Superconductivity and Beyond

Selected Papers of Alexei Abrikosov

World Scientific Series in 20th Century Physics

Published

Vol. 47 *Superconductivity and Beyond: Selected Papers of Alexei Abrikosov*
edited by Andrey Varlamov and Alexander Buzdin

Vol. 46 *The Essence of a Genius: A Tribute to Yoichiro Nambu*
by L. Brink and P. Ramond

Vol. 45 *The Finite Quantum Many-Body Problem: Selected Papers of Aage Bohr*
edited by Ricardo A. Broglia

Vol. 44 *Symmetries in Nature — The Scientific Heritage of Louis Michel*
edited by T. Damour, I. Todorov and B. Zhilinksii

Vol. 43 *Nambu: A Foreteller of Modern Physics*
edited by T. Eguchi and M. Y. Han

Vol. 42 *Selected Papers of K. C. Chou*
edited by Y.-L. Wu

Vol. 41 *Many-Body Theory of Molecules, Clusters, and Condensed Phases*
edited by N. H. March and G. G. N. Angilella

Vol. 40 *Murray Gell-Mann — Selected Papers*
edited by H. Fritzsch

Vol. 39 *Searching for the Superworld — A Volume in Honour of Antonino Zichichi on the Occasion of the Sixth Centenary Celebrations of the University of Turin, Italy*
edited by S. Ferrara and R. M. Mössbauer

Vol. 38 *Matter Particled — Patterns, Structure and Dynamics — Selected Research Papers of Yuval Ne'eman*
edited by R. Ruffini and Y. Verbin

Vol. 37 *Adventures in Theoretical Physics — Selected Papers with Commentaries*
by Stephen L. Adler

Vol. 36 *Selected Papers (1945–1980) with Commentary*
by Chen Ning Yang

Vol. 35 *A Career in Theoretical Physics (Second Edition)*
by P. W. Anderson

Vol. 34 *Quantum Mechanics in Phase Space: An Overview with Selected Papers*
edited by C. K. Zachos, D. B. Fairlie and T. L. Curtright

Vol. 33 *Formation and Evolution of Black Holes in the Galaxy — Selected Papers with Commentary*
edited by H. A. Bethe, G. E. Brown and C.-H. Lee

Vol. 32 *In Conclusion — A Collection of Summary Talks in High Energy Physics*
edited by J. D. Bjorken

Vol. 31 *From the Preshower to the New Technologies for Supercolliders — In Honour of Antonino Zichichi*
edited by B. H. Wiik, A. Wagner and H. Wenninger

Vol. 30 *Selected Papers of J. Robert Schrieffer — In Celebration of His 70th Birthday*
edited by N. E. Bonesteel and L. P. Gor'kov

Vol. 29 *Selected Works of Emil Wolf (with Commentary)*
edited by E. Wolf

For information on Vols. 1–28, please visit http://www.worldscientific.com/series/wsscp

World Scientific Series in 20th Century Physics Vol. 47

Superconductivity and Beyond

Selected Papers of Alexei Abrikosov

Editors

Andrey Varlamov
Institute of Superconductivity and Innovative Materials of the Italian Research Council (CNR-SPIN), Italy

Alexander Buzdin
University of Bordeaux, France

World Scientific

NEW JERSEY • LONDON • SINGAPORE • BEIJING • SHANGHAI • HONG KONG • TAIPEI • CHENNAI • TOKYO

Published by

World Scientific Publishing Co. Pte. Ltd.
5 Toh Tuck Link, Singapore 596224
USA office: 27 Warren Street, Suite 401-402, Hackensack, NJ 07601
UK office: 57 Shelton Street, Covent Garden, London WC2H 9HE

Library of Congress Control Number: 2024937302

British Library Cataloguing-in-Publication Data
A catalogue record for this book is available from the British Library.

World Scientific Series in 20th Century Physics — Vol. 47
SUPERCONDUCTIVITY AND BEYOND
Selected Papers of Alexei Abrikosov

ISBN 978-981-12-9178-4 (hardcover)
ISBN 978-981-12-9179-1 (ebook for institutions)
ISBN 978-981-12-9180-7 (ebook for individuals)

For any available supplementary material, please visit
https://www.worldscientific.com/worldscibooks/10.1142/13801#t=suppl

Desk Editor: Joseph Ang

Preface

When undergoing a magnetic resonance imaging examination in a clinic, reading about the discovery of the Higgs boson at the Large Hadron Collider, or learning about the setting of a new record for the speed of a superconducting magnetic cushion train, it's important to remind ourselves that behind all these fantastic achievements of modern technology lies the practical application of the phenomenon of superconductivity. This, in turn, is intricately connected with the figure of the outstanding theoretical physicist of the twentieth century, Alexei Alexeevich Abrikosov. We were fortunate to learn from him, work with him, and be his friends for many decades.

In this book, we aim to share memories of this distinguished man with our colleagues, both present and future, and to compile under one cover the best of his works. Most importantly, we want to highlight that in addition to his famous fundamental work on the vortex state of superconductors — work that brings us back to the immortal ideas of Leucippus and paves the way to the universe of vortex matter — Alexei Abrikosov made a huge contribution not only to superconductivity and the theory of metals and semiconductors but also to quantum electrodynamics and astrophysics.

Firstly, we express our gratitude to Professor David Khmelnitsky, who inspired us to undertake this work. Additionally, we extend our thanks to Alexey Abrikosov's widow and daughter, Svetlana and Natasha, his niece Natalia Abrikosova, and Professor Mikhail Chernikov, who provided us with rare photographs and other materials from their family archives. The book cover design was inspired by a painting which hangs in the living room of the Abrikosovs' house and evokes the discovery of superconducting vortices. We are also thankful to Dr. Ioulia Chikina, who assisted us in obtaining scarce copies of Alexei Abrikosov's works.

Alexander Buzdin
Andrey Varlamov

Contents

Part One

Memoirs of Disciples and Friends

Alexei Abrikosov: The life full of discoveries, twists, and turns

Alexander Buzdin, Andrey Varlamov

> "What Alyosha said has always been true; perhaps not every time it is suitable for public speaking. His assessments of people have always been harsh, but true. And his sharpness in his assessments became more understandable when I recognized that he simply compares others to himself and almost never finds an equal."
>
> M. Amusya (Shadows of the past, Memories of departed acquaintances and friends)

When you undergo an examination on a magnetic resonance imaging machine in a clinic, read about the discovery of the Higgs boson at the Large Hadron Collider or about a new speed record for a train on a superconducting magnetic cushion, then behind all these fantastic achievements of modern technology lies the practical use of the phenomenon of superconductivity, and hence, the figure of the outstanding theoretical physicist of the twentieth century Alexei Alekseevich Abrikosov.

The very phenomenon of superconductivity — the flow of current through wires without resistance, and therefore without heat release — was discovered at the beginning of the 20th century by the Dutch physicist Heike Kamerlingh Onnes. Despite the wide resonance in the scientific community and the high hopes associated with this discovery, for almost half a century it remained both mysterious from the point of view of its physical essence and worthless from the point of view of practical use. The point is that in the framework of classical physics, superconductivity should not have existed. And in practice it existed, but at very low temperatures and was very afraid of magnetic fields. The latter could not be avoided: why do you need a superconductor if you do not let strong currents through it? A strong current creates a noticeable magnetic field, which in turn destroys the superconducting state — the dog bites its tail.

Scientists from many countries, including the Soviet Union, have worked on the riddles of superconductivity. Back in 1950, future Nobel laureates Vitaly Ginzburg and Lev Landau published an article in which they propose amazing equations describing the properties of superconductors. The so-called phenomenological theory of superconductivity created by them made it possible to explain many unusual properties of existing superconductors but did not give an answer about the nature of this phenomenon.

A theoretical explanation of the nature of superconductivity — this one of the most mysterious phenomena of condensed matter physics — was found in 1957 by American physicists John Bardeen, Leon Cooper and Robert Schrieffer. It turned out that superconductivity is a macroscopic quantum phenomenon, i.e., the manifestation of the unusual properties of the microworld living according to the quantum laws in the everyday life around us. When a current flows in a normal metal, electrons drift along an electric field, moving chaotically, pushing like those running in a crowd. A superconductor through which a supercurrent flows is like a giant atom where electrons are soldiers united in pairs, with one gun has a bayonet pointing up, the other down. And these couples, in unison, in step, walk an endless march, thus transferring the charge and without disrupting their formation, without wasting energy. This beautiful picture brought peace to the souls of scientists but did not advance them in any way in understanding how to raise operating temperatures for using superconductors and rid them of their fear of magnetic fields.

It was at this time that a young disciple of Landau and a newly-minted researcher at the Institute of Physical Problems, Alexei Abrikosov, was actively working with experimenters to test the conclusions of the Ginzburg-Landau theory under various conditions. Everything is in perfect agreement as long as the samples are made at room temperature and only then cooled. But when superconducting films are deposited under conditions close to absolute zero (−273.15°C) and studied significant disagreements arise between the predictions of the theory and the data of experimental observations. Abrikosov thought about the restriction imposed by the authors on a certain parameter in the Ginzburg-Landau equations. The latter is connected precisely with the innate "fear" of the penetration of the magnetic field by the then known superconductors and allows, as long as the field is not too strong, to push it out of the sample volume. Abrikosov abandons this restriction and, purely formally, finds a completely unusual, hitherto unprecedented solution of the Ginzburg-Landau equations. This solution gives him a picture of a completely different scenario for the interaction of a superconductor with an external magnetic field: instead of completely pushing out the field and "break down" on this pretty soon, the superconductor can let the field into its volume in the form of single quantum vortices. These vortices are arranged in a periodic lattice, which is now called the Abrikosov vortex lattice. The stronger the field, the thicker its comb. In the cores of these vortices, the metal returns to its normal state, but the regions between them remain superconducting. And although the columns of soldiers — built in pairs of electrons — can no longer "march" across the entire battlefield, nevertheless, without disturbing their order, they flow around the vortex obstacles. Thus, reconciling the magnetic field and superconductivity, Abrikosov predicted the possibility of the existence of so-called type II superconductors, which can already withstand magnetic fields hundreds of times stronger than the superconductors known before his discovery. It is these superconductors in the overwhelming majority of cases that are used in practice today, and Alexei Abrikosov, without exaggeration, can be called the father of type II superconductivity. One of two fundamental obstacles to the use of superconductivity has fallen.

His discovery turned out to be so unusual and ahead of the concepts that existed in science that it did not immediately make its way. The work written in 1952 was published only five years later, and received its experimental confirmation eleven years later, in 1968. Half a century after the discovery (in 2003) A.A. Abrikosov was awarded the Nobel Prize in Physics for it.

Diligence of the future Nobel laureate manifested itself from childhood. From the archive of N.Yu. Abrikosova.

So, who is he, the man who showed mankind the way of practical use of superconductors, eliminated fundamental contradictions in quantum electrodynamics, developed elegant methods that have become classical in modern theoretical physics, left a deep mark on the theory of superconductivity and normal metals, according to whose text books students have been studying for more than half a century, wishing to become theoretical physicists all over the world studying the secrets of mastery? The diligence of the future Nobel laureate manifested itself from childhood.

A.I. Abrikosov with his son Alexei Abrikosov. From the archive of N.Yu. Abrikosova.

Alexei Alekseevich Abrikosov was born on June 25, 1928 into a famous Moscow family: his great-grandfather created the famous "Abrikosov and Sons" factory, which was famous throughout Russia for its confectionery, and, among other things, was the official supplier of the court of His Imperial Majesty. His father Aleksey Ivanovich Abrikosov, the vice-president of the Academy of Medical Sciences of the USSR, was a famous pathologist and author of monographs and textbooks on which generations of Soviet doctors were brought up. A lane in Moscow near the 1st Medical Institute was named after him, and a bust monument to A.I. Abrikosov was erected in front of the building of the Department of Pathological Anatomy. This was a man who knew one of the main secrets of Soviet power — from which Vladimir Lenin died. It was he who performed his autopsy and signed the death report.

"In the dreams of vortices". Alexei Abrikosov at the age 4 years on the stairs of family dacha. From the archive of N.Yu. Abrikosova.

The father's brother, the uncle of A.A. Abrikosov, Dmitrii Ivanovich Abrikosov, was a Russian diplomat. The February Revolution of 1917 found him in Japan, where he remained to represent the now defunct Russian Empire until 1925, when Japan established diplomatic relations with the Soviet Union. Providing documents to countless refugees from Bolshevik Russia, D.I. Abrikosov gave them the opportunity to start a new life outside their homeland.

Dmitrii Abrikosov with his sister Anna Abrikosova. Abrikosov family estate, 1895. From the archive of N.Yu. Abrikosova.

A.A.'s aunt Abrikosova, Anna Ivanovna, graduated from the first Moscow gymnasium with a gold medal and continued her history education at Girton College, Cambridge. In 1905, she married her cousin, Vladimir Abrikosov, a graduate of the Faculty of History and Philology of Moscow University, who also completed a university course at Oxford. The young intellectuals travelled to Europe and soon converted to Catholicism. Returning to Russia, the spouses were actively engaged in proselytizing Catholicism. Anna Abrikosova founded Catholic monasteries, translated the works of Catholic theologians into Russian, and financially supported poor Catholic children. While in Italy, she and her husband received an audience with the then Pope Pius X. After the October coup, she was repressed for her religious activities and for almost fifteen years, until her death in Butyrka prison, this woman of steel will support the spirit of her surrounding comrades (her name is mentioned in the "Gulag Archipelago" by A.I. Solzhenitsyn). In 1993 A.A. Abrikosov wrote a letter to John Paul II and drew attention to the fate of his aunt. As a result, the feat of A.I. Abrikosova is highly regarded by the Catholic Church: she is currently undergoing the procedure of beatification (i.e., canonization).

A.A. Abrikosov, I.M. Khalatnikov and L.D. Landau.

But let us return to the fate of A.A. Abrikosov. Already at the age of fifteen, he finished school. At twenty — with honors, Faculty of Physics, Moscow State University. At the same time, a year before that, he brilliantly passed Lev Landau's famous theoretical minimum: nine exams, which include the main sections of theoretical physics and the necessary (from Landau's point of view) sections of higher

Alexei Abrikosov presents his new work. From the archive of D.M. Chernikova.

mathematics. Landau took him to his postgraduate studies. This opened the door for him to one of the most famous theoretical physics clubs in the world: the Landau School.

The end of the forties — the beginning of the fifties of the twentieth century — the time of the work of prominent Soviet physicists on the atomic project. This specificity is reflected in the topics of A.A. Abrikosov.

Fanni Abrikosova (Wulf).

In 1951, the dissertation was successfully defended. Landau wanted to leave his student in the theoretical department of the IPP, but an unexpected obstacle arose. The head of the first department, a representative of the MGB at the institute without whose approval no personnel decisions are made, did not agree to sign the appointment of Abrikosov to the position of a researcher at the IFP. The fact is that the mother of A.A. Abrikosov, Fanni Wolf, was Jewish by nationality. Namely at this time in the Soviet Union with might and main there was a struggle against rootless cosmopolitans. The case of Abrikosov's hiring to work hung up for a long time. And then, on January 26, 1952, during his official visit, the commander-in-chief of the Mongolian People's Revolutionary Army and the head of the Mongolian government, Horlogiin Choibalsan, died in Moscow. "Pravda" on the first page, there appeared an official conclusion about his death, signed by the chief prosector of the Kremlin hospital Fanni Davidovna Wulf. This decides the fate of A.A. Abrikosov to work: "... since the Party has entrusted such a big deal to his mother, we can also trust your son."

The 1950s were a time of rapid development of quantum electrodynamics and solid state physics. Together with L.D. Landau and I.M. Khalatnikov, in 1954 A.A. Abrikosov published a work later called "Moscow zero"; it resolved the fundamental contradictions of the theory of elementary particles that was only being created in those years. The new calculation methods proposed in this work subsequently became an everyday working tool for thousands of scientists working in various fields of modern theoretical physics. They, together with some other beautiful ideas, such as the method of analytic continuation of Feynman diagrams, formed the basis of the one written by A.A. Abrikosov together with L.P. Gor'kov and I.E. Dzyaloshinsky monograph "Methods of Quantum Field Theory in Statistical Physics". In 1989, its authors were awarded the L.D. Landau Prize of the USSR Academy of Sciences. This book became a desktop for theoretical physicists in a number of countries, where it was translated and published.

As already mentioned, in 1957 A.A. Abrikosov published, perhaps, his most famous work, without which it is impossible to imagine the physics and technology of superconductivity today. It formulates the concept of superconductivity of the second kind, constructs a theory of the magnetic properties of such superconductors, which explains the accumulated experimental material, and reveals the existence of two critical fields and a phase of a mixed state between them, where the magnetic field partially penetrates into the superconductor in the form of quantum current vortices. This work is one of the most cited in the world scientific literature.

L.P. Gor'kov and A.A. Abrikosov in the Institute of Physical Problems. From the archive of D.M. Chernikova.

Following the discovery of type-II superconductivity, A.A. Abrikosov with coauthors obtained a number of important results in the newly created microscopic theory of superconductivity. These works lay the foundations for future applications: the prediction of the properties of superconducting alloys for the transmission of electricity over long distances, the creation of superconducting magnets for magnetic levitation, and the analysis of the high-frequency properties of superconductors, without which the most powerful accelerators working on superconducting resonators would not exist today. At the same time, together with L.P. Gor'kov, the properties of superconductors with magnetic impurities led to another fundamental discovery: the phenomenon of gapless superconductivity, which made it possible to abandon the already established dogmas and expanded the ideas of scientists about this most complex phenomenon.

L.A. Falkovsky and A.A. Abrikosov.

In subsequent years, A.A. Abrikosov's scientific interests shifted to the field of the theory of normal metals, semimetals and semiconductors. He studied the conductivity of metals with magnetic impurities (Abrikosov–Sula resonance), and together with his collaborators L.A. Falkovsky and S.D. Beneslavsky created the theory of semimetals and gapless semiconductors. Today, these works are especially relevant in connection to the discovery of graphene, topological insulators, and the prospects for the development of nano-electronics.

Along with his scientific activities, A.A. Abrikosov was actively involved in organizational and pedagogical work. His ability to explain complex problems clearly, lively, interestingly and with great humor manifested itself very early. When in 1955 P.L. Kapitsa, after finishing his disgrace (this followed the death of Stalin and the removal of Beria, Kapitsa's enemy), Kapitsa again became the director of the IFP. He decided to get acquainted with what the staff members of the institute were doing in his absence. Landau decided that the youngest employee of the department, Abrikosov, would talk about the work of theorists. Kapitsa liked Abrikosov's report so much that, despite his young age, he appointed him Scientific Secretary of the Institute. It is noteworthy that in this small institute (several dozen employees) work was carried out that brought three Nobel Prizes (Kapitsa, Landau, Abrikosov).

In 1964 A.A. Abrikosov was elected a corresponding member, and in 1987 a full member of the USSR Academy of Sciences (now the Russian Academy of Sciences). A.A. Abrikosov, together with other students of Landau — I.M. Khalatnikov, L.P. Gor'kov, and I.E. Dzyaloshinskii — created in 1965 the Institute of Theoretical Physics (which was later named after L.D. Landau as Department of Solid State Theory).

In addition to scientific work, A.A. Abrikosov taught almost all his life. At different periods of his life, he was a professor at Moscow State University, Nizhny Novgorod University, Moscow Institute of Physics and Technology (MIPT), and associate professor at the University of Illinois in Chicago.

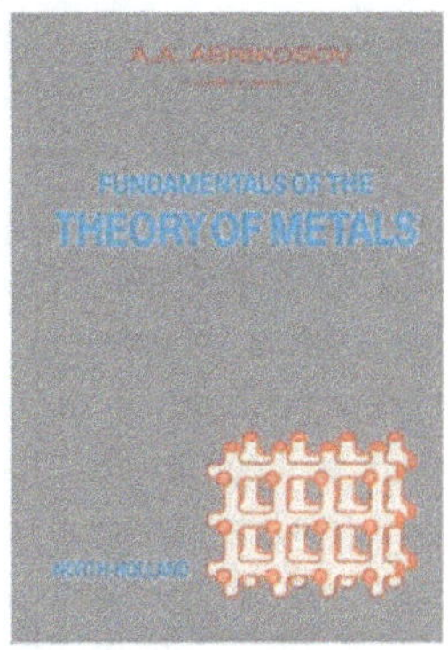

Since 1976, for fifteen years, A.A. Abrikosov headed the Department of Theoretical Physics at the Moscow Institute of Steel and Alloys (MISIS). In 1987, his textbook "Fundamentals of the Theory of Metals" was published, written on the basis of his brilliant courses of lectures given at MIPT and MISIS. In the same year, A.A. Abrikosov was elected director of the Institute for High Pressure Physics in Troitsk, Moscow Region.

Alexei Abrikosov with the authors of this memoir (A.I. Buzdin is at left and A.A. Varlamov is at right).

In 1991, A.A. Abrikosov accepted the offer of the Argonne National Laboratory (USA), moved to America and became a leading researcher at the Argonne Laboratory, head of the group of the theory of condensed matter. At this difficult time for science in Russia, Abrikosov was trying to help as many Russian physicists as possible by inviting them to work in Argonne. As a result, during this period, Russian dominated over English at his group in Argonne National Laboratory.

At the time of Abrikosov's invitation, there were no Nobel laureates at the Argonne National Laboratory and, undoubtedly, the laboratory's management hoped that this situation would soon change. The scale of Abrikosov's discoveries and contribution to the theory of superconductivity was such that this calculation was absolutely correct, and in 2003 A.A. Abrikosov, together with V.L. Ginzburg and A. Leggett, was awarded the Nobel Prize "for pioneering work on the theory of superconductivity and superfluidity".

Of course, the Nobel Prize was the crown of recognition of A.A. Abrikosov's contribution to physics, which was undeniable and was previously recognized: in Russia — Lenin (1966) and State (1982) prizes; the international scientific community — the Fritz London (1972), John Bardeen (1991) prizes, and subsequently the V.I. Vernadsky prize (2015). A.A. Abrikosov was an Honorary Foreign Member of the American Academy of Sciences and Arts, a member of the American Physical Society, a member of the Royal Society of London, an honorary doctor of many foreign universities.

A.A. Abrikosov worked at the Argonne National Laboratory for about twenty-five years. Immediately after his arrival there, he, in close cooperation with the experimenters of the laboratory, began to explain the phenomenon of high-temperature superconductivity. They discovered the existence of a specific feature in the spectrum of cuprate superconductors, which allowed A.A. Abrikosov to propose his own version of the theory of high-temperature superconductivity, capable of explaining a large number of existing experimental facts.

Abrikosov's work on quantum magnetoresistance, which was stimulated by experiments carried out at the Argonne National Laboratory and the University of Chicago, aroused great interest. Experiments have found a surprising linear dependence of resistance on the magnetic field. It turned out that a similar phenomenon was experimentally discovered back in 1928 by P.L. Kapitza, who mistook it for another effect. A.A. Abrikosov offered a beautiful explanation of this phenomenon — "quantum linear magnetoresistance" — based on the theory of semimetals with a linear electron dispersion law (the foundations of this theory were laid by him in the 1960s).

It should be noted that A.A. Abrikosov always paid great attention to experiment and stressed the importance for the theoretician to be in close contact with experimenters.

Despite the enormous contribution that A.A. Abrikosov made to various fields of physics, he believed that really good ideas rarely come to mind, and one should be able to appreciate them. Therefore, he did not like it when someone asked him what problem he should deal with: "Why am I going to tell him this? If I have an interesting problem, I will solve it myself". A.A. Abrikosov said that he would not remember that he could not solve any clearly formulated problem. It just took longer sometimes. But in the end, he always found a solution.

A.A. Abrikosov was a comprehensively gifted person; by the will of fate he chose physics as his field of activity. However, if it were another area of science, he would still have achieved outstanding results in it. A.A. Abrikosov will go down in history as a great scientist, whose work influenced many areas of physics and laid the foundation for a new understanding of the phenomenon of superconductivity. He was a very bright person with a brilliant sense of humor that helped him get through the difficult periods of his eventful life.

Aleksei A. Abrikosov

1928–2017

BIOGRAPHICAL *Memoirs*

A Biographical Memoir by
M. R. Norman

NATIONAL ACADEMY OF SCIENCES

ALEKSEI ALEKSEEVICH **ABRIKOSOV**

June 25, 1928–March 29, 2017

Elected to the NAS, 2000

By M. R. Norman

Shortly after the 2003 announcement that Aleksei Abrikosov had won the Nobel Prize in Physics, a number of colleagues took Alex to lunch at a nearby Italian restaurant. During lunch, one of the Russian visitors exclaimed that Alex should get a second Nobel Prize, this time in Literature for his famous "AGD" book with Lev Gor'kov and Igor Dzyaloshinskii (Methods of Quantum Field Theory in Statistical Physics.) Somewhat taken aback, I looked closely at this individual and realized that he was deadly serious. Although I could imagine the reaction of the Nobel Literature committee to such a book (for a lay person, perhaps analogous to trying to read Finnegan's Wake), I had to admit that my own copy of this book is quite dog-eared, having been put to good use over the years. In fact, you know you have made it in physics when your book gets a Dover edition. One of the most charming pictures I ever saw was a rare drawing in color that Alexei Tsvelik did (commissioned by Andrei Varlamov for Alex's 50th birthday) that was proudly displayed in Alex's home in Lemont, IL. It showed Alex with his fingers raised in a curled fashion as in the habit of medieval Popes. In his hand is an object resembling the Bible, but with the initials "AGD" on its front. Around this portrait are drawings of four of Alex's students, presumably learning at the feet of the Master.

But AGD is not the work Alex is best known for. In the early 1950s, he was trying to reconcile experimental data of his colleague Nikolay Zavaritskii with the recently published phenomenological theory of Ginzburg and Landau on superconductivity. He realized (1952) that Zavaritskii's data concerning the critical magnetic field at which superconductivity disappeared in thin films could be explained if he considered a case that Ginzburg and Landau had not considered, one where the surface energy between the normal and superconducting regions was negative. Alex referred to these as superconductors of the second group (now called type II superconductors). Afterwards, Alex realized that there were two critical fields, a lower one where magnetic flux first pene-

A drawing by Alexei Tsvelik that was commissioned by Andrey Varlamov for Alex's 50th birthday (1978). Note the copy of AGD Alex is holding. The students surrounding him are (from left to right) Alex Kovner, Ivan Ryzhkin, Andrey Varlamov, and Sergei Mukhin.

trated the sample, and a much higher one where superconductivity was destroyed. In between these fields, he predicted the existence of a lattice of quantized vortices. This discovery, one of the few where an analytic many-body wave function was derived, helped resolve puzzling data going back to the work of de Haas, Lev Shubnikov and others in the 1930s. This concept was so radical that Lev Landau initially dismissed it. But upon Richard Feynman's elucidation of vortices in liquid helium, Landau relented and allowed Alex to publish his paper (1957). As with many great papers, it was largely ignored in the beginning, but the discovery of the vortex lattice by neutron scattering along with its direct visualization by decoration experiments (sprinkling iron filings on the surface) in the 1960s brought the paper to the forefront. As Alex pointed out in his Nobel lecture, it is actually type I superconductors which are exotic, as the vast majority of important superconductors are type II. Perhaps most ubiquitously known for their role in MRI machines, type II superconductors power the Large Hadron Collider that was used to discover the Higgs boson, a concept which, itself, originated from superconductivity.

Alex was born in Moscow on June 25, 1928. He came from a well-known Russian family who owned a famous chocolate company (a poster of which you can find on Amazon), whose customers included the Czar's family. Alex's uncle, Dmitrii, was the last envoy of the Russian Empire in Japan, and his memoirs are available in English. Alex didn't even know of his uncle's existence until he was an older adult (but the KGB did, which was the reason Alex was not allowed to participate in the Soviet nuclear weapons program.) Alex's father was a world-famous pathologist, perhaps best known for leading the team that embalmed Lenin's body for display in the Lenin Mausoleum (he also did the autopsy for Lenin, a controversy given, claims, never proven, that he doctored the reports under orders to hide the fact that Lenin had syphilis). He was Vice President of the Academy of Medical Sciences and won a number of major awards (Stalin Prize, Hero of Socialist Labor, etc.). Alex's mother, Fanny, a former student of Alex's father, was a well-known pathologist in her own right.

Alex graduated from High School in 1943 during the midst of the German invasion, and in 1945 he transferred to the Physics Department at Moscow State University where he graduated three years later with an M.Sc. degree. He was then accepted to graduate school at the Institute for Physical Problems (later known as the Kapitza Institute), with his advisor being the future Nobel Laureate, Lev Landau. Alex passed the famous Landau minimum at the age of 19, and then obtained his PhD in 1951 with a thesis on thermal diffusion in plasmas. Landau wanted to keep Alex on at the institute, but there was a snag. A KGB official looked over Alex's file and noticed that not only was his mother Jewish, but her name was such as to make the KGB official suspect that Alex was Landau's nephew. Then fate intervened. The leader of Mongolia, Marshal Choibalsan, died while on a visit to Moscow, and Alex's mother performed the autopsy. When the KGB official found out about this, he was impressed enough to relent and allowed Alex to stay with Landau. The rest is history, so to speak.

The work for which Alex received his doctorate (equivalent to a habilitation) in 1955 was not his famous type II superconductivity work, but rather equally interesting work on quantum electrodynamics. This work was facilitated by Alex and Isaak Khalatnikov learning techniques developed in the West by Feynman and others, with important guidance coming from Landau about which diagrams to sum. Initially, they made a sign error in this work, and Landau realized they had discovered asymptotic freedom. When the sign error was pointed out to them (in one story, by Galanin and Tamm, in another by Galanin and Ioffe), they realized they had instead identified the so-called "Moscow zero," which can be interpreted as the renormalized charge vanishing in the point interaction limit due to complete screening. Asymptotic freedom was later discovered in the context of quantum chromodynamics by Gross, Politzer, and Wilczek, resulting in a

Nobel prize for them in 2004 (a fact that Isaak Khalatnikov ruefully acknowledged when I met him at the Landau Institute in Chernogolovka in 2007).

A famous picture of Landau's group in Moscow (1956). Sitting: Lyudmila Prozorova, Alex, Isaak Khalatnikov, Lev Landau, Evgeny Lifshitz. Standing: Semyon Gershtein, Lev Pitaevskii, Leonid Vainshtein, Robert Arkhipov, Igor Dzyaloshinskii.

During the same period as the quantum electrodynamics work, Alex did his Nobel-Prize-winning work on type II superconductivity. As in most truly profound work, the implications go far beyond what was originally intended. First, vortices are topological entities (the phase of the superconducting order parameter winds by 2π around the vortex core, i.e., the zero of the superconducting wave function). This led to consideration of such objects in other contexts, particularly in high energy physics. Perhaps the best-known example, though, is again in the field of superconductivity. By the Mermin-Wagner-Coleman theorem, a finite temperature XY phase transition (which superconductivity is equivalent to, since its order parameter is a complex scalar) is not allowed in two dimensions because of thermal fluctuations. It was later shown that a topological transition (binding/unbinding of vortex pairs) does occur, the famous Berezinskii-Kosterlitz-Thouless transition (for which the latter two won the Nobel Prize in 2016). To this can be added the prevalence of vortex-based ideas in microscopic theories of the fractional quantum Hall effect and quantum spin liquids, and the recent discovery of electric (as opposed to magnetic) vortices in certain transition metal oxides.

Besides the above, Alex also worked on the possibility of metallic hydrogen in the core of gas giants (much in vogue these days given recent high-pressure experiments claiming to have achieved the metallic phase), and then at the end of the 1950s, he turned to the microscopic theory of superconductivity after learning about the work of Bogoliubov. He and Khalatnikov generalized Bogoliubov's results to finite temperature, but their two papers were rejected by Evgeny Lifshitz because by then, the Russians had become aware of the papers by Bardeen, Cooper and Schrieffer (Nobel Prize, 1972) where these results were derived in a different manner (their papers, though, were later published after Valatin had independently derived some of their results). By then, Alex became interested in applying some of the techniques he had learned from the work in quantum electrodynamics to the problem of superconductivity. In a series of famous papers, he and Lev Gor'kov exploited an obscure formalism by Matsubara to extend zero temperature quantum field theory to finite temperatures. Their crucial insight was the realization that one could state the problem as a Fourier sum, with the inverse of the temperature playing the role of imaginary time. Since much was known analytically about such sums, hard problems, which would have previously involved complicated frequency integrals, became much easier to solve. With this formalism, they attacked a variety of problems, perhaps the most famous being superconductors in the presence of disorder. From their formalism, it became trivial to understand why non-magnetic impurities do not degrade superconductivity, whereas magnetic impurities do. Moreover, the latter degrade superconductivity in such a fashion that the energy gap, previously thought to be equivalent to the superconducting order parameter, actually vanishes before superconductivity disappears, leading to a novel region of gapless superconductivity. This body of work, as well as far more extensive work on "normal" materials, formed the basis of the famous textbook by the two of them and Igor Dzyaloshinskii, referred to in the introduction.

After this, Alex turned his attention to metals and semimetals. In 1965, he published a famous paper attacking the problem of magnetic impurities in a normal metal (the Kondo problem), again by summing selective (parquet) diagrams, where he ascertained the presence of a resonance (the Kondo, or Abrikosov-Suhl resonance, as Harry Suhl had independently derived the resonance the same year). A method Alex introduced was to express the spin operator as a Pauli matrix bracketed by so-called pseudo-fermions (now known as Abrikosov fermions). This was, to my knowledge, the first introduction of such auxiliary particles (commonly known as slave particles). Similar methods were later exploited by Piers Coleman and Nick Read in the 1980s to treat a lattice of Kondo impurities (the heavy fermion problem), and then in the 1990s by Phil Anderson, Patrick Lee, Nick Read, Subir Sachdev and Xiao-Gang Wen, among others, to address the problem of quantum spin liquids. These methods remain quite popular, particularly in the field of spin liquids, and led to Anderson's concept that spin ½ excitations could actually form a zero energy `spinon' Fermi surface, an idea going back to Pomeranchuk in 1941, and the existence of which is still being sought (several candidate materials have been proposed to have such surfaces).

It was also in the early 1960s that Alex first started working on semimetals (which are much in vogue these days with the discovery of topological varieties). With Leonid Falkovskii (with whom he had done an earlier paper on Raman scattering in superconductors), he considered the canonical example of bismuth (very recently discovered to be an ultra-low temperature superconductor), explaining the role of crystallographic deformation in yielding a semimetallic state (1962). In later work with Beneslavskii (1971), he considered the case of touching energy bands (with both linear and quadratic dispersion around the touching point), again of much relevance today with all of the excitement associated with Dirac and Weyl semimetals (the example they mentioned, HgTe, is quite prominent in recent literature). He also explored the possibility of an excitonic insulator phase in bismuth in a strong magnetic field (1973). For his work on semimetals and gapless superconductors, Alex would receive, along with some experimental colleagues, the State Prize of the USSR (1982). During the 1970s and early 1980s, he would also work on the electrical properties of quasi-one-dimensional metals (with Ryzhkin), including their superconducting properties.

During this time period, Alex moved around quite a bit. After Landau's car accident in 1962, the decision was made to found a theoretical institute in his name and move out of Moscow (from the Institute for Physical Problems that Alex had been at since 1948) to the village of Chernogolovka (a move which was obviously not relished by most). Alex, along with Isaak Khalatnikov, was instrumental in setting up this new institute. Although he was at the Landau Institute from 1965 to 1988, he also held positions in Moscow, first at the Moscow Institute of Physics and Technology (1972 to 1976), and then at the Moscow Institute of Steel and Alloys (1976 to 1991), and was also director of the Institute for High Pressure Physics from 1988 to 1991, at which time he departed for the United States. The lectures he gave in Moscow formed the basis of his classic textbook, *Fundaments of the Theory of Metals*, published in Russian in 1987, and in English the following year (a book that I again have put to great use over the years).

Two anecdotes from this period are perhaps worth sharing. The first was a story Alex related to me. During the Soviet era, foreign travel was particularly difficult to get permission for. Alex, being an avid mountaineer, along with some of his more enterprising colleagues, desired to climb Mount Kilimanjaro in Africa. But how to arrange it? They decided to call it a scientific expedition, and somehow got the authorities to sign off on it. Upon their return to Russia, the authorities wished to know what they had discovered during their expedition. After an embarrassing silence, one of the members suddenly spoke up: "It was cold." The other is related in Khalatnikov's memoirs. He and Alex were great champions of the practical joke. One took place during one of the symposia they held in Georgia. A shop that they kept passing had only a few items worth noting: framed portraits of Stalin (who was Georgian), long kitchen knives, and women's panties. The idea thus donned on them. They bought a knife and a pair of panties, and then hid them in Arkady Migdal's suitcase, along with a note reading, "you will answer for the outraged honor" appropriately translated to Georgian by Alex's graduate student, Rita (who was Georgian). They of course knew that Mrs. Migdal would be the one unpacking the bag when Arkady returned to Moscow, much to his embarrassment.

In the wake of perestroika, Alex decided that the future of science in Russia was not bright, and decided he would get away if he could. His initial thoughts were to go to France, given the presence there of his second son, but this prospect was not viewed well by some in France (Alex's second wife had been Philippe Nozieres' wife.) Word soon got out that Alex was looking around. Ivan Schuller at UCSD found out about this, and alerted Frank Fradin, who at that time was an associate laboratory director at Argonne. To throw off the authorities, Alex arranged a one month visit to Venezuela, and while he was on the visit, took a one week trip to Argonne to get acquainted with people there. When I first saw Alex during this visit, he reminded me of a farmer (he wore suspenders). But it was clear that he was an absolute treasure trove of information, and his presence would be of real benefit to the lab. So, Alex came in 1991, and six months later was appointed head of the Condensed Matter Theory Group, which he led from 1992 to 2000. This decision had its pluses and minuses. On the plus side, Alex was a great person to talk to, though you had to make sure you had plenty of time on your hands if you decided to venture into his office (he loved to tell stories). He also attracted many visitors, particularly Russians—Andrey Varlamov, Sasha Buzdin, Yuri Galperin, and Lev Pitaevskii were regular visitors, and Anatoli Larkin spent a six-month sabbatical (sadly, the Argonne administration moved too slowly and Anatoli left for Minnesota rather than staying at Argonne permanently.) One consequence of these visits was that Russian was commonly heard during seminars (sometimes from Alex himself, especially if he got bored with the speaker). One day, I arrived for a seminar late, only to be surprised to find out that the speaker, Sasha Finkelstein, was giving his lecture in Russian (his reaction was "ah, Mike is here,

Alex in his office at Argonne. Note the copy of AGD he is holding, and the picture of John Bardeen on the wall.

I must switch to English now.") This led to perhaps one of the more famous practical jokes during Alex's time at Argonne. Alex's postdoc, Tartakovskii, having seen the annoyance of some of the non-Russians to what was going on in the seminars, constructed an e-mail stating that it had come to Alex's attention that certain members of the theory group were deficient in the Russian language. Therefore, mandatory Russian language classes must be taken by these individuals. Now, there were two giveaways to this joke, first the date it was sent (April 1) and second the signature (Alex Apricot, "apricot" being the English translation of "abrikos"). Unfortunately, one of my colleagues did not notice these hints, and burst into my office saying that Alex had gone too far now, and he refused to take these language classes!

Although advanced in years, Alex did do some very interesting work while at Argonne. Perhaps his best-known work from this time was on quantum magnetoresistance (1998). A colleague at Argonne, Marie Louise Saboungi, had been providing samples of doped silver chalcogenides to Tom Rosenbaum's group at Chicago. A striking linear behavior with magnetic field was seen for the resistivity in these samples. Alex's explanation was drawn from his days working on semimetals, where he proposed that the material had a linear dispersion (nowadays, we would call this a Dirac semimetal), and with such a low carrier concentration, that with modest field one was already in the quantum limit where only one Landau level was relevant. Interest in his theory has since been renewed given all the recent excitement concerning topological semimetals (which this particular material has been conjectured to be).

Some of the participants from the symposium at Argonne in honor of Alex's 70th birthday (November 1998). From left to right, Helmut Brandt, Grisha Volovik, Dick Klemm, Boris Shklovskii, George Crabtree, Boris Altshuler, Lev Gor'kov, David Bishop, Alex, David Nelson, Mike Tinkham, Phil Anderson, Valerii Vinokur, Igor Dzyaloshinskii, and Dima Khmel'nitskii.

Alex pursued a number of other topics at Argonne. One was invoking van Hove singularities in the density of states as a mechanism to enhance superconductivity. As with most of his work, this project was motivated by experimental data, in this case that of Juan Carlos Campuzano's group (JC's office being next to Alex's), who found the presence of a flat band near the chemical potential in a cuprate superconductor. This flatness, Alex realized, leads to a power law divergence of the density of states (as opposed to the less singular logarithm associated with a normal van Hove singularity), and he derived the mean field consequences of such an electronic structure, particularly in regards to its influence on superconductivity. Alex also worked on how electronic transport occurs between layers (cuprates being a great example of weakly coupled planes), invoking the concept of resonant tunneling via defects to explain how the electrons moved from one layer to the next. Finally, based on some earlier work from 1978, he further pursued a potential mechanism of high Tc superconductivity where one replaces ions in the classic model for superconductivity by heavy holes in order to boost the energy scale.

In 2003, Alex won the Nobel Prize, giving him the opportunity to travel even more widely than before (a luxury he embraced given all the years he was denied travel while living in the Soviet Union). Unfortunately, six months later, he developed macular degeneration, completely losing all vision in one eye, and having very limited sight in the other. This began to limit both the travel and the work he could do. Alex, though, compensated for this. A way was found to blow up images on his computer monitor, and that, coupled with the jeweler glasses he wore, allowed him to do a certain amount of reading and writing. We offered him additional secretarial support (and a larger office), but Alex was too modest to accept such. Rather, he came in every day to the lab like clockwork (even after he officially retired in 2014), and for a long time even drove to the lab himself (we were rather terrified about this, and fortunately at some point, his wife, Svetlana, took over the driving duties)

Alex and Princess Madeleine at the Nobel Banquet (December 2003). I saw a picture of this scene in the Swedish papers the following day that cut Alex out!

Then, in the spring of 2016, Alex suffered a heart attack. His wife decided to move them to be with their daughter, Natalia, who was a physician living in Palo Alto. For a while, it seemed like he was doing better, but then he suffered another heart attack and died there on March 29, 2017.

Alex, married three times (Tatyana, Annie, and Svetlana, whom he was married to for forty years.) He had sons from his first two marriages (Alexey, a physicist as well, and Michel) along with a step daughter from his last (Natalia). Alex had a unique personality. He was opinionated, and shared more of his thoughts on people and events than he probably should have. He himself recognized this, which is the reason he never wrote his memoirs. Perhaps the most infamous example of this was an op-ed piece in the New York Times by Sergei Leskov from 1993, where Alex was quoted as saying that the best Russian scientists should be recruited away from Russia, and the rest should be ignored, which led to a howl of protests. On the other hand, he could be exceptionally kind. The reason I became a Fellow of the American Physical Society at such a young age was his nomination, among other kindnesses he bestowed on me over the years.

I will not dwell on the long list of prizes and honors Alex received during his long career (including induction into this Academy in 2000). Suffice it to say that the body of work that Alex left behind will be remembered for generations to come. I personally do not believe that AGD will ever go out of print. If you don't have a copy and you are a theoretical physicist, you should get one. There's no telling what gems you will find there.

ACKNOWLEDGEMENTS

The author would like to thank Andrey Varlamov, who not only checked over the original draft of this memoir, but also provided some of the information and one of the images I drew upon (the drawing by Tsvelik, now in his possession). One of the images (Alex with the Princess) was kindly provided to me by his daughter, Natalia. Some of the source material I used for this work can be found in the references below.

ALEKSEI ABRIKOSOV

REFERENCES

Abrikosov, A. A. 1965. *Academician Lev Landau.* Originally in Russian, and recently translated into English (2015) by Alexey Galda, Ivan Sadovskyy, and Andrey Varlamov, with Michael Norman as scientific editor Argonne, IL: Argonne.

Andreev, A. F., et al. 2008. Aleksei Alekseevich Abrikosov (on his 80th birthday). *Physics Uspekhi* 51:637-638.

Andreev, A. F., et al. 2017. In memory of Aleksei Alekseevich Abrikosov. *Physics Uspekhi* 60:638-639.

Dokshitzer, Yu. L. 2004. QCD phenomenology. In *Proceedings of the 2002 European School of High-Energy Physics.* Pp 1-33. Geneva: CERN.

Geshkenbein, B. V. 2001. Introducing Boris Ioffe. In *At the Frontiers of Particle Physics: Handbook of QCD, Vol. 1.* Edited by M. Shifman. Pp. 9-14. Singapore: World Scientific.

Gor'kov, Lev. P. 2010. Developing BCS Ideas in the former Soviet Union. *Intl. J. Modern Phys. B* 24:3835-3854.

Hargittai, Balazs and István Hargittai. 2005. Alexei A. Abrikosov. in *Candid Science V: Conversations with Famous Scientists.* Pp 177-197. Singapore: World Scientific Publishing.

Khalatnikov, Isaak M. 2007. *From the Atomic Bomb to the Landau Institute.* Heidelberg: Springer.

Leskov, Sergei. 1993. America's Soviet Scientists (*New York Times*, July 15, 1993).

Polyakov, A. 1992. A view from the island. https://arxiv.org/abs/hep-th/9211140

Varlamov, Andrey. 2017. Alex Abrikosov: life and scientific biography. XXIV Davydov Lectures (Kiev, Ukraine, Dec. 26, 2017).

Wikipedia. 2017. Alexei Alexeyevich Abrikosov, https://en.wikipedia.org/wiki/Alexei_Alexeyevich_Abrikosov

Alexei Abrikosov in Japan

Kazuo Kadowaki

I saw Dr. Alexei Abrikosov for the first time at the satellite conference of LT-19 (19th International Low Temperature Conference)[1] held in Cambridge, UK in 1990, but I had no chance to talk to him, unfortunately. At that time, he appeared to me so unusual because he made such a sharp contrast from other participants, always accompanied by a few tough-looking bodyguards beside him and he himself always taking movies by handy camera whenever and wherever he was going. He made such a strong impression on me, which still remains strongly in my memory.

Soon after LT-19, I heard that Abrikosov had moved to Argonne National Laboratory in USA. It was not so long time to have a chance to meet and to talk to him at Argonne National Laboratory in USA, because I often visited there to collaborate with Dr. George Crabtree's group in MSD (Materials Science Division) on the subject of High-T_c superconductors. I did not see Alexei Abrikosov so often at Argonne National Laboratory, but as soon as he saw me, he extended his hand in front of me for a shake by saying "*Hello*" with his full smile. He was so kind and gentle telling me many interesting jokes and stories related to all kinds of topics not only Physics, of course, but also world economy, politics, medicine, differences between scientific societies in USA from that in the former Soviet regime, *etc.* Sometimes he wanted to speak more without standing and talking in the corridor, so he would invite me to his office and often spent more than an hour with me.

At the end of 2003, or it may be in early 2004, I was at Argonne National Laboratory. I could have a chance to meet Abrikosov[2] and to say "Congratulations!", since it was soon after his Nobel Prize award in Physics that he shared with Ginzburg and Leggett. Sometime after that — I do not remember exactly when — he mentioned during our conversation, which I clearly remember, that he was interested in Japan very much, in particular Japanese culture, people, science, *etc.*

Soon after I came back to Japan, I started to think seriously to invite him to Japan. At the same time, I started to seek funding for that purpose. It was very fortunate that JSPS (Japan Society for the Promotion of Science) has such a program that funds Distinguished Research Fellows from foreign countries such as Nobel laureates to be invited to Japan for a few weeks. I decided to apply for funding in 2004 immediately, since this was a good timing, just after his Nobel Prize Award.

The application was smoothly accepted for the 2004 fiscal year, but the award was postponed to 2005, because Abrikosov had eye surgery and had difficulty to travel in 2004.

[1]The LT19 was held at 16–22 August 1990, Sussex, Brighton, UK. The satellite conference on "*High Temperature Superconductivity*" was held 13–15 August 1990 at Cambridge, UK.

[2]Hereafter, the title is dispensed.

We had to wait for his eye-sight recovery. Fortunately, the three consecutive operations resulted in success, so that his eye-sight conditions seem to have recovered in a year.

The plan was made for two weeks' visit starting from October 12 to 26, 2005. Abrikosov was accompanied by his daughter, Natalia Abrikosova, Alexander Buzdin and his wife Tatiana, George Crabtree and a few of organizing members.

I would like to mention few events which stayed in my memory and made strong and lasting impressions.

The day when Abrikosov had just arrived at Narita Airport, Tachiki, Abe (my administrative assistant), and I greeted him at the arrival lobby. Soon after we met each other, we moved to a large train car arranged by the University and left the airport for Tsukuba, about 60 km to the North from Narita airport. I was sitting near Abrikosov and asked him a few questions, despite that he looked quite tired from the long flight from Chicago. Among my questions I asked about his first impression of Japan. He immediately said "Green!", just one word. Then, he repeated that "Every piece of land is green!" I was a bit surprised to hear such phrases because I did not expect such an answer at all. That was his first voice and it impressed me very much. Then, he started to talk about many things that he experienced in the forests in Russia, stories about mountain climbing to Mt. Kilimanjaro, *etc.* He seemed to have a talent to tell a story dramatically and to attract the listener very much. A couple of stories from him were sufficient to help us pass away the hour of our train trip.

On Day 2, the University of Tsukuba organized the ceremony of Abrikosov's Honorary Degree (see Figure 1) and following the ceremony his memorial lecture was held. Four distinguished scientists in the field of superconductivity[3] were invited (see Figure 2). First, Abrikosov gave a very impressive talk entitled "*Type II Superconductors and the Vortex Lattice*". It was a story about the discovery of vortices that coexist in type II superconductor near the critical magnetic field H_{c2}. Although for Abrikosov the result was a natural mathematical consequence of Ginzburg-Landau equation on superconductivity discovered in 1954,[4] Landau did not consider it seriously until similar discovery of vortices in superfluid liquid by Feynman took place later.[5]

[3]The other four invited speakers were Akira Tonomura (Research Fellow at Hitachi Fundamental Laboratory, Japan, (deceased), the title of his talk was "*Real-Time Observation of Abrikosov Vortex Lattices Using Electron Waves*"), Alexander Buzdin (Professor at the University of Bordeaux, France, the title of his talk: "*Vortex Attraction and Vortex Intermediate State: Is It Possible?*"), Masashi Tachiki (Emeritus Professor at Tohoku University, the title of his talk was "*Emission of Terahertz Electromagnetic Waves by Means of Vortex Flow*"), and Shyoji Tanaka (President of the International Superconductivity Technology Center, (deceased), the title of his talk; "*Progress of Superconductivity Technology in Japan*").

[4]It was finally published in 1957. A.A. Abrikosov, Soviet Physics-JETP **5**, 1174 (1957). It is however noted that the calculated result by Abrikosov in the original paper was the vortex square lattice as a ground state, but it was pointed out later that the most stable lattice was hexagonal lattice. Therefore, it is nowadays called as the Abrikosov's hexagonal lattice of vortices.

[5]R.P. Feymnan: in "*Progress in Low Temperature Physics*", edited by D.F. Brewer, North Holland, Amsterdam, 1955, Vol. 1, Chap. 11.

Figure 1. Abrikosov receives the diploma of Doctor Honoris causa from Tsukuba University.

Figure 2. From left to right A. Abrikosov, A. Tonomura, A. Buzdin, K. Kadowaki, M.Tachiki at Tsukuba University.

The ceremony and associated lectures were open for students, young researchers, faculty members, *etc.* at the University and the event Hall was really full to capacity (more than 500 attendees).

On Day 3, we organized the round table panel discussion at the International Congress Hall at the University of Tsukuba.[6] The panel discussion consisted of the Chairman, Hiroshi Ezawa, the main speaker, Alexei Abrikosov, and six panelists: Masayasu Ishikawa, George Crabtree, Masashi Tachiki, Fumihiko Takano, Sadao Nakajima, Alexander Buzdin, and George Crabtree (see Figure 3). The program was divided into four parts: Part One was about the early history of Abrikosov and his surroundings entitled "From birth of Dr. Abrikosov to the BCS Theory". Abrikosov mostly talked about his family, his youth, student times, and how he chose physics instead of mathematics or chemistry which he really liked, why he became theoretical physicist and not experimentalist as he wanted, and how and when he met Landau. Abrikosov has passed the Landau minimum (famous graduate student entry examination by Landau) straightforwardly, and soon after he started to work in his group.

Part Two was devoted to the "*Glory Days of Superconductivity (BCS theory to mid 1970's)*", Part Three, "*New Development (mid 1970's to 1986)*" and Part Four, "*High-Temperature Superconductors in Oxide Cuprates and Future Perspectives (after 1986)*" Many wonderful stories may be found in the book "Abrikosov in Japan".

Throughout the whole event that lasted more than four hours, I realized a specific peculiarity about the scientific achievements in the former Soviet Union where so many excellent scientists were born. I finally came to think that complete isolation from the rest of the world may have resulted in the development of their own and unique Soviet scientific achievement. Furthermore, psychological pressure in communism might have helped to deepen the Soviet Union's societal focus on science, as the scientists lacked any other distractions. It is interesting to compare such a situation to the Edo era in Japan. Japan had been closed country for more than 250 years in Edo era. One may think that it might

[6]The detail of this panel discussion was recorded on tape and published in a scientific Journal in Japan "Kagaku" (which means "Science" in Japanese, Vol. 76, No. 6, pp. 655–642, 2006, translated by Hideki Matsumoto). The English version of this session will be published in a book "Abrikosov in Japan" (to be published), including the whole stories, events, *etc.* during his trip to Japan.

had been detrimental to society's development since there was no information and no stimulation from overseas countries. Actually, this was not true, and, to the contrary, Japan had made amazing progress in culture, science, arts, mathematics, *etc.,* and in all kind of human activities during the Edo era. The rate of literacy at the end of Edo era was more than 70%. There is a famous story that Perry[7] for the first time came to Japan in 1853, and shot a cannon to force the government to open the country. In his diary he wrote that "they all look so poor wearing flea-bitten poor kimono", but at the same time he was shocked to see that such Japanese children were reading books on the street! This may be a good lesson that the globalism that nowadays overwhelms the world by unifying the human activities, cultures, arts, music, business, *etc.* including the science and technologies all over the world, may be destroying individual unique development of human society. This looks totally ironic but seems to be universal.

Figure 3. Participants of the lively panel discussion with A. Abikosov at Tsukuba University.

At the end of the panel discussion, a memorial photo was taken with all the participants. (see Figure 3).

From October 15 on, Abrikosov visited Hitachi Central Laboratory, Science University of Tokyo, Kyoto University, and Tohoku University, returning back to Tokyo a day before his scheduled departure.

Abrikosov gave his four talks, all on different subjects, at the University of Tsukuba, at the Science University of Tokyo, Kyoto University and Tohoku University. I would like to add one more story that impressed me during the journey with Abrikosov in Japan.

Figure 4. Abrikosov's lecture at the crowded Yukawa lecture hall at Kyoto University.

In the morning of October 18th, all of us including Abrikosov moved from Tokyo to Kyoto by Shinkansen train. We planned

[7]Matthew Calbraith Perry (1794–1858).

to have a lunch at the faculty club at Kyoto. During lunch I received a phone call from Takeo Matsubara, who is the inventor of the famous Matsubara Green function. I wanted him to participate in Abrikosov's lecture. I was planning to set up a brief meeting of Abrikosov and him at Kyoto University, and I had informed him of Abrikosov's arrival from the Shinkansen by phone. "Now", he said with a quiet voice, "I am not able to come to Abrikosov's lecture in the afternoon". I was so shocked to hear Matsubara's quiet voice and his cancelation, because he told me earlier that morning that his intention was to attend Abrikosov's talk and meet Alexei. I asked him about the cause of his cancelation, and then I was shocked to hear about the reason. He continued to say that he was not able to come even on the campus of Kyoto University after he retired from Kyoto University many years ago. I could not believe that I am hearing such a story from him, and I was at complete loss of how to respond. He repeated the same phrase again and again with a quiet voice. I gently interrupted his sentence by saying that it is OK and proposing to him to join us for our banquet that night that was going to be held after the lecture at a restaurant near the University. He agreed. I was really taken by a feeling of relief. We agreed that I would come to his place by taxi and pick him up to bring us to the restaurant.

Abrikosov's lecture was a success, with a full auditorium and many people sitting even in the corridors and the backspace of the Yukawa lecture hall (see Figure 4). Abrikosov had a unique style of presentation: handwritten overhead projector slide sheets with full of names, words, equations, phrases, pictures, *etc.* I show one of them used in his talk at Kyoto University as an introduction. (see Figure 5) It looks like the last part of the brief history of superconductivity; I was so surprised when I saw it. I couldn't believe my eyes, because my name was written as a first discoverer of Y-123 compound $YBa_2Cu_3O_{7-x}$. Furthermore, hearing his talk, he knew much more details of our story that led to the discovery of high temperature superconductivity in $YBa_2Cu_3O_{7-x}$. I was wondering how did he obtain such a detailed information about our work, because we in fact worked only with a very few people and we never talked about it to Abrikosov. I had to think and believe that intelligence activities in Soviet time appear to be extremely good even for collecting the scientific information.

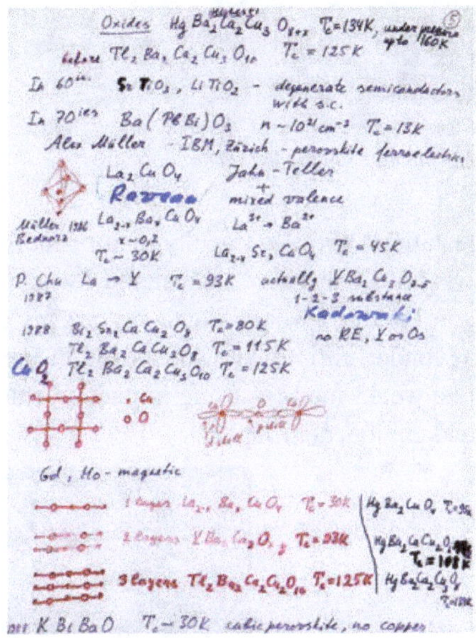

Figure 5. Overhead projector slide of Abrikosov's lecture on superconductivity at Kyoto University.

After Abrikosov's talk, it was surprising to see a long cue of students waiting for book signing of his book.

After Abrikosov's lecture in Kyoto University, we took a taxi and went to meet with Matsubara. Abrikosov and Matsubara met just on the street! Finally, Abrikosov and Matsubara could meet at the evening of 18 October on the streets in Kyoto. Such a dramatic meeting was somehow made possible. Then, a banquet was held at a traditional Japanese

Figure 6. Reception at the traditional Japanese restaurant in Kyoto in the honor of Abrikosov. Professor Matsubara is on the left side to Abrikosov.

restaurant with few faculty members of Kyoto University (Figure 6). After the banquet, we went to a bar where a Geisha girl was invited at Abrikosov's request (Figure 7).

Finally, I would like to express my deepest thank to JSPS who allowed us to have such a wonderful time with Abrikosov in Japan. I think that Abrikosov's visit in Japan in these two weeks made a strong impact on students, young professors as well as senior scientists and engineers in Japan.

Figure 7. Abrikosov talking to Maiko (an apprentice Geisha) at the Gion quarter of Kyoto.

Mountain hiking with Alexei Abrikosov

Yuri M. Galperin

In 1971, I — at that time a young postdoc — had luck to be accepted to a mountain hiking team planning a route over the Academician Obruchev Ridge located between Large and Small Yenisei — great Siberian rivers. The team consisted of three theoretical physicists, one biophysicist, two engineers, medical doctor, and architect. The oldest member of the team was Abrikosov. The route was rather complicated; it contained mountain passes and it was planned to spend about 40 days out of civilization. Consequently, we had to carry heavy backpacks, two rifles (for hunting and defense from bears — typical animals for that area) with ammunition. Under such rather severe conditions personal features of human beings clearly manifested themselves. In particular, Abrikosov always checked that his backpack was not lighter than other ones. He managed to keep a good mood under rather difficult conditions; in particular, when we were trapped in a valley by unexpected snowfall, he was ill and had high fever. By the way, on that day I learned an important lesson from him. The reason was that we wanted to prepare hot broth for Alex. Therefore, we took rifles and shot several ducks. Among them there was a duckling, which I shot by mistake. When Abrikosov saw that he was really angry with me: "One should never kill more than it is absolutely necessary". Since then I have never taken part in hunting. Alex Abrikosov was a perfect narrator. His stories near the camping fire were deeply impressive. Another his feature which impressed me was his neatness. His dress and camping tent were always clean, every morning he boiled water to be freshly shaved. By the way, his scientific drafts also were very accurate and clear allowing reproduction of the results. But this is another story.

Our team (left to right): Yuri Galperin, Alexei Abrikosov, Nina Rybakova, Vladimir Rybakov, Vadim Gurevich, Galina Raikhlina, Audronis Katulius, Vladimir Filov.

Part Two

Alexei Abrikosov on Himself and His Work

Autobiography

I was born in 1928, June 25, in Moscow, USSR (now Russia). My parents were physicians. I graduated from high school in 1943 and was accepted as student of the Institute for Power Engineers. In 1945 I transferred to the Physics Department of the Moscow State University, which I graduated in 1948 with a diploma (M. Sc. degree). After that I was accepted, as a Postgraduate (Ph. D. student) to the Institute for Physical Problems (now P. L. Kapitza Institute). My scientific adviser was L.D. Landau. After I defended in 1951 a thesis on thermal diffusion in completely and incompletely ionized plasmas, I got the Candidate of Science (Ph. D.) degree and was taken to the staff of the Institute, as Junior Scientist.

In 1951–1952 I worked with the experimentalist of the same institute, N.V. Zavaritskii on the experimental verification of the predictions of the recently published Ginzburg-Landau theory of superconductivity on the critical magnetic field of thin films. This resulted in our discovery of the "superconductors of the second group" (now *Type II superconductors*). After that I started to work on the magnetic properties of bulk Type II superconductors, and came to the conclusion that the transition from superconducting to the normal state happens gradually in increasing field with two limiting critical fields. Between these two values the field gradually penetrates the superconductor forming thin threads of magnetic flux surrounded by vortex currents. The array of these quantum vortices forms a regular structure (now referred in the literature, as *Abrikosov vortex lattice*). I compared my results with the magnetization curves obtained experimentally in the 1930s for superconducting alloys, and there was a perfect fit. The experimentalist explained their data, as due to inhomogenity of their samples. My paper was published in 1957 but the experimentalists accepted the vortex lattice only 10 years later, after it was demonstrated by decoration experiments.

In the middle of the 1950s I worked also on the transition from the insulating molecular phase into the atomic metallic phase in hydrogen and on the structure of hydrogen planets. Another topic of mine was quantum electrodynamics at high energies. The latter works became my Doctor of Science thesis (this degree is similar to Habilitation in Germany), which I defended in 1955.

In the end of the 1950s up to the early 1960s we worked with L. Gor'kov on the microscopic theory of superconductivity. We constructed the theory of superconductors in a high-frequency field (with I.M. Khalatnikov) and the theory of superconductors with magnetic impurities, where we discovered the possibility of the so-called *gapless superconductivity.* We also solved the mystery of the finite Knight shift at zero temperature, taking into account the spin-orbit scattering. Simultaneously we worked with I. M. Khalatnikov on the theory of nonsuperfluid He^3: thermodynamics, kinetics, sound dispersion, light and γ-ray scattering, etc. These works were based on the theory of a Fermi-liquid by L. Landau. I worked also during this time on the theory of strongly compressed matter.

In 1961 we published a book with L. Gor'kov and I. Dzyaloshinskii "Methods of the quantum field theory in statistical physics". Originally written in Russian, it was translated into English, German, Chinese, Japanese, and became (and still is) the main textbook on the subject.

In 1962–63 with my postgraduate L. Falkovsky we constructed the theory of semimetals of the Bi type. These substances have a very small number of charge carriers (in Bi ~ 10^{-5} per atom) and a very peculiar crystalline lattice differing from a simple cubic lattice by two small deformations. The resulting lattice has two atoms per unit cell, and in principle it could be an insulator. However, a simple cubic lattice has one atom per unit cell, and should be a "good" metal with the number of carriers of the order of one per atom, and small deformations cannot transform it into an insulator. This paradox can be resolved by constructing an artificial phase, which at zero deformation has an energy higher than a conventional metal but the energy decreases with deformation, so that eventually this phase becomes energetically favorable. This gives the opportunity to approach the (almost) insulating phase continuously. In this series of papers the energy spectrum was calculated, and the metal-insulator transition with the vanishing of the energy gap was predicted. The infrared properties were analyzed and the transparency thresholds in the frequency were established. All this was confirmed by experiments.

In 1962 our dear teacher, L.D. Landau, got in a car accident and suffered heavy injuries. His life was saved but his brain was damaged, and he never returned to science after that. He died in 1968 from remote consequences of the accident. After the accident the attitude towards theorists at the Institute for Physical Problems changed, and the Landau Group started to think about leaving.

In 1964 I was elected Corresponding Member of the Academy of Sciences, USSR (now Russian Academy of Sciences). In 1966 I was awarded the Lenin Prize together with V.L. Ginzburg and L.P. Gor'kov "for the theory of superconductivity in strong magnetic fields".

In 1965 I became the head of the Condensed Matter Theory Department in the newly organized Institute for Theoretical Physics (later named L. D. Landau Institute). I was one of its organizers.

In 1965–68 I published several papers on the Kondo effect at low temperatures, where I established the appearance of a resonance in the scattering amplitude of an electron from a magnetic impurity atom (now *Abrikosov-Suhl resonance*).

In 1971 I published a book "Introduction to the Theory of Normal Metals", which was translated into English. In 1972 I was awarded the International Fritz London Award for my works on low-temperature physics.

In 1970–75 I constructed the theory of gapless semiconductors, where I showed, that in substances of the type of HgTe, a strong interaction region close to the band matching point exists where the dependencies of various quantities on temperature and magnetic field are described by nontrivial power laws. At the same time I worked on the theory of an excitonic transition in Bi in strong magnetic fields. The predictions were in complete agreement with experimental data.

In 1975 I was awarded the title Doctor of Sciences Honoris Causa by the University of Lausanne (Switzerland).

In 1977–81 with my postgraduate, I.A. Ryzhkin, we constructed the theory of one-dimensional and quasi-one-dimensional metals. The main results were a) the probability distribution function of resistivity of a one-dimensional wire, where due to mesoscopic effects there was no self-averaging, and b) the conclusion that suppression of superconductivity in $(TMTSF)_2PF_6$ by nonmagnetic defects was an evidence of triplet pairing. Later this was confirmed. During the same years I worked on the theory of spin-glasses with short-range interaction, including semiconductor-based spin- glasses.

In 1982 with a group of experimentalists I was awarded the State Prize of USSR, for the works on semimetals and gapless semiconductors. In 1987 I was elected Full Member of the Academy of Sciences. In 1988 I published the book "Fundamentals of the Theory of Metals" on which I worked three years. It was translated into English and Japanese. The same year I was elected Director of the High Pressure Physics Institute in Troitsk, Moscow District. In 1989 L. Gor'kov, I. Dzyaloshinskii and myself were awarded the L. D. Landau award of the Academy of Sciences, USSR, for our book "Field Theory Methods in Statistical Physics".

In 1991 I accepted the offer of the Argonne National Laboratory, USA, and became Distinguished Argonne Scientist. Since then I continue to work at the same position. In the same year, together with V.L. Ginzburg and L.P. Gor'kov I was awarded the International John Bardeen Award and was elected as Foreign Honorary Member to the American Academy of Arts and Sciences. In 1992 I was elected Fellow of the American Physical Society.

Being at Argonne I became interested by the high-T_c layered cuprates. This interest resulted in a theory which was based on the Bardeen-Cooper-Scrieffer approach but took into account the specific features of the electron spectrum, namely the quasi-two-dimensionality and the existence of the "extended saddle point singularities", or "flat regions", which made the motion of quasiparticles in some regions of the Fermi surface quasi-one-dimensional. At the same time these regions had the maximal density of states. Another idea was the resonant tunneling mechanism of the electron transport between the CuO_2 planes. On the basis of these ideas I was able to explain almost all the unusual behavior of the high-T_c layered cuprates, including the isotope effect, neutron scattering, pseudogap and the metal-insulator transition.

In 1998 in connection with the experiments performed at Argonne and the University of Chicago I introduced a new phenomenon: "*Quantum Linear Magnetoresistance*". The analysis of experimental data showed that it was first discovered experimentally by Piotr Kapitza, as early, as 1928, but was confused with a different phenomenon. During these years in connection with experiments I studied also the effects of quantum interference on the magnetoresistance of layered substances and constructed a theory of an s-type superconductivity in UGe_2.

In 1999 I became a naturalized US citizen. In 2000 I was elected member of the National Academy of Sciences USA, and in 2001, as Foreign Member of the Royal Society of London, UK. In 2003 I received the title Doctor of Sciences Honoris Causa from the University of Bordeaux (France) and together with V. Ginzburg and A. Leggett was awarded the Nobel Prize in Physics "for pioneering work on the theory of superconductivity and superfluidity".

Apart from research, almost all my life I was teaching. First I was assistant, associated, full Professor at the Moscow State University, 1950–1969, then Professor at the Gorky (now Nizhniy Novgorod) University, 1970–72, and eventually Chair for theoretical physics at the Moscow Institute for Steel and Alloys (Technical University), 1976–1991. In USA I am Adjunct Professor at the University of Illinois at Chicago and the University of Utah. In addition I am a Leverhulm Adjunct Professor at the University of Loughborough, UK.

I am married and have two sons and one daughter.

My years with Landau

The discoverer of "type-II" superconductivity lets us in on the excitement of an important time for low-temperature physics

A. A. Abrikosov

During a decade or so of my life I had the opportunity to talk with Lev Davidovich Landau almost every day, and to profit from his advice. This may be why the period from about 1950 to 1960 was so successful for me, and I would like to share my memories of it with you.

In 1950 Vitali L. Ginzburg and Landau wrote their well known paper[1] on superconductivity. Without the microscopic theory, developed later by John Bardeen, Leon Cooper and J. Robert Schrieffer,[2] the meaning of several quantities entering the Ginzburg-Landau work remained unclear, above all the meaning of the "superconducting electron wave function" itself. Nevertheless this theory was the first to explain such phenomena as the surface energy of electrons at the superconducting-normal phase boundary and the dependence of the critical field and current in thin films on temperature and thickness.

A new type of superconductor

Experimental verification of the Ginzburg-Landau predictions concerning the critical fields of thin films was undertaken by my friend N. V. Zavaritski, who was then a young research student of A. I. Shalnikov. I often discussed the matter with Zavaritski. Generally his results fitted the theoretical predictions well, and he even managed to observe the change in the order of the phase transition with decreasing effective thickness (ratio of thickness to penetration depth at a given temperature). For that experiment he used the hysteresis of the dependence of the resistance $\rho(H)$ on the field. One day, Zavaritski slightly altered his technique of sample preparation. Usually he evaporated a metal drop on a glass plate and put the resulting mirror into a Dewar flask. This time instead, he began to do the evaporation inside the Dewar flask, with the glass plate at helium temperature.

We know now that in this case atoms reaching the plate are trapped at the place where they hit and are unable to move to form a regular structure. Therefore an amorphous substance is produced, which at every effective thickness will be a "type-II" superconductor. But at that time this was, of course, not known.

The dependence of critical field on thickness, as measured by Zavaritski,[3] did not follow the Ginzburg-Landau formulas. This seemed to be a paradox: Apart from its beauty, the theory really explained many things, and we were surprised to see that suddenly it had failed.

When Zavaritski and I discussed the possible origin of this discrepancy, we came to the idea that the approximation $\kappa \ll 1$ based on the surface-tension data (where κ is the adjustable Ginzburg-Landau parameter), could be incorrect for such new objects as the low-temperature films. In particular one could suppose that $\kappa > 1/2^{1/2}$. According to Ginzburg and Landau, the surface energy should be negative under these conditions. Intuitively, they felt that in this case the phase transition in the magnetic field would always be of second order, and this was in fact what Zavaritski observed.

When I calculated the dependence of critical field on effective thickness with $\kappa > 1/2^{1/2}$, the theory appeared to correspond to the experimental data. This gave me the courage, in a 1952 paper[4] containing this calculation, to state that apart from ordinary superconductors whose properties were familiar to everybody working on the subject, there exist in nature superconducting substances of another type, which I propose to call "superconductors of the second group" (now called type-II superconductors). The division between the first and the second group

Alexei Alexeivitch Abrikosov, the winner of the eighth Fritz London Award for low-temperature physics, is at the Landau Institute for Theoretical Physics in Moscow.

The theoretical group at the Institute tor Physical Problems in 1955: (seated, left to right) L. Prozhorova, A. A. Abrikosov, I. M. Khalatnikov, L. D. Landau, E. M. Lifshitz, (standing, left to right) S. Gerstein, L. P. Pitaevski, L. Veinstein, R. Arkhipov, I. Dzialoshinski.

was defined by the relation between the quantity k and its critical value $1/2^{1/2}$.

After that I tried to investigate the magnetic behavior of bulk type-II superconductors. The solution of the Ginzburg-Landau equation in the form of an infinitesimal superconducting layer in a normal sea of electrons was already contained in their paper. Starting from this solution I found that below the limiting critical field, which is the stability limit of every superconducting nucleation, a new and very peculiar phase arose, with a periodic distribution of the Ψ function, magnetic field and current. I called it the "mixed state."

Landau showed a notable interest in this work and wanted me to publish my results for the vicinity of the upper critical field, which I named H_{c2}. But I wanted to understand how the new mixed state looked in the total range of fields.

Landau has his doubts

At this time I became ill and had to stay in bed for almost three months. One day Landau visited me. The conversation, as in most cases, concerned everything but physics, and Landau sipped with great pleasure from a glass of *glühwein,* which was not at all like him. And then suddenly I destroyed all this paradise by telling him what I had wanted for the mixed state, namely the elementary vortices. As Landau's eyes fell on the London equation with a delta function on the right-hand side, he became furious. But then, remembering that a sick person should not be bothered, he took possession of himself and said: "When you recover we shall discuss it more thoroughly." He hastily bade me farewell and disapoeared.

He did not come to see me any more. When I felt better and appeared at the Institute and tried to tell him again about the vortices, he swore rather ingeniously. At that time I was still very young and did not know the temper of my teacher well enough. He had seen many kinds of pseudoscience in his life, and this made him suspicious toward unusual statements. However, by making some effort and disregarding the noise he made, one could always "drag through" him any reasonable idea. But at that time I sadly put my calculations in my table drawer until better times.

But in fact the idea was not so bad. Analyzing the solution that I got close to H_{c2}, I saw that in the plane perpendicular to the field there are points where Ψ becomes zero. The phase of the Ψ-function changes by 2π along a path around such a point. I thought about why such singularities should appear, and saw that it could not be otherwise. Indeed the Ginzburg-Landau equation contained not the magnetic field but the vector potential. If the magnetic field does not vary in sign over the whole sample, then the vector potential must increase with the coordinate. But the physical state in a uniform field (as is the situation close to H_{c2}) must be either uniform or, at most, vary periodically in space. So the increase of the vector potential must be compensated for by a change in the phase of the Ψ-function.

The figure on page 58 helps us to see this. Let the field be along the z-axis and let us choose the vector potential $\mathbf{A}$ such that $A_y = Hx$. Imagine the xy plane. The gray points are those I have mentioned, where Ψ becomes zero. If we want to have a unique determination of the phase we must draw cuts in the plane. We draw them through the gray points parallel to the y-axis. From the figure it is evident that when we go around the points the phase increases by $(\Delta\Psi)_1 = \pi y/a$ if we move along the lower path, and by $(\Delta\Psi)_1 = \pi y/a$ if we move along the upper one. That means that at every cut the gradient of the phase $\partial\Psi/\partial y$ undergoes a jump $2\pi/a$. If we use ordinary units (at that time I used the dimensionless Ginzburg-Landau units)

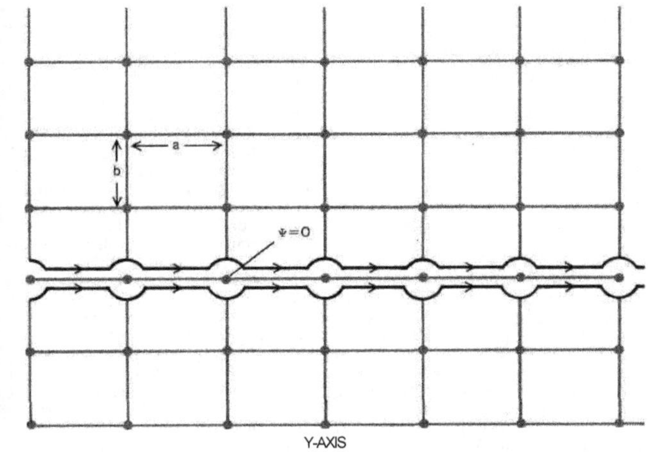

Vortex lattice of a type-II superconductor is seen in a schematic diagram with magnetic field along the z-axis and unit cell dimensions a and *b*. To determine the phase of the Ψ-function uniquely, draw a cut through the gray points parallel to the y-axis. Around each point, the phase changes by π if we follow the lower (clockwise) path and by $-\pi$ along the upper path, so that, with *y/a* the number of points, $(\Delta\Psi)_{lower}$ is $\pi y/a$ and $(\Delta\Psi)_{upper}$ is $-\pi y/a$. At every cut then, $\partial\Psi/\partial y$ jumps by $2\pi y/a$. In the limit, with decreasing magnetic field, one vortex exists, so that Ψ must equal zero along the z-axis.

we see that compensation for the increase in A_y demands

$$\frac{2e}{c} Hb = \frac{2\pi\hbar}{a}$$

or

$$Hab = \frac{\pi\hbar c}{e} = \Phi_0$$

which is the flux quantum. Because I used dimensionless quantities I did not mention the flux quantum on the right, but I understood that with a decreasing magnetic field the cell dimensions *"ab"* must increase, and as a limit one vortex must be considered, in which case the phase of Ψ changes by 2π in going around it. On the z-axis Ψ must equal zero. Otherwise the Ψ-function is not uniquely defined. Such a picture allowed me to find the lower critical field H_{c1} and the magnetization curve *M(H)*.

A worthwhile diversion

But, as I have noted (and this was in 1953), after the fuss my professor made I dropped the matter. There was also another reason. Interesting news appeared in a completely different field—quantum electrodynamics.

After the wonderful work of Julian Schwinger, Richard Feynman and Freeman Dyson, many people were interested in knowing whether it would be possible to sum up all the higher-order corrections and to find formulas for the Green's functions and physical phenomena without developing in powers of the fine-structure constant $e^2/\hbar c$. My friend Isaac Khalatnikov as well as myself had an old interest in the problem. At this time a paper[5] by S. F. Edwards was published, in which an attempt was made to sum up a ladder sequence of Feynman graphs for the electron-photon vertex part. We studied this paper and finally came to the conclusion that Edwards had done what he could but had not done what was really necessary, since he had no reason to choose this particular sequence. We tried to do something better and finally wrote some relation between the electron Green's function and the vertex part, which, as it soon became clear, was completely wrong. However when we substituted it into Dyson's equations we began to obtain various interesting consequences, as for example expressions for the electron mass and for the renormalized interaction.

Landau became extremely interested, but being busy with other problems he had no time to study the new technique of the quantum field theory. So he asked Khalat and me to teach him. I must confess that at that moment we were able to do calculations but had no true understanding of the fundamentals of the theory. Landau swore heavily, but after a month he said that he understood everything. He explained to us that we needed to find the main sequence of graphs having the highest power of the big logarithm with a given power of the interaction constant. This simple idea put everything in its place. We were indeed successful in calculating the asymptotic expressions for the Green's functions and various physical phenomena at high energies.[6] Moreover the principle of summing the main Feynman graphs proved afterwards to be extremely useful in various statistical problems.

Vortices from Feynman

Being occupied with such interesting things I of course did not turn my mind back to the work on type-II superconductors. But Landau had a long-standing interest in the state of He II in a rotating vessel. On the one hand, the helium should not be dragged along by the wall, but on the other hand, this was energetically favorable. In 1955 Landau and E. M. Lifshitz published a paper[7] in which they proposed a layer-type structure with velocity jumps on the layer boundaries. After a year they discovered Feynman's paper[8] in which it was shown that elementary vortices appear in rotating helium. Landau immediately said that Feynman was right, and that he and Lifshitz were wrong. Of course it was true. In terms of our superconducting notation, the He II could be considered as an extreme case of a type-II superconductor, with a correlation length of the order of interatomic distances and with an infinite penetration depth. But at that time this was not so evident.

When Landau began to praise Feynman's work I asked him: "Dau, why are you ready to accept the vortices from Feynman and flatly rejected the same idea from me." Landau answered: "You had something different." "Well then look, please," I said, and produced my calculations from the drawer. This time no objections followed. We discussed the subject very thoroughly and Landau's remarks were very useful.

When everything was put in order I remembered that I had already seen very similar magnetization curves with two critical fields, namely those of superconducting alloys. Digging for the corresponding experimental data I found the old work of 1937 by L. V. Shubnikov, W. I. Khotkevich, J. D. Shepeliov and J. N. Riabinin[9] on the magnetization curves of Pb-Tl alloys. They had prepared their samples very carefully, annealing them for a long time close to the melting temperature. So their samples were probably sufficiently uniform, and this was also confirmed by a rather small hysteresis. But at that time, and during the subsequent 25 years, everybody explained this form of the magnetization curve in terms of the formation of a "Mendelssohn sponge," that is, a nonuniform structure with a distribution of critical parameters.[10] It is worth mentioning that even many very good experimentalists finally believed in the mixed state only after they saw the powder figures of a vortex lattice obtained in 1966 by Uwe Essmann and Hermann Träuble.[11]

So my work was published in 1957 in *JETP*.[12] In the same year I reported it at a low temperature conference in Moscow in which some physicists from Oxford and Cambridge also took part. Nobody understood a single word. This could be explained however, by the fact that I had a terrible cold with high temperature and had hardly any idea myself of what I talked about. The translation of the paper was then published in the *Journal of Physics and Chemistry of Solids*,[13] but with more than one hundred errors in the formulas and text, and this of course did not improve the situation.

As you know, in the same year—1957—the famous paper by John Bardeen, Leon Cooper and J. Robert Schrieffer appeared. Everybody became enthusiastic about its ideas—we ourselves among the others. Therefore my paper did not attract attention. Of course it would be unjust for me to complain, because by using BCS theory we managed to get a lot of interesting results and were able to develop and improve greatly the methods of statistical calculations.

Then, in 1961 John Kunzler and his colleagues discovered that the alloy Nb_3Sn possesses a critical field of about 100 kOe.[14] Shortly after that, alloys with high critical fields began to be used for constructing superconducting coils. This drew attention to the theory of superconducting alloys.

Type-II rediscovered

In my work of 1957 I noted the connection between the quantity κ and the free path length; as I have said, nobody knew about this paper. However there did exist a series of papers by Brian Pippard[15] in which he qualitatively established the connection between the sign of the surface tension on the one hand and the ratio of the correlation length to the penetration depth on the other. He also mentioned the decrease of the correlation length with the free path length. In 1961, on the basis of Pippard's ideas, Bruce Goodman rediscovered that alloys with high critical fields have a negative surface tension.[16] Goodman calculated the magnetic properties of such alloys supposing a simple layer model for the distribution of the normal and superconducting phases. The results were in qualitative agreement with experiment.

I don't know how it happened, but probably somebody told Goodman about my work. What followed was completely incredible. In 1962 Goodman published another paper[17] in which he gave a short presentation of my theory and analyzed the experimental data for type-II superconductors, comparing them with the predictions of both theories, mine and his own. He came to the conclusion that the vortex model fits experiment much better than the laminar one. So the aim of Goodman's paper was to prove that his theory was worse than mine! I have never in my life seen another example of this kind, and took the first opportunity to express my admiration to Goodman.

After this paper by Goodman, physicists working on superconductivity finally developed an interest in my work, and I suppose that this was to a considerable extent the cause for the favor done to me by the London Award Committee. Therefore I would like to express once more my gratitude to Bruce Goodman.

Of course the fact that the award

Alexei Alexeivitch Abrikosov is seen here in a moment of relaxation. This photograph was taken in 1967 during a visit to the Kamchatka peninsula in the northeastern USSR.

A more formal portrait of Abrikosov is seen in this photograph, taken in 1970.

given me has the name of Fritz London is particularly pleasant for me, because he, together with Heinz London, invented the first phenomenological equations of superconductivity. As became apparent much later, they had described the electrodynamics of just the type-II superconductors. Also it was Fritz London who introduced the notion of the magnetic flux quantum, which has a direct relation to the subject.

Finally I would like to mention some of my other activity in the low-temperature field. With one exception it had a much quicker reception. My work with L. P. Gor'kov on the microscopic theory of superconductivity (high frequency behavior,[18] Knight shift[19] and the influence of impurities, particularly magnetic impurities,[20] on the properties of superconductors) was accepted by readers rather soon after publication, and has been developed further. It appeared, by the way, that the gapless superconductivity that we predicted for magnetic alloys could also exist under rather different conditions. My work on the Kondo effect[21] and the studies with Khalatnikov[22] of the properties of liquid He^3 had a lucky fate too.

The exception I have mentioned is the theory of semimetals of the bismuth type that L. A. Falkovski and I constructed ten years ago.[23] This theory explained such peculiarities as the strange crystal structure, the small number of free carriers and the large dielectric constant and gave formulas for the energy spectrum of bismuth that fitted experiment very well.

Recently I got some new results based on this theory and these have been published in the *Journal of Low Temperature Physics*.[24] I hope that they will be interesting to those who work on semimetals and on the metal-insulator transition. In conclusion I would like to express my gratitude to the Fritz London Award Committee for the honor it has given to me.

The author is the recipient of the 8th Fritz London Award for Distinguished Contributions in Low Temperature Physics. This article is adapted from his acceptance speech, delivered in his absence by Pierre Hohenberg at the 13th International Conference on Low Temperature Physics in Boulder, Colorado last August.

References

1. V. L. Ginzburg, L. D. Landau, Zh. Eksp. Teor.Fiz. 20, 1064(1950).
2. J. Bardeen, L. N. Cooper, J. R. Schrieffer, Phys. Rev. 106, 162 (1957); 108, 1175 (1957).
3. N. V. Zavaritski, Dokl. Akad. Nauk. SSSR86,501(1952).
4. A. A. Abrikosov, Dokl. Akad. Nauk, SSSR86,489(1952).
5. S. F. Edwards, Phys. Rev. 90, 284 (1953).
6. L. D. Landau, A. A. Abrikosov, I. M. Khalatnikov, Dokl. Akad. Nauk. SSSR 95,773(1954).
7. L. D. Landau, E. M. Lifshitz, Dokl. Akad. Nauk. SSSR 100, 669 (1955).
8. R. P. Feynman, *Progress in Low Temperature Physics,* Volume 1 (C. J. Gorter, ed.), Interscience, New York (1955), page 34.
9. L. V. Shubnikov, W. I. Khotkevich, J. D. Shepeliov, J. N. Riabinin, Zh. Eksp. Teor. Fiz. 7, 221 (1937).
10. K. Mendelssohn, Proc. Roy. Soc. A152, 34(1935).
11. H. Trimble, U. Essmann, J. Appl. Phys. 39,4052(1968).
12. A. A. Abrikosov, Zh. Eksp. Teor. Fiz. 32, 1442 (1957) [Sov. Phys.—JETP 5, 1174(1957)].
13. A. A. Abrikosov, J. Phys. Chem. Solids 2,199(1957).
14. J. E. Kunzler, E. Buehler, F. S. L. Hsu, J. H. Wernick, Phys. Rev. Lett. 6, 89 (1961).
15. A. B. Pippard, Proc. Roy. Soc. A216, 547 (1953).
16. B. B. Goodman, Phys. Rev. Lett. 6, 597 (1961).
17. B. B. Goodman, IBM J. Res. Devel. 6, 63(1962).
18. A. A. Abrikosov, L. P. Gor'kov, I. M. Khalatnikov, Zh. Eksp. Teor. Fiz. 35, 265 (1958) [Sov. Phys.—JETP 8, 182 (1959)].
19. A. A. Abrikosov, L. P. Gor'kov, Zh. Eksp. Teor. Fiz. 42, 1088 (1962) [Sov. Phys.—JETP 15, 752 (1962)].
20. A. A. Abrikosov, L. P. Gor'kov, Zh. Eksp. Teor. Fiz. 39, 1781 (1960) [Sov. Phys.—JETP 12, 1243 (1961)].
21. A. A. Abrikosov, Physics 2, 5 (1965).
22. A. A. Abrikosov, I. M. Khalatnikov, Rept. Progr. Phys. 22, 329 (1959).
23. A. A. Abrikosov, L. A. Falkovski, Zh. Eksp. Teor. Fiz. 43, 1089 (1962) [Sov. Phys.—JETP 16, 769 (1963)].
24. A. A. Abrikosov, J. Low Temp. Phys. 8, 315(1972). O

TYPE II SUPERCONDUCTORS AND THE VORTEX LATTICE

Nobel Lecture, December 8, 2003

by

ALEXEI A. ABRIKOSOV

Materials Science Division, Argonne National Laboratory, 9700 South Cass Ave., Argonne, IL 60439, USA.

In 1950, Vitalii Ginzburg and Lev Landau published their famous paper on the theory of superconductivity [1]. The approach was based on the general theory of the second order phase transitions proposed by Landau in 1937 [2]. There Landau introduced the main variable, the so called "order parameter" which was finite below the transition and zero above it. Different phase transitions had different order parameters, and whereas it was evident for, e. g., the ferromagnetic transition, namely, the spontaneous magnetization, it was far less evident for the superconducting transition. Ginzburg and Landau had a stroke of genius, when they chose, as the order parameter some sort of *wave function*. At that time nobody knew about Cooper pairs, and about their Bose condensate, where all particles become coherent, i. e. described by the same wave function. This assumption was the basis of the new theory, which managed to solve the main contradiction of the old theory by Fritz and Heinz London [3], namely, the positive surface energy. Besides it made many useful predictions, such as the critical magnetic field of thin films, the critical current in thin wires etc.

All these predictions required experimental verification, and my friend and University mate, Nikolay Zavaritskii, started to measure the critical field of thin films. Theory and experiment fitted perfectly, including the change of the nature of the transition: first order at larger thicknesses and second order at smaller ones. Everything seemed OK but Alexander Shalnikov, the boss of Zavaritskii was not satisfied. He said that the films used by Zavaritskii were bad, since they were prepared at room temperature. The atoms of the metal, evaporated on a glass substrate, could agglomerate, and therefore the film, actually consisted of small droplets. In order to avoid that, Shalnikov recommended to maintain the glass substrate at helium temperature during evaporation and until the measurements were finished. Then every metal atom hitting the surface would stick to its place, and the film would be homogeneous.

Zavaritskii followed this advice, and the result was a surprise: the dependence of the critical field on the thickness, or temperature (the theory contains the ratio of the thickness to the penetration depth, which depends on temperature), did not fit the predictions of the Ginzburg-Landau (GL) theory. Discussing these results with Zavaritskii, we couldn't believe that the theory was wrong: it was so beautiful, and fitted so well the previous data. Therefore we tried to find some solution in the framework of the theory itself, and we found it. The equations of the theory, where all entering quantities were expressed in corresponding units, depended only on the dimensionless "material" constant, κ, which was later called the Ginzburg-Landau (GL) parameter. The value of κ could be defined from the surface energy between the normal and superconducting phases. The latter, in its turn could be calculated from the period of the structure of the intermediate state. These data for conventional superconductors led to very small values of κ, and therefore the calculations in the paper by Ginzburg and Landau were done for this limiting case. It was also established that with increasing value of κ the surface energy between the superconducting and normal layers would become negative, and since this contradicted the existence of the intermediate state, such a case was not considered.

Therefore, I decided to look what would happen, if $\kappa > 1/\sqrt{2}$, when the surface energy became negative. The transition in this case became of the second order for any thickness. The theory fitted completely Zavaritskii's experimental data, and this led us to the conclusion that there exists a special kind of superconductors, which we called "superconductors of the second group", with $\kappa > 1/\sqrt{2}$ and negative surface energy. Now they are called Type II superconductors. I published my derivation in the Russian journal Doklady Akademii Nauk SSSR in 1954 [4]. This was the first time when the Type II superconductors were introduced. Since, however, this journal was never translated into English, there is a considerable confusion on this point, and the most common is just a statement that "there exist two types of superconductors...". In Russia the idea about Type II superconductors raised no objections, although such materials were considered, as exotic. In this connection it is worthwhile to mention that virtually all new superconducting compounds, discovered since the early 1960s up to the present time, are Type II superconductors. They include organics, A-15, Chevrel phases, heavy fermionic materials, fullerenes and high-temperature superconductors. One could say, that now Type I superconductors have become exotic.

After the work on films I decided to look what are the magnetic properties of bulk type II superconductors. It was definite that the transition to the normal state in magnetic field would be of the second order, and the transition point is defined by the condition of existence of a stationary infinitesimal nucleus. Such nucleation fields were actually defined in the GL paper. Their highest value corresponded in Type II superconductors to the so-called upper critical field H_{c2}:

$$H_{c2} = H_{cm}\kappa\sqrt{2} \quad , \tag{1}$$

where H_{cm} is the scale of magnetic fields, and it is defined, as the critical field of a first order transition of a Type I ($\kappa < 1/\sqrt{2}$) bulk cylinder in a longitudinal field.

At smaller magnetic fields one could imagine a linear combination of such nuclei centered at different points. Due to the homogeneity of space, the solution has to be periodic. Taking into account the necessity to renormalize the vector potential one arrived at the following general expression for the order parameter:

$$\Psi = \sum_{n=-\infty}^{\infty} C_n \exp\left[ikny - \frac{1}{2}\kappa^2\left(x - \frac{kn}{\kappa^2}\right)^2\right]. \tag{2}$$

Here and further the coordinates are measured in units of the penetration depth, λ, and k in 1/λ. The free energy becomes:

$$\frac{\Omega_s - \Omega_n^{(0)}}{H_{cm}^2/4\pi} = B^2 - \frac{\kappa - B}{1 + (2\kappa^2 - 1)\beta_A}, \tag{3}$$

where $\Omega_n^{(0)}$ is the free energy of a normal metal in zero field, B is the magnetic induction (average field), measured in units $H_{cm}\sqrt{2}$, and

$$\beta_A = \frac{\overline{|\Psi|^4}}{\left(\overline{|\Psi|^2}\right)^2}. \tag{4}$$

This dimensionless constant depends only on the geometry of the array, i.e., on the relative values of the coefficients C_n in (2).

According to (3), the choice must be such that β_A should be minimal. It can be shown that this minimal value is $\beta_A = 1.16$, and it corresponds to the following selection: $C_{n+4} = C_n, C_0 = C_1 = -C_2 = -C_3$ and $k = \kappa(\pi\sqrt{3})^{1/2}$. This function corresponds to a triangular lattice. A slightly larger value $\beta_A = 1.18$ characterizes the square lattice with equal coefficients $C_n = C$ and $k = \kappa(2\pi)^{1/2}$. In the latter case it is easier to illustrate the properties of the solution. It can be represented as a theta-function, namely,

$$\Psi = C\exp\left(-\frac{1}{2}\kappa^2 x^2\right)\vartheta_3\left[1;(2\pi)^{1/2}\kappa i(x + iy)\right]. \tag{5}$$

Using properties of the theta-function it can be shown that under rotation of the coordinate systems by $\pi/2$ the function Ψ is only multiplied by a phase factor $\exp(i\kappa^2 xy)$. Thus $|\Psi|^2$ has a symmetry of a square lattice.

At points $x = (\sqrt{2\pi}/\kappa)(m+1/2)$, $y = (\sqrt{2\pi}/\kappa)(n+1/2)$, m and n – integers, the function Ψ vanishes. Near these points in polar coordinates

$$\Psi \equiv |\Psi| e^{i\chi} \propto x + iy = \rho e^{i\varphi}. \tag{6}$$

The phase $\chi = \varphi$, and hence, it changes by 2π along a contour around the zero of Ψ. A similar situation takes place in the case of a triangular lattice. A natural question appears; how did it happen that the solution has these points? We just took a linear combination of the simple solutions, centered at different points, and the appearance of the zeroes with phases changing by 2π happened "by itself". In order to explain their appearance we have to take into account that in the GL equations the magnetic field is represented by the vector potential. If, on the average, the magnetic field is constant, the vector potential has to grow with coordinate. Since, however, the absolute value of the

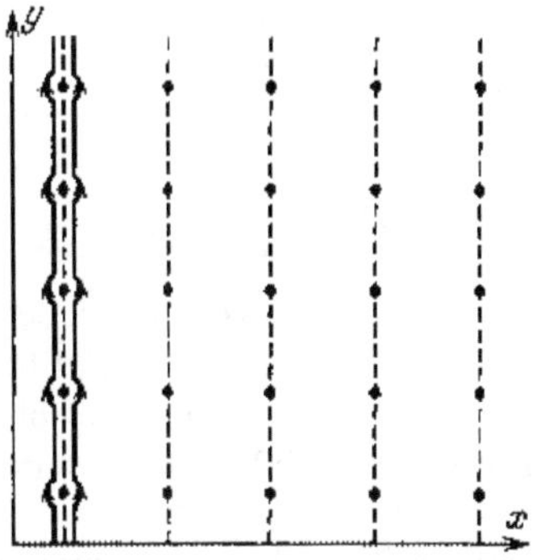

Figure 1. The dots correspond to zeros of the order parameter (square lattice). The dashed lines are the branch cuts introduced in order to make the phase single valued. The gradient of the phase has a discontinuity at every branch cut (see text).

order parameter cannot have a systematic growth, the growth of the vector potential has to be compensated. This can be done by the phase of the order parameter.

If the phase is taken into account, i.e.$\Psi = |\Psi|e^{i\chi}$, then χ enters the GL equations in the following combination with the vector potential:

$$\mathbf{A} - \frac{\hbar c}{2e}\nabla\chi \quad . \tag{7}$$

Consider the behavior of the complex order parameter in the coordinate plane (fig. 1). In order to define the phase unambiguously, branch cuts are introduced into this plane, going through the zeroes of the order parameter parallel to the y-axis.

If we move along the left shore of such a branch cut, the phase varies according to the formula

$$\chi_{left}(y) = \chi_{reg} - \pi\frac{y}{a} \quad ,$$

where the first term is the regular part, and the second is due to the rapid change of the phase in the vicinity of a zero of Ψ; a is the period. Along the right shore it changes according to the formula

$$\chi_{right}(y) = \chi_{reg} + \pi\frac{y}{a} \quad .$$

From these two expressions it can be concluded that the gradient of the phase has a discontinuity at every branch cut:

$$\Delta\left(\frac{\partial\chi}{\partial y}\right) = \frac{2\pi}{a} \quad . \tag{8}$$

If the magnetic field is directed along z, and we chose $A_y = Hx$, then the compensation of the growth of the vector potential, according to (7), can be achieved, if $Ha = \frac{\pi\hbar c}{ea}$, or

$$a = \sqrt{\frac{\pi \hbar c}{eH}} \, . \qquad (9)$$

It follows

$$Ha^2 = \frac{\pi \hbar c}{e} \equiv \Phi_0 \, . \qquad (10)$$

From these formulas two conclusions can be made: a) the period of the structure grows with decreasing magnetic field, and b) the flux of the magnetic field through one elementary cell is a universal constant, which is called "*magnetic flux quantum*". It is equal to 2.05×10^{-7} $Oe \times cm^2$ and was first introduced by F. London in 1950 [5].

The increase of the period with decreasing magnetic field happens not only close to H_{c2} but also at any field. Indeed, the reasoning, leading to fig. 1 and corresponding conclusions, remains justified, except that the vector potential is no more a linear function of the coordinate, and the compensation condition has to be reformulated. This leads to the substitution of the magnetic field by its average value $B = \frac{1}{a^2}\int_0^a\int_0^a Hdxdy$. Hence, we get the same result, as before, with B instead of H.

From this it can be concluded that even far from H_{c2} the period of the structure increases with decreasing magnetic field and its value, H_{c1}, at which $B = 0$, or $a = \infty$ is the boundary between a purely superconducting phase and the phase with a partial penetration of the magnetic field, which I called "*mixed state*". The boundary with the purely superconducting phase corresponds to the magnetic field

$$H_{c1} = \frac{H_{cm}}{\kappa\sqrt{2}} (\ln\kappa + 0.5) \qquad (11)$$

According to (1) with increasing κ the upper critical field H_{c2} grows, and simultaneously the lower critical field H_{c1} decreases.

Since the distance between the zeros of Ψ becomes infinite at H_{c1}, in its vicinity it is large, and only one such point can be considered. According to the GL theory the current can be written in a form

$$\mathbf{j} = \frac{\hbar e}{m}|\Psi|^2\left(\nabla\chi - \frac{2e}{\hbar c}\mathbf{A}\right) \, . \qquad (12)$$

In the vicinity of $\Psi = 0$ $\chi = \varphi$, and $\nabla\chi$ has only the φ-component, which is equal to $\frac{1}{\rho}\frac{\partial\chi}{\partial\varphi} = \frac{1}{\rho}$. Hence it is much larger than the second term in (12), and the current forms a vortex. In the general case these vortices from a lattice. The lines of the current in the vicinity of H_{c2} are presented at fig. 2.

A very similar structure has the triangular lattice, which for an isotropic model has a slightly lower energy. Since the energy difference is very small, in real substances the crystalline summetry can make the square lattice more favorable. Due to this structure, the mixed state is sometimes called "*the vortex lattice phase*".

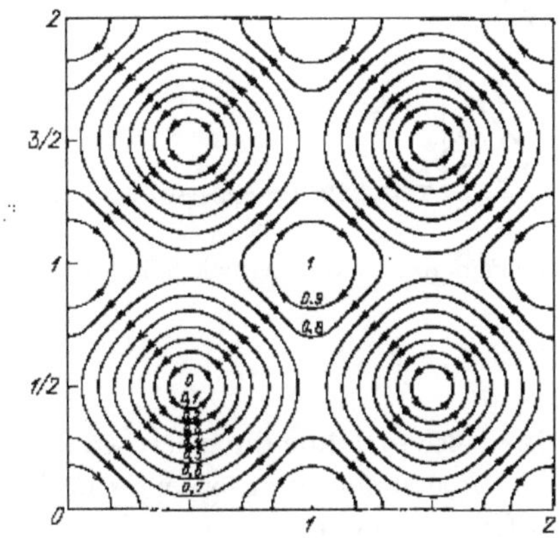

Figure. 2. The lines of current coinciding with the lines of constant $|\Psi|$ for a square lattice.

In the microscopic Bardeen-Cooper-Schrieffer (BCS) theory, as well as in the GL theory, which, as shown by Gor'kov [6], is the limiting case of the BCS theory at $T \to T_c$, there exist two characteristic lengths: the smaller "coherence length" ξ, which is the size of a Cooper pair, and the larger "penetration depth" λ. The GL parameter κ is, essentially, their ratio. For a pure superconductor at $T \to T_c$

$$\kappa = 0.96 \frac{\lambda_L}{\xi_0} \quad , \tag{13}$$

where $\lambda_L = (mc^2/4\pi ne^2)^{1/2}$ is the London penetration depth (n is the electron density), and $\xi_0 = 0.18\ (\hbar v/T_c)$is the coherence length at $T = 0$ (v is the electron velocity). In the case $\kappa >> 1$ $\lambda >> \xi$ (extreme type II, or London type, superconductor) every vortex has a "core" of the size ξ, where the order parameter varies rapidly, and the outer region of the size λ where the magnetic field decays to zero. According to formula (6), in the vicinity of the vortex axis the order parameter grows linearly with distance. The vanishing of Ψ at the center is due to the fact that this is the only way to avoid ambiguity of Ψ. Beyond distances of the order of ξ the order parameter approaches the equilibrium value at zero field. The overall shape of the behaviour of the order parameter and magnetic field in a vortex is presented at fig. 3.

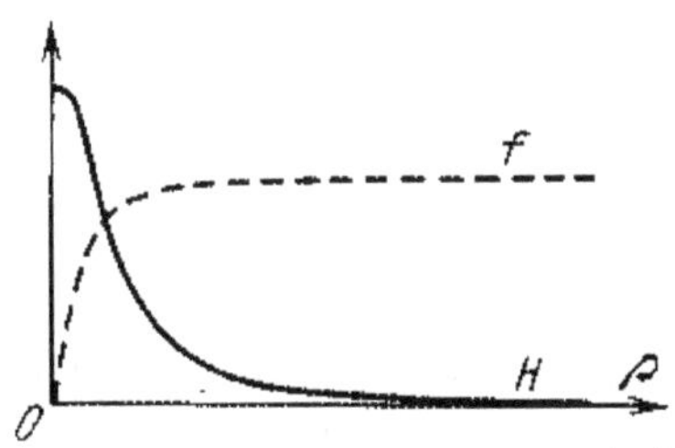

Figure 3. The plot of the magnetic field (bulk line) and $|\Psi|$ in a vortex.

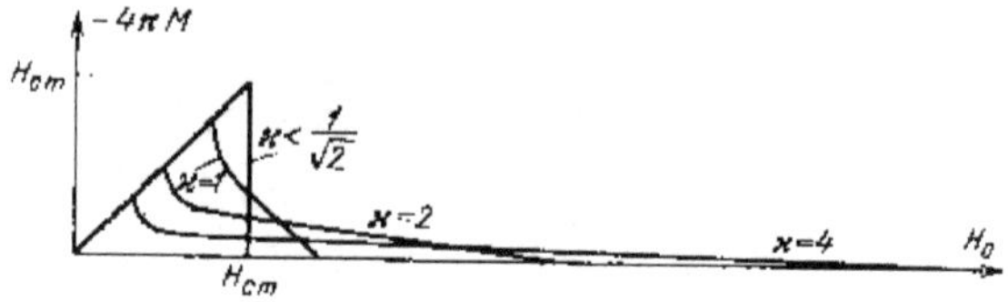

Figure 4. Dependence of magnetization on magnetic field for different values of κ.

The theory permitted also to define the macroscopic characteristics, namely, the dependence of the magnetization on external field. The latter is presented at fig. 4 for different values of κ.

For $\kappa < \frac{1}{\sqrt{2}}$ the dependence is a "triangle", reflecting an ideal diamagnetism below H_{cm} and no magnetization in the normal phase. At higher values of κ the vortex phase appears, and with increasing κ its lower boundary decreases, whereas its upper boundary increases. The limiting formula for the magnetization in the vicinity of the upper critical field is

$$-4\pi M = \frac{H_{c2} - H_0}{(2\kappa^2 - 1)\beta_A} \quad . \tag{14}$$

I compared the theoretical predictions about the magnetization curves with the experimental results obtained by Lev Shubnikov and his associates on Pb-Tl alloys in 1937 [7], and there was a very good fit.

Here I would like to describe the situation with experiments. The magnetization of superconducting alloys was first measured by De Haas und Casimir-Jonker in 1935 [8], and they got a gradual transition from the superconducting to the normal state with two critical fields. They explained it by inhomogenity of their samples. Shubnikov, who worked previously with De Haas, decided to make better samples, and his group annealed the alloys a long time at temperatures close to the melting point. After that the X-ray diffraction studies, performed at room temperature, did not show any inhomogenity. Since the authors could not imagine another explanation of the gradual transition, they wrote in their paper that the precipitation of another phase must happen at lower temperature. Unfortunately, L. V. Shubnikov was accused of an attempt to organize an "anti-Soviet strike", arrested and executed by the KGB the same year. I am sure that, given the opportunity, he would have discovered that there appeared a new phase, and that there exist a special kind of superconductors. I would like to pay here a tribute to Shubnikov, whose data gave me real inspiration. I have never met him but I heard about him from Landau, who was his close friend.

I made my derivation of the vortex lattice in 1953 but the publication was postponed since Landau first disagreed with the whole idea. Only after R. Feynman published his paper on vortices in superfluid helium [9], and Landau accepted the idea of vortices, he agreed with my derivation, and I

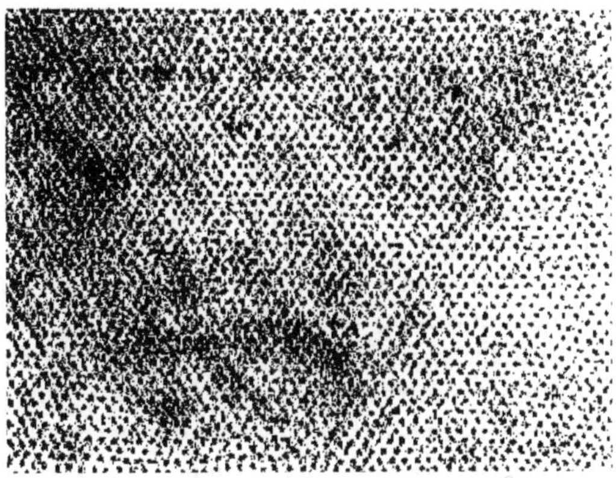

Figure 5. First decoration picture of vortices by Essmann and Traeuble (1967).

published my paper in 1957 [10]. Even then it did not attract attention, in spite of an English translation, and only after the discovery in the beginning of the 1960s of superconducting alloys and compounds with high critical magnetic fields there appeared an interest in my work. Nevertheless, even after that the experimentalists did not believe in the possibility of existence of a vortex lattice incommensurable with the crystalline lattice. Only after the vortex lattice was observed experimentally, first by neutron diffraction [11] and then by decoration [12], (fig. 5) they had no more doubts. Now there exist many different ways to get images of the vortex lattice. Apart from those, which I mentioned, there are electron holography, scanning tunneling microscopy (fig. 6) and magneto-optics.

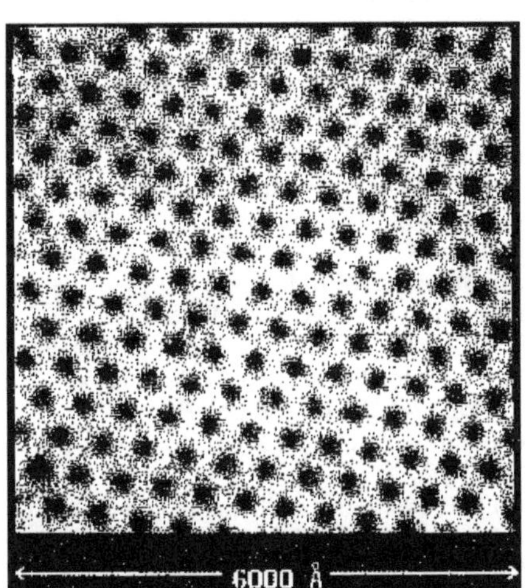

Figure 6. Vortices in $NbSe_2$ defined by scanning tunneling microscopy (STM).

After that I made only one work on vortices, namely I defined the lower critical field of thin films and the vortex lattice in its vicinity [13].

Although I worked in many different fields of theoretical physics afterwards, superconductivity was my favorite. In the beginning of the 1960s we did several works together with Lev Gor'kov. These were based on his Green functional presentation of the BCS theory, which permitted to extend the microscopic theory to spatially inhomogeneous problems. We studied the behaviour of superconductors in a high frequency field (with I. M. Khalatnikov) [14], the role of magnetic impurities [15], where we discovered the so called *"gapless"* superconductivity, and managed to solve the mystery of the finite Knight shift at low temperatures, introducing the spin-orbit scattering [16].

After the discovery of the high-temperature superconductivity in layered copper oxides by J. G. Bednorz and K. A. Mueller [17] I became interested in their properties. There existed many different approaches to these unusual substances and virtually all of them postulated some exotic mechanism of superconductivity. I based my approach on the BCS theory taking into account the specific features of the electron spectrum, mostly, the quasi-two-dimensionality and the so called "extended saddle point singularities", or "flat regions" in the electron spectrum [18]. Another idea was the resonant tunneling connection between the CuO_2, layers [19], which are responsible for conductivity and superconductivity. On this basis I was able to explain most of the experimental data about layered cuprates without dividing them into "good" ones, which should be mentioned on every possible occasion, and "bad" ones, which should be forgotten. As a result I can state that the so called "mystery" of high-T_c superconductivity does not exist.

REFERENCES

[1] V. L. Ginzburg and L. D. Landau, Zh. Eksp. Theor. Fiz **20,** 1064 (1950).
[2] L. D. Landau, Phys. Z. der Sowjet Union, **11**, 26 (1937); ibid. **11**, 129 (1937).
[3] F. London and H. London, Proc. Roy Soc. London, Ser. A **149,** 71 (1935).
[4] A. A. Abrikosov, Doklady Akademii Nauk SSSR **86**, 489 (1952).
[5] F. London, "Superfluids", V. 1, New York, 1950.
[6] L. P. Gor'kov, Soviet Phys. – JETP **9**, 1364 (1959); ibid. **10**, 998 (1960).
[7] L. V. Shubnikov *et al.*, Zh. Eksp. Teor. Fiz. **7.** 221 (1937).
[8] J. M. Kasimir-Jonker and W. J. De Haas, Physica **2**, 943 (1935).
[9] R. P. Feynman: in "Progress in Low Temperature Physics", ed. By D. F. Brewer, North-Holland, Amsterdam, 1955, V, 1, Ch. 11.
[10] A. A. Abrikosov, Sovjet Physics – JETP **5**, 1174 (1957).
[11] D. Cribier, B. Jacrot, L. M. Rao and B. Farnoux, Phys, Lett. **9**, 106 (1964).
[12] U. Essmann and H. Traeuble, Phys. Lett. A **24**, 526 (1967).
[13] A. A. Abrikosov. Sov. Phys. – JETP **19,** 988 (1964).
[14] A. A. Abrikosov, L. P. Gor'kov and I. M. Khalatnikov, Sov. Phys. – JETP **8**, 182 (1958); ibid. **10**, 132 (1959).
[15] A. A. Abrikosov and L. P. Gor'kov, Sov. Phys. – JETP **12**, 1243 (1961).
[16] A. A. Abrikosov and L. P. Gor'kov, Sov. Phys – JETP **15**, 752 (1962).
[17] J. G. Bednorz and K. A. Mueller, Zs. Physik. B **64**, 189 (1986).
[18] A. A. Abrikosov, Physica **C 341-348**, 97 (2000).
[19] A. A. Abrikosov, Physica **C 317-318**, 154 (1999).

Part Three

Selected Papers

SOVIET PHYSICS JETP VOLUME 3, NUMBER 1 AUGUST, 1956

The Infrared Catastrophe in Quantum Electrodynamics

A. A. ABRIKOSOV

Institute for Physical Problems, Academy of Sciences, USSR

(Submitted to JETP editor, October 28, 1954)

J. Exper. Theoret. Phys. USSR 30, 96-108 (January, 1956)

The infrared catastrophe is investigated by summing over diagrams. Expressions are obtained for $G(p)$ when $e^2/\pi \ln m^2/(p^2 - m^2) \gtrsim 1$ and for $\Gamma_\mu(p, q; l)$ when $p \cdot q \gg p^2 - m^2,\ q^2 - m^2$. The problem of radiation of additional quanta during the scattering of an electron of arbitrary energy by an external field is considered.

It is well known that the calculation of matrix elements of processes in quantum electrodynamics leads to infinities having various origins. Some of the infinities, which appear as a result of the divergence of integrals for large energies of virtual quanta and pairs, originate, as has already been pointed out[1], from an incorrect description of the interaction by means of the δ-function. In addition to these infinities (which are considered in references 1-4) there are also others, which result from integration over virtual quanta with k^2 close to zero, whenever the diagram under consideration includes a free electron with $p^2 = m^2$. This situation always occurs for matrix elements of real processes and has been named the infrared catastrophe. It is related to the fact that the very concept of a free line is a convention.

Actually, as shown by many authors[5], every process is accompanied by the radiation of a large number of low-energy quanta. For this reason, a properly formulated problem must take account of the possibility of such radiation with frequencies up to some maximum ω_{max}, corresponding to the fact that a real experimental apparatus always has a limited sensitivity to small changes in the energy of particles taking part in the process.

[1] L. D. Landau, A. A. Abrikosov and I. M. Khalatnikov, Dokl. Akad. Nauk SSSR 95, 497 (1954).

[2] L. D. Landau, A. A. Abrikosov and I. M. Khalatnikov, Dokl. Akad. Nauk SSSR 95, 773 (1954).

[3] L. D. Landau, A. A. Abrikosov and I. M. Khalatnikov, Dokl. Akad. Nauk SSSR 95, 1177 (1954)

[4] L. D. Landau, A. A. Abrikosov and I. M. Khalatnikov, Dokl. Akad. Nauk SSSR 96, 261 (1954).

[5] A. I. Akhiezer and V. B. Berestetskii, *Quantum Electrodynamics* GTTI (1953).

In practical calculations one usually uses the following device. One introduces a quantity λ_{min}, which serves as a fictitious photon mass. Then when we integrate over the momentum of the virtual photon the divergent result is replaced by a value proportional to $\ln(m/\lambda_{min})$. Next, one considers the same process, but replaces the virtual quantum by a real quantum (a process of the next higher order, in the preceeding approximation). It is not difficult to see that the cross section for the new process will be of the same order as the change in the first process which resulted from the introduction of the virtual quantum. We integrate over the momentum of the real quantum up to frequency ω_{max}, so that we again introduce the fictitious λ_{min}. When we combine the new cross section with the correction to the old cross section, the terms in $\ln \lambda_{min}$ cancel.

Such a procedure is completely valid for low energy processes and for large limiting frequencies ω_{max}. As we shall see later, the necessary conditions for applicability of this procedure are:

$$(e^2/\pi)\ln(m/\omega_{max}) \ll 1 \quad \text{for} \quad E \lesssim m,$$

$$(e^2/\pi)\ln^2(E/m),$$

$$(e^2/\pi)\ln(m/\omega_{max})\ln(E/m) \ll 1 \quad \text{for} \quad E \gg m,$$

where E is the order of magnitude of the energy of the process. When these conditions are violated, the procedure is invalid since, on the one hand, it becomes possible actually to radiate a large number of real quanta, and on the other hand virtual processes of arbitrary order begin to play an important role. The purpose of the present work is to investigate this question in general. In doing this, we shall here limit ourselves to considering the scattering of an electron by an external field. However, in principle, the results obtained are also applicable to other processes. In addition to the infrared catastrophe and the question of the radiation of soft quanta, we shall also consider the radiation of hard quanta during high energy scattering.

1. GREEN'S FUNCTION OF AN ELECTRON FOR $p^2 \approx m^2$

In reference 2 a calculation was made of the Green's function $G(p)$ and the vertex part $\Gamma_\mu(p, q; l)$ in the appropriate region. However, it is not hard to see that this calculation of $G(p)$ and Γ_μ applied only for $p^2 \gg m^2$. For the case where p^2 is so close to m^2 that $e^2/\pi \ln m^2/(p^2 - m^2) \gtrsim 1$, the calculation given in reference 2 is invalid. Since this region is directly related to the problem in which we are interested, we shall calculate $G(p)$ and the corresponding Γ_μ for this case.

The equations for Γ_μ and $G(p)$ are given in reference 1, Eq. (1.4). In solving them we can no longer (as was done in references 1,2) carry out a transition in the integrals to a four-dimensional Euclidean space by means of the substitution $k_0 \to ik_0$, since the vector p is essentially timelike. But for just this reason we can achieve this by the substitution $k_m \to -ik_m$ $(m = 1, 2, 3)$.

Let us consider the integral in the equation for $G(p)$. The region of integration where $k^2 \gg m^2$ gives a result already known from reference 2, and is of no special interest when $p^2 \approx m^2$. But now there is a new region which gives rise to a logarithmic integral. This is the region $p^2 - m^2 \ll (p-k)^2 - m^2 \ll m^2$.

For the case where $p^2 \approx m^2$, we shall try to find a $G(p)$ of the form

$$G(p) = \beta(m^2)\,\nu\left(\frac{m^2}{p^2 - m^2}\right)\frac{\hat{p}+m}{p^2 - m^2}, \tag{1}$$

where β was defined in reference 2, and ν is an unknown slowly varying function.

As for $\Gamma_\sigma(p, p-l; l)$, we make use of the fact that the Γ_μ always appears, in the cases which we need, with the factor $\hat{p}+m$ preceeding and following it (in the region of interest to us, $l^2 \ll m^2$, so that $\hat{p} \approx \hat{q}$), so we shall determine it up to terms which give a small contribution when bracketed in this way.

We shall show that $\Gamma_\sigma(p, p-l; l)$ should be of the form

$$(\hat{p}+m)\Gamma_\sigma(p, p-l, l)(\hat{p}+m) = (\hat{p}+m)\gamma_\sigma(\hat{p}+m)\,\alpha(m^2)\,\mu\left(\frac{m^2}{(p-l)^2 - m^2}\right), \tag{2}$$

where μ is a slowly varying function. In fact, looking at the integral in the equation for Γ_σ, we see that the region of integration $k^2 \gg m$ together with the term γ_σ in the equation gives us the known result $\gamma_\sigma \alpha(m^2)$. In this way all the terms in the equation can be absorbed into $\alpha(m^2)$. But a logarithmic integral is obtained not only from this region but also from the region $q^2 - m^2 \ll (p-\hat{k})^2 - m^2 \ll m^2$. Here we can neglect k compared to m and, since $k^2 \ll m^2$, we can replace d_l and d_t by d_l^0 and $d_t^0 = 1$.

From the spinor factors appearing in the numerator of the integral we easily obtain

$$(\hat{p}+m)\gamma_\mu(\hat{p}-\hat{k}+m) \quad (3)$$
$$\times\ \gamma_\sigma(\hat{p}-\hat{l}-\hat{k}+m)\gamma_\nu(\hat{p}+m)$$
$$\approx 4\,p_\mu p_\nu(\hat{p}+m)\gamma_\sigma(\hat{p}+m).$$

The form of the spinor agrees with that assumed in (2). Furthermore it is easy to see that, except for the the rapidly varying part $D_{\mu\nu}$, the integrand depends only on the projection of k along the direction of p ($(p-k)^2 - m^2 \approx -2k \cdot p$). Because of this we can integrate over the other components of k.

Going over to a Euclidean space we obtain, as in reference 2, the relation (in our case $d^4k = (2\pi)^{-2} dk_0\, dk_1\, dk_2\, dk_3$),

$$d^4k \rightarrow (i/\pi)\, dk_p \mathrm{K}^2 d\,|\mathrm{K}\,|, \quad (4)$$

where the vector **K** is perpendicular to k_p.

Resolving the k which appears in $D_{\mu\nu}$ along k_p and **K**, integrating over **K** and making a change of variables, we obtain for $\mu(\eta)$ the equation

$$\mu(\eta) = 1 - \frac{e^2}{2\pi}(3d_t^0 - d_l^0)\int_0^\eta \mu^3(z)\,\nu^2(z)\,dz, \quad (5)$$

where $\eta = \ln(m^2/[(p-l)^2 - m^2])$. In integrating over **K** we considered only those terms which give a logarithmic integral with respect to k_p. In addition we assumed that changing the sign of $q^2 - m^2$ in μ has no essential effect on its magnitude (as will be evident from the result). Equation (5) confirms the correctness of the form assumed for Γ_σ in Eq. (2).

We now look at the integral in the equation for $G(p)$[1,4] in the interesting region where $p^2 - m^2 \ll (p-k)^2 - m^2 \ll m^2$. It is not hard to show that the integral in this region is of order $\int k^{-3} d^4k$ so that, as in reference 2, we must include contributions to $G(p-k)$ and to Γ_μ in the integral of order $(p^2 - m^2)/[(p-k)^2 - m^2]$. The correction to $G(p)$ is obtained simply by expanding the denominator of formula (1) for $G(p-k)$. The correction to Γ_μ is gotten from the equation for Γ_μ, given in reference 1, Eq. (1). In the present case the important range in the integral in the equation for Γ_μ is $p^2 - m^2 \ll (p-k)^2 - m^2 \ll (p-l)^2 - m^2$. In this region the integral is of order $\int k^{-3}\, dk$, and therefore, as is easily seen, the main terms are those of order $(p^2 - m^2)/[(p-k)^2 - m^2]$ in the first $G(p-k)$ and the first $\Gamma_\mu(p, p-k; k)$. The corrections to the other terms in Γ and G are unimportant.

The correction to Γ_μ should be of the form

$$(\hat{p}+m)\,{}'\Gamma\,(p, p-l; l)(\hat{p}+m) \quad (6)$$
$$= \alpha(m^2)(\hat{p}+m)\gamma_\mu(\hat{p}+m)$$
$$\times \frac{p^2-m^2}{(p-l)^2-m^2}\,\delta[p^2-m^2,\ (p-l)^2-m^2].$$

Substituting in the equation for Γ_μ, we get an equation for δ:

$$\delta(\xi,\eta) = -\frac{e^2}{2\pi}(3d_t^0 - d_l^0)\,\nu(\eta)\,\mu^2(\eta) \quad (7)$$
$$\times \int_\eta^\xi [-\mu(z) + \delta(\xi, z)]\,\nu(z)\,dz.$$

Now substituting the corrections to G and Γ_μ in the equation for G and using previous results[2,4], we obtain an equation for ν:

$$\frac{1}{\nu(\xi)} = 1 + \frac{e^2}{2\pi}(3d_t^0 - d_l^0) \quad (8)$$
$$\times \int_0^\xi [-\mu(z) + \delta(\xi, z)]\,\nu(z)\,dz.$$

Equations (5), (7), (8), enable us to determine μ and ν. However there is no need to solve them directly. Actually, according to Ward's theorem[6], the relation:

$$\mu(\eta)\,\nu(\eta) = 1 \quad (9)$$

must hold. But then Eq. (5) gives us the function $\mu(\eta)$ in the form:

$$\mu(\eta) = \exp[-(e^2/2\pi)(3d_t^0 - d_l^0)\,\eta]. \quad (10)$$

The solution of equation (7) also presents no difficulties, and leads to the following form for $\delta(\xi, \eta)$:

$$\delta(\xi,\eta)/\mu(\eta) \quad (11)$$
$$= 1 - \exp\{(e^2/2\pi)(3d_t^0 - d_l^0)(\eta - \xi)\}.$$

Substituting this in equation (8), which can be written in the form $1/\nu(\xi) = 1 - \delta(\xi, 0)$, we get

[6] J. Ward, Phys. Rev. 78, 182 (1950).

$$\nu(\xi) = \exp[(e^2/2\pi)(3d_t^0 - d_l^0)\xi]. \tag{12}$$

This formula satisfies (9), as required.*

2. THE VERTEX PART WHEN $p \cdot q \gg p^2 - m^2$, $q^2 - m^2$.

In the preceding section we have already calculated one of the vertex parts of the type of interest to us here, namely $\Gamma_\mu(p,q;l)$ for $p^2 \approx m^2$, $q^2 \approx m^2$, $l^2 \ll m^2$. The other type of vertex $-\Gamma_\mu(p,q;l)$ for $l^2 \gg p^2$, q^2, $p^2 \gg m^2$, $q^2 \gg m^2$, was found in a paper by Sudakov.[7] Here we consider the general case, making no restrictions on the vectors other than $p \cdot q \gg p^2 - m^2$, $q^2 - m^2$. As before, we shall sum only over the principal terms, i.e., those containing the maximum logarithmic terms in e^2. In particular, we shall not include terms of order $e^2 \ln(l^2/m^2)$. For convenience of presentation, we shall use the symbols $p^2 - m^2 = 2m\Delta m_1$ and $q^2 - m^2 = 2m\Delta m_2$.

In the integrals representing the successive approximations to the vertex part, there are parts which include an integration over momenta of virtual quanta greater than the norms of the real momenta (in our case l^2 and m^2) which appear in the vertex part. If we separate this part in each of the integrals, it is not hard to see that together with the free term, the part containing only integrals over this region gives $\gamma_\mu \alpha(l^2)$ (just as in reference 2). In each of the remaining integrals, it turns out that the integration over $k \gg l$ (or m) must, if we neglect small terms, be carried out along a line lying closer to the principal vertex than that for the integrals with $k \ll l$ (or m). All this leads to the result that in the successive approximations for Γ_μ, we can integrate only over $k \ll l$ (or m), and take account of the integrals over the other region by a factor $\alpha(l)^2$ common to all terms, which determines the dependence of the integral on the "cutoff" limit. The dependence on l^2 given by this factor is unimportant in our case since, as we shall see later, there are terms in Γ_μ of order $[e^2 \ln^2(l^2/m^2)]^n$, which are much greater than the terms arising from $\alpha(l^2)$ which are of order $[e^2 \ln(l^2/m^2)]^n$. Because of this, we can always take $\alpha(l^2) \approx \alpha(m^2)$. In calculating the integrals for $k \ll l$, we use a method similar to that developed in the paper of Sudakov[7]. The principal part of the integral is gotten by resolving k along p, q, and in the plane perpendicular to these vectors. In integrating over the last component we must take the residue for $k^2 = 0$. Thus the additional vertex parts from the virtual quanta will be of the type considered in the preceding section, and each factor $\alpha(m^2)\mu$ from such a vertex will cancel against the factor $\beta(m^2)\nu$ from the factor G which is placed next to this Γ on the side toward the fundamental vertex. So, in calculating the vertex parts which we need, we can use the zeroth order result from neighboring Γ's and G's.

To make clear what the region of integration is, we shall first consider Γ_μ in first approximation. We begin with the part containing $d_t k^{-2}\delta_{\mu\nu}$. It has the the well-known form

$$\frac{e^2}{\pi i}\int \frac{\gamma_\mu(\hat{p} - \hat{k} + m)\gamma_\sigma(\hat{q} - \hat{k} + m)\gamma_\mu d_t(k^2)\, d^4k}{[(p-k)^2 - m^2][(q-k)^2 - m^2]k^2} \tag{13}$$

We introduce variables u, v, x:

$$k = p\left(\frac{a^2u - av}{a^2 - 1}\right) + q\left(\frac{a^2v - au}{a^2 - 1}\right) + k_\perp, \qquad x = -k_\perp^2, \quad a = (pq)/m^2. \tag{14}$$

In terms of these variables, the volume element in k-space is

$$d^4k = (m^2/4\pi)a^2(a^2 - 1)^{-1/2}\, dx\,du\,dv. \tag{15}$$

Expressed in terms of these variables, the important expressions which appear in the denominators are equal to

$$\begin{aligned} k^2 &= m^2a^2(a^2-1)^{-1}(2auv - u^2 - v^2) - x, \\ (p-k)^2 - m^2 &= 2m\Delta m_1 - 2(pq)v + k^2, \\ (q-k)^2 - m^2 &= 2m\Delta m_2 - 2(pq)u + k^2. \end{aligned} \tag{16}$$

Since x is positive (integration region $0 - \infty$), u and v must be chosen so that the point $k^2 = 0$ corresponds to $x > 0$. This requires that $2auv - u^2 - v^2 > 0$. We shall take $a > 0$, $a^2 - 1 > 0$. The contrary case is treated similarly, or even more simply by analytic continuation. Thus the integration must extend over the region between the two lines

$$u_1 = v(a + \sqrt{a^2 - 1}), \qquad u_2 = v(a - \sqrt{a^2 - 1}) = v/(a + \sqrt{a^2 - 1}). \tag{17}$$

---

* The product $\mu(\eta)\nu(\eta)$ does not depend on $p^2 - m^2$, so that at first glance it seems that the infrared catastrophe does not occur. Actually, the scattering matrix element contains terms in the exponent of order $\frac{e^2}{\pi}\frac{l^2}{m^2} \times \ln\frac{m^2}{p^2 - m^2}$ whose determination goes beyond the accuracy of this section. (cf. later sections).

[7] V. V. Sudakov, J. Exper. Theoret. Phys. USSR 30, 87, (1956). Soviet Phys. JETP 3, 65 (1956).

In addition it is clear from Eq. (16) that u and v are limited by the conditions

$$\Delta m_1/a \ll |v| \ll 1, \qquad (18)$$

$$\Delta m_2/a \ll |u| \ll 1.$$

The limit from above corresponds to the condition $k \ll l$. It is not hard to see that integration over positive and negative regions of u and v which satisfy our conditions gives the same contribution, and can be replaced by twice the integral over $u > 0$, $v > 0$. In the numerator, because of the smallness of u, v, and x, we can drop the k. When we bracket Γ on the left and right with $(\hat{p} + m)$ and $(\hat{q} + m)$ we get the expression $4\, p\cdot q(\hat{p} + m)\gamma_\sigma \times (\hat{q} + m)$ in the numerator.

We introduce the variables $\lambda = -\ln v$, $\mu = -\ln u$. In terms of these variables, the region of integration will have a form like that shown in Fig. 1 (the cross-hatched part). Of course, other relations between $\ln a$, $\ln(m/\Delta m_1)$ and $\ln(m/\Delta m_2)$ are possible. The integral will be proportional to the area cross-hatched in Fig. 1.

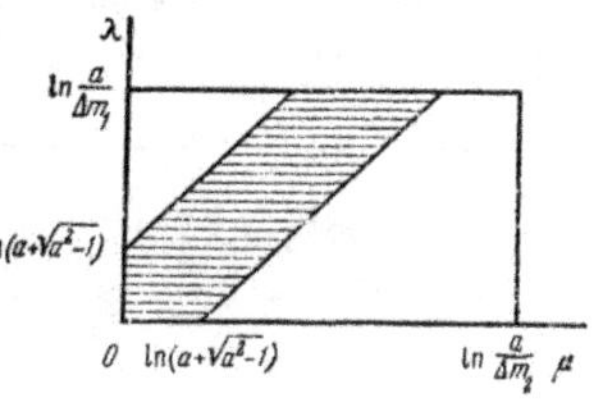

FIG. 1

We shall not yet give the corresponding expression, but will first consider the term in $(d_l - d_t)$. This term has the form

$$\frac{e^2}{\pi i}\int \frac{\hat{k}(\hat{p} - \hat{k} + m)\gamma_\sigma(\hat{q} - \hat{k} + m)\hat{k}[d_l(k^2) - d_t(k^2)]}{[(p-k)^2 - m^2][(q-k)^2 - m^2]k^4} d^4k\,. \qquad (19)$$

If we introduce in this integral the same variables as in (13), then the conditions (18) still hold. Then we can once more neglect $\hat{k}$ compared with $\hat{p}$ and $\hat{q}$ in the numerator, and upon multiplying on the left and right by $(\hat{p} + m)$ and $(\hat{q} + m)$ the numerator takes the form $4\, p{\cdot}k q{\cdot}k\, (\hat{p} + m)\gamma_\sigma(\hat{q} + m)$. Since the principal terms in square brackets of the expression in the denominator give the product $4\, p{\cdot}k q{\cdot}k$, upon cancelling these factors in numerator and denominator, we get in place of (19) the much simpler integral $\int k^{-4} d^4k$.

If we express this integral in terms of u, v, and x, it turns out that once more the important region in the integral is that in which $k^2 = 0$, but unlike what we had before, now both $k^2_{||}$ and x go to zero. As a result, the integral becomes proportional to the sum of the integrals $\int u^{-1}du$ (or, what is the same thing $\int v^{-1}dv$) taken along the two lines bounding the cross-hatched region of the type shown in Fig. 1. The particular form of the integral depends on the relation between Δm_1, Δm_2, and a.

Finally we obtain for the first approximation $\Gamma^{(1)}_\mu(p, q; l)$ the following formula:

$$(\hat{p} + m)\Gamma^{(1)}_\mu(p, q; l)(\hat{q} + m)$$
$$= (\hat{p} + m)\gamma_\mu(\hat{q} + m)(-f),$$

where the function f has the form:

$$f = (e^2/2\pi)\{a(a^2-1)^{-1/2} \times (3/2\eta^2 + 2\zeta\xi_1) + (1 - d_l^0)\xi_1\} \qquad (20a)$$

for $\xi_1 > 0$, $\xi_2 > \xi_1 + \eta$;

$$f = (e^2/2\pi)\{a(a^2-1)^{-1/2}[\eta^2 + \zeta(\xi_1 + \xi_2) - 1/2(\xi_1 - \xi_2)^2] + 1/2(1 - d_l^0)(\xi_1 + \xi_2)\} \qquad (20b)$$

for $\xi_1 > 0$, $\xi_2 > 0$, $\xi_1 + \eta > \xi_2 > \xi_1 - \eta$;

$$f = (e^2/2\pi)(3/2\eta^2 + 2\eta\xi_1 + 1/2\xi_1^2) \qquad (20c)$$

for $-\eta < \xi_1 < 0$, $\xi_2 > \xi_1 + \eta$;

$$f = (e^2/2\pi)[\eta^2 + \eta(\xi_1 + \xi_2) + \xi_1\xi_2 - 1/2\xi_2^2] \qquad (20d)$$

for $-\eta < \xi_1 < 0$, $0 < \xi_2 < \xi_1 + \eta$;

$$f = (e^2/2\pi)(\eta + \xi_1)(\eta + \xi_2) \qquad (20e)$$

for $-\eta < \xi_1 < 0$, $-\eta < \xi_2 < 0$,

where $\xi_1 = \ln(m/\Delta m_1)$, $\xi_2 = \ln(m/\Delta m_2)$, $\eta = \ln a$, $\zeta = \ln(a + \sqrt{a^2 - 1})$.

In all these terms, except those of the type $\zeta\xi$ in (20a) and (20b), we have taken $\zeta \approx \eta$, since taking account of their difference would go beyond our accuracy. For the same reason, the equations (20c), (20d) and (20e) should be taken for $a \gg 1$, and terms with one ξ or one η should be neglected.

We shall now assume that we are considering the n'th approximation for Γ_μ. In the corresponding diagram, n arbitrarily intersecting virtual lines can occur. We consider first the numerator of any diagram of this type. Upon multiplying on the left by $(\hat{p}+m)$ and on the right by $(\hat{q}+m)$ it will have the form

$$(\hat{p}+m)\gamma_\mu(\hat{p}+m)\gamma_\nu\ldots(\hat{p}+m)\gamma_\sigma(\hat{q}+m)\ldots\gamma_\xi(\hat{q}+m)\gamma_\eta(\hat{q}+m) \approx 2^{2n}p_\mu p_\nu\ldots q_\xi q_\eta\ldots(\hat{p}+m)\gamma_\sigma(\hat{q}+m). \tag{21}$$

Throughout all of this we have neglected $\hat{k}$ compared to $\hat{p}$ and $\hat{q}$, and the quantities $p^2 - m^2$ and $q^2 - m^2$ compared to $p_\mu q_\nu$. From formula (21) it is clear that the numerators of all diagrams of the same order are equal to one another. Differences can occur only in the denominators.

First we consider diagrams with a pair of lines. There can only be two such: one with parallel and one with intersecting lines. The integrand in both diagrams will be proportional to

$$\{(pk_1)[(pk_1)+(pk_2)][(qk_1)+(qk_2)](qk_1)\}^{-1} + \{(pk_1)[(pk_1)+(pk_2)][(qk_1)+(qk_2)](qk_2)\}^{-1}.$$

The integral with maximum degree of logarithmic divergence is obtained from the first term when $p\cdot k_2 \gg p\cdot k_1$, $q\cdot k_2 \gg q\cdot k_1$ and from the second if $p\cdot k_2 \gg p\cdot k_1$, $q\cdot k_1 \gg q\cdot k_2$ The expressions from both terms are the same, and so the integrations over $q\cdot k_1$ and $q\cdot k_2$ are independent. If we now make the change of variables $k_1 \rightleftarrows k_2$, and take half the sum of the expression thus obtained and the old expression, it turns out that $p\cdot k_1$ and $p\cdot k_2$ are also independent. Since the regions of integration are the same for both k's, we get the simple result

$$(\hat{p}+m)\Gamma_\mu^{(2)}(\hat{q}+m) = (\hat{p}+m)\gamma_\mu(\hat{q}+m)(-f)^2/2. \tag{22}$$

Generalizing this to a diagram of n'th order, we have

$$(\hat{p}+m)\Gamma_\mu^{(n)}(\hat{q}+m) = (\hat{p}+m)\gamma_\mu(\hat{q}+m)(-f)^n/n! \tag{23}$$

Summing all expressions of this type and noting, as pointed out at the beginning of this section, that the integrals over the momenta of virtual quanta in the region $k \gg l$ give a common factor $\alpha(m^2)$, we find the following result;

$$(\hat{p}+m)\Gamma_\mu(p,q;l)(\hat{q}+m) = \alpha(m^2)(\hat{p}+m)\gamma_\mu(\hat{q}+m)e^{-f} \tag{24}$$

This formula coincides with formula (10), if we take f from (20a) or (20b) and set $a \to 1$ $(\eta \to 0)$. If, however, we take f in the form (20e) we arrive at the result obtained by Sudakov[7].

If we consider the process of scattering of an electron by an external field, then to get the matrix element we must multiply (24) by the expression corresponding to a free line. It is well known that to each free line there corresponds $\{G(p) \times (\hat{p}-m)\}^{1/2}$, which in our case is equal to

$$\sqrt{\beta(m^2)}\left(\frac{m}{\Delta m}\right)^{(e^2/4\pi)(3-d_l^0)}$$

Later we shall show that only the cases (20b), (20d) and (20e) have a direct physical interpretation. The results for these cases are:

$$M_0 = (\bar{u}_1\gamma_\sigma u_2)e^{-f_1}, \tag{25}$$

$$f_1 = (e^2/2\pi)\{a(a^2-1)^{-1/2}[\eta^2 + \zeta(\xi_1+\xi_2) - 1/2(\xi_1-\xi_2)^2] - (\xi_1+\xi_2)\} \tag{26a}$$

for $\xi_1 > 0,\ \xi_2 > 0,\ \xi_1+\eta > \xi_2 > \xi_1 - \eta$;

$$f_1 = (e^2/2\pi)[\eta^2 + \eta\xi_1 + \eta\xi_2 + \xi_1\xi_2 - 1/2\,\xi_2^2] \tag{26b}$$

for $-\eta < \xi_1 < 0,\ 0 < \xi_2 < \xi_1 + \eta$;

$$f_1 = (e^2/2\pi)(\eta+\xi_1)(\eta+\xi_2) \tag{26c}$$

for $-\eta < \xi_1 < 0,\ -\eta < \xi_2 < 0$.

Generally speaking, when Γ_μ is an internal vertex, the relation between Δm_1 and Δm_2 is arbitrary. However, in this case we must consider lines which encompass several vertices, as we shall do for the case of one additional quantum.

3. GENERALIZATION OF THE FEYNMAN DIAGRAMS FOR OBTAINING PROBABILITIES OF MULTIPLE PROCESSES.

Before we consider the question of the radiation of additional quanta during the scattering of an electron, we shall present a method which considerably simplifies the calculation of processes involving a large number of real particles. In this section we shall consider the emission of real pairs as well as real quanta. In the literature, the rules for calculating matrix elements with virtual quanta are presented very well (cf., for example, reference 5). However the situation with regard to the rules for finding transition probabilities is not so satisfactory. We shall give a simple scheme which enables one to find such probabilities without difficulty.

First we consider processes in which the only real particles are photons. Aside from the appearance of their momenta in the propagation factors for virtual electrons, the presence of these photons manifests itself as follows. For each initial photon, there is a factor $2\pi/|\mathbf{K}|$ in the probability, where $\mathbf{K}$ is the photon wave vector. For all except one of the final photons there is a factor $(2\pi/|\mathbf{K}|) \times (2\pi)^{-3}\,\mathbf{K}^2 d|\mathbf{K}|\,d\Omega$ in the probability. The momentum of the remaining photon is determined by the conservation laws so that it gives only a factor $2\pi/|\mathbf{K}|$. In addition, each photon gives rise, in the appropriate place in the matrix element, to the factor $\hat{\epsilon}$ where ϵ is the polarization vector. The sum over polarizations is carried out by replacing $\hat{\epsilon}\cdot\cdot\cdot\hat{\epsilon}$ in the expression for the probability by $-\gamma_\mu\cdot\cdot\cdot\gamma_\mu$ ($\mu = 0, 1, 2, 3$). Finally there enters into the expression for the probability the quantity $2\pi\delta(E_{\text{in.}} - E_{\text{fin.}})$ where $E_{\text{in.}}$ and $E_{\text{fin.}}$ are the total energies of the initial and final states. The δ-function is eliminated by integrating over one of the $d|\mathbf{K}|$'s.

It is easy to see that the quantity $1/|\mathbf{K}|$ can be written in the form $2\int_{\omega>0} d\omega\,\delta(k^2)$, where $k^2 = \omega^2 - \mathbf{K}^2$. If we integrate with respect to the momentum of the corresponding photon, this replacement results in a great similarity between real and virtual photons. In fact, in place of $\int\cdots\gamma_\mu\cdots\gamma_\mu\cdots\delta_+(k^2) \times d^4k$ $(\delta_+(x) = \delta(x) - 1/i\pi x)$, which occurs for virtual photons, the expression

$$2\int_{\omega>0}\cdots\gamma_\mu\cdots\gamma_\mu\cdots\delta(k^2)\,d^4k$$

appears in the formula for the probability for real photons. The requirement $\omega > 0$ is relativistically invariant, and in particular signifies that the projection of the four-vector k on any time-like vector n with $n_0 > 0$ is positive. We note that in place of the factor $1/|\mathbf{K}|$ for the second photon, whose momentum is determined by the conservation laws and $\delta(E_{\text{in.}} - E_{\text{fin.}})$, we can simply put $2\delta_{\omega>0}(k^2)$. In doing this we still satisfy the energy conservation law and retain the correct factor, since $dE_{\text{fin}} = d\omega$.

We now go over to the case when the only real particles are electrons, and there are no photons or pairs. In this case each electron in the initial state contributes the expression $(1/2E)(\bar{u}\cdot\cdot\cdot u)$ to the probability, if we use the normalization $\bar{u}u = 2m$. Each of the electrons in the final state, except one whose momentum is determined by the conservation laws, gives the factor $(2E)^{-1}(\bar{u}\ldots u) \times (2\pi)^{-3}\,\mathbf{p}^2\,d|\mathbf{p}|\,d\Omega$, where E is the energy of the electron. The last of the final state electrons gives simply $(2E)^{-1}(\bar{u}\ldots u)$. If we carry out the sum over electron spins, we replace $\bar{u}u \to \hat{p} + m$. In addition there is a factor $2\pi\delta(E_{\text{in.}} - E_{\text{fin.}})$ in the probability. Just as in the case of photons, we can make the replacement

$$\frac{1}{E} = 2\int_{E>0}\delta(p^2 - m^2)\,dE.$$

If we are dealing with an electron for which we integrate over the momentum and sum over the spin, then we arrive at the result that each such electron gives a factor

$$\frac{\hat{p}+m}{2\pi}\int_{E>0}\delta(p^2 - m^2)\,d^4p.$$

As for the last of the final state electrons, after summing over its spin we can write for it $(\hat{p} + m)\,\delta_{E>0}(p^2 - m^2)$, which also includes $\delta(E_{\text{in.}} - E_{\text{fin}})$.

The similarity to virtual processes still does not appear explicitly in this form. In order to see the similarity, let us consider, for example, electron-electron scattering in zeroth approximation. From the conservation laws, it is clear that $d^4p = d^4q$, where q is the momentum of the virtual quantum which the electrons exchange with one another. If we make this substitution, then the expression for the probability suggests the matrix element for the interaction of a pair of electrons with exchange of a pair of identical virtual photons. A similar situation will occur in the more general case.

Now we consider the case when pairs are produced in a process. Since the momentum of one of the components of the pair is determined by the conservation laws if the momentum of the pair-producing photon is given, the integration need be taken over only one of the components, say the electron, and over the momentum of the photon producing the pair. In the integrand the factor $(\hat{p} + m)\,\delta_{E_p>0}(p^2 - m^2)$ refers to the electron, and the factor $(\hat{p} - \hat{k} + m)\,\delta_{E_{p-k}>0}((p-k)^2 - m^2)$ to the positron. The generalization to the case where real quanta, electrons, and pairs are present, presents no difficulty.

The rules we have given can be presented in the form of generalized Feynman diagrams. In fact we shall draw a diagram which shows a sort of doubled diagram for the matrix element of our process--a diagram together with its mirror image (the latter corresponds to the Hermitian adjoint matrix element). We now join those photon and electron lines for which there is a summation over spin and integration over momentum, provided the latter is not excluded by the conservation laws. We shall put a dash through such connected lines in order to distinguish them from the usual ones. After this the diagram will be completely analogous to the usual Feynman diagram for a matrix element. In particular, real pairs will be described by a loop of dashed electron lines. We note that if the matrix element is a sum of diagrams, then in the generalized diagrams each of the initial ones is joined in turn to the mirror images of all the others as well as to its own image.

In addition to virtual and intermediate lines, the fundamental elements of the generalized diagram will be:

1. A real photon line. The factor for it is

$$2\delta_{\omega>0}(k^2).$$

2. A real electron line. Its factor is

$$\frac{\hat{p}+m}{2\pi}\,\delta_{E>0}(p^2 - m^2).$$

3. Two identical photon lines between a pair of electrons, corresponding to the Coulomb interaction. The factor appearing for these lines is

$$(4\pi/q^2)^2.$$

The integration in the generalized diagram is taken over all closed photon lines, dashed as well as undashed, and over electron loops made up of ordinary and dashed lines. In the case of a loop, one also takes the spur. The factor $1/\pi i$ for photons and $-1/\pi i$ for a loop occurs only for virtual processes.

All of the rules given here refer to the case where the detailed characteristics of the emitted particles are not of interest (or are determined by conservation laws), i.e., we have closed lines. If this is not the case, then the usual factors for real particles occur. As examples, generalized diagrams are shown in Fig. 2 for: a – Compton effect, b – electron-electron scattering, c – materialization of a pair in the field of an electron. We show only one of the graphs for each effect.

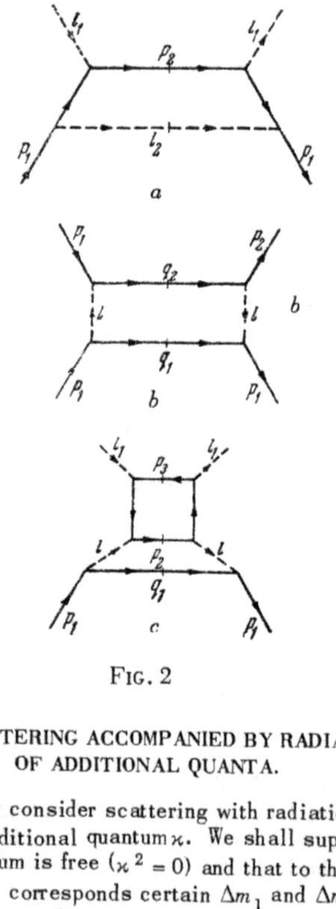

FIG. 2

4. SCATTERING ACCOMPANIED BY RADIATION OF ADDITIONAL QUANTA.

We now consider scattering with radiation of a single additional quantum $\varkappa$. We shall supose that the quantum is free ($\varkappa^2 = 0$) and that to the free line there corresponds certain Δm_1 and Δm_2, subject to the conditions $m\Delta m \ll q\varkappa \ll p\cdot q$. It is easy to see that the additional factor from the new vertex cancels against the factor from the internal electron line. There remain only lines which surround the fundamental vertex, or which circle both vertices. Since lines of these two types can be shifted arbitrarily, integrations over them are carried out independently, i.e., we must find the contribution from the lines surrounding both vertices, and then multiply by the already known factor for the fundamental vertex. As for lines surrounding both vertices, we see that, because of the appearance of a new denominator, we must impose the condition $|p \cdot qu| \ll q \cdot \varkappa$ in order to separate out the logarithmic part of the integral over u. We easily see that the new lines again give an exponential factor, while the sum of the exponents of the two exponentials does not depend on $q \cdot \varkappa$, but is determined only by the values of a, Δm_1 and Δm_2. Now considering an arbitrary number of additional quanta, we arrive at the conclusion that this same situation always occurs, i.e., the net effect of the virtual lines is a factor common to all the diagrams of scattering with radiation.

We now go on to consider the case of an arbitrary number of real quanta. We shall consider the probabilities of processes involving radiation of 1,2,...n quanta, making the following assumptions:

a). The total energy carried off by the quanta does not exceed ω_1 in the rest system of the initial electron, and does not exceed ω_2 in the rest frame of the final electron.

b). We associate with free electron lines, in all diagrams considered, certain values Δm_1 and Δm_2, such that $\Delta m_1 \ll m$, $\Delta m_2 \ll m$ and $a\Delta m_1 > \Delta m_2 > \Delta m_1/a$. In the present case the Δm are auxiliary quantities and can be chosen arbitrarily.

Let us consider the generalized diagrams corresponding to scattering with radiation. It is not hard to see that four groups of real lines occur (see Fig. 3): lines joining a--the q line on the left

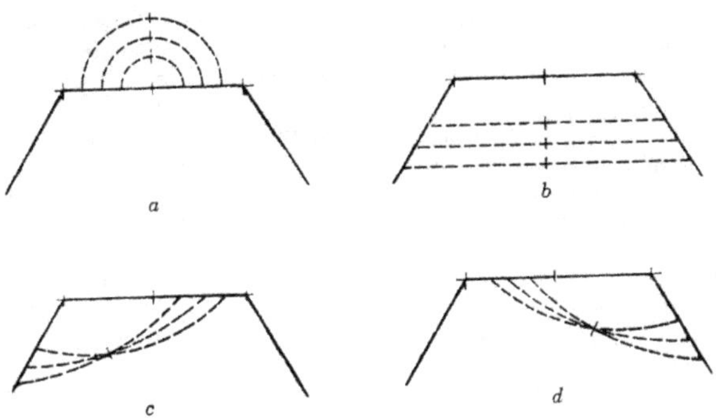

FIG. 3

to the one on the right, b--the p line on the left to the one on the right, c--the left q line to the right p line, d--the left p line to the right q line. If we consider the possible transpositions of lines we see that each group can be treated independently. For the same reason as before, we can use zeroth order values for the new vertices and electron lines. Consider a line of type a. There corresponds to it the factor

$$-2e^2 \int \frac{\gamma_\sigma(\hat{q}+\hat{k}+m)\gamma_\mu(\hat{q}+m)\gamma_\mu(\hat{q}+\hat{k}+m)\gamma_\tau \delta(k^2)\, d^4k}{[2m\Delta m_2 + 2(qk) + k^2]^2}. \tag{27}$$

If we choose the time axis of the vector k along the direction of q and keep only the logarithmic part of the integral, we see at once that it is equal to

$$-(e^2/\pi)\ln(\omega_2/\Delta m_2)\,\gamma_\sigma(\hat{q}+m)\gamma_\tau, \tag{28}$$

if $\omega_2 \ll m$. In the opposite case, ω_2 in (28) should be replaced by m.

If we consider successive lines of type a, it turns out that to get the maximum degree of logarithmic divergence in the integral, the lines must not intersect. Upon summing the integrals, we get the total result (for $\omega_2 \ll m$)

$$\gamma_\sigma(\hat{q}+m)\gamma_\tau(\omega_2/\Delta m_2)^{-e^2/\pi}. \tag{29}$$

Similar expressions also occur for type b.

Now we go to lines of type c and d. Both types of lines give the same contribution. If we consider one of the lines, for example of type c, then we find that the integral obtained differs from the integral for the case where the photon is virtual only by the fact that in case c we have $\hat{q}+\hat{k}$ in the denominator, in place of $\hat{q}-\hat{k}$ for a virtual photon, i.e., the difference is merely a sign change. Actually, just as for the case of a real photon, the main term for the virtual photon case comes from taking the δ-function in the photon propagation factor. The change in the domain of integration also produces no difference, since the double integral over positive frequencies corresponds to a double integral over $u>0$ and the corresponding v. But we know that the region $u<0$ gives the same contribution.

Thus we obtain for a single line of type c an expression proportional to the area cross-hatched in Fig.1, from which the lower part is cut off by the lines $\lambda = \ln(am/\omega_1)$, $\mu = \ln(am/\omega_2)$. This expression appears as the exponent in the sum over all lines. Multiplying the common exponential factor from all the real lines by the common exponential for all the virtual lines, we find that we get in the exponent an expression of the type of (26), with ξ_1 and ξ_2 replaced by $\ln(m/\omega_1)$ and $\ln(m/\omega_2)$. Now it is no longer difficult to explain the physical meaning of (20), and also to give the reasons why the other cases of (20) have no physical interpretation.

Upon comparing the formulas with the ω's with formula (26) it is evident the latter determine the matrix element for scattering with radiation of additional quanta with energy not exceeding Δm_1 in the coordinate system of the initial electron and no greater than Δm_2 in the frame of the final electron. If we assign the quantum frequency in one coordinate system, say ω_1, then upon transforming to the other system, we find that ω_2 lies within the limits

$$(a+\sqrt{a^2-1})\,\omega_1 > \omega_2 > \omega_1/(a+\sqrt{a^2-1}). \quad (30)$$

We note that the sloping lines bounding the integration region in Fig. 1 correspond to just this condition. For this reason, if we impose the condition $\omega \ll \omega_1$ in one coordinate system, then the maximum frequency in the other system cannot exceed $\omega_1(a+\sqrt{a^2-1})$, and conversely, i.e., it can actually be varied independently of ω_1 only within the interval given above. It is therefore clear that all the cases of (20) which are not contained in (26) have no physical meaning.

In a physical formulation of the problem, the frequency is limited in some one definite system. In general, such a bound would appear in Fig. 1. as some line bounding the cross-hatched region at its upper right corner. This line can be found by determining the maximum value of ω_2 for a given limiting frequency ω and a given ω_1. We shall not try to find the appropriate expressions for all cases which may arise, but merely point out that the resulting expression for the probability of a process with radiation of an arbitrary number of quanta with energy less than ω in the system of the initial electron, and of n additional quanta in the energy interval from ω to ω' is:

$$W = (W_0/n!)\, e^{-2f_1(\omega)}\,[f_1(\omega') - f_1(\omega)]^n, \quad (31)$$

where $f_1(\omega)$ is the f_1 of formula (26) with $\ln(m/\omega)$ substituted for ξ_1 and $\ln(m/\omega)+\eta$ in place of ξ_2.

Formula (31) is the familiar Poisson formula, and corresponds to the probability of n independent events. This is understandable, since we are dealing with quanta of such low energy that the radiation of these quanta essentially does not affect the electron's momentum, so that the radiation processes are actually independent. This result which we have obtained by a direct summation of diagrams was also demonstrated by Glauber[8] for the radiation of photons by an electron described by a given current.

We should point out that in order to get the correct result by such a method, we must impose restrictions on the domain of integration; these actually amount to a determination of the limits of energy and quantum frequency up to which the current can be regarded as classical. Strictly speaking, these limits can be obtained only through a quantum description of the current. This situation is not serious so long as one deals with quantum frequencies which are so small that $\ln(m/\omega) \gg \ln(E/m)$, but is very important for higher quantum energies.

In conclusion, I express my gratitude to Acad. L. D. Landau and V. V. Sudakov for many valuable comments.

---

[8] R. Glauber, Phys. Rev. **84**, 395 (1951).

Translated by M. Hamermesh
13

SOVIET PHYSICS JETP VOLUME 5, NUMBER 6 DECEMBER 15, 1957

On the Magnetic Properties of Superconductors of the Second Group

A. A. ABRIKOSOV

Institute of Physical Problems, Academy of Sciences, U.S.S.R.

(Submitted to JETP editor November 15, 1956)

J. Exptl. Theoret. Phys. (U.S.S.R.) **32**, 1442-1452 (June, 1957)

A study is made of the magnetic properties of bulk superconductors for which the parameter $\varkappa$ of the Ginzburg-Landau theory is greater than $1/\sqrt{2}$ (superconductors of the second group). The results explain some of the experimental data on the behavior of superconductive alloys in a magnetic field.

THE AUTHOR[1] has already noted that the quasi-microscopic Ginzburg-Landau theory[2] of superconductivity leads to the conclusion that there exist two groups of superconductors. For the first of these groups, the parameter $\varkappa$ entering into the Ginzburg-Landau theory is less than $1/\sqrt{2}$, and for the second group it is greater than $1/\sqrt{2}$.

This parameter $\varkappa$ determines to a great extent the surface energy at the normal – superconducting interface. It has already been mentioned,[2] in particular, that the calculated surface energy of a superconductor with $\varkappa > 1/\sqrt{2}$ is negative. Thus superconductors of the second group should have properties very different from those of the first group.

For pure metal, $\varkappa$ is found to be small. For instance for mercury, $\varkappa = 0.16$. In view of this, Ginzburg and Landau considered only the case in which $\varkappa \ll 1/\sqrt{2}$.

It has been shown by Zavaritskii,[3] however, that the properties of pure metal thin films condensed at liquid-helium temperatures are not described by such a theory. Zavaritskii and the present author have therefore suggested that such films correspond to $\varkappa > 1/\sqrt{2}$, and that superconductors can thus be divided into two groups. The critical field for superconductors of the second group has already been calculated[1] as a function of the film thickness. The agreement obtained with Zavaritskii's experimental data was not bad.

In the present work, a more detailed investigation of the magnetic properties of bulk superconductors of the second group (a cylinder in a longitudinal field) is undertaken. The results obtained show that $\varkappa > 1/\sqrt{2}$ for a large number of superconducting alloys whose magnetic properties had not previously been well understood.*

1. TRANSITION TO THE NORMAL STATE

It was shown by Ginzburg and Landau[2] that if $\varkappa > 1/\sqrt{2}$, the superconductivity is maintained at fields greater than H_{cm} at which equilibrium could exist between the normal and superconducting states. At fields higher than H_{cm}, a state with $\Psi = 0$ is unstable, and superconducting sections with $\Psi \neq 0$ may arise. It was shown[1] that this instability continues to some value H_{c2}, which for a bulk superconductor is $\varkappa\sqrt{2}H_{cm}$. At this value of the field the superconductor undergoes transition to the normal state by means of a second-order phase transition. In this section we shall investigate the properties of a superconductor in the neighborhood of the transition point, that is

$$\varkappa - H_0/\sqrt{2}H_{cm} \ll \varkappa. \tag{1}$$

The Ginzburg-Landau equations (in units previously used by the author,[2] and to be used henceforth),† can be written

$$\left(\frac{i\nabla}{\varkappa} + \mathbf{A}\right)^2\Psi = \Psi - \Psi|\Psi|^2, \tag{2}$$

$$-\operatorname{curl}\operatorname{curl}\mathbf{A} = |\Psi|^2\mathbf{A} + \frac{i}{2\varkappa}(\Psi^*\nabla\Psi - \Psi\nabla\Psi^*). \tag{3}$$

We shall assume that the superconductor fills all space, and that the external field H_0 is directed along the z axis. Let the potential A be directed along the y axis.

Close to the transition point $|\Psi|^2 \ll 1$, so that in the first approximation we may neglect the influence of Ψ on the field. We then obtain

$$H = \text{const} = H_0, \quad A = H_0 x. \tag{4}$$

We note that the point $x = 0$ may be located anywhere in space. Let us now go on to Eq. (2). Inserting (4) into the equation for Ψ, neglecting the term $|\Psi|^2\Psi$ and considering Ψ to be a function of x only in the first approximation, we obtain the oscillator type of equation

$$d^2\Psi/dx^2 - \varkappa^2(1 - H_0^2x^2)\Psi(x) = 0, \tag{5}$$

as found previously.[2] This equation has solutions when $H_0 = \varkappa/(2n+1)$. In particular, the largest value $H_0 = \varkappa$ corresponds to

$$\Psi = \exp\{-\varkappa^2x^2/2\}. \tag{6}$$

In addition to such solutions, Eq. (2) is satisfied by the function

$$\exp\left\{iky - \frac{\varkappa^2}{2}\left(x - \frac{k}{\varkappa^2}\right)^2\right\}. \tag{7}$$

Since the point $x = 0$ has no special properties, the conditions will be exactly the same at all points of space, and it is natural to pick Ψ in the form

$$\Psi = \sum_{n=-\infty}^{\infty} C_n e^{ikny}\psi_n(x),$$
$$\psi_n(x) = \exp\left[-\frac{\varkappa^2}{2}\left(x - \frac{kn}{\varkappa^2}\right)^2\right] \tag{8}$$

with arbitrary coefficients k and C_n.

This form of Ψ is a solution of the linear equation, and refers in facts to $H_0 = \varkappa$. In order to find a solution for $H_0 < \varkappa$, let us take into account the nonlinear terms in the equations. We find the approximation of interest by inserting A from Eq. (4) with $H_0 = \varkappa$, and Ψ from Eq. (8) into Eq. (3). This leads to

$$\frac{\partial^2 A}{\partial x^2} = \sum_{n,m} C_n C_m^* e^{i(n-m)ky}\left[\varkappa k - \frac{(n+m)k}{2\varkappa}\right]\psi_n(x)\psi_m(x),$$
$$-\frac{\partial^2 A}{\partial x\,\partial y} = \frac{ik}{2\varkappa}\sum_{n,m} C_n C_m^* e^{i(n-m)ky}(n-m)\psi_n(x)\psi_m(x) \tag{9}$$

It can be shown easily that

$$H = \partial A/\partial x = H_0 - |\Psi|^2/2\varkappa; \tag{10}$$

$$A = H_0 x - \frac{1}{2\varkappa}\int|\Psi|^2 dx, \tag{11}$$

is a solution of this equation, where Ψ is given by Eq. (8), and H_0 is some still undetermined constant. It will be shown later that this constant is equal to the external field strength.

Let us now consider Eq. (2) in the following approximation. We insert into it Eq. (11), and take account of the term $\Psi|\Psi|^2$. We shall assume in this approximation that the function is still of the form of Eq. (8), but we shall add small terms $\psi_n^{(1)}(x)$ to

---

* The suggestion that $\varkappa$ may be greater than $1/\sqrt{2}$ for alloys was first made by L. D. Landau.

† We note that in these units $H_{cm} = 1/\sqrt{2}$.

all the $C_n\psi_n(x)$. The equations for these small terms are of the form

$$\left(\frac{kn}{\varkappa}-\varkappa x\right)^2\psi_n^{(1)}-\frac{1}{\varkappa^2}\frac{d^2\psi_n^{(1)}}{dx^2}-\psi_n^{(1)}=2(\varkappa-H_0)\varkappa\left(x-\frac{nk}{\varkappa^2}\right)C_n\psi_n(x)+$$
$$+\sum_{p,m}C_pC_m^*C_{n-p+m}\left\{\left[x-\frac{k}{\varkappa^2}\left(n-\frac{p-m}{2}\right)\right]\psi_{n-p+m}(x)\int^x\psi_p(x')\psi_m(x')\,dx'-\psi_p(x)\psi_m(x)\psi_{n-p+m}(x)\right\}. \quad (12)$$

Thus $\psi_n^{(1)}$ is a solution of an inhomogeneous equation. But in order for such a solution to exist, it is necessary that the right side of the equation be orthogonal to the solution of the corresponding homogeneous equation. The solution of the homogeneous equation is just the function $\psi_n(x)$. Performing the necessary integration, we obtain the following condition:

$$\left(\frac{1}{2\varkappa^2}-1\right)\sum_{p,m}C_pC_m^*C_{n-p+m}\exp\left\{-\frac{k^2}{2\varkappa^2}[(p-n)^2+(m-n)^2]\right\}+\frac{\varkappa-H_0}{\varkappa}C_n=0. \quad (13)$$

Let us multiply this equation by C_n^* and sum over n. It is easily verified that we then obtain

$$\frac{\varkappa-H_0}{\varkappa}\overline{|\Psi|^2}+\left(\frac{1}{2\varkappa^2}-1\right)\overline{|\Psi|^4}=0, \quad (14)$$

where Ψ is the function of Eq. (8). Thus H_0 enters into Ψ in the form of a coefficient.

The quantity H_0 can be expressed in terms of the magnetic induction. According to (10) and (14), we have

$$B=\overline{H}=H_0-\overline{|\Psi|^2}/2\varkappa=H_0-(\varkappa-H_0)/(2\varkappa^2-1)\beta, \quad (15)$$

where

$$\beta=\overline{|\Psi|^4}/(\overline{|\Psi|^2})^2 \quad (16)$$

is independent of H_0.

Let us now calculate the free energy. According to Ginzburg and Landau,[2] the free energy is

$$\frac{F_{sH}-F_{s0}}{H_{cm}^2/4\pi}=\int dV\left\{\frac{1}{2}-|\Psi|^2+\frac{|\Psi|^4}{2}+H^2+\left(-\frac{i\nabla\Psi}{\varkappa}-\mathbf{A}\Psi\right)\left(\frac{i\nabla\Psi^*}{\varkappa}-\mathbf{A}\Psi^*\right)\right\}. \quad (17)$$

If we now make use of (2) and (3) and neglect the surface integral, which is of no importance to us, we obtain

$$\frac{F_{sH}-F_{s0}}{H_{cm}^2/4\pi}=\int\left(H^2-\frac{|\Psi|^4}{2}+\frac{1}{2}\right)dV. \quad (18)$$

The free energy per unit volume is proportional to the average of the integrand. Denoting this average by F_1 and writing it in terms of B, we obtain

$$F_1=\frac{1}{2}+B^2-\frac{(\varkappa-B)^2}{1+(2\varkappa^2-1)\beta}. \quad (19)$$

For a given B this quantity will be minimum when β has its lowest value.

We note that (13) can be written in the form

$$\partial\beta/\partial C_n^*=0, \quad (13')$$

becoming equivalent to the condition that β is a minimum. It is seen from (16) that β cannot be less than unity.

The condition that Ψ be periodic in x is the following simple recursion relation for the C_n:

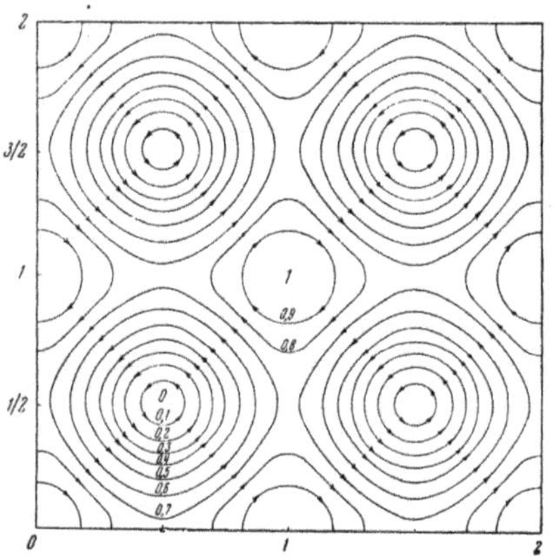

$$C_{n+N} = C_n. \tag{20}$$

The general analysis of the problem of choosing the coefficients is extremely complicated, but there is no reason to suppose that the actual value of N is a very large number. It would seem that the lowest energy corresponds to the simplest case* $N = 1$. In this case β is given by

$$\beta = \frac{k}{\varkappa\sqrt{2\pi}}\left(\sum_n \exp\left\{-\frac{k^2}{2\varkappa^2}n^2\right\}\right)^2.$$

The minimum value of β is obtained when $k = \varkappa\sqrt{2\pi}$, and is equal to

$$\beta = \left(\sum_n e^{-\pi n^2}\right)^2 = \vartheta_3^2(0, 1) = 1.18. \tag{21}$$

Then Ψ is

$$\Psi = Ce^{-\varkappa^2 x^2/2}\vartheta_3[\sqrt{2\pi}\,\varkappa i\,(x + iy);\; 1]. \tag{22}$$

It can be shown from the properties of the function ϑ_3 that rotation by 90° multiplies Ψ by the phase factor $e^{i\varkappa^2 xy}$, but does not change it otherwise.

* In addition to this case, which will be seen later to lead to a square lattice, we considered the example of a triangular lattice. The side of the triangle is $(1/\varkappa)\sqrt{8\pi/3\sqrt{3}}$. By numerical integration, however, we were able to show that this lattice corresponds to a larger value of β, namely $\beta = 1.32$ (see also the next section).

Thus $|\Psi|^2$ has the symmetry of a square lattice. When $x = y = (1/\varkappa)\sqrt{\pi/2}$, the function Ψ vanishes. According to (10), H attains its maximum value H_0 at these points.

We shall now show that H_0 is equal to the field strength. As is well known, the field strength is $4\pi\,\partial\bar{F}/\partial B$ (in the usual units). This leads to a relation between H and B which is identical with Eq. (15). In order to perform the comparison with experiment (this will be done in Sec. 3) more conveniently, the latter can be rewritten in terms of the function $M(H_0)$,

$$-4\pi M/(H_{c2} - H_0) = 1/1.18\,(2\varkappa^2 - 1), \tag{23}$$

where M and H can be taken in the usual units. We note that the magnetic moment is proportional to the first power of $H_{c2} - H_0$.

Equation (23) is the macroscopic characterization of the behavior of a bulk sample in an external field. As for the microscopic structure of the field penetration, collecting the results described above we obtain the following picture. In a plane perpendicular to the field direction, the structure is periodic with the symmetry of a square lattice. The field varies between its maximum value H_0 and its minimum value given by

$$H_{\min} = H_0 - (H_{c2} - H_0)\sqrt{2}\,/\,(2\varkappa^2 - 1). \tag{24}$$

When the applied field strength H_0 is increased, the field between the maxima increases up to values equal to $H = H_0 = \varkappa$.

The field and current distribution is shown in the figure. The lines correspond to $|\Psi|^2 = \text{const}$ (normalized to $|\Psi|^2_{\max} = 1$) or equivalently, to $H = \text{const}$. These are then the streamlines (their directions being given by the arrow).

2. THE BEGINNING OF FIELD PENETRATION INTO A SUPERCONDUCTOR

In the preceding section we concluded that in the neighborhood of the transition point to the normal state, superconductors of the second group are in a special state which is neither normal nor superconducting, and differs significantly from the intermediate state which may arise in noncylindrical bulk samples. Let us call this the mixed state. One naturally wonders how this state changes as the field is decreased, and where and how the transition to superconductivity occurs.

If we direct our attention to the experimental

$B(H_0)$ curves for alloys,[4] we see that this transition can be both a first and a second order phase transition. It would seem (see Sec. 3) that the first order phase transition occurs when $\varkappa$ is not very large, whereas the second order transition occurs when $\varkappa \gg 1$. It is extremely difficult to explain this on theoretical grounds, since this would require solving Eqs. (2) and (3) without any simplifying assumptions.

Let us consider only the simplest case of a second order phase transition, which would seem, as has already been mentioned, to take place when $\varkappa$ is not too small.

Let us start with a determination of the transition point. It is natural to assume that when the field first begins to penetrate, the structure of this penetration is again of the lattice type, although with a very large period. At the lattice points the field will have its maximum value, and will practically vanish between them. Thus the field penetration pattern is that of separated filaments. This pattern reminds one of the distribution of superfluidity in helium II by the creation of vortex filaments, as proposed by Onsager[5] and Feynman.[6] The similarity is even greater when one analyzes the behavior of the Ψ function in the neighborhood of these filaments in greater detail. When one considers the situation at fields close to H_{c2}, one sees easily that in the neighborhood of those points where Ψ vanishes and the field is maximum, the phase of the Ψ function is* θ, where θ is the polar angle, and thus changes by 2π in going once around such a point. It is natural to suppose that the phase behaves in the same way in the neighborhood of separate filaments when the separation between them is large. As will be seen later, the Ψ function vanishes also in the center of a filament. This behavior of the Ψ function, in particular of its phase, has been suggested by Feynman in application to vortices in helium II.

When the filaments are sufficiently separated, they can be considered independent of each other, and each filament can be treated separately. In this case, the problem has cylindrical symmetry. Let us introduce the modulus and phase of the Ψ function

$$\Psi = fe^{i\varphi} \tag{25}$$

* In the neighborhood of such a point with coordinates (x_1, y_1), Ψ is proportional to $(x - x_1) + i(y - y_1)$.

and consider the vector **A** directed perpendicular to the radius vector. We introduce the scalar quantity Q, which is equal to the absolute value of the vector $\mathbf{A} - \nabla\varphi/\varkappa$. Equations (2) and (3) then become

$$-\frac{1}{\varkappa^2 r}\frac{d}{dr}\left(r\frac{df}{dr}\right) + Q^2 f = f - f^3, \tag{26}$$

$$\frac{d}{dr}\left[\frac{1}{r}\frac{d}{dr}(rQ)\right] = Qf^2. \tag{27}$$

The field H is given by

$$H = -\frac{1}{r}\frac{d}{dr}(rQ). \tag{28}$$

The statement of the problem requires that $f \to 1$, $H \to 0$, $Q \to 0$ as $r \to \infty$. Further, we have assumed that for a separate filament $\varphi = \theta$. But this leads to $\nabla_\theta \varphi = \partial\varphi/r\partial\theta = 1/r$. In other words, $Q \to 1/\varkappa r$ as $r \to 0$. According to the theory of Onsager and Feynman the superfluid velocity v_s in helium II diminishes in the same way, with increasing distance from the center of the filament.

In addition, it is of course necessary to require that f be everywhere bounded. These conditions uniquely determine the solution of (26) and (27). In general, however, this solution can be obtained only by numerical integration, and this must be performed separately for each value of $\varkappa$.

When $\varkappa \gg 1$, the solution is considerably simplified. We note for this purpose that the distance over which Q should have a large variation is $r \sim 1$. As for f, it varies to a large extent at distances $r \sim 1/\varkappa$. In view of this, one may assume that f has already attained the value $f = 1$ in Eq. (26), from which we immediately obtain

$$Q = K_1(r)/\varkappa, \tag{29}$$

where K_1 is the Hankel function of imaginary argument. Further, we note that $Q = 1/\varkappa r$ when $r \ll 1$. Inserting this into the equation for f we obtain

$$\frac{1}{\varkappa^2}\left[\frac{1}{r}\frac{d}{dr}\left(r\frac{df}{dr}\right) - \frac{1}{r^2}f\right] = f^3 - f. \tag{30}$$

At distances $r \gg 1/\varkappa$, the solution of this equation is $f^2 = 1 - 1/\varkappa^2 r^2$, and when $r \ll 1/\varkappa$, the solution is $f = Cr$, where C is a constant which can be obtained by numerical solution of (30).

Let us now calculate ε, the free energy per unit length of a filament. To do this we must evaluate the integral

$$\varepsilon = 2\pi\int_0^\infty\left(H^2 + \frac{1-f^4}{2}\right) r dr$$
$$= \pi\int_0^\infty\left[(1-f^4)\, r - f^2 \frac{d}{dr}(r^2Q^2)\right] dr. \quad (31)$$

If we note that the integrand decays exponentially when $r \gg 1$, and assume that $\ln \varkappa \gg 1$, it is easily seen that the important contribution to the integral will be from the interval $1/\varkappa \ll r \ll 1$, and we then obtain

$$\varepsilon = (2\pi/\varkappa^2)\ln\varkappa.$$

Numerical integration of (30) and (31) makes it possible to obtain a correction to $\ln\varkappa$, so that we finally obtain

$$\varepsilon = (2\pi/\varkappa^2)(\ln\varkappa + 0.081). \quad (32)$$

In order that the formation of filaments be energetically possible, the field strength H_0 must be such that $F_1 - 2H_0B$ be negative. If the filament density, *i.e.*, the number of filaments crossing a unit area, be denoted by n, we have

$$F_1 = n\varepsilon, \qquad B = \overline{H} = n\int H dS = n\oint \mathbf{A}\, d\mathbf{l}. \quad (33)$$

In view of the fact that the contour of integration extends to $r = \infty$, and that at these large distances $Q = 0$, we conclude that $\mathbf{A} = \nabla\varphi/\varkappa$, which means that its absolute value is $1/\varkappa r$. Thus

$$B = 2\pi n/\varkappa. \quad (34)$$

It follows from this that the filaments start being formed when the field strength H_0 attains the value

$$H_{c1} = \varepsilon\varkappa/4\pi. \quad (35)$$

This formula is valid for arbitrary $\varkappa$ on the assumption that the phase transition is of second order. When $\varkappa \gg 1$, we obtain

$$H_{c1} = (1/2\varkappa)(\ln\varkappa + 0.08). \quad (36)$$

It is interesting to note that at the center of the filament the field is about twice H_{c1} in the given case. Indeed, according to Eq. (27) we have

$$H(0) = \int_0^\infty Qf^2 dr = (\ln\varkappa - 0{,}18)/\varkappa. \quad (37)$$

From (35) it is seen that H_{c1} is independent of n, and that an arbitrary number of filaments may be formed in this approximation for $H_0 > H_{c1}$. In order to find the function $n(H_0)$, and therefore $B(H_0)$, we must take account of the interaction between the filaments.

This is easily done if $\varkappa \gg 1$. Indeed, in this case the region in which f differs from unity ($r \sim 1/\varkappa$) is small compared with the distances at which the interaction between the filaments takes place ($r \sim 1$). In view of this, the filament centers will play the role of singularities in the solution, and their detailed structure will not be of importance. It is not difficult to see that the equation for the field together with the conditions at the centers of the filaments can be written in the form

$$\Delta H - H = -(2\pi/\varkappa)\sum_m \delta(\mathbf{r} - \mathbf{r}_m), \quad (38)$$

where the $\mathbf{r}_m$ are the coordinates of the centers of the filaments. The solution of this equation is

$$H = \frac{1}{\varkappa}\sum_m K_0(|\mathbf{r} - \mathbf{r}_m|). \quad (39)$$

The energy can be written in the form

$$F_1 = \overline{H^2 + (\nabla H)^2} = \overline{H(H - \Delta H)}$$
$$= \frac{2\pi n}{\varkappa^2}\sum_m K_0(|\mathbf{r}_m|), \quad (40)$$

where the $\mathbf{r}_m$ are calculated from one of the filament centers. The term corresponding to $\mathbf{r}_m = 0$ should be replaced by Eq. (32). The quantity n and the periodicity of the lattice, which enter into (40), can be expressed in terms of the magnetic induction B by Eq. (34). For a square lattice we obtain

$$F_1 = \frac{\varkappa B}{2\pi}\varepsilon + \frac{B}{\varkappa}\sum_{l^2+m^2\geqslant 1} K_0\left(\sqrt{\frac{2\pi(l^2+m^2)}{\varkappa B}}\right). \quad (41)$$

It is, however, possible that the lattice is actually not a square one or that its symmetry changes as the field is varied. Therefore the problem should be dealt with more carefully. The most probable structure which one can assume instead of the square one is the triangular one. For such a lattice one easily obtains

$$F_1 = \frac{\varkappa B}{2\pi}\varepsilon + \frac{B}{\varkappa}\sum_{l^2+m^2+lm\geqslant 1} K_0\left(\sqrt{\frac{4\pi(l^2+m^2+lm)}{\sqrt{3}\,\varkappa B}}\right). \tag{42}$$

By comparing the asymptotic form of this expression for small B with asymptotic form of (41),

$$F_{1\triangle} \approx \frac{\varkappa B}{2\pi}\varepsilon + 6\frac{B}{\varkappa}\sqrt{\frac{\pi}{2}}\left(\frac{\sqrt{3}\varkappa B}{4\pi}\right)^{1/4}\exp\left(-\sqrt{\frac{4\pi}{\sqrt{3}\,\varkappa B}}\right),$$

$$F_{1\square} \approx \frac{\varkappa B}{2\pi}\varepsilon + 4\frac{B}{\varkappa}\sqrt{\frac{\pi}{2}}\left(\frac{\varkappa B}{2\pi}\right)^{1/4}\exp\left(-\sqrt{\frac{2\pi}{\varkappa B}}\right), \tag{43}$$

one sees clearly that for sufficiently small B there should be a transition to the triangular modification. The value of H_0 at the transition point is determined by setting the free energy for a given field strength equal to $F_1 - 2H_0B$, and is

$$H_1' = H_{c1} + 0.0394/\varkappa. \tag{44}$$

This transition is clearly a first order phase transition. The induction undergoes a jump from $B_1 = 0.286/\varkappa$ to $B_2 = 0.294/\varkappa$. At fields greater than H_1', the square lattice is more favorable, which justifies to some degree the assumption made in the previous section.

We note that the transition takes place in the immediate neighborhood of H_{c1}, and that the discontinuity in the induction is so small (about 3%) that it would be extremely difficult to observe it on a $B(H_0)$ curve.

It is not difficult to see that when $H = H_{c1}$, the $B(H_0)$ curve has a vertical tangent. Indeed, when $H \approx H_{c1}$, the H_0-dependence of B is determined primarily by the term $H_0 \sim \exp[-(4\pi/\sqrt{3}\varkappa B)^{1/2}]$, and $dH_0/dB \to 0$ as $B \to 0$. In the other limiting case, when $\varkappa B \gg 1$, the various terms of the sum of Eq. (41) have arguments which lie close to each other, so that the sum can be replaced by an integral. This gives $F_1 \approx B^2$, and $H_0 \approx B$. Thus as H_0 is increased, the curve asymptotically approaches the straight line $B = H_0$. This goes on until the distance between the filaments becomes of the order of $1/\varkappa$, *i.e.*, until $H_0 \sim \varkappa$.

The dependence of $\varkappa B$ on $(H_0 - H_{c1})$ is given by the universal function

$$\varkappa(H_0 - H_{c1}) = \frac{1}{4}\sum_{x_{l,m}\neq 0}[2K_0(x_{l,m}) + x_{l,m}K_1(x_{l,m})],$$

$$\text{where } x_{l,m} = \sqrt{4\pi(l^2+m^2+lm)/\sqrt{3}\varkappa B} \quad \text{for } \varkappa(H_0 - H_{c1}) < 0.0394,$$

$$x_{l,m} = \sqrt{2\pi(l^2+m^2)/\varkappa B} \quad \text{for } \varkappa(H_0 - H_{c1}) > 0.0394. \tag{45}$$

We note that $(\partial B/\partial H_0)_T$ is always positive, which indicates that the situation discussed is stable.[7]

For very large values of $\varkappa$ Eq. (35) can be used to find H_{c1}. To do this one must solve (26) and (27) numerically, after which ε is calculated from (31). On the other hand, finding $B(H_0)$ in this case requires solving a separate and very complex problem.

To conclude this section, let us turn our attention to the intermediate state in noncylindrical speciments (or cylinders in nonparallel fields). It is most interesting that no intermediate state arises in specimens for which $\varkappa \gg 1$. One obtains instead a uniform pattern of a mixed state (in ellipsoidal samples). When the transition in the field H_{c1} is of the first order, however, an intermediate state does arise.

3. COMPARISON OF THEORY AND EXPERIMENT

No detailed data exist currently on the magnetic properties of superconducting alloys. The only work which is more or less suitable for comparison with the theory is that of Shubnikov, Khotkevich, Shepelev, and Riabinin,[4] performed in 1937. Let us compare the theoretical results with the data of this work.

If we are to speak of general qualitative characteristics, we must first note that starting at some alloy concentration, the $B(H_0)$ curve indicates a region of partial field penetration; this corresponds to the conclusion reached above on the existence of a mixed state. This region is bounded by the two field strengths H_{c1} and H_{c2}, and although the transition from the superconducting state to the mixed one can be either of first or of second order, the transition to the normal state must definitely be of first order. In the neighborhood of H_{c2}, the $B(H_0)$ curve is linear. All this is in agreement with the results of Sec. 1. Further, if the alloy concentration is sufficiently high, the transition at H_{c1} seems also to be of second order. According to the

Alloy composition	T°K	H_{c_2}	H_{cm}	$\varkappa\sqrt{2}$	$\tan\alpha$ (theory)	$\tan\alpha$ (exp.)	H_{c1} (theory)	H_{c1} (exp.)
Pure Pb	4.24		540	0.22÷0.36				
Pb + 2.5% Tl	4.22	704	596	1.18	2.12	2.2	—	510
	1.92	1100	846	1.3	1.21	1.2	—	740
Pb + 5% Tl	4.22	1010	561	1.8	0.38	0.38	—	430
	1.82	1730	854	2	0.27	0.27	—	600
Pb + 15% Tl	4.22	2170	534	4.07	0.054	0.052	150	230
	1.76	3760	762	4.9	0.036	0.036	210	310
Pb + 30% Tl	4.22	2840	575	4.94	0.036	0.033	150	170
	1.70	4460	756	5.9	0.025	0.023	190	240
Pb + 50% Tl	4.22	2270	370	6.14	0.023	0.02	90	110
	2.3	4270	520	8.2	0.013	0.014	120	150
Pb + 2% In	4.22	983	604	1.6	0.51	0.45	—	530
	1.95	1520	842	1.8	0.37	0.3	—	730
Pb + 8% In	4.22	2400	563	4.26	0.049	0.045	170	220
	1.75	3780	812	4.7	0.041	0.043	240	310

theory, at this point the $B(H_0)$ curve has a vertical tangent, and then begins gradually to approach the straight line $B = H_0$. This conclusion also more or less agrees with the experimental curves.

Going on to a quantitative comparison, we note that the two parameters H_{cm} and $\varkappa$ which enter into the theory must be determined from the experimental curves. This can be done by finding the upper critical field H_{c2} and the area under the curve of the magnetic moment $B(H_0) - H_0$, where H_0 is the field strength. In fact from the definition of H_{cm}, we have

$$\frac{1}{2}H_{cm}^2 = \int_0^\infty [H_0 - B(H_0)]\,dH_0 = \int_0^{H_{c2}} [H_0 - B(H_0)]\,dH_0. \quad (46)$$

Of course in order that H_{cm} defined in this way have meaning, it is necessary that the experimental $B(H_0)$ curve be that for equilibrium. When we know H_{cm}, we can find $\varkappa$ from the equation $H_{c2} = \varkappa\sqrt{2}H_{cm}$.

The values of H_{cm} and $\varkappa$ obtained in this way can be used to compare the magnetic moment in the neighborhood of the upper critical field H_{c2} with experiment. According to Sec. 1, the dependence of M on H_0 is given by Eq. (23).

We can compare with experiment, in addition to M, Eq. (36) for the first critical field.

The results of such comparison are given in the Table.* One should bear in mind that the accuracy with which the slope of the magnetic moment curve is determined in the neighborhood of H_{c2} (let us denote the ratio in Eq. (23) by $\tan\alpha$) and H_{c1} from the curves of Shubnikov, *et al.*[4] is no greater than 15%. It is seen from the table that the data for $\tan\alpha$ are in very good agreement with the theory. The values of $\tan\alpha$ vary from specimen to specimen by a factor as high as 150, and the theory always gives the correct value within the above-mentioned limits.

The situation with respect to H_{c1} is somewhat worse. Although the theoretical values are always of the same order of magnitude as the experimental ones, they lie too low in all cases. It should be borne in mind, however, that the values of $\varkappa$ dealt with here are not very large compared to unity, and this can lead to a systematic error in determining H_{c1} from (35), an equation which is valid only when $\varkappa \gg 1$. It is not difficult to see in which direction the theoretical value must be corrected in order to account for the fact that $\varkappa$ is not very large. This can be done by making the natural assumption that $H_{c1} = H_{cm}$ when $\varkappa = 1/\sqrt{2}$. Instead of this, insertion of $\varkappa = 1/\sqrt{2}$ into Eq. (36) leads to a negative value of H_{c1}. Thus for values of $\varkappa$ which are not

*It is significant here that the theory remains applicable for temperatures far from T_c. As fields close to H_{c2}, this conclusion is strictly correct, since Ψ is small in this region. Recently, however, Ginzburg[8] has presented arguments which tend to show that the theory is applicable also for arbitrary fields and temperatures.

very large, the actual theoretical value of H_{c1} should be greater than that given by (36). In principle, it is also possible to perform a comparison between experiment and Eq. (45) for $B(H_0)$ in the neighborhood of H_{c1}. This equation, however, is in principle valid only for $\varkappa \gg 1$, so that only a qualitative comparison is possible; this has already been done at the beginning of this section. Furthermore, the experimental curves in this region are measured with very low accuracy, so that a quantitative comparison is not really very meaningful. We note here that according to the experimental data, when $\varkappa$ is close to $1/\sqrt{2}$ the transition to the mixed state would seem to be a first order phase transition. In this case the concept of isolated filaments does not correspond at all to the mixed state at the transition point.

In order to investigate the dependence of H_{cm} and $\varkappa$ on the concentration, let us compare their values at a fixed temperature of 4.2°. For comparison, the table gives the data also for pure lead at the same temperature, taken from Lock.[9] It is seen from the table that H_{cm} hardly depends on the concentration (except for the 50% Pb-Tl alloy). The variation is no greater than 11% and has no apparent regularity. However, $\varkappa$ increases regularly with increasing concentration, and varies for Pb-Tl alloys from 1.18 at 2.5% Tl to 6.14 at 50% Tl.

Some years ago Pippard[10] made the interesting suggestion that there exists in superconductors a certain characteristic length (the coherence length) which determines the "range of order" of the superconducting electrons. This length characterizes the minimum distance at which the number of superconducting electrons may change significantly. It may, for instance, refer to the thickness of the transition layer between the normal and superconducting phases.

According to the theory of the previous sections, the quantity $\delta/\varkappa$ should play the role of such a length. This length determines the thickness of the transition layer in the intermediate state.[2] If we bear in mind that $\varkappa = (\sqrt{2}e/hc)H_{cm}\delta^2$, then it is clear that the characteristic length is proportional to $1/\sqrt{\varkappa}$. According to Pippard, this length should depend strongly on the impurity concentration. At high concentrations it should be of the order of the mean free path of an electron in the metal. It is seen from the table that this concept is not in disagreement with the experimental data.

The results described show that the quasi-microscopic theory is able to describe many of the regular properties of superconducting alloys. Thus these regular properties arise primarily not because macroscopic regions with relatively large values of T_c and H_{cm} arise in the alloy, as has been thought previously, but from the destruction of electron order at distances small compared with the penetration thickness δ.

It is true, of course, that macroscopic inhomogeneities may be of importance. Without even going into the fact that a strong inhomogeneity can change the overall pattern, we note that such a phenomenon as the remanent magnetic moment cannot be described by the present theory and is more probably a result of such inhomogeneities. The structure of the field penetration itself should be extremely sensitive to inhomogeneities, so that in real substances it would hardly be in the form of a regular lattice.

In conclusion I take this opportunity to express my gratitude to Academician L. D. Landau and N. V. Zavaritskii for valuable comments and discussion of the work.

[1] A. A. Abrikosov, Dokl. Akad. Nauk SSSR **86**, 489 (1952).

[2] V. L. Ginzburg, L. D. Landau, J. Exptl. Theoret. Phys. (U.S.S.R.) **20**, 1064 (1950).

[3] N. N. Zavaritskii, Dokl. Akad. Nauk SSSR **86**, 501 (1952).

[4] Shubnikov, Khotkevich, Shepelev, and Riabinin, J. Exptl. Theoret. Phys. (U.S.S.R.) **7**, 221 (1937).

[5] L. Onsager, Nuovo cimento **6**, Supp. 2, 249 (1949).

[6] R. P. Feynman, *Progress in Low Temperature Physics I* (North Holland Publish. Co., Amsterdam, 1955).

[7] V. P. Silin, J. Exptl. Theoret. Phys. (U.S.S.R.) **30**, 197 (1956); Soviet Phys. JETP **3**, 123 (1956).

[8] V. L. Ginzburg, J. Exptl. Theoret. Phys. (U.S.S.R.) **30**, 593 (1956); Soviet Phys. JETP **3**, 621 (1956).

[9] J. M. Lock, Proc. Roy. Soc. (London) A **208**, 391 (1951).

[10] A. B. Pippard, Proc. Cambridge Phil. Soc. **47**, 617 (1951); Proc. Roy. Soc. (London) A **216**, 547 (1953).

Translated by E. J. Saletan
285

SOVIET PHYSICS USPEKHI VOL. 66(1), NO. 1 SEPTEMBER-OCTOBER, 1958

THEORY OF THE FERMI FLUID

(The Properties of Liquid He^3 at Low Temperatures)

A. A. ABRIKOSOV AND I. M. KHALATNIKOV

Usp. Fiz. Nauk 56, 177-212 (October, 1958)

INTRODUCTION

THE study of the superfluidity of helium II gave rise to the question of the properties of quantum fluids, that is, of systems of interacting particles in whose behavior quantum effects play a major part. The theory of superfluidity given by L. D. Landau was at the same time the first theory of a quantum fluid. It described the properties of so-called Bose fluids, i.e., quantum systems of interacting particles for which the excitations obey Bose statistics.

Besides such fluids, however, there also exist others, forming a much more numerous class; these are the so-called Fermi fluids, whose excitations have spin 1/2 and obey Fermi statistics. These include liquid helium 3, the electrons in metals, and possibly heavy nuclei, although to be sure this last case is a quite special one. Until very recently there was no theory of the Fermi fluid. The theoretical calculations were usually confined to an approach analogous to that used for Bose fluids; some sort of energy spectrum was postulated, and this was then substituted into the Fermi distribution formula.

In view of the fact that in metals the situation was decidedly complicated by the anisotropy of the spectra, it was hard to test the correctness of such an approach. As for liquid helium 3 at low temperatures, the experimental data made it very clear that in this case such an approach could not be completely successful.

A consistent theory of Fermi fluids was developed by L. D. Landau in 1956.[1] He showed that the situtation is very different from that in Bose fluids, since in Fermi fluids a very important part is played by the interaction of the excitations, so that the excitations in a Fermi fluid cannot be regarded in all cases as an ideal gas.

In the present article we expound the theory of the Fermi fluid in its application to the isotropic model and show how the Landau theory provides an explanation of the various properties of liquid helium 3. In an appendix we present a recently completed research of L. D. Landau, which shows how the basic propositions of the theory of the Fermi liquid follow from a microscopic treatment of the interaction, and we also present there the theory of a rarefied Fermi gas. In this case one can obtain the basic quantities of the theory of the Fermi fluid by an actual calculation.

Before passing to the exposition of the theory, we recall for convenience in later applications, the main properties of liquid helium 3. The helium isotope of atomic weight 3 liquefies at 3.2°K (critical point $T_c = 3.3°$, $p_c = 845$ mm Hg), and remains liquid down to the lowest temperatures at which it has been studied (~ 0.2°K) and at pressures up to 30 atmos. The explanation is the same as in the case of helium 4: the weakness of the interaction between the atoms, and the smallness of the atomic mass. Owing to the latter circumstance the de-Broglie wavelength corresponding to the motion of helium 3 atoms at low temperatures can be larger than the distances between atoms; that is, the liquid becomes a quantum fluid.

Down to the lowest temperatures that have been used liquid helium 3 does not become a superfluid. Since according to the Landau theory every Bose liquid must possess superfluidity, while a Fermi liquid does not have this property, helium 3 must be regarded as a Fermi fluid. It must be stated that this fact is not trivial and does not follow automatically from the fact that helium 3 atoms have spin 1/2. A system of such atoms might have exclusively Bose excitations. Moreover, as we shall see later, such excitations actually exist, although they do not play any important part in determining the properties of helium 3. It is only the absence of superfluidity that enables us to conclude with assurance that helium 3 is a genuine Fermi fluid.

1. THE ENERGY OF THE EXCITATIONS

The possibility of describing an excited state of liquid helium 3 by means of a gas of quasi-particles and a corresponding distribution function in the energy scale is based on the fact that the interaction of the particles of a Fermi gas decreases rapidly as the temperature is lowered. In fact, it is clear from general considerations that the time between collisions is proportional to the square of the diffuseness of the Fermi distribution, that is, to the mean square of the temperature. At the same time the mean energy of the Fermi particles in the diffuse part of the distribution (measured from the limit energy) is itself proportional to the first power of the temperature, and it is these particles that determine the macroscopic properties of the system. Owing to this, the "energy uncertainty" arising from the collisions will, at low enough temperatures, be small in comparison with the mean energy itself. It can also be said that at sufficiently low temperatures the damping of the excitations, described by an imaginary term added to their energy, will be small in comparison with the excitation energy.

Two fundamental assumptions lie at the foundation of the Landau theory. The first is that the classification of the levels of the Fermi fluid corresponds to the classification of the levels of noninteracting atoms. This means that as one gradually turns on the interaction the atoms gradually go over into "quasi-particles", each of which has a definite energy. Thus the number of quasi-particles is equal to the number of atoms.

The second assumption is that the interaction of the quasi-particles can be taken into account by means of a self-consistent field of the surrounding particles, which manifests itself in the fact that the energy of the system is not the sum of the energies of the quasi-particles, but instead is a functional of their distribution function. The energy of the quasi-particles must be defined as the variational derivation of the energy density in momentum space, i.e.,

$$\delta E = \int \varepsilon \delta n \, d\tau, \tag{1.1}$$

where $d\tau = 2dp_x dp_y dp_z/(2\pi\hbar)^3$. The factor 2 in $d\tau$ arises from the fact that the quasi-particles have spin 1/2.

In some cases it is necessary to include a spin dependence of the particle energy. Owing to the fact that the spin is a quantum-mechanical quantity, we must regard the distribution function in this case as a statistical operator, and replace (1.1) by a definition of ϵ in the form

$$\delta E = \frac{1}{2} \mathrm{Sp}_\sigma \int \varepsilon \, \delta n \, d\tau. \tag{1.2}$$

The definition of the particle energy by Eq. (1.2) has the consequence that the equilibrium distribution function is in fact the Fermi function. To prove this it is most convenient to use the expression for the entropy*

$$S = -\frac{1}{2} \mathrm{Sp}_\sigma \int \{n \ln n + (1-n) \ln (1-n)\} \, d\tau, \tag{1.3}$$

since this formula is of purely combinatorial origin. From the condition that the entropy be a maximum for constant number of particles and constant energy,

$$N = \frac{1}{2} \mathrm{Sp}_\sigma \int n \, d\tau = \mathrm{const}, \quad E = \mathrm{const}$$

we can find the distribution function by taking the variation with respect to n:

$$n(\varepsilon) = n_F(\varepsilon) = \frac{1}{e^{\frac{\varepsilon - \mu}{T}} + 1}. \tag{1.4}$$

The energy ϵ, being a functional of n, depends on the temperature. This dependence can be put in the following form. If we denote by $\epsilon_0(\mathbf{p}, \sigma)$ the equilibrium energy of the quasi-particles for T = 0, then for a small departure from equilibrium or for small values of T it will be given by the formula

$$\begin{aligned} \varepsilon &= \varepsilon_0(\mathbf{p}, \sigma) + \delta\varepsilon(\mathbf{p}, \sigma) = \varepsilon_0(\mathbf{p}, \sigma) \\ &+ \frac{1}{2} \mathrm{Sp}_{\sigma'} \int f(\mathbf{p}, \sigma; \mathbf{p}', \sigma') \, \delta n(\mathbf{p}', \sigma') \, d\tau'. \end{aligned} \tag{1.5}$$

Here $\delta n = n - n_F$ (T = 0), and f is an operator depending on the momenta and spin operators of two particles. In view of its definition, as the second variational derivative of E with respect to δn, the function f must be symmetric under interchange of $\mathbf{p}, \sigma$ with $\mathbf{p}', \sigma'$. The function f is a very important quantity characterizing the Fermi fluid. As was shown by Landau, it is related to the forward scattering amplitude of two quasi-particles (cf. Appendix 2).

2. THE EFFECTIVE MASS

Because we are concerned here with very low temperatures, the energy ϵ_0 in the absence of an external field can be written in the form

$$\varepsilon_0 - \mu(0) = v(p - p_0), \tag{2.1}$$

where p_0 is the Fermi limiting momentum and v is the velocity at the Fermi limit. This velocity can be written in the form

$$v = \frac{p_0}{m^*}, \tag{2.2}$$

where m* is the effective mass. As has been shown by Landau, there is a definite relation between m*

*Here and throughout we use energy units for the temperature, i.e., k = 1.

and f, which is a consequence of the Galilean principle of relativity.

According to this principle the momentum per unit volume must be equal to the mass flux of the fluid. Because the velocity for the quasi-particles can be defined as the derivative $\partial\epsilon/\partial \mathbf{p}$, and the number of quasi-particles is equal to the number of atoms, we get:

$$\frac{1}{2}\mathrm{Sp}_\sigma \int \mathbf{p} n \, d\tau = \frac{1}{2}\mathrm{Sp}_\sigma \int m \frac{\partial \varepsilon}{\partial \mathbf{p}} n \, d\tau, \tag{2.3}$$

where m is the mass of an atom.

Let us take the variation of this relation with respect to n:

$$\frac{1}{2}\mathrm{Sp}_\sigma \int \mathbf{p} \delta n \, d\tau = \frac{1}{2}\mathrm{Sp}_\sigma \int m \frac{\partial \varepsilon}{\partial \mathbf{p}} \delta n \, d\tau$$
$$+ \frac{1}{2}\mathrm{Sp}_\sigma \mathrm{Sp}_{\sigma'} \int\int m n \frac{\partial}{\partial \mathbf{p}} f(\mathbf{p}, \sigma, \mathbf{p}', \sigma') \delta n' \, d\tau \, d\tau'.$$

Interchanging $\mathbf{p}$, σ and $\mathbf{p}'$, σ' in the last term and using the fact that δn is arbitrary, we get:

$$\frac{\mathbf{p}}{m} = \frac{\partial \varepsilon}{\partial \mathbf{p}} + \frac{1}{2}\mathrm{Sp}_{\sigma'} \int \frac{\partial f}{\partial \mathbf{p}'} n' \, d\tau' = \frac{\partial \varepsilon}{\partial \mathbf{p}} - \frac{1}{2}\mathrm{Sp}_{\sigma'} \int f \frac{\partial n'}{\partial \mathbf{p}'} d\tau. \tag{2.4}$$

If the system is not in an external magnetic field, the relation (2.4) does not lose its generality when we take its trace with respect to the unprimed spin variable $(1/2\,\mathrm{Sp}_\sigma)$. In the case of equilibrium at $T = 0$, ϵ in Eq. (2.4) can be taken in the form (2.1), and $\partial n/\partial p$ can be replaced by $-\delta(p - p_0)$. We then find:

$$\frac{1}{m} = \frac{1}{m^*} + \frac{1}{2}\frac{p_0}{(2\pi\hbar)^3}\mathrm{Sp}_\sigma \mathrm{Sp}_{\sigma'} \int f(\chi) \cos\chi \, d\Omega, \tag{2.5}$$

where $f(\chi)$ is the value of the function f for $|\mathbf{p}| = |\mathbf{p}'| = p_0$. Naturally, apart from a small correction, this relation also holds for temperatures near $T = 0$ and for small departures from equilibrium.

3. THE HEAT CAPACITY AND ENTROPY

Here we shall assume no external magnetic field, so that the quantities will not depend on the spin. A knowledge of the energy spectrum makes it possible to determine the heat capacity of the Fermi fluid.[2] It can be found in the usual way, by differentiating the energy with respect to the temperature for a prescribed number of particles.

An important point here is that the change of the energy density is given by Eq. (1.1); owing to this we can write

$$C = \left(\frac{\partial E}{\partial T}\right)_N = \int \varepsilon \left(\frac{\partial n}{\partial T}\right)_N d\tau \cong \int \varepsilon_0 \left(\frac{\partial n}{\partial T}\right)_N d\tau. \tag{3.1}$$

In the present case the replacement of ϵ by ϵ_0 is justified by the fact that whereas the main term in the heat capacity is linear in T the correction in Eq. (1.5) gives only cubic terms.

In fact, the δn in Eq. (1.5) can be put in the form

$$\delta n = n(T) - n(0) = \int_0^T \frac{\partial n}{\partial T} dT$$
$$= \int_0^T \frac{\partial n}{\partial \varepsilon}\left(-\frac{\varepsilon - \mu}{T} + \frac{\partial \varepsilon}{\partial T} - \frac{\partial \mu}{\partial T}\right) dT.$$

As is well known, at low temperatures the derivative of the Fermi function has the form

$$\frac{\partial n}{\partial \varepsilon} \approx -\delta(\varepsilon - \mu) - \frac{\pi^2}{6} T^2 \frac{\partial^2 \delta(\varepsilon - \mu)}{\partial \varepsilon^2}.$$

In view of this we have

$$\delta\varepsilon = \int f \delta n' \, d\tau' = \frac{\pi^2}{6} T^2 \left[\frac{\partial}{\partial \varepsilon'}\left(f \frac{d\tau'}{d\varepsilon'}\right)\right]_{\varepsilon' = \mu}$$
$$- \int_0^T \left[f\left(\frac{\partial \varepsilon}{\partial T} - \frac{\partial \mu}{\partial T}\right)\frac{d\tau'}{d\varepsilon'}\right]_{\varepsilon' = \mu} dT. \tag{3.2}$$

The derivative $\partial\mu/\partial T$ is equal to the entropy, which, as we shall see, is linear in T. As for the term in $\partial\epsilon'/\partial T$, its order of magnitude can be established by Eq. (3.2). Differentiating this formula with respect to T, we find without difficulty that $\partial\epsilon/\partial T$ is of the first order in T, and consequently $\delta\epsilon$ is of the second order.

Because of the replacement ϵ by ϵ_0 the energy can be calculated simply from the integral

$$E = \int \varepsilon_0 n \, d\tau \quad \text{with} \quad N = \int n \, d\tau = \text{const},$$

where, again to within an error of cubic terms in T, we can replace ϵ by ϵ_0 in the functions n appearing in the integrals for E and N. After this the calculation of the integrals does not differ from that for the usual case of a Fermi gas.

In this way we find:

$$c = \frac{C}{N} = \gamma T; \quad \gamma = \frac{\pi^2}{3N}\left(\frac{d\tau}{d\varepsilon}\right)_\mu = \left(\frac{\pi}{3N}\right)^{\frac{2}{3}} \frac{m^*}{\hbar^2}. \tag{3.3}$$

Because of the linear dependence on the temperature the entropy is equal to the heat capacity.

By comparing Eq. (3.3) with the experimental data on the entropy of liquid He^3 at low temperatures we can find the parameter γ, and consequently can also determine the effective mass. Unfortunately, at present the experimental curve has only been found as far as the beginning of the linear part.[3] From these data we can obtain only an approximate value of γ, which turns out to be about 3 cal mole^{-1} deg^{-2}. Because of the equality of the number of quasi-particles to the number of particles, we can find the Fermi limit momentum from the density of liquid He^3 (ρ = 0.078 g/cm³). In this way we get:

$$m^* = 1.43 m_{He^3}, \quad \frac{p_0}{\hbar} = 0.76 \cdot 10^8 \text{ cm}^{-1}. \tag{3.4}$$

4. THE MAGNETIC SUSCEPTIBILITY

Because particles possessing spin necessarily have a magnetic moment, the Fermi fluid must be magnetic. In the isotropic case there are only two possibilities. The fluid can be paramagnetic or ferromagnetic. Antiferromagnetism in an isotropic Fermi fluid is impossible. We present here only the case of the paramagnetic Fermi liquid, which corresponds to liquid He^3. A calculation of the susceptibility of such a liquid was given by Landau in reference 1 (cf. also reference 2).

When the system is in a magnetic field H, the particles naturally have an additional energy depending on their spins. If they were free particles, the change of the energy would be $-\beta(\boldsymbol{\sigma}\cdot\mathbf{H})$. In the Fermi fluid, however, we have to take into account also the fact that there is a change in the distribution function. Thus we get

$$\delta\epsilon = -\beta(\boldsymbol{\sigma}\cdot\mathbf{H}) + \frac{1}{2}\mathrm{Sp}_{\sigma'}\int f\delta n' d\tau'. \tag{4.1}$$

In the present case the spin dependence of the function f is important. This dependence can be written in the general case in the form

$$f(\mathbf{p}, \sigma; \mathbf{p}', \sigma') = f(\mathbf{p}, \mathbf{p}') + \zeta_{ik}(\mathbf{p}, \mathbf{p}')\sigma_i\sigma_k'. \tag{4.2}$$

It is natural to assume, however, that the spin interaction is mainly of exchange origin. In this case f takes the form

$$f(\mathbf{p}, \sigma; \mathbf{p}', \sigma') = f(\mathbf{p}, \mathbf{p}') + \zeta(\mathbf{p}, \mathbf{p}')\,\boldsymbol{\sigma}\cdot\boldsymbol{\sigma}'. \tag{4.3}$$

The change of the distribution function appearing in Eq. (4.1), δn, is due only to the change of the energy ϵ, since the chemical potential changes only in the second order in H. Because of this we can set $\delta n = \partial n/\partial\epsilon \;\; \delta\epsilon$ in Eq. (4.1). This gives

$$\delta\epsilon = -\beta(\boldsymbol{\sigma}\cdot\mathbf{H}) + \frac{1}{2}\mathrm{Sp}_{\sigma'}\int f\frac{\partial n'}{\partial\epsilon'}\delta\epsilon' d\tau'. \tag{4.4}$$

This is the equation for $\delta\epsilon$, which can be assumed to have the form

$$\delta\epsilon = -\xi(\boldsymbol{\sigma}\cdot\mathbf{H}). \tag{4.5}$$

Substituting this into Eq. (4.4), we get the equation

$$\xi = \beta + \frac{1}{4}\int\zeta\frac{\partial n'}{\partial\epsilon'}\xi' d\tau'. \tag{4.6}$$

In view of the fact that $\partial n'/\partial\epsilon' \cong -\delta(\epsilon' - \mu)$, the integration is taken over the surface of the Fermi sphere. We shall see below that only the value of ξ taken on the surface of the Fermi sphere is of importance for the magnetic susceptibility. In this case both the arguments of ζ (p, p′) have the absolute value p_0, and ζ depends only on the angle between them. Denoting $\int\zeta(\theta)\, d\Omega/4\pi$ by $\bar{\zeta}$, we get from Eq. (4.6):

$$\xi = \frac{\beta}{1 + \frac{1}{4}\bar{\zeta}\left(\frac{d\tau}{d\epsilon}\right)_\mu}. \tag{4.7}$$

The susceptibility is found from the formula

$$\chi = \frac{\partial}{\partial\mathbf{H}}\frac{1}{2}\mathrm{Sp}\int n\beta\boldsymbol{\sigma}\, d\tau.$$

The value of this expression is of course determined by just the spin-dependent change of n. Thus we find:

$$\chi = \frac{1}{4}\beta\xi\left(\frac{d\tau}{d\epsilon}\right)_\mu = \frac{\frac{1}{4}\beta^2\left(\frac{d\tau}{d\epsilon}\right)_\mu}{1 + \frac{1}{4}\bar{\zeta}\left(\frac{d\tau}{d\epsilon}\right)_\mu}. \tag{4.8}$$

The quantity $(\partial\tau/\partial\epsilon)_\mu$ can be expressed in terms of the coefficient γ in the linear law of the heat capacity; this gives

$$\frac{1}{\chi} = \beta^{-2}\left(\bar{\zeta} + \frac{4\pi^2}{3\gamma N}\right). \tag{4.9}$$

Equation (4.9) can be compared with the experimental data on the magnetic susceptibility of He^3;[4] this shows that $\bar{\zeta}$ is negative and is of magnitude 0.85 times the second term. Thus the exchange interaction has a decided effect on the magnitude of the magnetic susceptibility of liquid He^3. The sign of the effect is such that the exchange interaction facilitates parallel orientation of the spins. It does not lead to ferromagnetism, however, since the Fermi tendency toward an antiparallel arrangement of the spins prevails. It is possible that at higher pressures the parameters change in such a way that ferromagnetism becomes possible. There are no experimental data up to the present, however, to support such a conclusion.

5. THE KINETIC EQUATION

The kinetic equation for the distribution function in the absence of a magnetic field has the usual form

$$\frac{\partial n}{\partial t} + \frac{\partial n}{\partial \mathbf{r}}\cdot\frac{\partial\epsilon}{\partial\mathbf{p}} - \frac{\partial n}{\partial\mathbf{p}}\cdot\frac{\partial\epsilon}{\partial\mathbf{r}} = I(n), \tag{5.1}$$

where I(n) is the collision integral. Here, however, we have to take into account the fact that the energy ϵ is a functional of the distribution function, and thus also depends on the coordinates. By means of this kinetic equation Landau[1] obtained expressions for the fluxes of energy and momentum.

To find the momentum flux one must multiply (5.1) by p_i and integrate over the phase space. Conservation of momentum in the collisions makes the integral $\int p_i I(n) d\tau$ vanish, and we have the equation

$$\frac{\partial}{\partial t}\int p_i n\, d\tau + \int p_i\left(\frac{\partial n}{\partial x_k}\cdot\frac{\partial \varepsilon}{\partial p_k} - \frac{\partial n}{\partial p_k}\cdot\frac{\partial \varepsilon}{\partial x_k}\right) d\tau = 0.$$

This expression in parentheses in the second integral can be written

$$\frac{\partial}{\partial x_k}\left(n\frac{\partial \varepsilon}{\partial p_k}\right) - \frac{\partial}{\partial p_k}\left(n\frac{\partial \varepsilon}{\partial x_k}\right).$$

Substituting this into the integral and integrating the second term by parts, we get:

$$\frac{\partial}{\partial x_k}\int p_i\frac{\partial \varepsilon}{\partial p_k} n\, d\tau + \int n\frac{\partial \varepsilon}{\partial x_i} d\tau.$$

The last term can be put in the form

$$\frac{\partial}{\partial x_i}\int n\varepsilon\, d\tau - \int \varepsilon\frac{\partial n}{\partial x_i} d\tau.$$

But in view of the fact that $\delta E = \int \varepsilon \delta n d\tau$, the last term here is equal to $\partial E/\partial x_i$. Thus we get the equation

$$\frac{\partial}{\partial t}\int p_i n\, d\tau + \frac{\partial \Pi_{ik}}{\partial x_k} = 0. \tag{5.2}$$

This equation expresses the law of conservation of momentum. The tensor Π_{ik} is the flux of momentum. It is given by

$$\Pi_{ik} = \int p_i\frac{\partial \varepsilon}{\partial p_k} n\, d\tau + \delta_{ik}\left[\mathrm{Sp}\int n\varepsilon\, d\tau - E\right]. \tag{5.3}$$

In a similar way we also get the law of conservation of energy. To do this we multiply the kinetic equation by ϵ and integrate with respect to $d\tau$. Because of the conservation of energy in the collisions the integral $\int \epsilon I(n) d\tau$ vanishes, and we get:

$$\int \varepsilon\frac{\partial n}{\partial t} d\tau + \int \varepsilon\left(\frac{\partial n}{\partial \mathbf{r}}\frac{\partial \varepsilon}{\partial \mathbf{p}} - \frac{\partial n}{\partial \mathbf{p}}\cdot\frac{\partial \varepsilon}{\partial \mathbf{r}}\right) d\tau = 0.$$

Writing the term in $\partial n/\partial \mathbf{r}$ in the form

$$\frac{\partial}{\partial \mathbf{r}}\int n\varepsilon\frac{\partial \varepsilon}{\partial \mathbf{p}} d\tau - \int n\frac{\partial \varepsilon}{\partial \mathbf{r}}\frac{\partial \varepsilon}{\partial \mathbf{p}} d\tau - \int n\frac{\partial^2 \varepsilon}{\partial \mathbf{r}\,\partial \mathbf{p}}\varepsilon\, d\tau$$

and integrating by parts with respect to $\mathbf{p}$ in the term in $\partial n/\partial \mathbf{p}$, we get the law of conservation of energy

$$\frac{\partial E}{\partial t} = \int \varepsilon\frac{\partial n}{\partial t} d\tau = -\operatorname{div}\mathbf{Q}, \tag{5.4}$$

where the flux of energy is given by

$$\mathbf{Q} = \int n\varepsilon\frac{\partial \varepsilon}{\partial \mathbf{p}} d\tau. \tag{5.5}$$

6. THE VISCOSITY

From (5.1) and the expressions (5.2) for the momentum flux and (5.5) for the energy flux we can determine the coefficients of viscosity and thermal conductivity of the Fermi fluid.[5]

We begin by finding the viscosity. To do this, suppose that there occurs a motion in the liquid with a certain velocity $\mathbf{u}$ which is a slowly varying function of position. In this case the distribution function will differ only slightly from the equilibrium value:

$$n = n_0 + \delta n, \tag{6.1}$$

where

$$n_0 = \frac{1}{e^{\frac{\varepsilon_0 - \mathbf{p}\cdot\mathbf{u} - \mu}{T}} + 1}; \quad |\delta n| \ll n_0. \tag{6.2}$$

The quantity δn is found from the kinetic equation (5.1). As usual, we must insert the function n_0 in the left member of the kinetic equation. Furthermore we shall assume that at the point considered $\mathbf{u} = 0$. Substituting Eq. (6.2) into (5.1) we find:

$$\frac{\partial n_0}{\partial t} - \frac{1}{3}\frac{\partial n_0}{\partial \varepsilon}\mathbf{p}\frac{\partial \varepsilon}{\partial \mathbf{p}}\operatorname{div}\mathbf{u} - \frac{1}{2}\frac{\partial n_0}{\partial \varepsilon}\left(p_i\frac{\partial \varepsilon}{\partial p_k} - \frac{1}{3}p_l\frac{\partial \varepsilon}{\partial p_l}\delta_{ik}\right) \times\left(\frac{\partial u_i}{\partial x_k} + \frac{\partial u_k}{\partial x_i} - \frac{2}{3}\delta_{ik}\frac{\partial u_l}{\partial x_l}\right) = I(n). \tag{6.3}$$

We now transform $\partial n_0/\partial t$ and show that this expression also does not depend on the term in f in Eq. (1.5). According to Eq. (6.2) we can write

$$\delta n_0 = \frac{\partial n_0}{\partial \varepsilon_0}\left(\delta\varepsilon_0 - \delta\mu - \frac{\varepsilon_0 - \mu}{T}\delta T - \mathbf{p}\cdot\delta\mathbf{u}\right). \tag{6.4}$$

Since the derivative $\partial n_0/\partial \epsilon_0$ in Eq. (6.4) is different from zero in a small neighborhood of the point $\epsilon_0 = \mu$, and is a rapidly changing function in this region, we can suppose the quantities in the parentheses evaluated at that point (the error will be of the relative order $(T/\mu)^2$). The variations $\delta\epsilon_0$ and $\delta\mu$ are arbitrary and are by no means equal to each other. On the other hand, the distribution function is normalized by the relation

$$\int n_0 d\tau = N,$$

where N is the number of atoms in unit volume. Taking the variation of this relation, we find:

$$\begin{aligned}\delta N &= \int \frac{\partial n_0}{\partial \varepsilon_0}\left(\delta\varepsilon_0 - \delta\mu - \frac{\varepsilon_0 - \mu}{T}\delta T - \mathbf{p}\cdot\delta\mathbf{u}\right) d\tau \\ &\approx -\left[(\delta\varepsilon_0 - \delta\mu - \mathbf{p}\cdot\delta\mathbf{u})\frac{d\tau}{d\varepsilon_0}\right]_{\varepsilon_0=\mu},\end{aligned} \tag{6.5}$$

since $\int(\partial n_0/\partial \epsilon_0)\, d\epsilon_0 = -1$. Comparing (6.5) and (6.4), we get:

$$\frac{\partial n_0}{\partial t} = -\frac{\partial N}{\partial t}\cdot\frac{\partial n_0}{\partial \varepsilon}\cdot\left(\frac{d\varepsilon_0}{d\tau}\right)_{\varepsilon_0=\mu}.$$

The number of atoms N satisfies the equation of continuity

$$\frac{\partial N}{\partial t} + N\operatorname{div}\mathbf{u} = 0. \tag{6.6}$$

Thus the term in $\partial n_0/\partial t$ gives an addition to the term in div u, which as a result now takes the form

$$-\frac{\partial n_0}{\partial \varepsilon}\left(\frac{1}{3}\mathbf{p}\frac{\partial \varepsilon_0}{\partial \mathbf{p}} - N\left(\frac{d\varepsilon_0}{d\tau}\right)_{\varepsilon_0=\mu}\right)\operatorname{div}\mathbf{u}. \tag{6.7}$$

Using Eq. (2.1) and the equality of the number of excitations to the number of particles, which makes $p_0/h = (3\pi^2 N)^{1/3}$, we find that the expression just written vanishes. Thus apart from small corrections of the order $(T/\mu)^2$ there are no terms in div u in the left member of the kinetic equation.

The right member of Eq. (5.1) is the collision integral

$$I(n) = -\int w\{n_1 n_2(1-n_1')(1-n_2') - (1-n_1)(1-n_2)n_1' n_2'\}$$
$$\times\, \delta(\mathbf{p}_1+\mathbf{p}_2-\mathbf{p}_1'-\mathbf{p}_2')\,\delta(\varepsilon_1+\varepsilon_2-\varepsilon_1'-\varepsilon_2')\,d\tau_2 d\tau_1' d\mathbf{p}_2'. \tag{6.8}$$

Here we must keep in mind that the δ function of the energies has as its arguments the actual energies, i. e., the energies as given by Eq. (1.5). Owing to this, substitution of the function $n_0(\epsilon_0)$ into the collision integral does not make it vanish. The integral vanishes only when we substitute in it the true equilibrium function, i. e., $n_0(\epsilon)$. The total distribution function can be written

$$n = n_0(\varepsilon) - \frac{\partial n_0}{\partial \varepsilon}\int f\,\delta n\,d\tau + \delta n. \tag{6.9}$$

We introduce the notation

$$\delta n = -\frac{n_0(1-n_0)}{T}\nu = \frac{\partial n_0}{\partial \varepsilon_0}\nu. \tag{6.10}$$

Substituting this in Eq. (6.9), we find that the term added to the equilibrium distribution function also has the form (6.10). Instead of ν however, it contains the quantity

$$\phi = \nu + \left(\frac{d\tau}{d\varepsilon_0}\right)_{\mu}\int f(\chi)\,\nu\,\frac{d\Omega}{4\pi}. \tag{6.11}$$

Substitution of Eq. (6.9) in the collision integral, with neglect of terms quadratic in ψ, gives:

$$I(n) = \frac{1}{T}\int w n_{01} n_{02}(1-n_{01}')(1-n_{02}')(\psi_1+\psi_2-\psi_1'-\psi_2')$$
$$\times\, \delta(\mathbf{p}_1+\mathbf{p}_2-\mathbf{p}_1'-\mathbf{p}_2')\,\delta(\varepsilon_1+\varepsilon_2-\varepsilon_1'-\varepsilon_2')\,d\tau_2 d\tau_1' d\mathbf{p}_2'. \tag{6.12}$$

In general the collision probability w depends on all four momenta. The momenta of importance here, however, are those with magnitudes close to the limit value on the Fermi surface. We can therefore assume that w depends only on θ, the angle between $\mathbf{p}_1$ and $\mathbf{p}_2$, and φ, the angle between the planes $(\mathbf{p}_1, \mathbf{p}_2)$ and $(\mathbf{p}_1', \mathbf{p}_2')$.

Let us make use of the fact that the momenta of the particles in the main region of the integration differ little in magnitude from the limit momentum p_0. If we rotate the plane of the vectors $(\mathbf{p}_1', \mathbf{p}_2')$

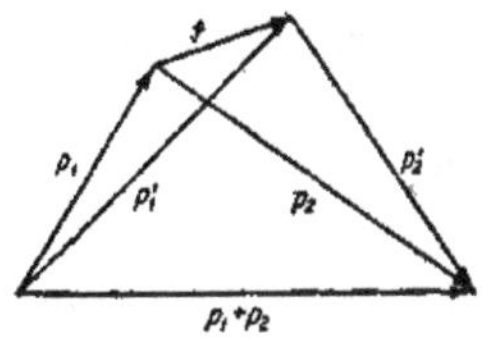

FIG. 1

through the angle φ relative to the axis in the direction $\mathbf{p}_1 + \mathbf{p}_2$, so that this plane coincides with the plane $(\mathbf{p}_1, \mathbf{p}_2)$, we get the diagram shown in Fig. 1. It is clear that the vector f will be small in magnitude and that the angles between all the momenta and the axis $\mathbf{p}_1 + \mathbf{p}_2$ will be approximately equal to $\theta/2$ or $-\theta/2$. Thus we get

$$p_1' \approx p_1 + f_z\cos\frac{\theta}{2} + f_r\sin\frac{\theta}{2}, \quad p_2' \approx p_2 - f_z\cos\frac{\theta}{2} + f_r\sin\frac{\theta}{2}, \tag{6.13}$$

where f_z is the component of f along the axis $\mathbf{p}_1 + \mathbf{p}_2$ and f_r is the perpendicular component. We get rid of the δ function by integrating over $d\mathbf{p}_2'$, and replace the integral over $d\tau_1'$ by that over $df_r df_z d\varphi$, introducing a system of cylindrical coordinates with axis along $\mathbf{p}_1 + \mathbf{p}_2$:

$$d\tau_1' = 2p_0\sin\frac{\theta}{2}\,df_r df_z d\varphi\,\frac{1}{(2\pi\hbar)^3}. \tag{6.14}$$

We now introduce the following notation:

$$x = \frac{\varepsilon_1' - \mu}{T}, \quad t = \frac{\varepsilon_1 - \mu}{T}, \quad \varkappa = \frac{\mu}{T}, \quad y = \frac{\varepsilon_2' - \mu}{T}; \tag{6.15}$$

with the values of ϵ_i from Eq. (2.1). The change from the variables f_r and f_z to x and y is accomplished very easily by means of Eq. (6.13). Then we get:

$$d\tau_1'\int d\tau_2\,\delta(\varepsilon) = \frac{m^{*3}T^2 dx\,dy\left(\frac{d\Omega}{2\pi}\right)\left(\frac{d\varphi_2}{2\pi}\right)}{8\pi^4\hbar^4\cos\left(\frac{\theta}{2}\right)}, \tag{6.16}$$

where we have denoted the angular differential $\sin\theta d\theta d\varphi$ by $d\Omega$, which involves the arguments of $w(\theta, \varphi)$. We note that because of the indistinguishibility of the particles the angle φ varies only from zero to π.

From considerations of symmetry it is clear that ψ must have the form

$$\psi = \frac{1}{2}q(t)\left(p_i\frac{\partial\varepsilon}{\partial p_k} - \frac{1}{3}p_l\frac{\partial\varepsilon}{\partial p_l}\delta_{ik}\right)\cdot\left(\frac{\partial u_i}{\partial x_k} + \frac{\partial u_k}{\partial x_i} - \frac{2}{3}\delta_{ik}\frac{\partial u_m}{\partial x_m}\right). \tag{6.17}$$

When this expression is substituted in the collision integral the second factor can be transformed by means of the addition theorem for spherical harmonics. After the integration over the angle φ_2 only the first terms remain, i.e.,

$$P_2(\theta_1+\theta_2) \to P_2(\theta_1)P_2(\theta_2), \text{ where } P_2(\theta) = \frac{3}{2}\cos^2\theta - 1.$$

As the result we get from Eq. (6.3) the following equation for q:

$$n_0(t)[1-n_0(t)] = \frac{m^{*3}T^2}{8\pi^4\hbar^6\cos\frac{\theta}{2}}\int\frac{d\Omega}{2\pi}\int\limits_{\substack{-\infty \\ x+y>t-\infty}}^{\infty} dx \int\limits_{-\infty}^{\infty} dy\, w(\theta,\varphi)$$
$$\times n_0(t)\, n_0(x+y-t)[1-n_0(x)][1-n_0(y)] \tag{6.18}$$
$$\times [q(t)+q(x+y-t)P_2(\theta)-q(x)P_2(\theta')-q(y)P_2(\theta_2')].$$

In view of the fact that in the range of temperatures we are considering only values of x, y, and t much smaller than κ are of importance, we can take the lower limit of the integration in Eq. (6.12) to be $-\infty$. Under these conditions, assuming that q is a symmetric function (as is confirmed by the results) we easily verify that all the terms with different q's can be reduced to the same form as regards the dependence on x and y. Thus the bracketed expression takes the simple form

$$q(t)+q(x)[P_2(\theta)-P_2(\theta_1')-P_2(\theta_2')].$$

The change from the angles θ_1' and θ_2' to the angles θ and φ is easily obtained from Fig. 1. Integrating over the variable y, on which q no longer depends, after some manipulation we get the following equation:

$$\frac{8\pi^4\hbar^6}{m^{*3}T^2} = \overline{\frac{w(\theta,\varphi)}{\cos\frac{\theta}{2}}\cdot\left[\frac{3}{4}(1-\cos\theta)^2\sin^2\varphi-1\right]}$$
$$\times\left\{\int\limits_0^\infty \frac{dx\, x[q(x+t)+q(x-t)]}{e^x-1}+2\int\limits_0^\infty\frac{dx\, xq(x)}{e^x+1}+\int\limits_0^t dx\cdot xq(x-t)\right\}$$
$$+\overline{\frac{w(\theta,\varphi)}{\cos\frac{\theta}{2}}}\,q(t)\cdot\frac{\pi^2+t^2}{2}, \tag{6.19}$$

where the bar denotes averages over the solid angle.

The equation that has been obtained is complicated, but analysis shows that for arbitrary assumptions about the form of $w(\theta,\varphi)$ the error in the coefficient of viscosity will be smaller than 10 percent if we simply assume that the values we need satisfy $t^2 \ll \pi^2$. Then q is a constant, given by

$$q = \frac{64\pi^2\hbar^6}{3m^{*3}T^2}\left[\overline{\frac{w(\theta,\varphi)}{\cos\left(\frac{\theta}{2}\right)}(1-\cos\theta)^2\sin^2\varphi}\right]^{-1}. \tag{6.20}$$

The flux of momentum is given by (5.3). Inserting in it $n = n_0 + \delta n$ and Eq. (6.10), and using the fact that the energy is also a functional of the distribution function, we get:

$$\Pi_{ik} = \int p_i\frac{\partial\varepsilon(p)}{\partial p_k}\left[\delta n-\frac{\partial n_0}{\partial\varepsilon_0}\int f(\mathbf{p},\mathbf{p}')\,\delta n'\,d\tau'\right]d\tau$$
$$=\int p_i\frac{\partial\varepsilon(p)}{\partial p_k}\frac{\partial n_0}{\partial\varepsilon_0}\left[\nu+\int f\nu\left(\frac{d\tau'}{d\varepsilon_0'}\right)_{\varepsilon_0'=\mu}\frac{d\Omega'}{4\pi}\right]d\tau \tag{6.21}$$
$$=\int p_i\frac{\partial\varepsilon(p)}{\partial p_k}\frac{\partial n_0}{\partial\varepsilon_0}\,\psi\, d\tau.$$

Thus it is seen that the expression for the momentum flux involves the same function ψ that appears in the kinetic equation. Consequently, the function f does not enter the expression for the momentum flux, and it has the same form as for a Fermi gas of particles with the mass m^* and the scattering law described by the function $w(\theta,\varphi)$.*

Substituting Eqs. (6.17) and (6.20) in Eq. (6.21) and defining the coefficient of viscosity as the proportionality coefficient between Π_{ik} and $-(\partial u_i/\partial x_k + \partial u_k/\partial x_i - 2/3\, \partial u_l/\partial x_l\, \delta_{ik})$, we find

$$\eta = \frac{64}{45}T^{-2}\frac{\hbar^3 p_0^5}{m^{*4}}\left[\overline{\frac{w(\theta,\varphi)}{\cos\frac{\theta}{2}}(1-\cos\theta)^2\sin^2\varphi}\right]^{-1}. \tag{6.22}$$

The viscosity is proportional to T^{-2}. This dependence was earlier predicted by Pomeranchuk[6] on the basis of qualitative considerations. As for the numerical values of the viscosity, they depend on the definite form of the averaged function $w(\theta,\varphi)$ and therefore cannot be found precisely. But Eq. (6.22) makes it possible to determine the order of magnitude of η. To do this we make use of the fact, which will be proved in Appendix 2, that the function f is of the order of magnitude of the scattering amplitude of the quasi-particles. Consequently $w \sim (2\pi/\hbar) f^2$, and the value of f can be determined from experimental data on the speed of sound (cf. Sec. 8). Using the numerical values of m^* and p_0 from Eq. (3.4), we get for He^3

$$\eta = \frac{a}{T^2}\ (a \sim 10^{-6} \div 10^{-5}\ \text{poise};\ T\ \text{in}\ {}^\circ\text{K}). \tag{6.23}$$

This order of magnitude corresponds to the experimental results found by K. N. Zinov'eva.[7] Because her measurements went down only to 0.35°K, it is indeed impossible to check the temperature dependence. Nevertheless the experimental results show an increase of the viscosity with decreasing temperature.

As already noted early in this section, apart from terms of order $(T/\mu)^2$ there is no part pro-

*In reference 5 the second term in the formula (6.9) was not included, and incorrect formulas were thus obtained for the viscosity and thermal conductivity coefficients. The correct values can be obtained by taking f = 0 in the expressions (26) and (30) of reference 5.

portional to div **u** on the left side of the kinetic equation. This means that the coefficient of second (dilatational) viscosity is smaller than η by a factor $(T/\mu)^2$ and plays no part in physical effects (for example, the absorption of sound).

7. THE THERMAL CONDUCTIVITY

We now go on to the calculation of the thermal conductivity.[5] Assuming a small temperature gradiant, we get in the left side of the kinetic equation:

$$-\frac{\partial n_0}{\partial \varepsilon_0}\left(\frac{\varepsilon_0-\mu}{T}-s\right)\frac{\partial \varepsilon_0}{\partial \mathbf{p}}\nabla T. \tag{7.1}$$

At the beginning the transformations of the collision integral are the same as those made in the preceding section [Eqs. (6.8) to (6.16)]. In the present case it is reasonable to try to find the function ψ in the form

$$\psi = q(t)\frac{\partial \varepsilon_0}{\partial \mathbf{p}}\nabla T. \tag{7.2}$$

Instead of Eq. (6.8) we now get a similar equation, but with $\cos\theta$ instead of $P_2(\cos\theta)$ in the collision integral and the factor $(\epsilon_0-\mu)/T - s$ in the left member. The presence of such a factor shows that the unknown function must contain parts both symmetric and antisymmetric in t, namely:

$$q(t) = q_s(t) + q_a(t). \tag{7.3}$$

Substitution in the collision integral gives the two equations

$$\frac{8\pi^4\hbar^6}{m^{*3}T^2}s = \frac{\overline{w(\theta,\varphi)}}{\cos\frac{\theta}{2}}\Big[q_s(t)\frac{\pi^2+t^2}{2} - \int_0^\infty dx\,\frac{x[q_s(x+t)+q_s(x-t)]}{e^x-1} - 2\int_0^\infty \frac{dx\,xq_s(x)}{e^x+1} - \int dx\,xq_s(x-t)\Big], \tag{7.4}$$

$$\frac{8\pi^4\hbar^6}{m^{*3}T^2} = \frac{\overline{w(\theta,\varphi)(1+2\cos\theta)}}{\cos\frac{\theta}{2}}\Big[\int_0^\infty dx\,\frac{x[q_a(x-t)-q_a(x+t)]}{e^x-1} + 2t\int_0^\infty\frac{dx\,q_a(x)}{e^x+1} + \int_0^t dx\cdot xq_a(x-t)\Big] + \frac{\overline{w(\theta,\varphi)}}{\cos\frac{\theta}{2}}q_a(t)\frac{\pi^2+t^2}{2}. \tag{7.5}$$

In the present case, however, the kinetic equation does not completely determine the solution of the problem. One must use along with it a supplementary condition, which expresses the conservation of the flux of matter. This is

$$\delta\int n\frac{\partial \varepsilon}{\partial \mathbf{p}}d\tau = \int \delta n\frac{\partial \varepsilon_0}{\partial \mathbf{p}}d\tau + \int n_0\frac{\partial}{\partial \mathbf{p}}\Big[\int f\delta n'\,d\tau'\Big]d\tau$$
$$= \int\frac{\partial \varepsilon_0}{\partial \mathbf{p}}\Big[\delta n - \frac{\partial n_0}{\partial \varepsilon_0}\int f\delta n'd\tau'\Big]d\tau = 0. \tag{7.6}$$

Here we have performed an integration by parts and have replaced ϵ by ϵ_0 in the integrals that already contain δn. Substituting the expressions (6.10), (6.11), and (7.2) in Eq. (7.6), we get

$$\int\frac{\partial n_0}{\partial \varepsilon_0}\left(\frac{\partial \varepsilon_0}{\partial \mathbf{p}}\right)^2 q(t)\,d\tau = 0. \tag{7.7}$$

To find $q_a(t)$ we have only to solve (7.5) in a way like that used in the treatment of the viscosity. The situation is different for the even part $q_s'(t)$. First, we must note that q_s = constant makes the right member of Eq. (7.4) vanish identically. Therefore the constant term $q_s(0)$ is to be determined not from this equation, but from the condition (7.7). Furthermore it is easy to show that the subsequent terms in $q_s(t)$, namely $a_2t^4 + a_4t^4 + \ldots$, make no contribution to the thermal conductivity. In fact, the occurrence of such a term as a_mt^{2m}, for example, must change the constant term in $q_s(t)$ by a quantity a_m^0 which is such that

$$\int\frac{\partial n_0}{\partial \varepsilon_0}\left(\frac{\partial \varepsilon}{\partial \mathbf{p}}\right)^2[a_mt^{2m}+a_m^0]\,d\tau = 0.$$

From this condition we get:

$$a_m^0 = -a_m(2m)!\,R_m;\quad R_m = \int_0^\infty\frac{z^{2m-1}}{e^z+1}dz.$$

To find the energy flux we have to calculate integrals of the form

$$\int\frac{\partial n_0}{\partial \varepsilon_0}F(\varepsilon_0)(a_mt^{2m}+a_m^0)\,d\tau$$
$$=\left(F\frac{d\tau}{d\varepsilon_0}\right)_{\varepsilon_0=\mu}[a_m(2m)!\,R_m + a_m^0] = 0.$$

Thus we get the result that to find the thermal conductivity coefficient we need only solve (7.5) and then choose the constant term so as to satisfy the condition (7.7). As in the case of the viscosity, it is enough for practical purposes to find the solution on the assumption $t^2 \ll \pi^2$. We then get:

$$q = \frac{24\pi^2\hbar^6}{m^{*3}T^2}\left(t - \frac{\pi^2T_m^*}{p_0^2}\right)\left[\frac{\overline{w(\theta,\varphi)(1-\cos\theta)}}{\cos\frac{\theta}{2}}\right]^{-1}. \tag{7.8}$$

The energy flux is given by Eq. (5.5). Substituting Eq. (6.1) in that formula, we have:

$$Q = \int\varepsilon_0(\mathbf{p})\frac{\partial \varepsilon_0(\mathbf{p})}{\partial p}\delta n\,d\tau - \int\varepsilon_0(\mathbf{p})\frac{\partial \varepsilon_0(\mathbf{p})}{\partial \mathbf{p}}\cdot\frac{\partial n}{\partial \varepsilon_0}\int f\delta n'\,d\tau'\,d\tau. \tag{7.9}$$

Substituting Eqs. (6.10), (6.11), (7.2), and (7.8) we find the value of the thermal conductivity coefficient

$$\varkappa = \frac{8}{3}\frac{\pi^2\hbar^3 p_0^3}{m^{*4}T}\cdot\left[\overline{\frac{w(\vartheta,\varphi)(1-\cos\theta)}{\cos\frac{\theta}{2}}}\right]^{-1}. \qquad (7.10)$$

As in the case of the viscosity, the function f does not appear in the coefficient κ.

The temperature dependence of the thermal conductivity coefficient is given by the T^{-1} law, which also agrees with the qualitative prediction of Pomeranchuk.

Estimating the numerical value of κ for He^3 in the same way as we did that of the viscosity, we get the relation

$$\varkappa = \frac{\beta}{T}(\beta \sim 10^2 \div 10^3\ \text{erg cm}^{-1}\,\text{sec}^{-1}\,\text{deg}^{-1};\ T \text{ in } {}^\circ\text{K}). \qquad (7.11)$$

In any case the expressions obtained for the viscosity and thermal conductivity coefficients cease to be valid for temperatures $T \sim \mu$. In addition to this, however, there is a limitation that shifts the region of applicability of the theory toward considerably lower temperatures. This is that the range of excitation energies with which we are concerned, of the order of T, must be much larger than the quantum uncertainty in the energy that arises owing to the collisions, i.e.,

$$\tau \gg \frac{\hbar}{T}, \qquad (7.12)$$

where τ is the time between collisions. We note that the condition (7.12) is required not only for the calculation of the kinetic coefficients as it was done above, but also for the validity of the entire theory of the Fermi fluid. For He^3 the quantity τ can be determined from the experimental values of the viscosity[7] [cf. Eq. (9.12)]. If we use the point at the very lowest temperature, i.e., $\eta = 4.8 \times 10^{-5}$ poise at $T = 0.35°K$, we get from Eq. (7.12)

$$T \ll 0.35^\circ\,\text{K}. \qquad (7.13)$$

It is clear, however, that this value is too high, since at $T = 0.35°K$ the viscosity is still not proportional to T^{-2}, but follows a weaker law of variation.

8. THE PROPAGATION OF SOUND

The propagation of sound in a Fermi fluid has a number of quite specific peculiarities. If we consider sound of a prescribed frequency, then at not too low temperatures its propagation takes place according to the laws of ordinary hydrodynamics. In this case the damping of the sound is proportional to τ. When the temperature is lowered the collision time will increase proportional to T^{-2}, and finally, at temperatures such that $\omega\tau \sim 1$, the sound ceases to be propagated at all. It turns out, however, that when the temperature is lowered still further the possibility of propagation of sound reappears. The velocity will then be different, and the sound is no longer simply a wave of compression and rarefaction. This phenomenon was also predicted by Landau,[8] and was called by him the "zeroth sound". Because only the relation between ω and τ is essential in fixing the nature of the sound, these two kinds of sound can be described as low-frequency sound ($\omega\tau \ll 1$) and high-frequency sound ($\omega\tau \gg 1$).

The speed of sound at low frequencies or not too low temperatures, for which the condition $\omega\tau \ll 1$ holds, is determined by the compressibility in the usual way. It was shown by Landau[1] that there is an important dependence of the speed on the function f.

It is convenient to express the compressibility in terms of the derivative $\partial\mu/\partial N$. Here N is the total number of particles. Using the fact that the chemical potential μ depends only on N/V, we find:

$$\frac{\partial\mu}{\partial N} = -\frac{V\frac{\partial\mu}{\partial V}}{N} = -\frac{V^2}{N^2}\frac{\partial p}{\partial V}. \qquad (8.1)$$

As is well known, the speed of sound is given by the relation

$$c^2 = \frac{\partial p}{\partial \rho} = \frac{\partial p}{\partial\left(\frac{mN}{V}\right)} = \frac{1}{m}\left(N\frac{\partial\mu}{\partial N}\right). \qquad (8.2)$$

The derivative $\partial\mu/\partial N$ is calculated in the following way. Because $\mu = \epsilon(p_0)$, the change of μ occurs both on account of the change of p_0 and also because of the change of the form of the function $\epsilon(p)$:

$$\delta\mu = \int f\,\delta n'\,d\tau' + \frac{\partial\epsilon_0}{\partial p_0}\delta p_0. \qquad (8.3)$$

The changes δN and δp are connected by the relation

$$\delta N = 8\pi p_0^2\,\delta p_0\frac{V}{(2\pi\hbar)^3}. \qquad (8.4)$$

Since only changes δn close to the Fermi limit are important in the integral of Eq. (8.3), the integration over the absolute value of the momentum can be performed. This gives

$$\int f\,\delta n'\,d\tau' = \frac{\delta N}{4\pi V}\int f\,d\Omega. \qquad (8.5)$$

Substitution in Eq. (8.3) gives [cf. also Eqs. (2.1) and (2.2)]:

$$\frac{\partial\mu}{\partial N} = \frac{1}{4\pi V}\int f\,d\Omega + \frac{(2\pi\hbar)^3}{8\pi p_0 m^* V}. \qquad (8.6)$$

Equation (2.5) enables us to express the effective mass m* appearing here in terms of the actual mass m of the atoms. Substituting the resulting expression in Eq. (8.2), and using the fact that

$$N = \frac{8\pi p_0^3 V}{3(2\pi\hbar)^3},$$

we find:

$$c^2 = \frac{p_0^2}{3m^2} + \frac{2}{3m}\left(\frac{p_0}{2\pi\hbar}\right)^3 \int f(\chi)(1-\cos\chi)\,d\Omega. \quad (8.7)$$

The formulas (2.5) and (8.7) make it possible to get some information about the function f(χ). If we introduce the notation

$$F(\chi) = f(\chi)\left(\frac{d\tau}{d\varepsilon}\right)_{\varepsilon=\mu} = f(\chi)\frac{p_0 m^*}{\pi^2\hbar^3} = \sum_k F_k P_k(\cos\chi), \quad (8.8)$$

the first coefficients of the expansion in Legendre polynomials can be found from Eqs. (2.5) and (8.7). Indeed, it is not hard to get the relations

$$\frac{1}{m} = \frac{1}{m^*}(1+\overline{F\cos\chi}), \quad c^2 = \left(\frac{p_0^2}{3m^2}\right)\frac{1+\overline{F}}{1+\overline{F\cos\chi}}. \quad (8.9)$$

Substituting the values of the parameters for He^3 from Eq. (3.4) and the value of the speed of sound,[9] $c_{T=0} = 183$ m/sec, we find: d:

$$F_0 = \overline{F} = 4.65, \quad F_1 = 3\overline{F\cos\chi} = 1.3. \quad (8.10)$$

In view of the fact that the later coefficients cannot be determined in such a simple way, hereafter we shall confine ourselves in specific cases to the first two harmonic contributions to the function F:

$$F = F_0 + F_1\cos\chi. \quad (8.11)$$

We now go on to the study of the "zeroth sound", i.e., the case in which $\omega\tau \gg 1$. The kinetic equation has the form (5.4). We substitute in it the distribution function in the form

$$n = n_0 + \delta n, \quad (8.12)$$

where the small deviation $\delta n \sim e^{i(\mathbf{k}\cdot\mathbf{r} - \omega t)}$. Here, of course, we have to take into account the fact that, according to Eq. (1.5), the energy of the excitations also undergoes a change. Inasmuch as the collision integral has the behavior $I(n) \sim \delta n/\tau$, and $\omega\tau \gg 1$, we can neglect I(n).

Thus we find that

$$(\mathbf{k}\cdot\mathbf{v} - \omega)\delta n - \mathbf{k}\cdot\mathbf{v}\frac{\partial n_0}{\partial\varepsilon}\int f\delta n'\,d\tau' = 0.$$

From the form of this equation it follows that δn is proportional to $\partial n_0/\partial\epsilon$. Writing $\delta n = \partial n_0/\partial\epsilon\,\nu$, we get:

$$(\mathbf{k}\cdot\mathbf{v} - \omega)\nu + \mathbf{k}\cdot\mathbf{v}\int F\nu'\frac{d\Omega'}{4\pi} = 0. \quad (8.13)$$

If we take **k** as the polar axis and introduce the symbols $u = \omega/k$ for the speed of propagation of the wave and $s = u/v$, then Eq. (8.13) takes the form:

$$(s-\cos\theta)\nu(\theta,\varphi) = \cos\theta\int F(\chi)\nu(\theta',\varphi')\frac{d\Omega}{4\pi}. \quad (8.14)$$

From this we can determine $\nu(\theta,\varphi)$ and the characteristic value s. In view of the fact that the change of the distribution function, δn, is proportional to $\partial n_0/\partial\epsilon$, it is clear that the essential point is the deformation of the Fermi surface. The shape of this deformation is determined by the function ν. Equation (8.14) depends essentially on the function F.

Let us consider first as an example the simplest case, namely $F = F_0 = $ const. With this we get from Eq. (8.14) (without the periodic exponential factor)

$$\nu = \frac{\text{const}\cdot\cos\theta}{s-\cos\theta}. \quad (8.15)$$

A displacement of the Fermi surface as a whole (this corresponds to ordinary sound) would be given by $\nu \sim \cos\theta$. Thus in the present case we have to do with an anisotropic deformation of the Fermi surface. As we shall see latter, s must be larger than unity. Thus the Fermi surface is stretched out in the direction of the motion.

Substituting Eq. (8.15) in Eq. (8.14) with $F = F_0$, we find the equation for s. Carrying out the elementary integration, we find:

$$\frac{s}{2}\ln\frac{s+1}{s-1} - 1 = \frac{1}{F_0}. \quad (8.16)$$

From this it can be seen that if s is real (which corresponds to undamped waves) it must be larger than 1, i.e.,

$$u > v. \quad (8.17)$$

From Eq. (8.14) it can be seen that this condition remains valid for any function F. Furthermore, since the left member of Eq. (8.16) is always positive, it is clear that the condition for the existence of zeroth sound is $F_0 > 0$.

If the function F_0 is large, s is also large. From Eq. (8.16) we get. $s \to (F_0/3)^{1/2}$ for $F_0 \to \infty$. On the other hand, for $F_0 \to 0$ we have $s \to 1$, i.e., $u \to v$. This is the case of an almost free Fermi gas.

It is not hard to see that the conclusion that $s \to 1$ for $F \to 0$ does not depend on the form of F. In fact, it follows from Eq. (8.14) that for $F \to 0$, $s \to 1$, and ν is different from zero only for small θ. According to Eq. (8.7), in a weakly nonideal Fermi gas $c^2 \approx p_0^2/3m$, i.e., $c \cong 3^{-1/2}u$. Thus the speed of the zeroth sound will exceed the ordinary speed of sound by a factor $3^{1/2}$.

Let us now consider the general case F(χ).[10] We represent this function as a sum of spherical harmonics as in Eq. (8.8). We substitute Eq. (8.8) in Eq. (8.14) and use the addition theorem of Legendre polynomials

$$P_n(\chi) = \sum_{m=-n}^{n} P_n^m(\theta) P_n^m(\theta') e^{im(\varphi-\varphi')} \frac{(n-|m|)!}{(n+|m|)!},$$

where $P_n^m = P_n^{-m}$ are the associated Legendre polynomials. After the indicated substitution we get:

$$(\cos\theta - s)\nu + \cos\theta \sum \frac{(n-|m|)!}{(n+|m|)!} P_n^m(\theta) F_n e^{im\varphi} \times \int P_n^m(\theta')\nu(\theta',\varphi') e^{-im\varphi'} \frac{d\Omega'}{4\pi} = 0. \tag{8.18}$$

We introduce the notation

$$F_n \frac{(n-|m|)!}{(n+|m|)!} \int P_n^m(\theta')\nu(\theta'\varphi') e^{-im\varphi'} \frac{d\Omega'}{4\pi} = \Phi_{nm} \tag{8.19}$$

and solve Eq. (8.18) for ν:

$$\nu = \frac{\cos\theta}{\cos\theta - s} \sum \Phi_{nm} P_n^m(\theta) e^{im\varphi}. \tag{8.20}$$

Substituting this expression into Eq. (8.19) and carrying out the integration over φ we get:

$$F_n \frac{(n-|m|)!}{(n+|m|)!} \int \sum_k P_k^m(\theta') \frac{\cos\theta'}{\cos\theta' - s} \cdot P_n^m(\theta') \frac{d\Omega'}{4\pi} \Phi_{km} = \sum_k \Phi_{km}\delta_{kn}. \tag{8.21}$$

Thus we have obtained a system of homogeneous equations for the quantities Φ_{km}. This system separates into independent subsystems corresponding to different values of m. From Eq. (8.21) it follows that in the Fermi fluid at absolute zero vibrations of several different types can be propagated, characterized by different dependences of the amplitude on the angles θ, φ. The value m = 0 corresponds to vibrations in which ν is isotropic in the plane perpendicular to **k**. For m $\neq$ 0 the vibrations are polarized in a definite way in this plane. The number of types of vibrations is determined by the number of possible values of m ($|m| \leq n$). The speeds of propagation of the vibrations are determined by the requirement that the determinants of the corresponding subsystems of equations be equal to zero:

$$\| \delta_{kn} + F_n \Omega_{kn}^m(s) \| = 0 \quad (N \gg n,\ k \gg |m|)$$
$$\Omega_{kn}^m(s) = \frac{(n-|m|)!}{(n+|m|)!} \int P_k^m(\theta') \frac{\cos(\theta')}{\cos\theta' - s} P_n^m(\theta') \frac{d\Omega'}{4\pi}. \tag{8.22}$$

Because $P_n^{-m} = P_n^m$, the coefficients Ω_{kn}^m do not depend on the sign of m, so that vibrations that differ only by the sign of m are propagated with the same speed.

From Eq. (8.22) it can be seen that the equations for the speeds are transcendental equations. In the general case they do not always have real roots. Cases are also possible, however, in which there are several real roots. Then there are several types of vibrations with the same polarization in the plane perpendicular to **k**.

As an example let us consider the case in which the function $F(\chi)$ contains only the zeroth and first harmonics [the function (8.11)]. Here the coefficients Ω_{kn}^m are given by

$$\Omega_{00}^0 = \frac{1}{2}\int_{-1}^{1} \frac{x\,dx}{x-s} = 1 - \frac{s}{2}\ln\frac{s+1}{s-1} = -w,$$

$$\Omega_{10}^0 = \Omega_{01}^0 = \frac{1}{2}\int_{-1}^{1} \frac{x^2dx}{x-s} = -sw,$$

$$\Omega_{11}^0 = \frac{1}{2}\int_{-1}^{1} \frac{x^3\,dx}{x-s} = \frac{1}{3} - s^2w,$$

$$\Omega_{11}^1 = \frac{1}{4}\int_{-1}^{1} \frac{(1-x^2)}{x-s}\,dx = \frac{1}{2}\left[(s^2-1)w - \frac{1}{3}\right].$$

On substituting in the determinant (8.22) we get for the speed of propagation of vibrations of type m = 0 the equation

$$w = \frac{1+F_1/3}{F_0 + F_0F_1/3 + F_1s^2}. \tag{8.23}$$

For the case m = 1 we get the equation

$$w = \frac{F_1 - 6}{3F_1(s^2-1)}. \tag{8.24}$$

This equation has one real root for $F_1 > 6$.

Let us now look at the application of our formulas to the case of liquid He^3. According to Eq. (8.10) there is one speed for the vibrations of type m = 0 [the root of Eq. (8.23)]

$$s = \frac{u}{v} = 1.72, \quad u = 192 \text{ m/sec}.$$

The vibrations of type m = 1 (and also all with m > 1) are absent. It is of course possible that this conclusion is a consequence of the crudeness of the approximation we have taken for $F(\chi)$, but we see no reasons for thinking so.

In addition to the vibrations that have been considered, waves of an entirely different type can be propagated in a Fermi fluid at absolute zero; these can be called spin waves.[8] In the treatment of these vibrations one must take into account the spin part of the function f, which is given by Eq. (4.3). We introduce a new symbol

$$\Phi(\chi) = f(\chi)\left(\frac{d\tau}{d\varepsilon}\right)_{\varepsilon=\mu} = F(\chi) + Z(\chi)\sigma\sigma'. \tag{8.25}$$

The equation for zeroth sound now takes the form

$$(s-\cos\theta)\,\nu = \cos\theta\cdot\frac{1}{2}\,\mathrm{Sp}_{\sigma'}\,\frac{1}{4\pi}\int \Phi\nu'\,d\Omega'. \qquad (8.26)$$

If we assume that ν does not depend on the spin, this goes over into Eq. (8.14). It is also possible, however, to assume a different form for the function ν, namely to take

$$\nu = \boldsymbol{\nu}\cdot\boldsymbol{\sigma} \qquad (8.27)$$

For this function we get the equation

$$(s-\cos\theta)\,\boldsymbol{\nu} = \cos\theta\int z\boldsymbol{\nu}'\,\frac{d\Omega'}{16\pi}. \qquad (8.28)$$

Thus we have for the components of the vector $\boldsymbol{\nu}$ an equation that differs from Eq. (8.14) only by the replacement of F by Z/4. Therefore all the further developments must be the same as those carried out above.

According to the results of Sec. 4, for the case of He^3 the quantity Z_0 is approximately -3.4, that is, a negative quantity. Although it is of course impossible to draw any rigorous conclusions from this, still it is most probable that no spin waves can be propagated in He^3.

The possibility of the propagation of sound waves at $T = 0$ means that the Fermi fluid has a Bose branch of the energy spectrum, in the form of phonons with energy $\epsilon = up$. But the contributions from this branch to the thermodynamic quantities involve higher powers of T (heat capacity $\sim T^3$), which were not included in the approximate theory that has been considered.

9. THE DISPERSION AND ABSORPTION OF SOUND

In the preceding section we considered the two limiting cases $\omega\tau \ll 1$ and $\omega\tau \gg 1$. Let us now examine how the transition occurs from ordinary sound to zeroth sound, i.e., let us consider the dispersion of sound.[10] Because the region of the dispersion, $\omega\tau \sim 1$, is at the same time the region of strong absorption, our treatment will also enable us to deal with the problem of the absorption of sound.

In the present case we have to find the solution of the kinetic equation, including the collision integral. Because the interaction law of the excitations is still unknown, however, we shall simplify the form of the collision integral.

For this purpose we could introduce a certain effective time τ and replace the collision integral I in Eq. (5.1) by the expression $-\delta n/\tau$. With this replacement, however, the kinetic equation will not yield any conservation laws for the number of excitations, the momentum, and the energy, and this makes the transition to hydrodynamics impossible. Since both the second (dilatational) viscosity and the effects of thermal conductivity (see below) are negligibly small in our case, the collision integral contains no terms in the zeroth and first spherical harmonics. We therefore replace the collision integral by the following expression:*

$$I(n) \to -\tau^{-1}\left(\delta n - \overline{\delta n} - \overline{3\,\delta n\cos\theta}\,\cos\theta\right). \qquad (9.1)$$

It is easy to see that the integration over $d\tau_p$ makes this expression go to zero. It also goes to zero after multiplication by ϵ or by $p\cos\theta$ and integration over $d\tau_p$.† Thus the equations for the conservation of the number of particles, the momentum, and the energy are automatically satisfied. According to Eqs. (8.14) and (9.1) the basic kinetic equation now takes the following form:

$$(kv\cos\theta-\omega)\,\nu + kv\cos\theta\int F\nu'\,\frac{d\Omega'}{4\pi} = -\frac{1}{i\tau}\left(\nu-\overline{\nu}-\overline{3\nu\cos\theta}\cdot\cos\theta\right). \qquad (9.2)$$

Having in mind the application of the results to He^3, in order not to complicate the problem we shall write the function F in the two-term form of Eq. (8.11).

We introduce the notation

$$\overline{\nu}=\nu_0;\quad \overline{3\nu\cos\theta}=\nu_1;\quad \sigma=-i\tau kv;\quad \xi=\frac{i\omega\tau-1}{i\tau kv}, \qquad (9.3)$$

after which we get without difficulty from Eq. (9.2):

$$(\cos\theta-\xi)\,\nu+\cos\theta\left(F_0\nu_0+\frac{1}{3}F_1\nu_1\cos\theta\right)=\frac{1}{\sigma}\left(\nu_0+\nu_1\cos\theta\right). \qquad (9.4)$$

We now solve the equation for ν and calculate $\overline{\nu} = \nu_0$ and $\overline{\nu\cos\theta} = \nu_1/3$. We thus find two equations for the two quantities ν_0 and ν_1:

$$\nu_0 = F_0\nu_0 w + \frac{1}{3}\,wF_1\nu_1 - \frac{1}{\sigma}\,\frac{w+1}{\xi}\,\nu_0 - \frac{\nu_1}{\sigma}\,w, \qquad (9.5)$$

$$\frac{1}{3}\nu_1 = F_0\nu_0\xi w - \frac{1}{3}\left(\frac{1}{3}-\xi^2 w\right)F_1\nu_1 - \frac{w}{\sigma}\,\nu_0 - \frac{\xi w}{\sigma}\,\nu_1. \qquad (9.6)$$

where $w = \frac{1}{2}\ln\frac{\xi+1}{\xi-1} - 1$.

From the condition for compatibility of (9.5) and (9.6) we get an equation for the complex speed of sound

$$\left(1+\frac{1}{\xi\sigma}\right)\left(1+\frac{F_1}{3}\right) - w\left\{\left(1+\frac{F_1}{3}\right)\left(F_0-\frac{1}{\xi\sigma}\right) + \xi^2\left(F_1-\frac{3}{\xi\sigma}\right)\cdot\left(1+\frac{1}{\xi\sigma}\right)\right\} = 0. \qquad (9.7)$$

This is the desired equation for the frequency

*Here and in what follows the bar denotes averaging over the angles.

†In this integration only the range of values of momentum close to p_0 is of importance, since the function δn contains a δ-function singularity for $\varepsilon = \mu$.

dependence of the speed of sound or, in other words, for the dispersion of sound in the Fermi fluid. We first consider two extreme cases.

a) Low frequencies, $\omega\tau \ll 1$. Here we have $\sigma \to 0$, $\xi\sigma \to -1$, and $\xi \to \infty$. Expansion of w in powers of $1/\xi$ gives

$$w = \frac{1}{3\xi^2} + \frac{1}{5\xi^4}$$

and after simple manipulations Eq. (9.7) takes the form

$$\left(1+\frac{1}{\xi\sigma}\right)^2 = \frac{1}{3\xi^2}\cdot\left(1+\frac{F_1}{3}\right)\cdot\left(F_0-\frac{1}{\xi\sigma}\right) + \frac{1}{5\xi^4}\left(1+\frac{1}{\xi\sigma}\right)\cdot\left(F_1-\frac{3}{\xi\sigma}\right). \tag{9.8}$$

Recalling the relations (9.3), we have:

$$\left(\frac{1+\xi\sigma}{\sigma}\right)^2 = \left(\frac{\omega}{kv}\right)^2.$$

From Eqs. (9.8) and (9.3) we find in first order in $i\omega\tau$

$$\left(\frac{\omega}{kv}\right)^2 = \frac{1}{3}(1+F_0)\left(1+\frac{F_1}{3}\right) - \frac{4}{15}i\omega\tau\left(1+\frac{F_1}{3}\right). \tag{9.9}$$

The first term corresponds to the speed of ordinary sound in the Fermi fluid. The absorption of sound in the region $\omega\tau \ll 1$ is found by the elementary rule from Eq. (9.9) as the imaginary part of the wave vector:

$$\gamma = \operatorname{Im} k = \frac{2\omega^2\tau v^2}{15c^3}\left(1+\frac{F_1}{3}\right). \tag{9.10}$$

In view of the fact that the present case corresponds to ordinary sound, we can use instead of Eq. (9.10) the more precise ordinary formula

$$\gamma = \frac{\omega^2}{2\rho c^3}\left\{\left(\frac{4}{3}\eta+\zeta\right)+\frac{\varkappa}{c_p}\left(\frac{c_p}{c_v}-1\right)\right\}.$$

Noting that $\zeta \ll \eta$ and also that κ/c_p, which is a quantity of the order of η, is multiplied by the small factor $(c_p/c_v) - 1$ [it is not hard to verify that it is of the order $(T/\mu)^2$], we find:

$$\gamma = \frac{2\omega^2}{3\rho c^3}\eta. \tag{9.11}$$

Comparing this expression with Eq. (9.10), we can determine the time τ from experimental data:

$$\eta = \frac{1}{5}\rho\tau v^2\left(1+\frac{1}{3}F_1\right). \tag{9.12}$$

From the data of K. N. Zinov'eva on the viscosity[7] we get

$$\tau \cong 2.3\cdot 10^{-12}\cdot T^{-2}\ \text{sec}\ (T \text{ in } {}^\circ\text{K}) \tag{9.13}$$

The corresponding value for the absorption coefficient is

$$\gamma \sim 1\cdot 10^{-17}\left(\frac{\omega}{T}\right)^2 \text{cm}^{-1}. \tag{9.14}$$

b) We now consider the second extreme case, that of high frequencies and low temperatures, $\omega\tau \gg 1$. In this case

$$\sigma \to \infty;\ \xi\sigma \to \infty;\ \xi = s + i\xi',\ |\xi'| \ll s.$$

Equation (9.7) takes the form

$$\left(1+\frac{F_1}{3}\right) - w(s)\left\{\left(1+\frac{F_1}{3}\right)F_0 + s^2F_1\right\} = 0. \tag{9.15}$$

This equation agrees exactly with the equation (8.23) that determines the speed of the zeroth sound.

As for the absorption of the zeroth sound, to calculate it we must find the imaginary part ξ' of the speed of sound. From Eq. (9.9) we get the equation

$$\xi'\left\{\frac{1}{w(s)}\left(1+\frac{F_1}{3}\right)\left(\frac{s}{s^2-1}-\frac{w(s)+1}{s}\right) - 2sw(s)F_1\right\} - \frac{1}{\omega\tau}\left\{\left(1+\frac{F_1}{3}\right)(1+w(s)) + w(s)s^2(3-F_1)\right\} = 0. \tag{9.16}$$

By means of Eq. (9.3) we find the absorption coefficient

$$\gamma = \operatorname{Im} k = \frac{1}{s\tau v} = \frac{\omega\xi'}{s^2 v}. \tag{9.17}$$

If we substitute here the values of the parameters for He^3 ($s = 1.84$, $v = 1.13 \times 10^4$ cm/sec), we get:

$$\xi' = \frac{1.5}{\omega\tau},\quad \gamma \sim 3\cdot 10^6\cdot T^2 \text{cm}^{-1}. \tag{9.18}$$

Thus the absorption of the zeroth sound does not depend on the frequency and increases with increasing temperature as $1/\tau$, i.e., proportional to T^2.

In this calculation the speed of sound has been treated classically. The validity of this treatment depends on the inequality $\hbar\omega \ll T$. For the case $\hbar\omega \gtrsim T$ it is necessary to treat the problem by quantum theory. The decrease of the number of sound quanta per unit time, due to the collisions of the Fermi particles, is given by

$$\int w\{n_1n_2(1-n_1')(1-n_2') - n_1'n_2'(1-n_1)(1-n_2)\} \times\delta(\mathbf{p}'+\mathbf{p}_2'-\mathbf{p}_1-\mathbf{p}_2-\hbar\mathbf{k}) \times\delta(\varepsilon_1'+\varepsilon_2'-\varepsilon_1-\varepsilon_2-\hbar\omega)\,d\tau_1\,d\tau_2\,d\tau_1'\,d\mathbf{p}_2'. \tag{9.19}$$

The function w, which depends on the four momenta, is unknown. It is possible, however, to carry out the calculation in such a way that the absorption coefficient is expressed in terms of its classical value (9.17); this has been done in a paper by Landau.[8]

Making use of the fact that in the region in which the Fermi distribution falls off (which is the important one for the problem) the only rapidly varying function is the occupation number $n(\epsilon)$, we can assume that the angular integrals give a certain constant, which does not change as we go from the classical region $\hbar\omega \ll T$ to the quantum region $\hbar\omega \gg T$. There then remains only the integral over the energies

$$I = \int \{n_1 n_2 (1 - n_1')(1 - n_2') - n_1' n_2' (1 - n_1)(1 - n_2)\} \times \delta(\varepsilon_1' + \varepsilon_2' - \varepsilon_1 - \varepsilon_2 - \hbar\omega)\, d\varepsilon_1\, d\varepsilon_2\, d\varepsilon_1'\, d\varepsilon_2'. \tag{9.20}$$

Substituting the Fermi function for n and extending the integration from $-\infty$ to ∞ (this is permissible because of the rapid convergence of the integrals), after a rather lengthy calculation we get

$$I = T^3 \frac{2\pi^2}{3} \hbar\omega \left[1 + \left(\frac{\hbar\omega}{2\pi T}\right)^2\right]. \tag{9.21}$$

The absorption coefficient is proportional to the integral I. From the classical limiting case we can determine the proportionality factor. Since the classical value of γ must be obtained for $\hbar\omega \ll T$, we have in the general case

$$\gamma = \gamma_{cl} \left[1 + \left(\frac{\hbar\omega}{2\pi T}\right)^2\right], \tag{9.22}$$

where γ_{cl} is given by Eq. (9.17). In view of the fact that γ_{cl} is proportional to T^2, in the quantum limiting case $\hbar\omega \gg T$ the absorption coefficient γ must be proportional to ω^2.

10. THE SCATTERING OF LIGHT (FLUCTUATIONS OF THE DISTRIBUTION FUNCTION)

According to the foregoing, a particular condition for the possibility of the propagation of "zeroth sound" is the inequality

$$\omega_{sound}\tau \gg 1, \tag{10.1}$$

where τ is the time between collisions. For He^3 this is of the order of $10^{-12}T^{-2}$ sec. If we take even a temperature $\sim 0.01°$K, a frequency higher than 10^8 cps is required for the direct observation of zeroth sound, and this makes the performance of such an experiment very difficult.

Besides the direct experiment, an indirect method can be proposed, which consists of the observation of the Rayleigh scattering of light in liquid He^3.* As is well known, in Rayleigh scattering there appear, in addition to the main line, satellites that differ from it in frequency by

$$\Delta\omega = \pm 2 \frac{u}{c} \omega \sin \frac{\theta}{2},$$

where u is the speed of sound and θ is the scattering angle. The speed of the zeroth sound in He^3 is of the order of 2×10^4 cm/sec, that is, $\Delta\omega \sim 10^{-6}\omega$. Thus the observation of the frequency distribution of the scattered light provides in principle a possibility for measuring the speed of the zeroth sound. The condition (10.1) can be satisfied owing to the high frequency of visible light.

In addition to this side of the matter, the scattering of light in a Fermi fluid at sufficiently low temperatures has a number of specific features which give interest to the theoretical study of this phenomenon, in particular the frequency distribution of the intensity.*

As is well known, the frequency and angular distributions for the Rayleigh scattering of unpolarized light are given by the formula[12]†

$$dh = \frac{\omega^4}{6\pi c^4} \cdot \frac{1}{2\pi V} \overline{\left| \int \delta D_{\Delta\omega}(\mathbf{r})\, e^{-i\mathbf{q}\cdot\mathbf{r}} dV \right|^2} \frac{3}{4} (1 + \cos^2\theta) \frac{d\Omega}{4\pi} d\Delta\omega, \tag{10.2}$$

where ω is the frequency of the incident light, θ is the scattering angle, $\mathbf{q}$ is the change of the wave vector of the light, of magnitude $(2\omega/c) \sin(\theta/2)$, and $\delta D_{\Delta\omega}$ is the Fourier component of the fluctuation $\delta D(t)$ of the dielectric constant:

$$\delta D_{\Delta\omega} = \frac{1}{\sqrt{t_0}} \int_0^{t_0} \delta D(t)\, e^{i\Delta\omega t} dt, \tag{10.3}$$

where t_0 is a certain large value, which will be let go to infinity in the final formula.

The bar in Eq. (10.2) means averaging over the fluctuations. Hereafter we shall for simplicity take the volume of the system to be unity.

Because of the very small polarizability of helium atoms we can assume that the change of the dielectric constant occurs as a result of density fluctuations, i.e., $\delta D = (\partial D/\partial N)\delta N$, where N is the number of particles in unit volume. But according to the general theory of the Fermi fluid the number of excitations is equal to the number of atoms in the fluid. Thus we can write

$$\int \delta D_{\Delta\omega}(\mathbf{r})\, e^{-i\mathbf{q}\cdot\mathbf{r}} dV = \frac{\partial D}{\partial N} \int \delta n_{\mathbf{q},\Delta\omega}(\mathbf{p})\, d\tau_p, \tag{10.4}$$

where $\delta n_{\mathbf{q},\Delta\omega}(\mathbf{p})$ is the Fourier component with respect to $\mathbf{r}$ and t [the latter in the sense of Eq. (10.3)] of the fluctuation of the distribution function of the excitations.

Before going on to further calculations, let us

*The idea of using the Rayleigh scattering was first proposed by S. P. Kapitza.

*We note that in the case of high temperatures, for which $\omega_{sound}\tau \ll 1$, the scattering of light will be described by the usual formulas (cf. reference 12).

†Here dh is the so-called differential extinction coefficient. The integral of dh with respect to $d\Omega$ and $d\Delta\omega$ gives the total extinction coefficient h, which is the damping decrement of the photon flux density in the medium.

note one important fact. In Eq. (10.2) the averaging is taken over all possible fluctuations. In the range of temperatures and frequencies in which $\hbar\Delta\omega \gtrsim T$ we must take account of quantum effects in carrying out this averaging. For this, however, (cf. reference 12) it is enough to know the result for the purely classical case, that is, for $T \gg \hbar\Delta\omega$, and then introduce a certain correction factor. For scattering with increase of the frequency by the amount $\Delta\omega$ (anti-Stokes scattering) we must introduce the factor $(h\Delta\omega/T)N(\Delta\omega)$, and for scattering with decrease of the frequency by $\Delta\omega$ (Stokes scattering) the factor is $(h\Delta\omega/T)[N(\Delta\omega) + 1]$, where $N(\Delta\omega)$ is the Bose distribution function. If we use negative values of $\Delta\omega$ for the description of the Stokes scattering, then owing to the relation $N(-\Delta\omega) + 1 = -N(\Delta\omega)$ it turns out that the correction factor for both cases has the form

$$\frac{\hbar\Delta\omega}{T}\left[e^{\frac{\hbar\Delta\omega}{T}} - 1\right]^{-1}. \tag{10.5}$$

Thus we shall take $T \gg \hbar\omega$. To find the fluctuations of the distribution function we shall use the method proposed by Rytov[13] and by Landau and Lifshitz[14] for calculating fluctuations in electrodynamics and hydrodynamics.* By this method we find the fluctuation of the "random force" appearing in the kinetic equation; then by solving this equation we can also get the fluctuation of the distribution function.

For the case of the Fermi fluid we shall start with the kinetic equation, which we write in the form

$$\frac{\partial\delta n}{\partial t} + \frac{\partial\delta n}{\partial \mathbf{r}}\cdot\frac{\partial\varepsilon}{\partial \mathbf{p}} - \frac{\partial n_0}{\partial \mathbf{p}}\int f(\mathbf{p}, \mathbf{p}')\frac{\partial\delta n(\mathbf{p}')}{\partial \mathbf{r}}\,d\tau_{p'} = I(\delta n) + y(\mathbf{p}, \mathbf{r}, t). \tag{10.6}$$

Except for the "random force" $y(\mathbf{r}, \mathbf{p}, t)$ this equation is an approximation to the kinetic equation (5.1) that is linear in δn.

In what follows we shall be concerned only with the case of frequencies and temperatures satisfying the relation (10.1), i.e., with the case in which the collisions can be neglected. The detailed form of the collision integral is then of no importance, since it plays the part of an auxiliary quantity in the calculations and can be set equal to zero in the final result. In view of this we set

$$I(\delta n) = -\frac{\delta n}{\tau}, \tag{10.7}$$

where τ is a large quantity. We have also to find the rate of change of the entropy. Recalling that the number of particles and the total energy are prescribed, and using the relation

$$\delta\varepsilon(\mathbf{p}) = \int f(\mathbf{p}, \mathbf{p}')\,\delta n(\mathbf{p}')\,d\tau_{p'}, \tag{10.8}$$

we find:

$$S = -\left\{\int \frac{\delta n\,[I(\delta n) + y]}{n_0(1 - n_0)}\,d\tau_p\,dV + \frac{1}{T}\int f(\mathbf{p}, \mathbf{p}')\,\delta(\mathbf{r} - \mathbf{r}')\,\delta n I(\delta n')\,d\tau_p\,dV\,d\tau_{p'}\,dV'\right\}. \tag{10.9}$$

Noting that $n_0(1 - n_0) \approx T\delta(\epsilon - \mu)$, where μ is the chemical potential, we readily see that $\delta n(\mathbf{p})$ must have the form

$$\delta n(\mathbf{p}) = \nu(\theta, \varphi)\cdot\delta(\varepsilon - \mu), \tag{10.10}$$

where θ and φ are the polar angles of the vector $\mathbf{p}$. This formula means that the fluctuations of the distribution function occur only in the neighborhood of the Fermi surface.

It is natural to take a similar form for $\mathbf{y}$, i.e.,

$$y(\mathbf{p}) = y^{\varepsilon}(\theta, \varphi)\,\delta(\varepsilon - \mu). \tag{10.11}$$

We now introduce the notation

$$F(\chi) = \left[f(\mathbf{p}, \mathbf{p}')\frac{d\tau}{d\varepsilon}\right]_{\varepsilon=\varepsilon'=\mu},$$

where χ is the angle between $\mathbf{p}$ and $\mathbf{p}'$, and expand δn, y, and F in spherical harmonics

$$\left.\begin{aligned} \nu(\theta, \varphi) &= \sum_{n=0}^{\infty}\sum_{m=-n}^{n} A_n^m P_n^m(\cos\theta)\,e^{im\varphi}, \\ y^{\varepsilon}(\theta, \varphi) &= \sum_{n=0}^{\infty}\sum_{m=-n}^{n} y_n^m P_n^m(\cos\theta)\,e^{im\varphi}, \\ F(\chi) &= \sum_{n=0}^{\infty} F_n P_n(\cos\chi). \end{aligned}\right\} \tag{10.12}$$

Since ν and y^{ϵ} are real quantities, we have

$$A_n^m = (A_n^{-m})^*, \quad y_n^m = (y_n^{-m})^*.$$

Using the expression (10.7) for the collision integral, we get the following formula for the rate of change of the entropy:

$$\dot{S} = \int dV\left(\frac{d\tau_p}{d\varepsilon}\right)_{\varepsilon=\mu}\frac{1}{T}\sum_{n=0}^{\infty}\sum_{m=-n}^{n}\left(\frac{F_n}{2n+1} + 1\right) \times\frac{1}{2n+1}\left(\frac{(n+|m|)!}{(n-|m|)!}\right)\left(\frac{A_n^m}{\tau} - y_n^m\right)A_n^{-m}. \tag{10.13}$$

Let us now introduce the notation

$$\dot{x}_n^m = -\frac{A_n^m}{\tau} + y_n^m. \tag{10.14}$$

Then in order for the expression (10.13) to take the form (cf. reference 14)

$$\dot{S} = -\sum_i X_i\cdot\dot{x}_j$$

---

*The writers express their gratitude to L. P. Gor'kov, I. E. Dzyaloshinskii, and L. P. Pitaevskii, who called attention to the possibility of applying this method to the kinetic equation.

we must take as the generalized forces X_i the expressions (10.15)

$$X_n^m = \frac{1}{T}\left(\frac{d\tau_p}{d\varepsilon}\right)_{\varepsilon=\mu}\left(\frac{F_n}{2n+1}+1\right)\frac{1}{2n+1}\cdot\frac{(n+|m|)!}{(n-|m|)!}A_n^{-m}.$$

In the expression (10.14) the quantity y_n^m plays the role of the "random force." Writing this relation in the form

$$x_n^m = -\sum_{n'm'}\gamma_{n,n'}^{m,m'}X_{n'}^{m'} + y_n^m,$$

where the coefficients γ can be determined without difficulty from Eq. (10.15), we have according to the general theory of fluctuations:

$$\overline{y_n^m(\mathbf{r},t)\,y_{n'}^{m'}(\mathbf{r}',t')} = (\gamma_{n,n'}^{m,m'}+\gamma_{n',n}^{m',m})\,\delta(t-t')\,\delta(\mathbf{r}-\mathbf{r}')$$

$$= \frac{2}{\tau}\delta_{n,n'}\delta_{m,-m'}\delta(\mathbf{r}-\mathbf{r}')\,\delta(t-t') \qquad (10.16)$$

$$\times\left[T\left(\frac{d\tau_p}{d\varepsilon}\right)_{\varepsilon=\mu}\left(\frac{F_n}{2n+1}+1\right)\frac{1}{2n+1}\frac{(n+|m|)!}{(n-|m|)!}\right]^{-1}.$$

Finally, using Eq. (10.12) and the relation

$$\sum_n (2n+1)P_n(\cos\chi) = 2\delta(\cos\chi-1),$$

we get after some transformations the general relation

$$\overline{y(\mathbf{p},\mathbf{r},t)\,y(\mathbf{p}',\mathbf{r}',t')} = \frac{2T}{\tau}\delta(\mathbf{r}-\mathbf{r}')\,\delta(t-t') \qquad (10.17)$$

$$\times\left\{\frac{(2\pi\hbar)^3}{2}\delta(\mathbf{p}-\mathbf{p}')\,\delta(\varepsilon-\mu)+\delta(\varepsilon-\mu)\,\delta(\varepsilon'-\mu)\left(\frac{d\varepsilon}{d\tau_p}\right)_{\varepsilon=\mu}\right.$$

$$\left.\times\sum_{n=0}^{\infty}\frac{F_nP_n(\cos\chi)}{1+F_n/2n+1}\right\}.$$

By means of this formula and the kinetic equation (10.16) we can calculate the fluctuations of the distribution function, with which we are concerned here. Since in the general case of an arbitrary function f this is a rather complicated task, we confine ourselves to the case f = const.

Making use of the fact that the fluctuations occur only at the Fermi surface, and referring to Eq. (10.6), we express $\nu_{\mathbf{q},\Delta\omega}(\theta,\varphi)$ in terms of the corresponding Fourier component of $y^{\epsilon}(\theta,\varphi)$. This gives

$$\int \nu_{\mathbf{q},\Delta\omega}(\theta,\varphi)\frac{d\Omega}{4\pi}$$

$$= \int\frac{y^{\varepsilon}_{\mathbf{q},\Delta\omega}(\theta,\varphi)\frac{d\Omega}{4\pi}}{-i\omega+\frac{1}{\tau}+i\mathbf{q}\mathbf{v}}\Bigg/\left(1+F_0\int\frac{i(\mathbf{q}\mathbf{v})\frac{d\Omega}{4\pi}}{-i\omega+\frac{1}{\tau}+i\mathbf{q}\mathbf{v}}\right), \qquad (10.18)$$

where $\mathbf{v} = (\partial\epsilon/\partial\mathbf{p})_{\epsilon=\mu}$. Averaging the square of the magnitude of this expression by means of Eq. (10.17)[here we have to remember that the Fourier components with respect to t are taken according to Eq. (10.3)], we find:

$$\overline{\left|\int\nu_{\mathbf{q},\Delta\omega}(\theta,\varphi)\frac{d\Omega}{4\pi}\right|^2} = 2\left(\frac{d\varepsilon}{d\tau_p}\right)_{\varepsilon=\mu}\frac{T}{\tau}\left[\frac{1}{2}\int_{-1}^{1}\frac{dx}{\left|qvx-\omega+\frac{i}{\tau}\right|^2}\right.$$

$$\left.-\frac{F_0}{1+F_0}\left|\frac{1}{2}\int_{-1}^{+1}\frac{dx}{qvx-\omega+\frac{i}{\tau}}\right|^2\right]\Bigg/\left|1+\frac{F_0}{2}\int_{-1}^{+1}\frac{qvx\,dx}{qvx-\omega+\frac{i}{\tau}}\right|^2. \qquad (10.19)$$

We are interested in the limit of the value of this expression for $\tau\to\infty$. In the case in which $qv > |\Delta\omega|$, the denominator does not have any poles, and the important quantity is the residue in the integral in the numerator. We get as the result:

$$\frac{1}{2\pi}\overline{|\nu_{\mathbf{q},-\omega}(\mathbf{p})\,d\tau_p|^2} = T\left(\frac{d\tau_p}{d\varepsilon}\right)\frac{1}{qv}\left\{\left[1+F_0\left(1\right.\right.\right.$$

$$\left.\left.\left.-\frac{\Delta\omega}{2qv}\ln\frac{qv+\Delta\omega}{qv-\Delta\omega}\right)\right]^2+\left(\frac{F_0\Delta\omega\pi}{2qv}\right)^2\right\}^{-1}. \qquad (10.20)$$

In the opposite case, i.e., for $qv \ll \Delta\omega$, the pole in the denominator of the expression (10.19) is important. It is not hard to see that such a pole occurs if $F_0 > 0$ and is given by $\Delta\omega = \pm sqv$, where s satisfies the equation

$$1+F_0\left[1-\frac{1}{2}\ln\left(\frac{s+1}{s-1}\right)\right] = 0. \qquad (10.21)$$

This equation is identical with Eq. (8.16), which determines the speed of the zeroth sound. Using the relation

$$\frac{1}{\pi}\lim_{\tau\to\infty}\frac{\frac{1}{\tau}}{(\omega-\omega_0)^2+\tau^{-2}} = \delta(\omega-\omega_0),$$

we get without difficulty:

$$\frac{1}{2\pi}\overline{\left|\int\nu_{\mathbf{q},\Delta\omega}(\mathbf{p})\,d\tau_p\right|^2} = T\left(\frac{d\tau_p}{d\varepsilon}\right)_{\varepsilon=\mu}\frac{2(s^2-1)}{F_0(1+F_0-s^2)}$$

$$\times[\delta(\Delta\omega-sqv)+\delta(\Delta\omega+sqv)]. \qquad (10.22)$$

Thus the distribution of the scattered light in direction and frequency has the form [we have also introduced the quantum factor (10.5)]:

$$dh = \frac{\omega^4}{4\pi c^4}\left(\frac{\partial D}{\partial N}\right)^2\left(\frac{d\tau_p}{d\varepsilon}\right)_{\varepsilon=\mu}(1+\cos^2\theta)\cdot\frac{\hbar\Delta\omega}{e^{\hbar\Delta\omega/T}-1}$$

$$\times\left[\frac{\theta(qv-|\Delta\omega|)}{2qv}\Bigg/\left\{\left[1+F_0\left(1-\frac{\Delta\omega}{2qv}\ln\frac{qv+\Delta\omega}{qv-\Delta\omega}\right)\right]^2\right.\right.$$

$$\left.+\left(\frac{F_0\Delta\omega\pi}{2qv}\right)^2\right\}+\frac{s^2-1}{F_0(1+F_0-s^2)}[\delta(\Delta\omega-sqv) \qquad (10.23)$$

$$\left.+\delta(\Delta\omega+sqv)]\right]\frac{d\Omega}{4\pi}d\Delta\omega,$$

where

$$\theta(y) = \begin{cases}1 & y>0,\\ 0 & y<0.\end{cases}$$

This result has a simple physical meaning. As

can easily be seen, the frequency spectrum consists of a central part $-qv < \Delta\omega < qv$ and two narrow lines at $\Delta\omega = \pm sqv$. The central part corresponds to the Doppler broadening of the main line. Comparison of Eq. (10.21) with Eq. (8.16) shows that the side lines are the satellites of the Rayleigh scattering, which appear because of the possibility of the propagation of zeroth sound ($sv = u$). The ratio between the intensities of the central part and the satellites in general depends on the scattering angle. In the two extreme cases, however, of high temperatures ($T \gg \hbar\omega u/c$) and low temperatures ($T \ll \hbar\omega u/c$) this ratio does not depend on the angle. One can carry out numerical computations for He^3, using unknown (sic) parameters. It then turns out that for the high-temperature case about 20 percent of the whole intensity is in the central part, and about 40 percent in each of the side lines. In the low-temperature case, owing to the quantum factor, the distribution is cut off on the side of positive $\Delta\omega$. In particular, the only one of the two satellites left is the Stokes line with $\Delta\omega = -uq$, which gets 90 percent of the intensity. In this case the central part receives only 10 percent of the whole intensity.

The total scattered intensity is obtained by integrating Eq. (10.23) over $d\Delta\omega$ and $d\Omega$. For the high-temperature case

$$\left(T \gg \hbar\Delta\omega \sim \hbar\omega\frac{u}{c}\right)$$

it is

$$h = \frac{\omega^4 T}{6\pi c^4}\left(\frac{\partial D}{\partial N}\right)^2\left(\frac{d\tau_p}{d\varepsilon}\right)_{\varepsilon=\mu} J_1, \qquad (10.24)$$

where J_1 is a numerical integral, which in the case of He^3 is about 0.5. In the quantum limiting case ($T \ll \hbar\omega u/c$) we have:

$$h = \frac{\hbar\omega^5 v}{6\pi c^4}\left(\frac{\partial D}{\partial N}\right)\left(\frac{d\tau_p}{d\varepsilon}\right)_{\varepsilon=\mu} J_2. \qquad (10.25)$$

Here J_2 is a numerical integral, which in the case of He^3 is about 0.2. To carry out quantitative calculations from these expressions for the case of He^3, we need to know the value of $\partial D/\partial N_0$. There are no measurements of this quantity; therefore we take $D - 1$ to be proportional to N, and use the proportionality factor as given by the data on liquid He^4. For wavelength $\lambda = 5461$ A the index of refraction of liquid He^4 is 1.027. This gives $\partial D/\partial N = 2.5 \times 10^{-24}$. Substitution in Eqs. (10.24) and (10.25) gives (T in °K):

$$h(\mathrm{He}^3)\ 10^{-69}\omega^4 T \ \mathrm{cm}^{-1} \quad \text{for} \quad \omega \ll 2\cdot 10^{17} T \ \mathrm{sec}^{-1},$$

$$h(\mathrm{He}^3)\ 10^{-87}\omega^5 \ \mathrm{cm}^{-1} \quad \text{for} \quad \omega \gg 2\cdot 10^{17} T \ \mathrm{sec}^{-1}.$$

It must be remembered that the frequency must satisfy the relation (10.1) i.e., $\Delta\omega \gg 1/\tau$, or

$$\omega \gg 10^{18} T^2 \ \mathrm{sec}^{-1}.$$

If this condition is not fulfilled, the line width will be too large. Thus in the visible range of frequencies we need temperatures below 0.05°K. It is not hard to see that at temperatures of the order of 0.01° and frequencies in the visible region we have $h(He^3) \sim 10^{-9}\,\mathrm{cm}^{-1}$, which is of course too small for the effect to be measurable.* But owing to the fact that ω occurs raised to a high power in the expression for h, it is possible that one may be able to measure the scattering in the ultraviolet region.

APPENDICES

MICROSCOPIC THEORY OF THE FERMI FLUID

A 1. The Rarefied Fermi Gas

In this section we consider the properties of a non-ideal Fermi gas in which the dimensions of the particles are small in comparison with the mean free path.[16] This model, a number of whose properties have been examined recently by Yang and Huang,[17] and also by Yang and Lee,[18] was called by those writers the hard-sphere model. In reality it possesses a more general character, and corresponds to a system of Fermi particles with arbitrary short-range forces having a radius of action small in comparison with the mean wavelength.

Such a model makes it possible to obtain expressions for the various quantities describing the Fermi fluid, for example the energy, the effective mass, and the function f, in the form of expansions in powers of $(a/\lambda\!\!\!^-)$ (a is the radius and $\lambda\!\!\!^-$ is the wavelength). We shall carry out the calculation to within terms of the order $(a/\lambda\!\!\!^-)^2$. By a procedure similar in principle one could determine also terms of several higher orders, but this is of no special interest. We begin with the calculation of the energy.

We use perturbation theory to take into account the interaction energy of the particles, which we write in the form (with the volume of the gas taken as unity for simplicity)

$$V = 2U \sum_{\substack{n_1 n_2 m_1 m_2 \\ n_1 < n_2}} a^+_{m_1} a^+_{m_2} a_{n_2} a_{n_1}, \qquad (A1.1)$$

where a_i^+ and a_i are creation and annihilation

*For liquid He^4 in the visible region $h \sim 10^{-8}\,\mathrm{cm}^{-1}\psi^{15}$ (about the same value should hold for He^3 for $\Delta\omega\tau \ll 1$). Although measurements of h for He^4 have been carried out, they are at the limit of what is experimentally possible.

operators for the particles. The summation is taken with the restriction that the total momentum is conserved; furthermore the spin component for the state m_1 is equal to that for the state n_1, and similarly for m_2 and n_2. The position of U outside the sign of summation corresponds to the fact that the interaction between all pairs of particles is the same, and the scattering amplitude does not depend on the angle. In first approximation the quantity U is connected with this amplitude by the relation

$$U = \frac{4\pi a\hbar^2}{m} \tag{A1.2}$$

(a is the scattering amplitude).

The first-order correction to the energy is the diagonal matrix element of V:

$$E_n^{(1)} = 2U \sum_{n_1<n_2} N_{n_1} N_{n_2} Q_{n_1 n_2}, \tag{A1.3}$$

where the N_i are occupation numbers.

The factor Q_{ik} in Eq. (A1.3) takes account of the fact that Fermi particles with a scattering amplitude independent of the angle do not interact in the case of parallel spins. Thus we shall take

$$Q_{ik} = \frac{1}{4} - \sigma_i \sigma_k \tag{A1.4}$$

(σ_i is the spin operator of the i-th particle).

Substituting Eqs. (A1.2) and (A1.4) in Eq. (A1.3), we get

$$E^{(1)} = (2\pi a\hbar^2/m)\, N^2/2. \tag{A1.5}$$

To find the second-order correction we use the expression from perturbation theory

$$E_n^{(2)} = \sum_{m\neq n} |V_{nm}|^2/(E_n - E_m). \tag{A1.6}$$

Substituting the expression (A1.1) in this formula, we get the following sum:

$$4U^2 \sum_{\substack{n_1 n_2 m_1 m_2 \\ n_1<n_2}} \frac{N_{n_1} N_{n_2} (1-N_{m_1})(1-N_{m_2})\, Q_{n_1 n_2}}{E_{n_1}+E_{n_2}-E_{m_1}-E_{m_2}}, \tag{A1.7'}$$

where N_i are the equilibrium occupation numbers and E_i the energies of the particles.

Because our purpose is to get the expansion of the energy in powers of a, we have to take into account the fact that the relation (A1.2) between U and the scattering amplitude is not exact, but is valid only to first order. When second-order terms are included we get instead of (A1.2) the following relations:

$$2U + 4U^2 \sum_{m_1 m_2} Q_{n_1 n_2}/(E_{n_1}+E_{n_2}-E_{m_1}-E_{m_2}) = \frac{8\pi a\hbar^2}{m}. \tag{A1.2'}$$

If we get from this the expression of U in terms of a and substitute it in Eq. (A1.3), the resulting expression for $E^{(1)}$ contains terms proportional to a^2, which must naturally be assigned to the second-order correction. Taking this into account, we get the following value of the second-order correction to the energy:

$$E^{(2)} = 2U^2 \sum_{n_1 n_2 m_1 m_2} \left\{ \frac{N_{n_1} N_{n_2} (1-N_{m_1})(1-N_{m_2}) \cdot Q_{n_1} Q_{n_2}}{E_{n_1}+E_{n_2}-E_{m_1}-E_{m_2}} - \frac{N_{n_1} N_{n_2} Q_{n_1 n_2}}{E_{n_1}+E_{n_2}-E_{m_1}-E_{m_2}} \right\}. \tag{A1.7}$$

In view of the fact that the expression in brackets is symmetric in n_1 and n_2, we replace the restriction $n_1 > n_2$ by the factor 1/2.

The meaning of the operation just performed lies in the fact that in actuality an expansion in powers of U is not what is used. The existence of a constant U would simply lead to an infinity in the energy, as can be seen directly from the formula (A1.7). In the present case the essential point is that the scattering amplitude a has a finite, and moreover a small, value, which makes possible an expansion in powers of this quantity.

In the first term of the expression (A1.7) the term with the product of four N_i's is equal to zero because of the fact that the denominator is antisymmetric under the interchange $n_1 n_2 \rightleftarrows m_1 m_2$, while its numerator is symmetric, and all the ranges of summation are identical. The two surviving terms with products of three N_i's are identical. Thus we get finally:

$$E^2 = -4U^2 \sum_{n_1 n_2 m_1 m_2} \frac{N_{n_1} N_{n_2} N_{m_1} Q_{n_1 n_2}}{E_{n_1}+E_{n_2}-E_{m_1}-E_{m_2}}. \tag{A1.8}$$

This is the expression for the energy that is valid on the basic assumption $(a/\lambda) \ll 1$.

Our purpose is to obtain the characteristics of the degenerate Fermi gas. From the expression (A1.8) we find the energy of the ground state

$$E^{(2)} = -\frac{4U^2}{(2\pi\hbar)^9} \int\limits_{|\mathbf{p}_1|<p_0;} d\mathbf{p}_1 \int\limits_{|\mathbf{p}_2|<p_0;} d\mathbf{p}_2 \int\limits_{|\mathbf{p}_3|<p_0;} d\mathbf{p}_3 \int d\mathbf{p}_4 \frac{\delta(\mathbf{p}_1+\mathbf{p}_2-\mathbf{p}_3-\mathbf{p}_4)}{(\mathbf{p}_1^2+\mathbf{p}_2^2-\mathbf{p}_3^2-\mathbf{p}_4^2)\frac{1}{2m}}, \tag{A1.9}$$

where p_0 is the limiting momentum.

According to Sec. 1 the energy of the excitations is given by the relation

$$\varepsilon_i = \frac{\delta E}{\delta N_i}. \tag{A1.10}$$

Variation of the expressions (1.3) and (1.8) with respect to N_i gives

$$\varepsilon(\mathbf{p}) = \frac{\mathbf{p}^2}{2m} + \frac{UN}{2} + \frac{2U^2}{(2\pi\hbar)^6} \int d\mathbf{p}_1 \int d\mathbf{p}_2 \int d\mathbf{p}_3 \tag{A1.11}$$

$$\times \left[\frac{\delta(\mathbf{p}_1+\mathbf{p}_2-\mathbf{p}-\mathbf{p}_3)}{(\mathbf{p}^2+\mathbf{p}_3^2-\mathbf{p}_1^2-\mathbf{p}_2^2)/2m} - 2\frac{\delta(\mathbf{p}_1+\mathbf{p}-\mathbf{p}_2-\mathbf{p}_3)}{(\mathbf{p}^2+\mathbf{p}_1^2-\mathbf{p}_2^2-\mathbf{p}_3^2)/2m} \right].$$

Thus the problem of calculating the energy of the ground state and the effective mass of the excitations reduces to the calculation of the integrals (A1.9) and (A1.11). The integration is rather cumbersome because of the high multiplicity of the integrals and the awkwardness of the regions of integration.

Instead of this we can use a simpler method, based on the introduction of the function f. If we introduce this function

$$f_{ik}=\delta^2E/\delta N_i\delta N_k, \qquad (A1.12)$$

then according to Eqs. (2.5) and (8.7) we can determine the effective mass and the speed of low-frequency sound. By a suitable integration we can find the energy of the ground state from the speed of sound.

Thus the problem reduces to the determination of the quantity f. Taking the variations of the expressions (A1.3) and (A1.8) with respect to N_1, and then with respect to N_k, we find the following expression:

$$f=2UQ_{\sigma,\sigma'}-\frac{8U^2}{(2\pi\hbar)^3}\int d\mathbf{p}_1\int_{|\mathbf{p}_1|<p_0} d\mathbf{p}_2 \times\Big[Q_{\sigma\sigma'}\frac{\delta(\mathbf{p}+\mathbf{p}'-\mathbf{p}_1-\mathbf{p}_2)}{(\mathbf{p}^2+\mathbf{p}'^2-\mathbf{p}_1^2-\mathbf{p}_2^2)/2m} \qquad (A1.13)$$

$$+\frac{1}{4}\frac{\delta(\mathbf{p}+\mathbf{p}_1-\mathbf{p}'-\mathbf{p}_2)}{(\mathbf{p}^2+\mathbf{p}_1^2-\mathbf{p}'^2-\mathbf{p}_2^2)/2m}+\frac{1}{4}\frac{\delta(\mathbf{p}'+\mathbf{p}_1-\mathbf{p}-\mathbf{p}_2)}{(\mathbf{p}'^2+\mathbf{p}_1^2-\mathbf{p}^2-\mathbf{p}_2^2)/2m}\Big].$$

In the calculation we shall at once set $|\mathbf{p}| = |\mathbf{p}_1| = p_0$. The integration in the second term of f is considerably simpler than those in Eqs. (A1.9) and (A1.11). We get as the result:

$$f(\chi)=\frac{2\pi a\hbar^2}{m}\left[1+2\left(\frac{3}{\pi}\right)^{\frac{1}{3}}aN^{\frac{1}{3}}\left(2+\frac{\cos\chi}{2\sin\frac{\chi}{2}}\ln\frac{1+\sin\frac{\chi}{2}}{1-\sin\frac{\chi}{2}}\right)\right]$$

$$-\frac{8\pi a\hbar^2}{m}(\sigma_1\sigma_2)\left[1+2\left(\frac{3}{\pi}\right)^{\frac{1}{3}}aN^{\frac{1}{3}}\left(1-\frac{\sin\frac{\chi}{2}}{2}\right)\ln\frac{1+\sin\frac{\chi}{2}}{1-\sin\frac{\chi}{2}}\right]. \qquad (A1.14)$$

A peculiarity of this expression deserves attention. For angles χ close to π the function f has a logarithmic singularity,

$$f(\chi)\sim\left(\frac{1}{4}-\sigma_1\sigma_2\right)\ln\frac{1}{\pi-\chi}. \qquad (A1.15)$$

It is clear that in this case the approximation we have used is, strictly speaking, not legitimate. The singularity of the function f at $\chi = \pi$ is a manifestation of the singularity in the scattering amplitude of excitations colliding at the angle π. The corresponding expression, obtained by summing the main terms of the perturbation theory, is proportional to

$$\left[1+2\left(\frac{3}{\pi}\right)^{\frac{1}{3}}aN^{\frac{1}{3}}\ln\frac{p_0^2}{\epsilon}\right]^{-1}, \qquad (A1.16)$$

where $\epsilon = p^2 + p'^2 - 2p_0^2$.

In the case of positive a this expression goes to zero for $p^2 = p'^2 = p_0^2$.

If, however, $a < 0$ (this is possible for Fermi systems), the scattering amplitude has a pole near the Fermi surface. This corresponds to the possibility of the formation of bound pairs of excitations with opposite momenta, which was pointed out by Cooper[19] and is the main cause of the phenomenon of superconductivity in metals.[20]

Thus the expression found for f is not valid for angles close to π. But in view of the fact that the singularity is only logarithmic, it manifests itself only in the immediate neighborhood of the singular point. And since only integrals of f together with regular functions occur in the expressions with which we are concerned, the logarithmic singularity of the function f is of no importance.

Substituting Eq. (A1.4) in Eq. (2.5), we find the value of the effective mass

$$m/m^*=1-(8/15)(3/\pi)^{\frac{2}{3}}(7\ln 2-1)a^2N^{\frac{2}{3}}. \qquad (A1.17)$$

We note that if we insert here the value $m^* = 1.43$ m, which corresponds to liquid He^3, the resulting value of a is 1.6×10^{-8}; that is, it is of the order of magnitude of the gas-kinetic value of the diameter of the helium atom. Such a comparision of course has no meaning in a strict sense. The model under consideration cannot describe liquid He^3. This can already be seen from the fact that the quantity $(m^*-m)/m^*$, which in the theory should be a quantity of second order in a, is 1/3 for He^3.

Substituting the formula for f in the expression for the speed of sound, we get:

$$c^2=\frac{\pi^{\frac{4}{3}}}{3^{\frac{1}{3}}}N^{\frac{2}{3}}\frac{\hbar^2}{m^2}+2\frac{\pi a\hbar^2}{m^2}N\left[1+\frac{4}{15}\left(\frac{3}{\pi}\right)^{\frac{1}{3}}aN^{\frac{1}{3}}(11-2\ln 2)\right]. \qquad (A1.18)$$

From the value of c^2 just found it is not hard to obtain the energy of the ground state of the Fermi fluid. To do this we use the relation (8.2), $c^2 = N/m(\partial\mu/\partial N)$. This gives:

$$E=\int\mu\, dN=E^{(0)}+\frac{\pi a\hbar^2}{m}N^2\left[1+\frac{6}{35}\left(\frac{3}{\pi}\right)^{\frac{1}{3}}aN^{\frac{1}{3}}(11-2\ln 2)\right]. \qquad (A1.19)$$

Equation (A1.2) agrees with the result of Lee and Yang.[15]

The same result can also be obtained by direct integration in Eq. (A1.9). To calculate the integral in Eq. (A1.9) it is convenient to introduce new variables

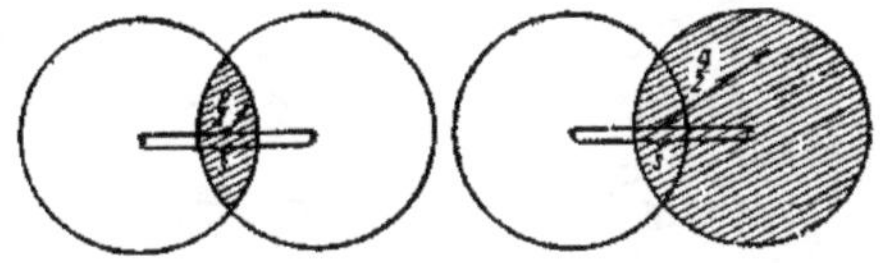

FIG. 2

$$\mathbf{p} = \mathbf{p}_1 - \mathbf{p}_3; \quad \mathbf{q} = \mathbf{p}_3 - \mathbf{p}_4; \quad \mathbf{s} = \mathbf{p}_1 + \mathbf{p}_2 = \mathbf{p}_3 + \mathbf{p}_4.$$

In these variables $E^{(2)}$ takes the form

$$E^{(2)} = -\frac{mU^2}{4(2\pi\hbar)^9} \int ds \int dp \int dq \frac{1}{p^2 - q^2},$$

where the region of integration for the vector s is $0 < |s| < 2p_0$, and the regions of values for the vectors p and q are shown in Fig. 2. If we introduce the variables $x_q = \cos(\mathbf{q}, \mathbf{s})$ and $x_p = \cos(\mathbf{p}, \mathbf{s})$, $E^{(2)}$ is given by

$$E^{(2)} = -\frac{8mU^2\pi^3}{(2\pi\hbar)^9} \int_0^{2p_0} s^2 ds \int_0^1 dx_p \int_0^{z(x_p)} p^2 dp \int_{-1}^1 dx_q \int_0^{z(x_q)} q^2 dq \frac{1}{p^2 - q^2},$$

where z(x) satisfies the relation

$$z^2 + 2zsx + s^2 = 4p_0^2.$$

By a series of transformations and integrations by parts with respect to dx_p and dx_q we then find:

$$E^2 = \frac{2mU^2p_0^7}{\pi^6\hbar^9} \int_0^1 s^2 ds \left[\int_0^{1+s} p^2 dp \int_0^{1-s} q^2 dq \right.$$

$$\left. + \frac{1}{4s^2} \int_{\sqrt{1-s^2}}^{1+s} p\,dp(1 - p^2 - s^2) \int_{\sqrt{1-s^2}}^{1-s} q\,dq(1 - q^2 - s^2) \right] \frac{1}{p^2 - q^2}.$$

Proceeding further to integrate by parts with respect to ds and then performing the remaining integration, we get:

$$E^{(2)} = (6/35)(3/\pi)^{\frac{1}{3}}(11 - 2\ln 2)\,aN^{\frac{1}{3}}E^{(1)}.$$

Here we have inserted the expression (A1.2) for U and have set $p_0 = \hbar(3\pi^2 N)^{1/3}$. This result is identical with the second-order term in Eq. (A1.19).

A.2. Microscopic Theory of the Fermi Fluid for T = 0

Here we present the general microscopic theory of the Fermi fluid at T = 0, as developed recently by Landau;[18] this theory is valid for arbitrary interactions. As we shall see, this treatment provides a basis for the phenomenological theory and an understanding of the physical meaning of the function f that appears in it.

The microscopic treatment is based on methods taken from the quantum theory of fields. As is well known, in quantum field theory one works with vacuum expectation values of chronological products of field operators (i.e., products of operators taken in the order of succession of their times). These quantities are the so-called Green's functions. In our case, instead of vacuum expectation values we shall consider expectation values for the ground state of the system. It is not hard to verify that such expectation values have the same properties as the Green's functions of the field theory, and in particular that in principle they can be calculated by means of the Feynman-diagram technique.

For definiteness we shall first give the name of Green's function to the quantity

$$G_{1,2} = -i\langle T(\psi_1\psi_2^+)\rangle, \tag{A2.1}$$

where the brackets < > denote the expectation value for the ground state, ψ and ψ^+ are the field operators of the Fermi particles, the indices 1, 2 denote sets of space, time, and spin coordinates, and T, as usual, denotes the time-ordered product with change of sign when ψ_1 and ψ_2^+ are interchanged. In the absence of external fields the Green's function depends only on the differences of the coordinates and the times. As for the spin indices, if the interaction of the particles does not depend on their spins, then $G_{\alpha\beta} \sim \delta_{\alpha\beta}$.

We denote the Fourier transform of the Green's function with respect to the coordinate and time differences by $G(p)\,\delta_{\alpha\beta}$ [here p is the four-dimensional vector $(\mathbf{p}, \epsilon)$]. The poles of G(p) determine the energies of the quasi-particles. If the magnitude of p is close to p_0 and ϵ is close to the limiting energy μ, then G(p) has the form

$$G(p) = \frac{a}{\epsilon - \mu - v(p - p_0) + i\delta}. \tag{A2.2}$$

Here a is a positive constant, and δ is a small quantity which agrees in sign with the sign of $\epsilon - \mu$, or, close to a pole, with the sign of $p - p_0$ (δ is introduced to fix the rule for going around the poles, so that it will correspond to taking the T-product in the x, t representation).

Let us now introduce the expectation value of the time-ordered product of four operators

$$\Phi_{12,34} = \langle T(\psi_1\psi_2\psi_3^+\psi_4^+)\rangle. \tag{A2.3}$$

We give the name of vertex part $\Gamma_{12,34}$ to the quantity defined by the following relation:

$$\Phi_{12,34} = G_{1,3}\,G_{2,4} - G_{1,4}\,G_{2,3}$$
$$+ i \sum_{1',2',3',4'} G_{1,1'}\,G_{2,2'}\,G_{3,3'}\,G_{4,4'}\,\Gamma_{1'2',3'4'}. \tag{A2.4}$$

The Fourier transform of the vertex part has the form

$$(2\pi h)^4 \Gamma_{\alpha\beta,\gamma\delta}(p_1 p_2, p_3 p_4)\delta(p_1 + p_2 - p_3 - p_4).$$

It is clear that the function $\Gamma_{\alpha\beta\ \gamma\delta}(P_1 P_2, P_3 P_4)$ changes sign on the interchange $\alpha p_1 \rightleftarrows \beta p_2$. As in the quantum field theory, the function Γ with a definite relation between the components of each of the vectors P [for example, near the Fermi surface $\epsilon - \mu = v(p - p_0)$], multiplied by a^2, plays the role of the scattering amplitude of the quasi-particles.

Let us consider Γ in the case in which p_1 is close to p_3, and consequently p_2 is close to p_4, and introduce the definition

$$\Gamma(P_1 P_2, P_1 + K\ P_2 - K) \equiv \Gamma(P_1 P_2, K),$$

where $K = (k, \omega)$ is a small four-vector.

If it were possible to calculate such a vertex part by perturbation theory, then in first order one would have to take the diagrams shown in Figs. 3, I. The corresponding expressions contain integrals of products of two Green's functions. Whereas for diagrams (a) and (b) the case $K = 0$ is not in any way singled out, in case (c) the poles of the two Green's functions approach each other for $K = 0$, which, as we shall see, leads to the appearance of a singularity of Γ.

If we denote by $\Gamma^{(1)}$ the set of all possible diagrams for Γ that do not have "singular elements" [i.e., integrals of G(p)G(p) +K)], it is not hard to see that the total Γ is obtained by summation of the "ladders" shown in Fig. 3, II. This can be accomplished by solving the integral equation

$$\Gamma_{\alpha\beta,\gamma\delta}(P_1 P_2; K) = \Gamma^{(1)}_{\alpha\beta,\gamma\delta}(P_1 P \qquad\qquad \text{(A2.5)}$$
$$- \frac{i}{(2\pi h)^4}\int \Gamma^{(1)}_{\alpha\epsilon,\gamma\zeta}(P_1 Q)\, G(Q)\, G(Q + K)\, \Gamma_{\zeta\beta,\epsilon\delta}(QP_2; K)\, d^4Q.$$

Here, in view of the fact that $\Gamma^{(1)}$ has no singularity at $K = 0$, we simply set $K = 0$ everywhere in $\Gamma^{(1)}$.

Let us now examine the product in the integrand. The integral over Q will consist of the principal value and a term arising from the passage around the poles. Because the arguments of the two G functions are nearly equal, we can assume that all the other parts of the integral have the same values at these poles. The passage around the poles then gives a contribution only in the case in which the detours are on opposite sides of the real axis. This means either $q < p_0$, $|q + k| > p_0$, or vice versa. Because of the smallness of k it is not hard to see that here $q \approx p_0$ and $\epsilon_q \approx \mu$. Thus in the part of the integral over Q which comes from the passage around the poles the product G(Q)G(Q +K) can be replaced by $A\,\delta(\epsilon - \mu)\,\delta(q - p_0)$.

The coefficient A can be found by integrating G(Q)G(Q + K), and is found to be $(2\pi i a^2 k \cos\theta)/(\omega - vk\cos\theta)$, where θ is the angle between q

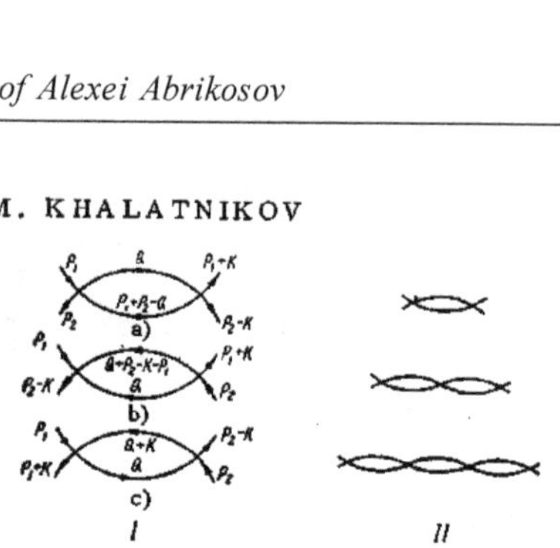

FIG. 3

and k. Thus the product G(Q)G(Q + K) can be written in the form

$$G(Q)\,G(Q + K) = 2\pi i a^2 \frac{k\cos\theta}{\omega - vk\cos\theta}\,\delta(\epsilon - \mu)\,\delta(q - p_0) + \varphi(Q), \qquad \text{(A2.6)}$$

where φ corresponds to the principal value of the integral and has no singularities (therefore one puts $K = 0$ in this term).

If we consider the expression (A2.6) in the limit $K = 0$, then it makes a great difference how this limit is taken, i.e., what the ratio of ω and k approaches. The same is true about Γ in the limit $\omega \to 0$, $k \to 0$.

First of all let us consider Γ in the limit $k \to 0$, $(k/\omega) \to 0$. For this limit, which we denote by Γ^{ω}, we get from Eqs. (A2.5) and (A2.6):

$$\Gamma^{\omega}_{\alpha\beta,\gamma\delta}(P_1 P_2) = \Gamma^{(1)}_{\alpha\beta,\gamma\delta}(P_1 P_2) \qquad\qquad \text{(A2.7)}$$
$$- \frac{i}{(2\pi h)^4}\int \Gamma^{(1)}_{\alpha\epsilon,\gamma\zeta}(P, Q)\,\varphi(Q)\,\Gamma^{(\omega)}_{\zeta\beta,\epsilon\delta}(QP_2)\, d^4Q.$$

It is not hard to eliminate $\Gamma^{(1)}$ from the two equations (A2.5) and (A2.7). Using Eq. (A2.6), we find after simple transformations

$$\Gamma_{\alpha\beta,\gamma\delta}(P_1 P_2; K) = \Gamma^{(\omega)}_{\alpha\beta,\gamma\delta}(P_1 P_2) \qquad\qquad \text{(A2.8)}$$
$$+ \frac{a^2 p_0^2}{(2\pi h)^3}\int \Gamma^{\omega}_{\alpha\epsilon,\gamma\zeta}(P, Q)\,\Gamma_{\zeta\beta,\epsilon\delta}(QP_2, K)\,\frac{k\cos\theta}{\omega - vk\cos\theta}\, d\Omega.$$

Let us now take the other limit, namely $K \to 0$, $(\omega/k) \to 0$. This limiting value, multiplied by a^2, corresponds physically to the forward scattering amplitude for quasi-particles with energy on the Fermi surface. We denote this quantity by Γ^k. From Eq. (A2.8) we find the relation between $a^2\Gamma^k$ and $a^2\Gamma^{\omega}$

$$a^2\Gamma^{k}_{\alpha\beta,\gamma\delta}(P_1 P_2) = a^2\Gamma^{\omega}_{\alpha\beta,\gamma\delta}(P_1 P_2) \qquad\qquad \text{(A2.9)}$$
$$- \frac{p_0^2}{v(2\pi h)^3}\int a^2\Gamma^{\omega}_{\alpha\epsilon,\gamma\zeta}(P, Q)\, a^2\Gamma^{k}_{\zeta\beta,\epsilon\delta}(QP_2)\, d\Omega.$$

Let us now study the poles of the function $\Gamma(P_1 P_2; K)$. Just as the poles of the function G(P) give the relation between the energies and momenta of the Fermi excitations, the poles of the function Γ characterize the "two-particle" excitations, in other words the Bose excitations. In view of the fact that in the neighborhood of a pole $\Gamma(P_1 P_2; K) \gg \Gamma^{\omega}(P_1 P_2)$, we can neglect the term Γ^{ω} in the right member. Furthermore it must be noted that the variable P_2, and also the indices β and δ play the part of parameters in the equation. The function Γ can be represented in the form of a product $\chi_{\alpha\beta}(P_1; K)\,\chi_{\beta\delta}(P_2)$, and $\chi(P_2)$ cancels on the two sides of the equation. If we also introduce the notation

$$\nu_{\alpha\gamma}(\mathbf{n}) = \frac{\mathbf{n}\cdot\mathbf{k}}{\omega - v\mathbf{n}\cdot\mathbf{k}}\,\chi_{\alpha\gamma}(P_1; K),$$

where $\mathbf{n}$ is the unit vector in the direction of P_1, we get for $\nu_{\alpha\gamma}$ the equation

$$(\omega - v\mathbf{n}\cdot\mathbf{k})\,\nu_{\alpha\gamma}(\mathbf{n}) = \mathbf{n}\cdot\mathbf{k}\,\frac{p_0^2 a^2}{(2\pi\hbar)^3}\int \Gamma^{\omega}_{\alpha\delta,\gamma\zeta}(\mathbf{n}, \mathbf{l})\,\nu_{\zeta\delta}(\mathbf{l})\,d\Omega. \quad \text{(A2.10)}$$

Comparing this equation with Eqs. (8.14) and (10.4), we see without difficulty that it is the same as the equation for the zeroth sound and the spin waves, with the quantity $a^2\Gamma^{\omega}$ playing the role of the function f. This confirms the equation for the zeroth sound obtained in the phenomenological theory, and thus confirms the hypothesis made in that theory about the functional dependence of the energy of the excitations on the distribution function. On the other hand, owing to Eq. (A2.9), we have established a connection between the function f and the scattering amplitude of the quasi-particles at the angle 0°.

Let us examine this relation in greater detail. Denoting the amplitude multiplied by $(d\tau/d\epsilon)_{\epsilon=\mu}$ by $A(\mathbf{n}_1\cdot\sigma_1, \mathbf{n}_2\cdot\sigma_2)$, we get:

$$A(\mathbf{n}_1\cdot\sigma_1, \mathbf{n}_2\cdot\sigma_2) = \Phi(\mathbf{n}_1\cdot\sigma_1, \mathbf{n}_2\cdot\sigma_2) - \frac{1}{2}\,\mathrm{Sp}_{\sigma'}\int \Phi(\mathbf{n}_1\cdot\sigma_1, \mathbf{n}'\cdot\sigma')\,A(\mathbf{n}'\cdot\sigma', \mathbf{n}_2\cdot\sigma_2)\,\frac{d\Omega}{4\pi}. \quad \text{(A2.11)}$$

As regards their spin dependence, Φ and A each contain two terms: one independent of the spins, and one proportional to $\sigma_1\cdot\sigma_2$. If we write

$$A(\mathbf{n}_1\cdot\sigma_1, \mathbf{n}_2\cdot\sigma_2) = B(\mathbf{n}_1\cdot\mathbf{n}_2) + \sigma_1\cdot\sigma_2\cdot C(\mathbf{n}_1, \mathbf{n}_2), \quad \text{(A2.12)}$$

B and C satisfy the equations

$$B(\mathbf{n}_1, \mathbf{n}_2) = F(\mathbf{n}_1, \mathbf{n}_2) - \int F(\mathbf{n}_1, \mathbf{n}')\,B(\mathbf{n}', \mathbf{n}_2)\frac{d\Omega}{4\pi}, \quad \text{(A2.13)}$$

$$C(\mathbf{n}_1, \mathbf{n}_2) = Z(\mathbf{n}_1, \mathbf{n}_2) - \frac{1}{4}\int Z(\mathbf{n}_1, \mathbf{n}')\,C(\mathbf{n}', \mathbf{n}_2)\frac{d\Omega}{4\pi}.$$

Being scalars, the quantities B and C, and also F and Z, can depend only on $\cos\chi$. If we introduce expansions in spherical harmonics, $B(\chi) = \Sigma B_l P_l(\cos\chi)$, and so on, then we can easily get relations between the coefficients B_l and F_l, and between C_l and Z_l:

$$B_l = \frac{F_l}{1 + \frac{F_l}{2l+1}}, \quad C_l = \frac{Z_l}{1 + \frac{Z_l}{4(2l+1)}}. \quad \text{(A2.14)}$$

For example, let us find the scattering amplitude of the excitations for a rarefied Fermi gas. From Eqs. (A1.16) and (A2.13) we get:

$$A(\chi, \sigma_1\cdot\sigma_2) = 8\left(\frac{3}{\pi}\right)^{\frac{1}{3}} aN^{\frac{1}{3}}\left(\frac{1}{4} - \sigma_1\cdot\sigma_2\right) \quad \text{(A2.15)}$$

$$+ 16\left[\left(\frac{3}{\pi}\right)^{\frac{1}{3}} aN^{\frac{1}{3}}\right]^2\left[\frac{1}{4} - \sigma_1\cdot\sigma_2 + \frac{\cos\chi}{8\sin\frac{\chi}{2}}\ln\frac{1+\sin\left(\frac{\chi}{2}\right)}{1-\sin\left(\frac{\chi}{2}\right)}\right.$$

$$\left. - \sigma_1\cdot\sigma_2\left(1 - \frac{\sin\left(\frac{\chi}{2}\right)}{2}\ln\frac{1+\sin\left(\frac{\chi}{2}\right)}{1-\sin\left(\frac{\chi}{2}\right)}\right)\right].$$

We note that in the case of parallel spins this expression goes to zero for $\chi = 0$, as it must according to the Pauli principle (for f this does not occur).

[1] L. D. Landau, J. Exptl. Theoret. Phys. (U.S.S.R.) **30**, 1058 (1956), Soviet Phys. JETP **3**, 920 (1956).

[2] I. M. Khalatnikov and A. A. Abrikosov, J. Exptl. Theoret. Phys. (U.S.S.R.) **32**, 915 (1957), Soviet Phys. JETP **5**, 745 (1957).

[3] Abraham, Osborne, and Weinstock, Phys. Rev. **98**, 551 (1955).

[4] Fairbank, Ard, and Walters, Phys. Rev. **95**, 567 (1954).

[5] A. A. Abrikosov and I. M. Khalatnikov, J. Exptl. Theoret. Phys. (U.S.S.R.) **32**, 1083 (1957), Soviet Phys. JETP **5**, 887 (1957).

[6] I. Ia. Pomeranchuk, J. Exptl. Theoret. Phys. (U.S S.R.) **20**, 919 (1950).

[7] K. N. Zinov'eva, J. Exptl. Theoret. Phys. (U.S.S.R.) **34**, 609 (1958), Soviet Phys. JETP **7**, 421 (1958).

[8] L. D. Landau, J. Exptl. Theoret. Phys. (U.S.S.R.) **32**, 59 (1957), Soviet Phys. JETP **5**, 101 (1957).

[9] Lagner, Sydoriak, and Roberts, Program of the Fifth International Conference on Low-Temperature Physics and Chemistry, p. 41. University of Wisconsin 1957.

[10] I. M. Khalatnikov and A. A. Abrikosov, J. Exptl. Theoret. Phys. (U.S.S.R.) **33**, 110 (1957), Soviet Phys. JETP **6**, 84 (1958).

[11] A. A. Abrikosov and I. M. Khalatnikov, J. Exptl. Theoret. Phys. (U.S.S.R.) **34**, 198 (1958), Soviet Phys. JETP **7**, 135 (1958).

[12] L. D. Landau and E. M. Lifshitz, Электродинамика сплошных сред (Electrodynamics of Continuous Media), Gostekhizdat, Moscow 1957.

[13] S. M. Rytov, Теория электрических флуктуаций

и теплового излучения (Theory of Electrical Fluctuations and Thermal Radiation), Acad. Sci. Press, 1953.

[14]L. D. Landau and E. M. Lifshitz, J. Exptl. Theoret. Phys. (U.S.S.R.) **32**, 618 (1957), Soviet Phys. JETP **5**, 512 (1957).

[15]V. L. Ginzburg, J. Exptl. Theoret. Phys. (U.S.S.R.) **13**, 243 (1943).

[16]A. A. Abrikosov and I. M. Khalatnikov, J. Exptl. Theoret. Phys. (U.S.S.R.) **33**, 1154 (1957), Soviet Phys. JETP **6**, 888 (1958).

[17]K. Huang and C. N. Yang, Phys. Rev. **105**, 767 (1957).

[18]T. D. Lee and C. N. Yang, Phys. Rev. **105**, 1119 (1957).

[19]L. N. Cooper, Phys. Rev. **104**, 1189 (1956).

[20]Bardeen, Cooper, and Schrieffer, Phys. Rev. **106**, 162 (1957).

[21]L. D. Landau, J. Exptl. Theoret. Phys. (U.S.S.R.) (in press).

Translated by W. H. Furry

SOVIET PHYSICS JETP VOLUME 35(8), NUMBER 6 JUNE, 1959

ON THE THEORY OF SUPERCONDUCTING ALLOYS

I. THE ELECTRODYNAMICS OF ALLOYS AT ABSOLUTE ZERO

A. A. ABRIKOSOV and L. P. GOR' KOV

Institute for Physics Problems, Academy of Sciences, U.S.S.R.

Submitted to JETP editor July 16, 1958

J. Exptl. Theoret. Phys. (U.S.S.R.) 35, 1558-1571 (December, 1958)

In this paper we give the theory of superconductors containing impurities at the absolute zero. The dependence of the penetration depth on the impurity concentration is considered for small atomic concentrations. We have obtained the electrodynamic equations in an alternating field for superconductors with a mean free path which is smaller than the correlation length.

It is well known that Bardeen, Cooper, and Schrieffer[1] succeeded in formulating the electrodynamics of superconductors with the aid of their theory of superconductivity. They obtained equations connecting the current with the vector potential which replaced the old equations of the phenomenological theory of H. and F. London. Of most interest was the non-local character of the connection between the current and the field for the majority of pure superconductors.

The non-local character is put into the very basis of the theory which is based on the physical idea of Cooper's[2] about the formation of bound singlet pairs of electrons near the Fermi surface under the action of the phonon interaction. The dimensions of such a pair causes the existence of a correlation between electrons at distances of the order $\xi_0 \sim 10^{-4}$, which becomes apparent in the form of a non-local connection between the field and the current, if the penetration depth of the field is less than ξ_0. The majority of pure superconductors belong to that class.

The creation of a theory of superconductivity made it possible to consider a whole series of phenomena. With its aid they could construct the thermodynamics and electrodynamics of superconductors[1] and, in particular, determine the behavior of superconductors in a high-frequency field.[3] A very interesting problem is that of "alloys," i.e., of superconductors with impurity atoms of other elements or other lattice defects. In the normal state these defects cause the so-called residual resistance.

At very small concentrations the role of impurities is small. Increasing the concentration, however, leads to a decrease of the spatial correlation of the electrons in the superconductor. For a sufficiently concentrated alloy the role of the correlation parameter goes over from ξ_0 to the electron mean free path. Thanks to this, it is natural to expect the appearance of a local connection between the current and the vector potential for such concentrations where the mean free path becomes less than the penetration depth. The difference with the Londons' theory will lie in a change in the constant of proportionality of $\mathbf{j}$ to $\mathbf{A}$. Below we shall consider this problem, assuming, however, that the atomic impurity concentration is small.

The problem of the penetration of a weak electromagnetic field into a medium is solved if we know the Green function of the field in the medium

$$D_{ik}(x-x') = \frac{4\pi i}{c}\langle T(A_i(x)A_k(x'))\rangle. \quad (1)$$

The function D_{ik} is directly connected with experimentally observed quantities. Let $D_{ik}(\mathbf{k}, \omega_0)$ be the Fourier transform of $D_{ik}(x-x')$:

$$D_{ik}(x-x') = \frac{4\pi}{(2\pi)^4}\int \exp\{i\mathbf{k}\cdot(\mathbf{x}-\mathbf{x}') - i\omega_0(t-t')\}D_{ik}(\mathbf{k}, \omega_0)d^3\mathbf{k}\,d\omega_0.$$

In view of the symmetry of the tensor D_{ik} we have

$$D_{ik}(\mathbf{k}, \omega_0) = D_1(\mathbf{k}^2, \omega_0)(\delta_{ik} - k_ik_k/k^2) + D_2(\mathbf{k}^2, \omega_0)k_ik_k/k^2.$$

The function D_2 which determines the longitudinal part of D_{ik} is connected with the gauge of the vector potential and is arbitrary. The physical quantities depend only on the transverse part, i.e., on D_1.

With the aid of the general theory of the anomalous skin effect we can arrive at the following formula for the surface impedance in the case of diffuse scattering at the surface*

$$Z = \frac{4\pi^2 i\omega_0}{c^2} \Big/ \int_0^\infty \ln[\mathbf{k}^2 D_1(\mathbf{k}^2, \omega_0)]\, d|\mathbf{k}|. \tag{2}$$

We recall that the penetration depth of the field is

$$\delta = \operatorname{Re}\left[\frac{c^2}{4\pi\omega_0} iZ\right]. \tag{3}$$

Usually, instead of finding the function D_{ik} one finds the connection between the current and the vector potential

$$j_i = -Q_{ik}(\mathbf{k}, \omega_0) A_k(\mathbf{k}, \omega_0). \tag{4}$$

To find the connection between the tensors D_{ik} and Q_{ik}. It is sufficient to substitute (4) into the Maxwell equation and to write down with its aid an equation for D_{ik}. This gives

$$(D^{-1})_{ik} = (\mathbf{k}^2\delta_{ik} - k_i k_k) + \frac{4\pi}{c} Q_{ik}. \tag{5}$$

The first term corresponds to the "zero" value of the tensor D_{ik}, obtained by neglecting the interaction of the field with the electrons. In Eqs. (2) to (5) we have implied $\omega_0 > 0$.

From Eq. (5) it is clear that the tensor Q_{ik} must be symmetric. Also, since the current is conserved ($k_i j_i = 0$) it follows from Eq. (4) that it must be transverse.† Equation (5) determines thus only the transverse part of D_{ik} and its longitudinal part remains arbitrary, as it should be.

We evaluate below the function $D_{ik}(\mathbf{k}, \omega_0)$ for a superconductor with impurities, using the techniques of the quantum field theory‡ (see reference 4, which in contradistinction to other methods[1,5] makes it possible to state the problem very clearly). We assume that we have already taken into account the interaction of the electrons with the impurities and consider their interaction with the electromagnetic field. The Hamiltonian of this interaction is

$$H_1' = \frac{e^2}{mc^2}\psi^+(x)\psi(x)\mathbf{A}^2(x) + \frac{ie\mathbf{A}(x)}{2mc}(\nabla_x - \nabla_{x'})\psi^+(x')\psi(x)\Big|_{x\to x'}. \tag{6}$$

Dyson's equation for the function D will be of the form

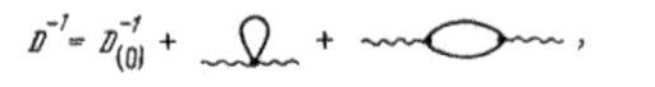 (7)

where a solid line corresponds to an electron and a wavy one to a photon. More complicated diagrams with internal photon lines can be neglected in the non-relativistic approximation.

Comparing Eqs. (7) and (5) we see that, apart from a factor $4\pi/c$ the term added to $D_{(0)}$ corresponds to the tensor Q_{ik}. The first term of this addition gives simply $(Ne^2/mc)\delta_{ik}$ in Q_{ik}. The second diagram is more complicated. For the following it is convenient to separate, as is usually done, one of the photon vertices, for instance p_i, from the remaining part which, apart from a coefficient, can be considered as the result of including the photon vertex p_k in the electron line. To find the coefficient is not difficult (the factor 2 comes from taking the trace over the spin). In this way we get

$$Q_{ik}(\mathbf{k}, \omega_0) = \frac{Ne^2}{mc}\delta_{ik} - \frac{2ie^2}{m^2c}\int p_i \Pi_k(p_+, p_-)\frac{d^4p}{(2\pi)^4}, \tag{8}$$

where $p_\pm = (\mathbf{p} \pm \mathbf{k}/2,\ \omega \pm \omega_0/2)$.

As far as the interaction of the electrons with the impurities is concerned, the corresponding term in the Hamiltonian is of the form

$$H_2' = \sum_a H_a, \qquad H_a = u(\mathbf{x} - \mathbf{x}_a)\psi^+(x)\psi(x), \tag{9}$$

where $u(\mathbf{x} - \mathbf{x}_a)$ is the interaction potential with an impurity atom at the point $\mathbf{x}_a$. We shall assume that the impurity atoms are randomly distributed in the metal and that their distance apart is large compared to the lattice period. We assume, of course, that the impurity concentration is so small that the interaction between the electrons is not changed. We shall denote the impurity vertex in the Feynman diagrams by a cross. It corresponds to a factor

$$u(\mathbf{q})\, e^{i\mathbf{q}\cdot\mathbf{x}_a}\,\delta(\omega_1 - \omega_2), \tag{10}$$

where $u(\mathbf{q})$ is the Fourier component of the potential $u(\mathbf{x})$ and $\mathbf{q}$ is the momentum transferred.

For the sake of clarity we first give the calculations for the application to a normal metal with a residual resistance (Sec. 1). We then consider superconducting alloys (Sec. 2). All calculations refer only to the absolute zero.

*The correspondence of this equation with the usual expression is easily established using Eqs. (4) and (5).

†The method of finding the tensor Q_{ik} usually employed in various problems of the theory of superconductivity (see references 1 and 3) leads to an expression which is not purely transverse. One can, however, show by means of a more rigorous calculation based on the equations of reference 4 that to get the correct result for the given problem it is sufficient to retain only the transverse part of the tensor Q_{ik} obtained by the usual method, or to use a vector potential satisfying the condition div $\mathbf{A} = 0$.

‡We must emphasize that the method used here is suitable only at $T = 0$. In the case where $T \neq 0$ there will occur, instead of simple time-ordered products, more complicated combinations which make this method practically innapplicable to the problem under consideration.

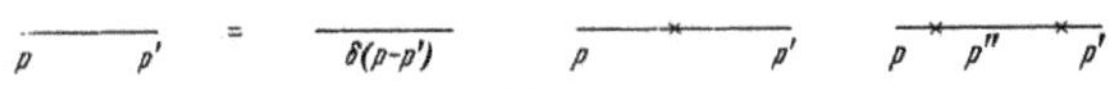

FIG. 1

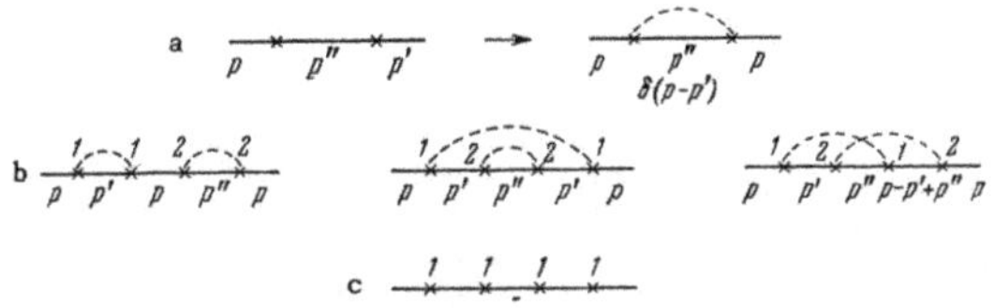

FIG. 2

1. RESIDUAL RESISTANCE OF A NORMAL METAL

We introduce the Green function of a system of fermions

$$G_{\alpha\beta}(x, x') = -i\langle T(\psi_\alpha(x)\,\psi^+_\beta(x'))\rangle. \qquad (11)$$

For a pure metal we have:*

$$G_{0\alpha\beta}(x - x') = \delta_{\alpha\beta}(2\pi)^{-4}\int G_0(p)\,e^{ip(x-x')}d^4p, \qquad (12)$$
$$G_0(p) = (\omega - \xi + i\delta\,\mathrm{sign}\,\omega)^{-1},$$

where $\xi = v(|\mathbf{p}| - p_0)$, $v = p_0/m$, $\delta \to +0$. When impurities are present the Green function (11) is not the same as expression (12). We shall write it in the form

$$G_{\alpha\beta}(x, x') = \delta_{\alpha\beta}(2\pi)^{-7}\int G(\mathbf{p}, \mathbf{p}', \omega)$$
$$\times \exp\{i\mathbf{p}\cdot\mathbf{x} - i\mathbf{p}'\cdot\mathbf{x}' - i\omega(t - t')\}\,d^3\mathbf{p}\,d^3\mathbf{p}'\,d\omega. \qquad (13)$$

The function $G(\mathbf{p}, \mathbf{p}', \omega)$ is expressed as a sum of the diagrams given in Fig. 1 according to the well-known rules of field theory. Each line corresponds to $G_0(\mathbf{p})$.

Summing over the diagrams gives an integral equation for $G(\mathbf{p}, \mathbf{p}', \omega)$:

$$G(\mathbf{p}, \mathbf{p}', \omega) = \delta(\mathbf{p} - \mathbf{p}')\,G_0(p) \qquad (14)$$
$$+ (2\pi)^{-3}\int G_0(p)\sum_a u(\mathbf{p} - \mathbf{p}'')\,e^{i(\mathbf{p}-\mathbf{p}'')\mathbf{x}_a}\,G(\mathbf{p}'', \mathbf{p}', \omega)\,d^3\mathbf{p}''$$

We are not interested in the exact solution of (14), since the impurity atoms are randomly distributed over the metal. We must average all expressions over the positions of all impurity atoms. It is then essential that by virtue of our assumption about the smallness of the atomic concentration the average distance apart of the impurity atoms is much larger than the atomic distances in the metallic lattice, so that the averaging can take place over volumes with dimensions large compared to the interatomic distances. After such an averaging, the Green function $G(\mathbf{p}, \mathbf{p}', \omega)$ is clearly of the form

$$\overline{G(\mathbf{p}, \mathbf{p}', \omega)} = G(\mathbf{p}, \omega)\,\delta(\mathbf{p} - \mathbf{p}'). \qquad (15)$$

The absolute magnitude of the momenta $\mathbf{p}$, $\mathbf{p}'$ which are of interest to us are of the order of magnitude of the momentum p_0 at the Fermi surface which in turn is of the order of the inverse of the interatomic distances. This fact makes it at once easy to carry out the averaging.

We shall perform the calculation in the Born approximation, i.e., we shall assume that $p_0^3\int u(\mathbf{x})\,d^3\mathbf{x} \ll E_{ph}$. Once can show that the final results expressed in terms of the collision time will also be correct in the general case.

The simplest diagram for G contains only one cross. The value obtained by averaging over the position of the impurity atom is a constant $\overline{u(\mathbf{q})\,e^{i(\mathbf{q}\cdot\mathbf{x}_\alpha)}} = u(0)$, which can be included in the ground state energy and which in the following we assume to be equal to zero. The next diagram, as far as complexity is concerned, contains two crosses referring to one atom (Fig. 2a). The average value of these diagrams (without external G_0) is equal to

$$\frac{1}{V}\int |u(\mathbf{p} - \mathbf{p}')|^2\,G_0(p')\,\frac{d^3\mathbf{p}'}{(2\pi)^3}, \qquad (16)$$

where V is the volume of the system. In the following we shall be interested in values of $|\mathbf{p}|$ which in absolute magnitude are close to p_0. The integral in Eq. (16) can be split into two parts: an integral over values of $|\mathbf{p}'|$ far from the Fermi surface and one over values near the Fermi surface (the limits of the second integral over $|\mathbf{p}'|$ can be taken symmetric with respect to $|\mathbf{p}'| = p_0$). The integral over the far region gives a real constant which together with $u(0)$ is a renormalization of the chemical potential and need not be considered. One may assume that in the second integral $u(\mathbf{p} - \mathbf{p}')$ is a slowly-varying function. Substituting Eq. (12) for $G_0(\mathbf{p})$ and summing over the impurity atoms (which simply means multiplying by the number of atoms) we get the essen-

*Here and henceforth we use units with $\hbar = 1$.

tial contribution to the G-function:

$$i \operatorname{sign} \omega/2\tau,$$

where

$$\frac{1}{\tau} = n\frac{p_0 m}{(2\pi)^2}\int |u(\theta)|^2 d\Omega \tag{17}$$

is the time between collisions, n the number of impurity atoms per unit volume. From this it is clear that the region near the Fermi surface plays the leading role in the integrals.

From this point of view not all diagrams are equivalent. Let us compare, for instance, the three diagrams of Fig. 2b. (Dotted lines connect crosses referring to one atom.) We see easily that in the first two expressions the integration over $\mathbf{p}'$ and $\mathbf{p}''$ can be performed over arbitrary angles between the momenta near the Fermi surface. On the other hand, in the third integral the requirement that all arguments of the G-functions are near the Fermi surface leads to a limitation of the angles. As a result the contribution from such a graph turns out to be less by a factor $(vp_0G)^{-1}$ as compared to the others. Since in the following we need values of ω and $\xi \sim 1/\tau$ where τ is the time between collisions, the smallness of the "intersecting" graphs can be estimated to be $(p_0 l)^{-1}$ where $l = v\tau$ is the mean free path.

One can easily show that a small contribution is also given by diagrams containing more than two crosses from one impurity atom. Let us, for instance, compare the total contribution (from all impurity atoms) of diagrams of the kind of the first one of Fig. 2b with diagrams of the type 2c. The first ones give a quantity of the order of magnitude

$$\frac{1}{\tau^2}G_0 \sim \frac{1}{\tau}.$$

The second ones give, as far as order of magnitude is concerned

$$\frac{1}{\tau}\frac{u^2(\mathbf{q})\,p_0^4}{v^2} \sim \frac{1}{\tau}\left[\frac{\int u dV p_0^3}{E_{\text{ph}}}\right]^2 \ll \frac{1}{\tau}.$$

It is thus clear that we need consider only diagrams containing up to two crosses on one impurity atom.

Summing all essential diagrams (i.e., only "pair" diagrams and those which do not contain "intersections" like the third diagram of Fig. 2b) we get the following expression for the G-function:

$$G(p) = G_0(p) + nG_0(p)\int |u(\mathbf{p}-\mathbf{p}')|^2 G(p')\frac{d^3p'}{(2\pi)^3}G(p). \tag{18}$$

If we do not wish to use the Born approximation, it is necessary to take into account diagrams with many crosses at one impurity atom. One can show

a b c

FIG. 3

that all changes which this entails are taken into account by replacing the Born amplitude $u(\theta)$ by the total scattering amplitude. This remains true also for all calculations in the following. We can therefore imply in all formulae by $u(\theta)$ the total scattering amplitude.

The solution of Eq. (18) is

$$G(p) = \frac{1}{\omega - \xi - \overline{G}(\omega)},$$

where $\overline{G}(\omega)$ satisfies the equation

$$\overline{G}(\omega) = n\int |u(\mathbf{p}-\mathbf{p}')|^2 \frac{1}{\omega - \xi - \overline{G}(\omega)}\frac{d^3\mathbf{p}'}{(2\pi)^3}$$

Assuming $\overline{G}(\omega)$ to be purely imaginary ($\overline{G}(\omega) = -i\beta$) and applying to the integral on the right hand side the same approach as was applied to (16) we find

$$\beta = \operatorname{sign}\beta/2\tau,$$

where τ is determined by Eq. (17). Comparing G which is obtained in this manner with the case of small amounts of impurities ($G \to G_0$) we find that $\beta = \operatorname{sign}\omega/2\tau$ or

$$G = (\omega - \xi + i \operatorname{sign}\omega/2\tau)^{-1}. \tag{19}$$

Going over to the x-representation we see easily that the whole change in G as compared to G_0 is obtained by multiplying the latter by an exponential screening factor

$$G = G_0 e^{-|\mathbf{x}-\mathbf{x}'|/2l}, \tag{20}$$

where $l = v\tau$.

We now proceed to evaluate Q_{ik}. Putting the photon vertex into the electron Green function $G(\mathbf{p}, \mathbf{p}', \omega)$ we get:

$$\Pi(p_+, p_-) = (2\pi)^{-3}\int G\left(\mathbf{p}_+, \mathbf{p}'_+, \omega + \frac{\omega_0}{2}\right)\mathbf{p}'G\left(\mathbf{p}'_-, \mathbf{p}_-, \omega - \frac{\omega_0}{2}\right)d^3\mathbf{p}'. \tag{21}$$

The function $G(\mathbf{p}, \mathbf{p}', \omega)$ entering here corresponds to the sum of the diagrams in Fig. 1 and is described by Eq. (14). To average over the positions of the impurity atoms we must take into account that the average of two Green functions is not equal to the product of the two average values.

For a pure metal expression (21) corresponds to the diagram of Fig. 3a. After averaging over the positions of the impurity atoms, it turns out that apart from the simple diagrams corresponding to the transition from the zeroth Green functions $G_0(p)$ to the function $G(p)$ of (19) the essential diagrams for the quantity (21) are those of Fig. 3b. The large contribution from these corrections is connected with the fact that the photon momentum occurring in the vertex $|\mathbf{k}| \ll p_0$. Diagrams of another kind, for instance the graph of Fig. 3c, are far smaller since one of the integrations in them is over a region of momenta far from the Fermi surface. The averaging of the quantity (21) is thus performed by the summation of the "ladder" diagrams of Fig. 3b.

The integral equation for $\Pi(p_+, p_-)$ is

$$\Pi(p_+, p_-) = G(p_+)G(p_-)\left[\mathbf{p} + \frac{n}{(2\pi)^3}\int |u(\mathbf{p}-\mathbf{p}')|^2\, \Pi(p'_+, p'_-)\, d^3\mathbf{p}'\right]. \quad (22)$$

There are two limiting cases possible:

(a) The skin effect is anomalous, $(kv) \gg \tau$; it is easy to verify that the integral on the right hand side of (22) will be negligibly small in that case $(\sim 1/kv\tau)$.

(b) The skin effect is normal, $(kv) \ll 1/\tau$; this case is just of the greatest interest to us. In Eq. (22) we can then assume $p_+ = p_-$.

The vector obtained from the integral on the right hand side of (22) will clearly be parallel to $\mathbf{p}$. We introduce the notation

$$\mathbf{p}\Lambda(\omega, \omega_0) = \frac{n}{(2\pi)^3}\int |u(\mathbf{p}-\mathbf{p}')|^2\, \Pi(p'_+, p'_-)\, d^3\mathbf{p}'. \quad (23)$$

Since $|\mathbf{p}| \approx p_0$, we can assume $\Lambda(\omega, \omega_0)$ to be independent of $|\mathbf{p}|$. We multiply (22) by $n(2\pi)^{-3} \times |u(\mathbf{1}-\mathbf{p})|^2$ and integrate over $d^3\mathbf{p}$:

$$\mathbf{l}\Lambda(\omega, \omega_0) = \frac{n}{(2\pi)^3}\int |u(\mathbf{l}-\mathbf{p})|^2 \mathbf{p} G(p_+)G(p_-)\left[1 + \Lambda(\omega, \omega_0)\right] d^3\mathbf{p}. \quad (24)$$

Replacing $G(p)$ by Eq. (19) we find easily that $\Lambda(\omega, \omega_0)$ is only different from zero if $|\omega| < \omega_0/2$, since in the opposite case both poles in (24) for the integration over ξ according to (20) lie in the same half-plane. In that integral Λ does not depend on ω. Integrating over ξ and using the relation

$$\cos\theta = \cos\theta'\cos\theta'' + \sin\theta'\sin\theta''\cos(\varphi' - \varphi''), \quad (25)$$

we get

$$\Lambda(\omega, \omega_0) = \begin{cases} \dfrac{i}{\tau_1}\dfrac{1}{\omega_0 + i/\tau_{tr}} & \omega^2 < \omega_0^2/4, \\ 0, & \omega^2 > \omega_0^2/4, \end{cases} \quad (26)$$

where

$$\frac{1}{\tau_1} = \frac{1}{\tau} - \frac{1}{\tau_{tr}}; \quad \frac{1}{\tau_{tr}} = \frac{nmp_0}{(2\pi)^2}\int |u(\theta)|^2(1-\cos\theta)\, d\Omega. \quad (27)$$

After substituting (26) into (22) and (7) and integrating, assuming that $\omega_0\tau \ll 1$, we find

$$Q_{ik} = -\frac{4\pi i\omega_0\sigma}{c^2}\delta_{ik}. \quad (28)$$

The "transport" collision time enters into the conductivity $\sigma = Ne^2\tau_{tr}/m$, as should be the case.

2. SUPERCONDUCTING ALLOYS

The method given above will now be applied to a study of a superconductor with impurities. The main difference as compared to what was given above lies in the fact that a superconductor is described by three Green functions:

$$\begin{aligned} \delta_{\alpha\beta}G(x, x') &= -i\langle T(\psi_\alpha(x)\psi_\beta^+(x'))\rangle; \\ \hat{I}_{\alpha\beta}F^+(x, x') &= \langle T(\psi_\alpha^+(x)\psi_\beta^+(x'))\rangle; \\ -\hat{I}_{\alpha\beta}F(x, x') &= \langle T(\psi_\alpha(x)\psi_\beta(x'))\rangle, \end{aligned} \quad (29)$$

where $\hat{I}_{\alpha\beta} = -\hat{I}_{\beta\alpha}$, $\hat{I}^2 = -E$. These functions were introduced in reference 4 for a pure superconductor and are equal to

$$\begin{aligned} G(x - x') &= (2\pi)^{-4}\int G(p)\, e^{ip(x-x')}\, d^4p, \\ F^+(x - x') &= (2\pi)^{-4}\int F^+(p)\, e^{ip(x-x')}\, d^4p, \\ F(x - x') &= F^+(x - x'), \end{aligned}$$

where

$$G_0(p) = \frac{\omega + \xi}{\omega^2 - \xi^2 - \Delta^2 + i\delta}; \quad F_0^+(p) = \frac{i\Delta}{\omega^2 - \xi^2 - \Delta^2 + i\delta} \quad (\delta \to +0). \quad (30)$$

The occurrence of three Green functions makes it necessary to change the diagram technique.* The functions G, F^+, and F can be represented by the lines with arrows given in Fig. 4a. If vertices are included in an electron line it is necessary to put an arrow next to the vertex. If the interaction Hamiltonian includes products $\psi_\alpha^+(y')\psi_\alpha(y)$ (this always occurs in our problem) the arrows around the vertex must be aimed in the same direction. There are then two possibilities: both arrows can be directed towards the left or towards the right. From Fig. 4c it is clear what will be the result of including vertices $\psi^+\psi$ in the functions G, F^+, and F. The result can be written in the form

$$\begin{aligned} G(x, x') &\to G(x, y')G(y, x') + F(x, y)F^+(y', x'); \\ F^+(x, x') &\to F^+(x, y')G(y, x') + G(y, x)F^+(y', x'); \\ F(x, x') &\to G(x, y')F(y, x') + F(x, y)G(x', y'). \end{aligned} \quad (31)$$

*The technique with three Green functions was worked out by S. T. Beliaev[6] in an application to the Bose gas.

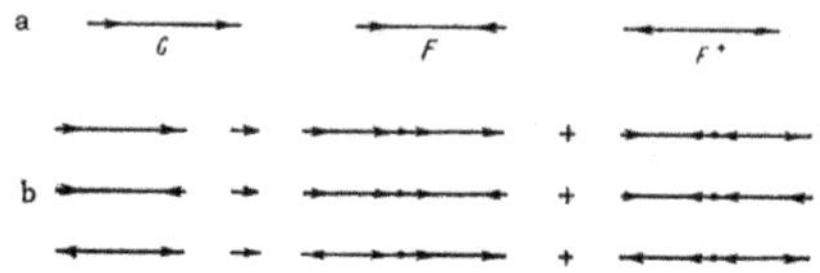

FIG. 4

The inclusion of vertices leads in this way to the appearance of F and F^+ lines in the diagrams for G and similarly to G lines in the diagrams for F and F^+. The equations for G, F^+, and F are therefore more complicated than Eq. (14), obtained by summing the diagrams of Fig. 1 with one G-function.

The derivation of these equations and their solution are given in the Appendix. It turns out that the averaging over the positions of the impurity atoms of the functions G(p) and $F^+(p) = F(p)$ can be obtained from the expressions (30) for a pure superconductor by the substitution

$$\omega;\ \Delta \to \omega\eta;\ \ \Delta\eta, \tag{32}$$

where

$$\eta = \begin{cases} 1 + i/2\tau\sqrt{\omega^2 - \Delta^2}; & \omega^2 > \Delta^2, \\ 1 + 1/2\tau\sqrt{\Delta^2 - \omega^2}; & \omega^2 < \Delta^2; \end{cases}$$

τ is the collision time for the metal in its normal state. One can easily verify that, as in Eq. (18) of Sec. 1, these equations signify in coordinate space a multiplication of the function $G_0(x)$ and $F_0^+(x)$ by $e^{-|\mathbf{x}|/2l}$

We now turn to the determination of the function Q_{ik}. This time we have two diagrams corresponding to the simple loop for a normal metal. They are given in Fig. 5. This means that instead of a product of two G's we have now the combination $GG - FF^+$ (the sign is due to the fact that the operator $\nabla_\mathbf{y} - \nabla_{\mathbf{y}'}$ occurs in the electromagnetic interaction, which changes sign if the direction of the arrows at the vertex is changed). For the following it will again be convenient for us, taking the two loops of Fig. 5 together, and taking one of the photon vertices p_i separately, to consider the remaining factor as the result of including the photon vertex p_k in the function G. After this Q_{ik} will be of the form (8) and the problem lies in evaluating the vector $\mathbf{\Pi}$.

To average $\mathbf{\Pi}$ over the impurities we note at once that since superconductivity changes the Green functions only near the Fermi surface we need again only sum over ladder graphs of the type given in Fig. 3b. Owing to the modified diagram technique, we must consider, as can easily be seen from Fig. 6, not one, but four different vectors $\mathbf{\Pi}^{(i)}$ $(i = 1, \ldots, 4)$ which occur as a result of including photon vertices in G, F^+, $G^{(-)}$, and F (where $G^{(-)}$ denotes G with the arguments interchanged). The first of these quantities enters directly into the integral (8) for Q_{ik}.

FIG. 5

We obtain easily for the $\mathbf{\Pi}^{(i)}$ a set of four integral equations. The first one of them is obtained with the aid of Fig. 6 and has the form:

$$\begin{aligned} \mathbf{\Pi}^{(1)}(p_+, p_-) &= \mathbf{p}\,[G(p_+)G(p_-) - F(p_+)F(p_-)] \\ &+ G(p_+)G(p_-)\Lambda^{(1)} + F(p_+)G(p_-)\Lambda^{(2)} \\ &+ F(p_+)F(p_-)\Lambda^{(3)} + G(p_+)F(p_-)\Lambda^{(4)}. \end{aligned} \tag{33}$$

The quantities $\Lambda^{(i)}$ $(\omega, \omega_0, \mathbf{k}, \mathbf{p})$ are the integrals (23) of the $\mathbf{\Pi}^{(i)}(p_+, p_-)$. The remaining equations are obtained in a similar way. We shall not write them down since, as was pointed out above, we only need $\mathbf{\Pi}^{(1)}(p_+, p_-)$.

As was already noted, for high impurity concentrations the role of the correlation length is assumed here not by $\xi_0 = v/\Delta$ but by the mean free path l. As the impurity concentration increases this length decreases. We shall consider here the case where l is much less than the penetration depth. It is then natural to expect a local connection between the current and the vector potential. The case considered is characterized by the condition $kl \ll 1$, which leads to a considerably simplified equation.

Indeed, neglecting $\mathbf{k}$, we obtain $\mathbf{p}_+ = \mathbf{p}_-$. It is easy to check that in that case $\Lambda^{(i)}(\omega, \omega_0, \mathbf{k}, \mathbf{p}) \approx \mathbf{p}\Lambda^{(i)}(\omega, \omega_0)$ and

$$\Lambda^{(3)}(\omega, \omega_0) = -\Lambda^{(1)}(\omega, \omega_0); \quad \Lambda^{(2)}(\omega, \omega_0) = \Lambda^{(4)}(\omega, \omega_0). \tag{34}$$

The set of equations for $\Lambda^{(1,2)}$ takes the form

$$\begin{aligned} \mathbf{p}\Lambda^{(1)}(\omega, \omega_0) &= \frac{n}{(2\pi)^3}\int |u(\mathbf{p} - \mathbf{p}')|^2\,\mathbf{p}'\,\{[G(p'_+)G(p'_-) \\ &- F(p'_+)F(p'_-)] \times [1 + \Lambda^{(1)}(\omega, \omega_0)] \\ &+ [F(p'_+)G(p'_-) + G(p'_+)F(p'_-)]\,\Lambda^{(2)}(\omega, \omega_0)\}\,d^3\mathbf{p}; \\ \mathbf{p}\Lambda^{(2)}(\omega, \omega_0) &= \frac{n}{(2\pi)^3}\int |u(\mathbf{p} - \mathbf{p}')|^2\,\mathbf{p}'\,\{[F(p'_+)G(p'_-) \\ &- G(-p'_+)F(p'_-)]\,[1 + \Lambda^{(1)}(\omega, \omega_0)] \\ &+ [G(-p'_+)G(p'_-) + F(p'_+)F(p'_-)]\,\Lambda^{(2)}(\omega, \omega_0)\}\,d^3\mathbf{p}. \end{aligned} \tag{35}$$

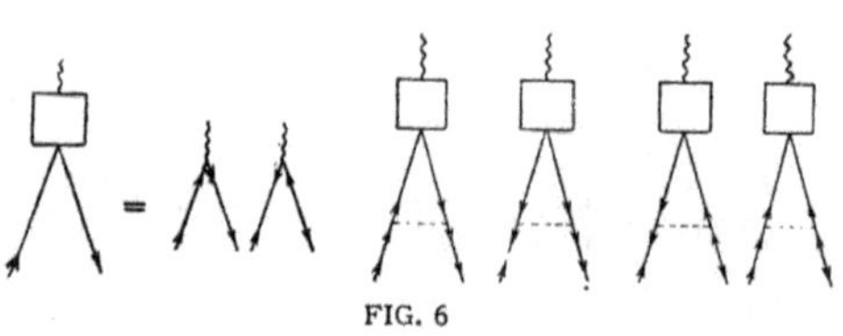

FIG. 6

Substituting into (35) the expressions (30) and (32) for the functions G and F, integrating over ξ, and using the relation (25) we arrive at the following expressions for the quantities of interest to us:

$$\Lambda^{(1)}(\omega,\omega_0)=\frac{i}{2\tau_1}\frac{1-(\omega_+\omega_-+\Delta^2)/\sqrt{\omega_+^2-\Delta^2}\sqrt{\omega_-^2-\Delta^2}}{\sqrt{\omega_+^2-\Delta^2}+\sqrt{\omega_-^2-\Delta^2}+i/\tau_{tr}};$$

$$\Lambda^{(2)}(\omega,\omega_0)=\frac{\Delta}{2\tau_1}\frac{2\omega}{\sqrt{\omega_+^2-\Delta^2}\sqrt{\omega_-^2-\Delta^2}\left[\sqrt{\omega_+^2-\Delta^2}+\sqrt{\omega_-^2-\Delta^2}+i/\tau_{tr}\right]}. \tag{36}$$

Substituting (34) and (36) into (33) we find $\Pi^{(1)}(p_+, p_-)$ and substituting this into Eq. (8) we get an expression for Q:

$$Q_{ik}(\omega_0)=\delta_{ik}\left\{\frac{Ne^2}{mc}-\frac{ie^2N}{2\pi mc}\int d\xi\int_C d\omega[((\omega_+\omega_-+\Delta^2)\eta_+\eta_-+\xi^2)\right.$$

$$\times(1+\Lambda^{(1)}(\omega,\omega_0))+2i\Delta\omega\Lambda^{(2)}(\omega,\omega_0)] \tag{37}$$

$$\times\frac{1}{[\xi^2-(\omega_+^2-\Delta^2)\eta_+^2][\xi^2-(\omega_-^2-\Delta^2)\eta_-^2]}.$$

The contour around the singular points in the integral over ω is given in Fig. 7a for the case $\omega_0 < 2\Delta$ and in Fig. 7c for the case $\omega_0 > 2\Delta$. Before performing the integration (37) we note that the integral of the first term in (33) diverges. One must therefore observe the well-known caution in the problem of the order of integration over ω and ξ. A closer inspection shows that one must first integrate by parts.

We first find the penetration depth of a static magnetic field into a superconductor. The latter is connected with Q_{ik} by the relation

$$\delta=\sqrt{\frac{c}{4\pi Q(0)}};\quad \text{if } Q_{ik}=\delta_{ik}Q). \tag{38}$$

The contour C in Fig. 7a can in the case $\omega_0 = 0$ conveniently be deformed so that the integration over ω is performed along the imaginary axis (this is possible because of the decrease of the expression under the integral sign as $\omega\to\infty$):

$$Q_{ik}(0)=\delta_{ik}Q(0)=\frac{Ne^2}{mc}\delta_{ik}\left\{1\right.$$

$$+\frac{1}{2\pi}\iint_{-\infty}^{\infty}d\xi\,d\omega\,\frac{1+\dfrac{\Delta^2}{2\tau_1(\omega^2+\Delta^2)\left(\sqrt{\omega^2+\Delta^2}+1/2\tau_{tr}\right)}}{\xi^2+(\sqrt{\omega^2+\Delta^2}+1/2\tau_{tr})^2}$$

$$\left.-\frac{1}{\pi}\iint_{-\infty}^{\infty}\frac{d\xi\,d\omega\,\omega^2\left(1+\dfrac{1}{2\tau\sqrt{\omega^2+\Delta^2}}\right)^2}{[\xi^2+(\sqrt{\omega^2+\Delta^2}+1/2\tau_{tr})^2]^2}\right\}.$$

Integrating by parts over ω in the first term of the first integral makes it possible formally to cancel the diverging terms in that expression. In the remaining double integral one can change the order of integration. The integral over ξ is at once performed:

$$Q(0)=\frac{Ne^2}{mc}\left\{1-\frac{\Delta^2}{4\tau_{tr}}\int_{-\infty}^{\infty}\frac{d\omega}{(\omega^2+\Delta^2)^{3/2}(\sqrt{\omega^2+\Delta^2}+1/2\tau_{tr})}\right\}.$$

After that we integrate over ω and find

$$Q(0)=\frac{Ne^2}{mc}\tau_{tr}\Delta\begin{cases}\pi-\dfrac{4\tau_{tr}\Delta}{\sqrt{1-(2\tau_{tr}\Delta)^2}}\cosh^{-1}\dfrac{1}{2\tau_{tr}\Delta}; & 2\tau_{tr}\Delta<1,\\[2ex] \pi-\dfrac{4\tau_{tr}\Delta}{\sqrt{(2\tau_{tr}\Delta)^2-1}}\arccos\dfrac{1}{2\tau_{tr}\Delta}; & 2\tau_{tr}\Delta>1.\end{cases} \tag{39}$$

Substituting this into Eq. (38) we can from this obtain the penetration depth. If $\tau_{tr}\Delta \ll 1$ it is (in the usual units) of the form

$$\delta=\frac{c}{2\pi}\sqrt{\hbar/\sigma\Delta}, \tag{40}$$

where $\sigma = Ne^2\tau_{tr}/m$ is the conductivity of the metal in the normal state. In the case $\tau_{tr}\Delta \gg 1$ the expression for the penetration depth goes over into the usual London formula:

$$\delta=\sqrt{mc^2/4\pi Ne^2}.$$

It is well known[1] that pure superconductors fall into two classes: in one we have $\delta > v/\Delta$ for all temperatures; they can be called London superconductors. In that case there is no limit whatever for the quantity $\tau_{tr}\Delta$. For the other superconductors, which can be called Pippard superconductors, we have $\delta < v/\Delta$ at $T = 0$. In that case from the condition $l \ll \delta$ follows always that $\tau_{tr}\Delta \ll 1$. The Pippard superconductors are thus described by Eq. (40) for $l \ll \delta$. In the case when the mean free path $l \gg \delta$ the theory developed in reference 1 is clearly correct for Pippard superconductors,

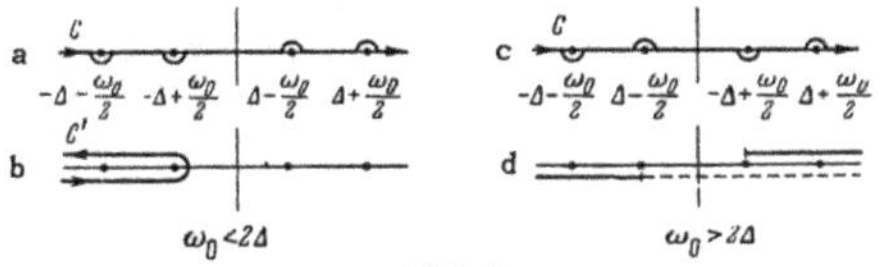

FIG. 7

independent of the relation between l and v/Δ. From this theory we obtain

$$Q(k, 0) = \frac{3\pi^2}{4mc}\frac{Ne^2\Delta}{vk}. \tag{41}$$

Comparison of (40) and (41) shows that Eq. (40) must be used, if

$$\delta / l_{tr} \gg 3\pi / 4. \tag{42}$$

We now proceed to the case of an alternating field. It is of interest to follow the transition from a superconductor to a normal metal as the frequency ω_0 increases from zero to $\omega_0 \gg \Delta$. We consider only the case $\tau_{tr}\Delta \ll 1$. We subtract from the expression under the integral sign its value at $\omega_0 = 0$. The remainder can then be integrated first over ξ. In the case when $\omega_0 < 2\Delta$ we go over to the contour C′ (Fig. 7b) for the integration over ω. We get

$$Q(\omega_0) = \frac{Ne^2}{mc}\left\{\int_{\Delta+\omega_0/2}^{\infty} d\omega\left[1 - \frac{\omega_+\omega_- + \Delta^2}{\sqrt{\omega_+^2 - \Delta^2}\sqrt{\omega_-^2 - \Delta^2}}\right]\right.$$
$$\times \frac{\sqrt{\omega_+^2 - \Delta^2} + \sqrt{\omega_-^2 - \Delta^2}}{\left(\sqrt{\omega_+^2 - \Delta^2} + \sqrt{\omega_-^2 - \Delta^2}\right)^2 + 1/\tau_{tr}^2}$$
$$\left. + \tau_{tr}\int_{\Delta-\omega_0/2}^{\Delta+\omega_0/2} \frac{\omega_+\omega_- + \Delta^2}{\sqrt{\omega_+^2 - \Delta^2}\sqrt{\omega_-^2 - \Delta^2}}\right\}.$$

The contribution from the first integral is of the order $(\tau\Delta)^2$ and we can neglect it. The second integral leads to a complete elliptic integral.[7] The final expression has the following form for the case $\omega_0 < 2\Delta$

$$Q(\omega_0) = \frac{2\sigma\Delta}{c}E\left(\frac{\omega_0}{2\Delta}\right). \tag{43}$$

In particular, we get for $\omega_0 \ll 2\Delta$

$$Q(\omega_0) = \frac{\pi\sigma\Delta}{c}\left[1 - \left(\frac{\omega_0}{4\Delta}\right)^2\right]. \tag{44}$$

In the case $\omega_0 > 2\Delta$ it is convenient to complete the integral to go from $-\infty$ to $+\infty$ in the lower half-plane (Fig. 7c). This part can then be discarded, since there are no poles in the expression under the integral sign. This makes it possible to simplify the expressions considerably. As a result, Q can be expressed as the sum of elliptic integrals, if $\tau\Delta \ll 1$:

$$Q(\omega_0) = \frac{\sigma}{c}\left\{\omega_0 E(\kappa) - \frac{\omega_0^2 - (2\Delta)^2}{\omega_0}K(\kappa)\right.$$
$$\left. - i\left[\omega_0 E(\kappa_1) - \frac{(2\Delta)^2}{\omega_0}K(\kappa_1)\right]\right\}, \tag{45}$$

where $\kappa = 2\Delta/\omega_0$; $\kappa_1 = \sqrt{1-\kappa^2}$.

In particular if $\omega_0 \gg 2\Delta$

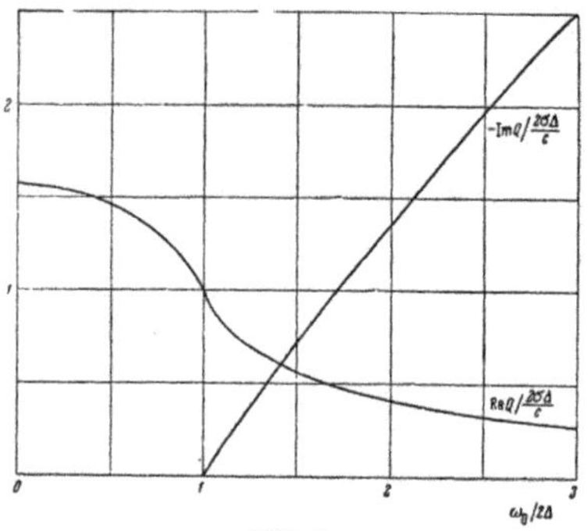

FIG. 8

$$Q(\omega_0) = -\frac{\sigma\omega_0}{c}\left\{i\left[1 - \frac{2\Delta^2}{\omega_0^2}\left(\ln\frac{2\omega_0}{\Delta} + \frac{1}{2}\right)\right] - \frac{\pi\Delta^2}{\omega_0^2}\right\}. \tag{46}$$

From Eq. (46) it is clear that if $\omega_0 \gg \Delta$ the superconducting properties disappear; the usual formula for the normal skin effect turns up. It is of interest that the imaginary part of the current corresponding to absorption occurs only for $\omega_0 > 2\Delta$; the connection of this threshold energy and the binding energy of the pairs is evident. Near the threshold $|\omega_0 - 2\Delta| \ll \Delta$ we have:

$$Q = \frac{\sigma}{c}\left[2\Delta + \left(\Delta - \frac{\omega_0}{2}\right)\left(\ln\frac{16\Delta}{2\Delta - \omega_0} - 1\right)\right]; \quad \omega_0 < 2\Delta, \tag{47}$$
$$Q = \frac{\sigma}{c}\left[2\Delta - \left(\frac{\omega_0}{2} - \Delta\right)\left(\ln\frac{16\Delta}{\omega_0 - 2\Delta} - 1 + i\pi\right)\right]; \quad \omega_0 > 2\Delta.$$

In Fig. 8 we have given the complete curves of the dependence of the real and imaginary parts of Q on ω_0. Along the ordinate axis are put $\mathrm{Re}\left[Q/\frac{2\Delta\sigma}{c}\right]$ and $-\mathrm{Im}\left[Q/\frac{2\sigma\Delta}{c}\right]$ in their dependence on $\omega_0/2\Delta$. The impedance of a bulk specimen with a plane surface is obtained with the aid of Eq. (4) and is of the form

$$Z = -\frac{i\omega_0}{c}\sqrt{\frac{4\pi}{cQ(\omega_0)}}; \tag{48}$$

If Q is complex the radical means the analytical continuation of that quantity which is positive for real Q.

The Eqs. (39), (40), and (43) to (48) describe completely the behavior of alloys in a constant or alternating weak electromagnetic field at $T = 0$. Apart from alloys, these expressions can also be applied to films of a pure substance with a thickness less than the penetration depth. Such films are as a rule polycrystalline and the dimensions of the crystallites are not more than the film thickness. The presence of boundaries between the crystallites will lead to a break-up of the correlation. The local character of the connection in the case $\delta \gg l$ makes it possible to apply the results obtained above to films without any changes. It may then turn out to be convenient to use the complex dielectric susceptibility which is connected with Q through the relation

$$\varepsilon = -4\pi c Q/\omega^2 \tag{49}$$

In conclusion the authors use this opportunity to express their gratitude to Academician L. D. Landau for his constant attention and valuable remarks.

APPENDIX

In the case of the superconducting alloy the equations replacing (14) are easily found by drawing the corresponding diagrams. They are of the form

$$G(\mathbf{p}, \mathbf{p}', \omega) = G_0(p)\,\delta(\mathbf{p}-\mathbf{p}')$$
$$+\frac{1}{(2\pi)^3}\Big\{G_0(p)\int u(\mathbf{p}-\mathbf{p}'')\sum_a e^{i(\mathbf{p}-\mathbf{p}'')\mathbf{x}_a}\,G(\mathbf{p}'', \mathbf{p}', \omega)\,d^3\mathbf{p}''$$
$$+F_0(p)\int u(\mathbf{p}-\mathbf{p}'')\sum_a e^{i(\mathbf{p}-\mathbf{p}'')\mathbf{x}_a}F^+(\mathbf{p}'', \mathbf{p}', \omega)\,d^3\mathbf{p}''\Big\};$$

$$F^+(\mathbf{p}, \mathbf{p}', \omega) = F_0^+(p)\,\delta(\mathbf{p}-\mathbf{p}') \tag{A.1}$$
$$+\frac{1}{(2\pi)^3}\Big\{F_0(p)\int u(\mathbf{p}-\mathbf{p}'')\sum_a e^{i(\mathbf{p}-\mathbf{p}'')\mathbf{x}_a}\,G(\mathbf{p}'', \mathbf{p}', \omega)\,d^3\mathbf{p}''$$
$$+G_0(-p)\int u(\mathbf{p}-\mathbf{p}'')\sum_a e^{i(\mathbf{p}-\mathbf{p}'')\mathbf{x}_a}F^+(\mathbf{p}'', \mathbf{p}', \omega)\,d^3\mathbf{p}''\Big\}$$

In principle it is necessary to consider also the equations for the function $-I_{\alpha\beta}F(x, x') = \langle T(\psi_\alpha(x)\psi_\beta(x'))\rangle$.

For a pure superconductor $F(x, x') = F^+(x, x')$. This property will be true also for alloys after Eq. (A.1) is averaged over the positions of the impurity atoms.

The averaging is performed exactly as in Sec. 1. Using the equations[4] that are satisfied by the functions $G_0(p)$ and $F_0(p)$, this set can be put into a very simple form:

$$(\omega-\xi-\overline{G_\omega})\,G(p)+(i\Delta-\overline{F_\omega^+})\,F^+(p)=1,$$
$$(\omega+\xi-\overline{G_\omega})\,F^+(p)-(i\Delta-\overline{F_\omega^+})\,G(p)=0.$$

The solution of this set is

$$G(p)=\frac{\omega-\overline{G_\omega}+\xi}{(\omega-\overline{G_\omega})^2+(i\Delta-\overline{F_\omega^+})^2-\xi^2};$$
$$F^+(p)=\frac{i\Delta-\overline{F_\omega^+}}{(\omega-\overline{G_\omega})^2+(i\Delta-\overline{F_\omega^+})^2-\xi^2}. \tag{A.2}$$

Remembering the definition of $\overline{G}_\omega$ we see that as before $\overline{G}_\omega$ contains a constant term which means a renormalization of the chemical potential and which arises from the integration far from the Fermi surface:

$$\delta\mu=\frac{n}{(2\pi)^3}\int\frac{|u(\mathbf{p}-\mathbf{p}')|^2\,\xi d\xi}{(\omega-G_\omega)^2+(i\Delta-F_\omega^+)^2-\xi^2}.$$

After subtracting this term, the remaining part of $\overline{G}_\omega$ is apart from a coefficient determined by the same integral as $\overline{F_\omega^+}$, as is clear from (A.2). In this way

$$\overline{G_\omega}/\omega=\overline{F_\omega^+}/\Delta.$$

After this it is easy to find that the solution of the set (A.2) gives Eqs. (30) and (32).

[1] Bardeen, Cooper, and Schrieffer, Phys. Rev. **108**, 1175 (1957).
[2] L. N. Cooper, Phys. Rev. **104**, 1189 (1956).
[3] Abrikosov, Gor'kov, and Khalatnikov, J. Exptl. Theoret. Phys. (U.S.S.R.) **35**, 265 (1958), Soviet Phys. JETP **8**, 182 (1959); D. C. Mattis and J. Bardeen, Phys. Rev. **111**, 412 (1958).
[4] L. P. Gor'kov, J. Exptl. Theoret. Phys. (U.S.S.R.) **34**, 735 (1958), Soviet Phys. JETP **7**, 505 (1958).
[5] N. N. Bogoliubov, J. Exptl. Theoret. Phys. (U.S.S.R.) **34**, 58, 73 (1958), Soviet Phys. JETP **7**, 41, 51 (1958); V. V. Tolmachev and S. V. Tiablikov, J. Exptl. Theoret. Phys. (U.S.S.R.) **34**, 66 (1958), Soviet Phys. JETP **7**, 46 (1958).
[6] S. T. Beliaev, J. Exptl. Theoret. Phys. (U.S.S.R.) **34**, 417 (1958), Soviet Phys. JETP **7**, 289 (1958).
[7] E. Jahnke and F. Emde, Tables of Functions, Dover, New York, 1945.

Translated by D. ter Haar
322

SUPERCONDUCTING ALLOYS AT FINITE TEMPERATURES

A. A. ABRIKOSOV and L. P. GOR' KOV

Institute for Physical Problems, Academy of Sciences, U.S.S.R.

Submitted to JETP editor July 16, 1958

J. Exptl. Theoret. Phys. (U.S.S.R.) **36**, 319-320 (January, 1959)

WE have recently[1] formulated the electrodynamics of superconductors that contain small atomic concentrations of impurities at T = 0. The method of reference 1 is not applicable at finite temperatures.

The authors of the present paper obtained, together with I. E. Dzyaloshinskii, a generalization of the method applied at T = 0. It was basically founded upon the formulation of the thermodynamic theory proposed by Matsubara.[2] An exposition of this method will be given in a separate paper. We note here only that the evaluation of the basic functions at $T \neq 0$ is formally very similar to the calculations at T = 0. In the case of equilibrium practically the only change involved is the replacement of the integrals over the frequency by sums over the discrete variable

$$\frac{1}{2\pi}\int_{-\infty}^{\infty} d\omega f(\omega) \to \frac{iT}{\hbar}\sum_{n=-\infty}^{\infty} f(i\omega_n), \quad (1)$$

where $\omega_n = (\pi T/\hbar)(2n + 1)$, and T is the temperature in energy units.

This method was applied by us to study the equilibrium properties of superconducting alloys at finite temperatures. All calculations are exactly the same as the evaluations performed[1] for T = 0 apart from the transformation (1).

The evaluation of the functions G(x, x′) and F(x, x′) leads to the conclusion that in alloys, as at T = 0, these functions are simply multiplied by an exponential factor

$$G(x, x') = G_0(x, x')\exp\{-|\mathbf{x}-\mathbf{x}'|/2l\},$$
$$F(x, x') = F_0(x, x')\exp\{-|\mathbf{x}-\mathbf{x}'|/2l\}, \quad (2)$$

where l is the mean free path in the normal state.

To find the thermodynamic functions it is sufficient to determine the particle density N(μ, T) as a function of the chemical potential and temperature:

$$N = \langle\psi^+(x)\psi(x)\rangle = -i\,[G(x, x')]_{x=x';\,t=t'-0}. \quad (3)$$

One sees easily that the function N(μ, T) is the same as in the case of a pure superconductor. In the model under consideration, where one assumes that the impurities do not influence the interaction between the electrons, but only scatter them, the occurrence of impurities will therefore not change the thermodynamic functions. In particular, the critical temperature T_c is not shifted. We emphasize once again that this result is, of course, valid only if the impurity concentration is supposed to be small.

We have considered the behavior of alloys in a constant magnetic field. The connection between the current and the vector potential is local in the London case. At finite temperatures it is of the form

$$\mathbf{j} = -Q\mathbf{A}, \quad (4)$$

$$Q = \frac{Ne^2}{mc}2\pi T\Delta^2\sum_0^\infty 1/[(\hbar\omega_n)^2+\Delta^2]\left(\sqrt{(\hbar\omega_n)^2+\Delta^2}+\hbar/2\tau_{tr}\right) \quad (5)$$

where $\omega_n = (\pi T/\hbar)(2n+1)$; Δ(T) the temperature dependent "gap" occurring in the paper by Bardeen, Cooper, and Schrieffer;[3] $\tau_{tr} = [\int w(\theta)(1-\cos\theta)\,d\Omega]^{-1}$ is the transport collision time for the metal in its normal state, and N is the electron density.

From Eq. (4) together with the Maxwell equation one obtains the connection between Q and the penetration depth:

$$\delta = \sqrt{c/4\pi Q}. \quad (6)$$

In the limiting case of large free paths, $\Delta\tau_{tr}/\hbar \gg 1$, this formula goes over into the usual result for a pure London superconductor, i.e., a superconductor with $\delta \gg \hbar v/\Delta$:

$$\delta = \sqrt{mc^2/4\pi N_s e^2}, \quad (7)$$

where N_s is the number of superconducting electrons (see Ref. 3),

$$N_s/N_0 = 2\pi T\Delta^2\sum_n[(\hbar\omega_n)^2+\Delta^2]^{-3/2} = \frac{1}{4T}\int_\Delta^\infty\frac{\varepsilon\,d\varepsilon}{\cosh^2\frac{\varepsilon}{2T}\sqrt{\varepsilon^2-\Delta^2}}. \quad (8)$$

In the opposite limiting case, $\Delta\tau_{tr}/\hbar \ll 1$,

$$\delta = (c/2\pi)\,[\hbar/\Delta\sigma\tanh(\Delta/2T)]^{1/2}, \quad (9)$$

where $\sigma = Ne^2\tau_{tr}/m$ is the conductivity of the alloy in the normal state.

It is well known that superconductors fall into two classes. For London superconductors $\delta > \hbar v/\Delta$. In that case the connection between Δ and $\hbar/\tau_{tr}$ can be arbitrary. In the case of the Pippard superconductors, for which $\delta < \hbar v/\Delta$, the condition $\delta \gg l$ leads to $\Delta\tau_{tr}/\hbar \ll 1$. Hence, only Eq. (9) refers to Pippard superconductors.

As was noted in reference 1, all formulae obtained for δ can by analogy with this be applied for the characteristics of films of thickness $d \ll \delta$.

In that case the quantity δ no longer plays the role of the penetration depth, but this quantity goes over into the effective dielectric constant for low frequencies and determines the magnetic susceptibility. Since the connection between $\mathbf{j}$ and $\mathbf{A}$ is local, ϵ and χ, expressed in terms of δ, have the usual, London form:

$$\varepsilon = -c^2/\omega^2\delta^2, \tag{10}$$

$$\chi = \frac{1}{4\pi}\left(1 - \frac{2\delta}{d}\tanh\frac{d}{2\delta}\right) \approx -\frac{1}{12\pi}\left(\frac{d}{2\delta}\right)^2. \tag{11}$$

In conclusion the authors express their gratitude to Academician L. D. Landau for discussions of this paper.

[1] A. A. Abrikosov and L. P. Gor'kov, J. Exptl. Theoret. Phys. (U.S.S.R.) **35**, 1558 (1958), Soviet Phys. JETP **8**, 1090 (1959).

[2] T. Matsubara, Progr. Theoret. Phys. **14**, 351 (1955).

[3] Bardeen, Cooper, and Schrieffer, Phys. Rev. **108**, 1175 (1957).

Translated by D. ter Haar
48

SOVIET PHYSICS JETP VOLUME 36 (9), NUMBER 3 SEPTEMBER, 1959

ON THE APPLICATION OF QUANTUM-FIELD-THEORY METHODS TO PROBLEMS OF QUANTUM STATISTICS AT FINITE TEMPERATURES

A. A. ABRIKOSOV, L. P. GOR' KOV, and I. E. DZYALOSHINSKIĬ

Institute of Physical Problems, Academy of Sciences, U.S.S.R.

Submitted to JETP editor December 4, 1958

J. Exptl. Theoret. Phys. (U.S.S.R.) **36**, 900-908 (March, 1959)

A formulation of thermodynamic perturbation theory is proposed which makes it possible to make full use of quantum-field-theory methods in quantum statistics at finite temperatures. The method is a generalization of the Matsubara technique and is based on the expansion of the Green's functions in Fourier series in the "imaginary time" variable. The technique so obtained differs from the usual diagram technique for $T = 0$ by the replacement of integration over the frequencies by summation over discrete values of the imaginary frequency. The analytic properties of the Fourier components of the Green's functions are examined. It is shown that owing to the possibility of analytic continuation a knowledge of the corresponding equilibrium Green's functions is sufficient for the solution of various kinetic and nonstationary processes.

1. INTRODUCTION

THE methods of quantum field theory have recently been successfully applied to problems of statistics.[1-4] The application is based on the fact that the apparatus of quantum field theory is developed through the wide use of the "diagram technique," which provides a very intuitive representation of the structure and character of any approximation. Earlier methods of quantum statistics, which started from ideal-gas approximations, could scarcely take the interaction between the particles into account beyond one or two approximations because of the complexity of the older quantum perturbation theory. This is utterly inadequate for the actual many-body problem; to obtain any approximation of physical significance it is necessary to sum over an infinite set of different terms of the perturbation-theory series. The diagram technique is extraordinarily useful in such problems, since it formulates simple rules by which any term of the perturbation theory can be written down.

The basis of the quantum-field-theory methods is the calculation of the so-called Green's function of a particle, which for the case of temperatures different from zero is defined as

$$G(1,2) = -i\,\mathrm{Sp}\,\{\exp[(\Omega + \mu N - \hat{H})/T]\cdot T(\tilde{\psi}(x_1)\tilde{\psi}^+(x_2))\}, \quad (1)$$

where $\tilde{\psi}$, $\tilde{\psi}^+$ are second-quantization operators in the Heisenberg representation. Knowledge of the Green's function makes it possible easily to calculate all the thermodynamic quantities of the system. For example, the particle number density as a function of the chemical potential μ and the temperature is connected with G by the relation

$$N(\mu, T) = \mp iG(x, x)|_{t'=t+0}$$

(the minus sign is for Fermi statistics, the plus for Bose statistics).

At the absolute zero of temperature the Green's function is calculated by going over to the so-called "interaction representation." In this representation G(1, 2) has the form

$$G(1,2) = -i\,\langle T(\psi_1\psi_2^+ S)\rangle / \langle S\rangle, \quad (1a)$$

where $\langle\ldots\rangle$ means the average over the ground state of the system, and S is the well known S matrix of the quantum field theory. Expanding S in powers of the interaction constant, we obtain the usual Feynman diagram technique.

For temperatures other than zero it is already impossible to represent G in a form like Eq. (1a), and consequently it is impossible to calculate G by means of the diagram technique.

Several years ago Matsubara[4] proposed a new formulation of the thermodynamic perturbation theory which is almost completely analogous to the diagram technique of field theory. The present paper is devoted to the description of a technique which is a further development of the Matsubara method. This technique makes it possible to calculate the thermodynamic quantities at finite temperatures with almost the same simplicity with which this is done for $T = 0$. It will further be shown that by establishing the connection of the quantities that figure in this technique with the corresponding quantities in the usual technique one can determine all the kinetic characteristics of the system.*

2. THE MATSUBARA METHOD

The basic idea of the Matsubara method is the passage from the time t to a certain imaginary "time" τ. We note that the ordinary Green's function $G(1, 2)$ can be represented in the form

$$G(1,2)=\begin{cases} -i\mathrm{Sp}\left\{\exp\left[\frac{\Omega+\mu N-\hat{H}}{T}\right]\right. \\ \times\exp\left[\frac{i(\hat{H}-\mu N)}{\hbar}(t_1-t_2)\right]\psi(\mathbf{r}_1) \\ \left.\times\exp\left[-\frac{i(\hat{H}-\mu N)}{\hbar}(t_1-t_2)\right]\psi^+(\mathbf{r}_2)\right\},\ t_1>t_2, \\ \pm i\mathrm{Sp}\left\{\exp\left[\frac{\Omega+\mu N-\hat{H}}{T}\right]\right. \\ \times\exp\left[-\frac{i(\hat{H}-\mu N)}{\hbar}(t_1-t_2)\right]\psi^+(\mathbf{r}_2) \\ \left.\times\exp\left[\frac{i(\hat{H}-\mu N)}{\hbar}(t_1-t_2)\right]\psi(\mathbf{r}_1)\right\},\ t_1<t_2, \end{cases} \tag{2}$$

where ψ, ψ^+ are the second-quantization operators in the Schrödinger representation.

In the Matsubara method one introduces a temperature-dependent Green's function $\mathfrak{G}$, which is obtained from G by the replacement $t \to -i\tau\hbar$:

$$\mathfrak{G}(1,2)=\begin{cases} -\mathrm{Sp}\left\{\exp\left[\frac{\Omega+\mu N-\hat{H}}{T}\right]\right. \\ \times\exp\left[(\hat{H}-\mu N)(\tau_1-\tau_2)\right]\psi(\mathbf{r}_1) \\ \left.\times\exp\left[-(\hat{H}-\mu N)(\tau_1-\tau_2)\right]\psi^+(\mathbf{r}_2)\right\},\ \tau_1>\tau_2, \\ \pm\mathrm{Sp}\left\{\exp\left[\frac{\Omega+\mu N-\hat{H}}{T}\right]\right. \\ \times\exp\left[-(\hat{H}-\mu N)(\tau_1-\tau_2)\right]\psi^+(\mathbf{r}_2) \\ \left.\times\exp\left[(\hat{H}-\mu N)(\tau_1-\tau_2)\right]\psi(\mathbf{r}_1)\right\},\ \tau_1<\tau_2. \end{cases} \tag{3}$$

Matsubara showed that in the formula (3) one can go over to a sort of "interaction representation". Namely, let us represent the statistical matrix $\hat{\rho} = \exp\{(\mu N - \hat{H})/T\}$ in the form

$$\exp\left[\frac{\mu N-\hat{H}}{T}\right]=\exp\left[\frac{\mu N-\hat{H}_0}{T}\right]\cdot\hat{S}(1/T) \tag{4}$$

and introduce the new operators $\bar{\psi}(\mathbf{r}, \tau)$, $\psi(\mathbf{r}, \tau)$ in the "interaction representation" by the formulas

$$\psi(\mathbf{r},\tau)=\exp\left[\tau(\hat{H}_0-\mu N)\right]\psi(\mathbf{r})\exp\left[-\tau(\hat{H}_0-\mu N)\right],$$
$$\bar{\psi}(\mathbf{r},\tau)=\exp\left[\tau(\hat{H}_0-\mu N)\right]\psi^+(\mathbf{r})\exp\left[-\tau(\hat{H}_0-\mu N)\right]. \tag{5}$$

Then, using the formal resemblance of the equation satisfied by the matrix $\hat{S}(1/T)$, $-\partial\hat{S}(\tau)/\partial\tau = H(\tau)\hat{S}(\tau)$, with the Schrödinger equation, we write $\hat{S}(1/T)$, in complete analogy with the usual S matrix, in the form

$$\hat{S}(1/T)=T_\tau\exp\left\{-\int_0^{1/T}\hat{H}_{\mathrm{int}}(\tau)\,d\tau\right\},$$

where T_τ is the time-ordering operator with respect to the imaginary "time" τ. The interaction Hamiltonian $H_{\mathrm{int}}(\tau)$ now depends on the operators $\psi(\mathbf{r}, \tau)$ and $\bar{\psi}(\mathbf{r}, \tau)$ in the same way that it formerly depended on $\psi(\mathbf{r})$ and $\psi^+(\mathbf{r})$.

Now, using the possibility of cyclic permutation of the operators under the sign Sp in Eq. (2), we can show that $\mathfrak{G}$ can be written in a form analogous to that used in the field theory:

$$\mathfrak{G}(1,2)=-\frac{\mathrm{Sp}\{\exp[(\mu N-\hat{H}_0)/T]T_\tau[\psi_1\bar{\psi}_2\hat{S}(1/T)]\}}{\mathrm{Sp}\{\exp[(\mu N-\hat{H}_0)/T]\hat{S}(1/T)\}}, \tag{6}$$

It is not hard to see that for the potential Ω we can write the following formula:

$$\Omega=\Omega_0-T\ln\left[\frac{\mathrm{Sp}\left\{\exp\left[\frac{\mu N-\hat{H}_0}{T}\right]\hat{S}(1/T)\right\}}{\mathrm{Sp}\left\{\exp\left[\frac{\mu N-\hat{H}_0}{T}\right]\right\}}\right]. \tag{6a}$$

The expressions (6) and (6a) have the forms of the corresponding formulas of field theory:

$$\mathfrak{G}(1,2)=-\langle T(\psi_1\bar{\psi}_2 S)\rangle/\langle S\rangle,\ \Omega=\Omega_0-T\ln\langle S\rangle, \tag{7}$$

if we understand the averaging symbol $\langle\ldots\rangle$ to mean the operation

$$\mathrm{Sp}\left\{\exp\left[\frac{\mu N-\hat{H}_0}{T}\right]\ldots\right\}\Big/\mathrm{Sp}\left\{\exp\left[\frac{\mu N-\hat{H}_0}{T}\right]\right\}.$$

Matsubara showed that for such averages of T-products of several ψ-operators there exists a "Wick's theorem," according to which the average in question breaks up into a sum of products of averages by pairs of operators:

*We have learned that similar results have been obtained by E. S. Fradkin (J. Exptl. Theoret. Phys. (U.S.S.R.), in press).

$$\mathfrak{G}_0(1,2) = -\langle T_\tau(\psi_1\bar{\psi}_2)\rangle.$$

Of course this does not apply to Bose systems below the point of the Bose condensation. Similarly, the Wick's theorem is not applicable to Fermi systems in which superconductivity exists. For the construction of a thermodynamic perturbation theory in this range of temperatures one must use in the first case an obvious generalization of some work of Belyaev,[5] and in the second case, work of Gor'kov.[6]

Expanding the expression for $S(1/T)$ appearing in Eq. (7) in a series of powers of the interaction Hamiltonian and using Wick's theorem, we get the usual diagram technique, in which each line corresponds to a zeroth-order Green's function $\mathfrak{G}_0(1, 2)$. For example, in the case of the four-fermion interaction with

$$\hat{H}_{\rm int} = \frac{g}{2}\Lambda_{\alpha\beta;\gamma\delta}\bar{\psi}_\alpha(x)\bar{\psi}_\beta(x)\psi_\gamma(x)\psi_\delta(x),$$

where $x = \{\mathbf{r}, \tau\}$ and $\Lambda_{\alpha\beta;\gamma\delta}$ is some function of the spin indices antisymmetric in $\alpha\beta$ and $\gamma\delta$,

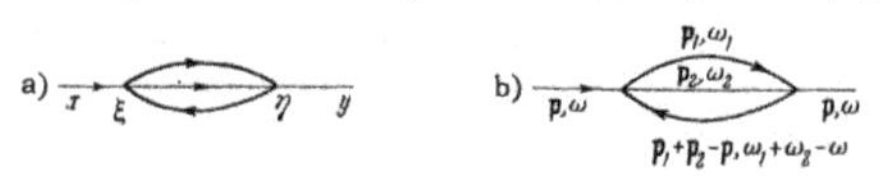

FIG. 1

we find for the second-approximation correction to the Green's function, corresponding to the diagram of Fig. 1a

$$2g^2\Lambda_{\lambda\mu\nu\tau}\Lambda_{\rho\pi\kappa\omega}\int_0^{1/T} d\tau_\xi \int_0^{1/T} d\tau_\eta \int d\xi \int d\eta\, \mathfrak{G}^0_{\alpha\lambda}(x-\xi)$$
$$\times\, \mathfrak{G}^0_{\nu\rho}(\xi-\eta)\,\mathfrak{G}^0_{\tau\pi}(\xi-\eta)\mathfrak{G}^0_{\kappa\mu}(\eta-\xi)\,\mathfrak{G}^0_{\omega\beta}(\eta-y). \quad (8)$$

In the case of the electron-phonon interaction[1,4]

$$\hat{H}_{\rm int} = g\bar{\psi}\psi\varphi,$$

the first correction to the electron Green's function, corresponding to the diagram of Fig. 2a, has

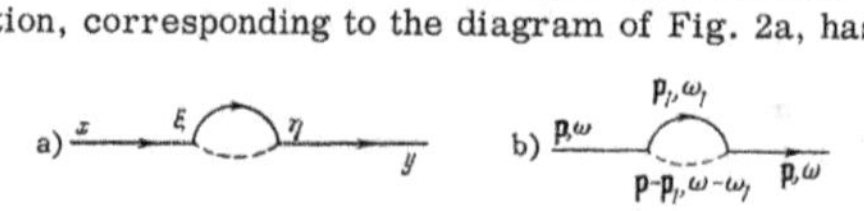

FIG. 2

the form (the dotted line indicates the phonon Green's function $\mathfrak{D}_0 = -\langle T_\tau(\varphi_1\varphi_2)\rangle$, which is the analogue of the ordinary D function of the phonon):

$$-g^2\int_0^{1/T} d\tau_\xi \int_0^{1/T} d\tau_\eta \int d\xi \int d\eta\, \mathfrak{G}^0$$
$$\times (x-\xi)\,\mathfrak{G}^0(\xi-\eta)\,\mathfrak{D}^0(\xi-\eta)\,\mathfrak{G}^0(\eta-y). \quad (9)$$

The actual calculation of expressions of types (8), (9) and of the more complicated expressions in the higher approximations is, however, very laborious in the Matsubara technique. The point is that the success of field-theory methods in statistics for $T = 0$ is due to the highly automatic nature of the calculations, which is achieved by the use of Fourier expansions of all the quantities with respect to all four coordinates. It is obvious that in the Matsubara method the automatic feature is lacking because of the finiteness of the range of τ, from zero to $1/T$. Indeed in the coordinate representations (as regards τ) $\mathfrak{G}^0$ (and $\mathfrak{G}$) is a discontinuous function of the variable τ; therefore in fact all the integrals over τ break up into integrals over a very large number of regions. The number of such regions obviously increases rapidly with increase of the order of the approximation.

3. THE EXPANSION IN FOURIER SERIES

The Matsubara technique can be decidedly improved by the exploitation of certain general properties of the thermodynamic Green's functions.

As we have shown, by Eq. (3) the Green's function $\mathfrak{G}(1, 2)$ is a function of the difference $\tau_1 - \tau_2$, specified in the interval from $-1/T$ to $1/T$. It is therefore expedient to continue it periodically and expand it in Fourier series in the variable τ:*

$$\mathfrak{G}(\tau) = T\sum_n e^{-i\omega_n\tau}\mathfrak{G}(\omega_n),\; \omega_n = \pi T n,$$

$$\mathfrak{G}(\omega_n) = \frac{1}{2}\int_{-1/T}^{1/T} e^{i\omega_n\tau}\mathfrak{G}(\tau)\,d\tau. \quad (10)$$

One major property of the function $\mathfrak{G}$ is essential for the transformation of the perturbation-theory series. It follows from Eq. (3) that the function $\mathfrak{G}$ for negative values of τ is simply related to the $\mathfrak{G}$ for $\tau > 0$, namely we have the relation

$$\mathfrak{G}(\tau) = \mp\mathfrak{G}(\tau + 1/T),\; \tau < 0, \quad (11)$$

where the minus sign is for Fermi statistics and the plus for Bose statistics. The formula (11) is easily derived if we use the fact that one may cyclically permute the operators under the sign of the trace. The relation (11) is of course also valid for the free Green's functions. For the functions as periodically continued this relation is valid for any value of τ.

*Recently I. M. Khalatnikov called our attention to the fact that the idea of expanding the temperature-dependent Green's functions of Matsubara in Fourier series is also contained in a paper by Ezawa, Tomozawa, and Umezawa.[9] These writers, however, being interested only in the application of the Matsurbara method to the problem of multiple production, confined themselves to the case of zero chemical potention. Meanwhile the construction of the technique is nontrivial just in the case $\mu \neq 0$.

If we note further that at each vertex of a Feynman diagram an even number of fermion lines come together, it is easy to see that all the integrals $\int_0^{1/T} \ldots d\tau$ in formulas of the type of Eqs. (8) and (9) can be replaced by $\frac{1}{2}\int_{-1/T}^{1/T} \ldots d\tau$. When this is done the Fourier transformation (10) is easily carried out in all terms of the perturbation-theory series.

The relation (11) also has the consequence that in the Fourier expansion of a fermion Green's function only the components with $\omega_n = \pi T(2n+1)$ are different from zero, and the expansion of a boson Green's function has only components with $\omega_n = \pi T \cdot 2n$. We shall present the expressions for the zeroth-order Green's functions $\mathfrak{G}_0(\omega_n, \mathbf{p})$ (here the Fourier integral transformation for the space coordinates has also been carried out).

Fermions:

$$\mathfrak{G}_0(\omega_n, \mathbf{p}) = [i\omega_n + \mu - \varepsilon(\mathbf{p})]^{-1}, \qquad \omega_n = (2n+1)\pi T$$

(here $\epsilon(\mathbf{p}) = p^2/2m$, and μ is the chemical potential); bosons above the Bose condensation point:

$$\mathfrak{G}_0(\omega_n, \mathbf{p}) = [i\omega_n + \mu - \varepsilon(\mathbf{p})]^{-1}, \qquad \omega_n = 2n\pi T;$$

phonons:

$$\mathfrak{D}_0(\omega_n, \mathbf{k}) = -\omega^2(\mathbf{k}) / [\omega^2(\mathbf{k}) + \omega_n^2], \quad \omega_n = 2n\pi T$$

($\omega(\mathbf{k})$ is the energy of a phonon).

Carrying out the Fourier transformation in all the terms of the expansion of the Green's function in the perturbation-theory series, we can verify that the technique so obtained is entirely equivalent to the diagram technique in the momentum space at absolute zero. To each line of a diagram there corresponds a zeroth-order Green's function $\mathfrak{G}_0(\omega_n, \mathbf{p})$, and to each vertex a δ function expressing the conservation laws $\Sigma p = 0$, $\Sigma\omega_n = 0$. The only difference is that in our case we have instead of an integration over frequency a summation over discrete imaginary "frequencies" $i\omega_n$. For example, the formula (8) for the correction to the electron Green's function from the four-fermion interaction is transformed into the form

$$\Lambda_{\alpha\mu\nu\tau}\Lambda_{\nu\tau\mu\beta}\frac{2g^2T^2}{(2\pi)^6}\sum_{\omega_1,\omega_2}\int d^3\mathbf{p}_1 d^3\mathbf{p}_2\,[i\omega_1 + \mu - \varepsilon(\mathbf{p}_1)]^{-1}$$
$$\times [i\omega_2 + \mu - \varepsilon(\mathbf{p}_2)]^{-1}[i(\omega_1 + \omega_2 - \omega) + \mu - \varepsilon(\mathbf{p}_1 + \mathbf{p}_2 - \mathbf{p})]^{-1}$$
$$\times [i\omega + \mu - \varepsilon(\mathbf{p})]^{-2},$$

corresponding to the diagram of Fig. 1b.

The expression (9) takes the form

$$\delta\mathfrak{G}(\omega, \mathbf{p}) = \frac{g^2T}{(2\pi)^3}\Big\{\sum_{\omega'}\int d^3\mathbf{p}'\,[i\omega' + \mu - \varepsilon(\mathbf{p}')]^{-1}\omega^2(\mathbf{p} - \mathbf{p}')$$
$$\times [\omega^2(\mathbf{p} - \mathbf{p}') + (\omega - \omega')^2]^{-1}\Big\}[i\omega + \mu - \varepsilon(\mathbf{p})]^{-2},$$

which corresponds to the diagram of Fig. 2b.

On passage to the absolute zero of temperature, large values of n play the main part in the sums. Therefore in the perturbation-theory formulas we must make the replacement

$$T\sum_{\omega_n} \rightarrow \frac{1}{2\pi}\int_{-\infty}^{\infty} \ldots d\omega.$$

Furthermore, it is obvious that when one replaces all frequencies by imaginary values, $\omega \rightarrow -i\omega'$, integrals with respect to $d\omega$ from $-\infty$ to $+\infty$ of products of several factors of the form

$$(i\omega + \mu - \varepsilon(\mathbf{p}))^{-1}$$

go over into integrals of products of factors

$$(\omega' + \mu - \varepsilon(\mathbf{p}))^{-1}, \tag{12}$$

which corresponds to a rotation of the path of integration to real ω. In the case of Fermi Green's functions $\epsilon(p) - \mu$ can be either positive or negative. Therefore after the indicated transformation of the contour of integration the path around the poles is chosen as shown in Fig. 3a. For Bose particles the chemical potential μ of an ideal gas

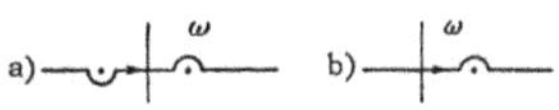

FIG. 3

is always negative; therefore the poles in Eq. (11) lie in the lower half-plane of ω' (Fig. 3b). Conversely, it can be seen from this how from the well known rules of the diagram technique for $T = 0$ one can write down the corresponding term of the perturbation theory for $T \neq 0$. To do this we first make the replacement $\omega' \rightarrow i\omega_n$ ($n = (2m+1)$ for Fermi particles and $n = 2m$ for Bose systems) in the corresponding expressions for $T = 0$, and must then go from integration over the frequencies to summation:

$$\int_{-\infty}^{+\infty} d\omega \rightarrow -2\pi T i \sum_{\omega_n}.$$

Everything said up to now has referred to calculations by perturbation theory. In quite a number of problems, however, calculations by perturbation theory have been found to be inadequate. For these purposes it is essential to know certain exact equations connecting the quantities with each other. One such relation is the Dyson equation. As is well known, in field theory one can write an equation connecting the Green's function with the so-called vertex part (the Dyson equation). The analogue of this equation can of course be obtained in the technique proposed here. For the case of the four-fermion interaction it is shown schematically, in the language of the diagrams, in Fig. 4. Here heavy

FIG. 4

lines mean the complete Green's function $\mathfrak{G}$, and the rhomb means the complete vertex part, i.e., the sum of all connected diagrams having four external fermion lines. In the momentum representation the Dyson equation corresponding to this diagram has the form:

$$\mathfrak{G}^{-1}_{\alpha\beta}(\omega, \mathbf{p}) = \mathfrak{G}^{0-1}_{\alpha\beta}(\omega, \mathbf{p}) - \frac{2gT}{(2\pi)^3}\sum_{\omega'}\int d^3\mathbf{p}\,\mathfrak{G}_{\nu\mu}(\omega', \mathbf{p}')\,\Lambda_{\alpha\mu;\nu\beta}$$
$$-\frac{2gT^2}{(2\pi)^6}\Lambda_{\alpha\lambda,\mu\nu}\sum_{\omega_1\omega_2}\int d^3\mathbf{p}_1 d^3\mathbf{p}_2\,\mathfrak{G}_{\mu\tau}(\omega_1, \mathbf{p}_1)\,\mathfrak{G}_{\nu\rho}(\omega_2, \mathbf{p}_2)$$
$$\times\mathfrak{G}_{\varkappa\lambda}(\omega_1+\omega_2-\omega; \mathbf{p}_1+\mathbf{p}_2-\mathbf{p})$$
$$\times\mathfrak{T}_{\tau\rho,\varkappa\beta}(\omega_1, \mathbf{p}; \omega_2, \mathbf{p}_2; \omega_1+\omega_2-\omega, \mathbf{p}_1+\mathbf{p}_2-\mathbf{p}; \omega, \mathbf{p}).$$

Here $\mathfrak{T}$ is the complete vertex part already mentioned. It is connected with the Fourier components of the two-particle Green's function

$$\mathfrak{G}_{\alpha\beta,\gamma\delta}(x_1, x_2; x_3, x_4) = \langle T(\psi_\alpha(x_1)\psi_\beta(x_2)\bar{\psi}_\gamma(x_3)\bar{\psi}_\delta(x_4)\hat{S})\rangle/\langle\hat{S}\rangle$$

by the relation

$$\mathfrak{G}_{\alpha\beta,\gamma\delta}(\omega_1, \mathbf{p}_1, \omega_2, \mathbf{p}_2; \omega_3, \mathbf{p}_3, \omega_4, \mathbf{p}_4)$$
$$= \frac{(2\pi)^3}{T}\left\{\frac{(2\pi)^3}{T}\left[\mathfrak{G}_{\alpha\delta}(\omega_1, \mathbf{p}_1)\mathfrak{G}_{\beta\gamma}(\omega_2, \mathbf{p}_2)\right.\right.$$
$$\left.\times\delta_{\omega_1\omega_4}\delta(\mathbf{p}_1-\mathbf{p}_4) - \mathfrak{G}_{\alpha\gamma}(\omega_1, \mathbf{p}_1)\,\mathfrak{G}_{\beta\delta}(\omega_2, \mathbf{p}_2)\,\delta_{\omega_1\omega_3}\delta(\mathbf{p}_1-\mathbf{p}_3)\right]$$
$$-\frac{1}{2}\mathfrak{G}_{\alpha\lambda}(\omega_1, \mathbf{p}_1)\,\mathfrak{G}_{\beta\mu}(\omega_2, \mathbf{p}_2)\,\mathfrak{T}_{\lambda\mu,\nu\tau}(\omega_1, \mathbf{p}_1; \omega_2, \mathbf{p}_2; \omega_3, \mathbf{p}_3; \omega_4, \mathbf{p}_4)$$
$$\left.\times\mathfrak{G}_{\nu\gamma}(\omega_3, \mathbf{p}_3)\mathfrak{G}_{\tau\delta}(\mathbf{p}_4, \omega_4)\right\}\delta_{\omega_1+\omega_2-\omega_3-\omega_4}\delta(\mathbf{p}_1+\mathbf{p}_2-\mathbf{p}_3-\mathbf{p}_4).$$

We shall not write out the analogous Dyson equation for the case of the electron-phonon interaction.

Knowing the thermodynamic Green's function $\mathfrak{G}$ one can find thermodynamic functions of the system. For this purpose one can use, for example, the following simple relation

$$\frac{\partial\Omega}{\partial g} = T\frac{n}{g}\sum_{\omega}\int\frac{d^3\mathbf{p}}{(2\pi)^3}\left[\mathfrak{G}^{0-1}_{\alpha\beta}(\mathfrak{G}_{\beta\alpha} - \mathfrak{G}_{\beta\alpha}{}^0)\right].$$

Here n is a coefficient depending on the form of the interaction. For the interaction with phonons n = 1; for the four-fermion interaction n = ½.

4. THE ANALYTICAL PROPERTIES OF THE GREEN'S FUNCTIONS AND THE TRANSITION TO TIME-DEPENDENT QUANTITIES

As has already been mentioned, a knowledge of the Green's function is sufficient for the determination of the thermodynamic characteristics of a system. But for the study of various kinetic phenomena, such as scattering or absorption of light or sound, diffusion, and so on, one must know the ordinary time-dependent functions. We shall show how, knowing the thermodynamic one-particle function $\mathfrak{G}(\omega_n, \mathbf{p})$, one can find the time-dependent $G(\omega, \mathbf{p})$.

Beginning with the coordinate space, and comparing Eqs. (2) and (3), we see that in the range of τ from $-1/T$ to $1/T$ the transition from $\mathfrak{G}$ to G is accomplished by replacing τ by it. It is much more interesting, however, to find the connection between the Fourier components $G(\omega, \mathbf{p})$ and $\mathfrak{G}(\omega_n, \mathbf{p})$.

As has been shown by Landau,[7] when the function $G(\omega, \mathbf{p})$ is expanded as a Fourier integral, its Fourier transform is a nonanalytic function of ω. Its real and imaginary parts are connected by the relation

$$G'(\omega, \mathbf{p}) = \frac{1}{\pi}\int_{-\infty}^{+\infty}\coth\frac{x}{2T}\frac{G''(x, \mathbf{p})}{x-\omega}dx$$

for the Fermi statistics, and the relation

$$G'(\omega, \mathbf{p}) = \frac{1}{\pi}\int_{-\infty}^{+\infty}\tanh\frac{x}{2T}\frac{G''(x, \mathbf{p})}{x-\omega}dx$$

for Bose statistics.

It can be shown that the Fourier transform of the so-called retarded Green's function $G^R(\mathbf{r}_1-\mathbf{r}_2, t_1-t_2)$,

$$G^R(1,2) = \begin{cases} -i\,\mathrm{Sp}\left\{\exp\left[\frac{\Omega+\mu N-\hat{H}}{T}\right]\right. & \\ \left.\times[\tilde{\psi}(x_1)\tilde{\psi}^+(x_2) \pm \tilde{\psi}^+(x_2)\tilde{\psi}(x_1)]\right\}, & t_1>t_2, \\ 0, & t_1<t_2. \end{cases}$$

is analytic in the upper half-plane. Here the plus sign is for Fermi statistics, the minus for Bose statistics. It turns out that the Fourier transform $G^R(\omega, \mathbf{p})$ of this function is simply related to $G(\omega, \mathbf{p})$. Namely, their real parts are identical, and

$$G''^R(\omega, \mathbf{p}) = \coth(\omega/2T)\,G''(\omega, \mathbf{p})$$

for the Fermi statistics, and

$$G''^R(\omega, \mathbf{p}) = \tanh(\omega/2T)\,G''(\omega, \mathbf{p})$$

for the Bose statistics. For $G^R(\omega, \mathbf{p})$ we can write an expansion of the type used by Lehmann[8]

$$G^R(\omega, \mathbf{p}) = \int_{-\infty}^{+\infty} \frac{\rho(x, \mathbf{p})}{x - \omega - i\delta} dx, \tag{13}$$

where

$$\rho(\omega, \mathbf{p}) = -(2\pi)^3 \sum_{n,m} \exp\left[\frac{\Omega + \mu N_n - E_n}{T}\right] \times |\psi_{mn}(0)|^2 \left(1 \pm \exp\left[-\frac{\omega_{mn}}{T}\right]\right) \times \delta(\mathbf{p} - \mathbf{p}_{mn})\,\delta(\omega - \omega_{mn}). \tag{14}$$

Here $\omega_{mn} = E_m - E_n - \mu$, $\mathbf{p}_{mn} = \mathbf{p}_m - \mathbf{p}_n$; E_n, $\mathbf{p}_n$ are the energy and momentum in the n-th state of the system. It can be seen from Eq. (14) that $\rho(\omega, \mathbf{p})$ is a real function, for which the integral $\int_{-\infty}^{+\infty} \rho(\omega, \mathbf{p})\, d\omega$ is finite.

Let us write the analogous Lehmann expansion for $G(\omega_n, \mathbf{p})$. By the use of Eq. (6) it is easy to show, by a method analogous to that of reference 7, that

$$\mathfrak{G}(\omega_n, \mathbf{p}) = \int_{-\infty}^{\infty} \frac{\rho(x)\, dx}{x - i\omega_n}. \tag{15}$$

Comparing Eqs. (13) and (15), we see that

$$\mathfrak{G}(\omega_n) = G^R(i\omega_n), \quad \omega_n > 0. \tag{16}$$

On the other hand, let us use the integral (15), regarded formally as a function of the complex variable $i\omega_n$, to determine a function $\mathfrak{G}(\omega)$ analytic in the upper half-plane. In virtue of Eq. (16) and the well known theorem of the theory of functions of a complex variable regarding the analytic continuation of a function given at an infinite set of discrete points possessing a point of concentration, we find from the above that G^R is, apart from a constant factor, the analytic continuation of the function $\mathfrak{G}(-i\omega_n)$ into the upper half-plane. Namely, we have

$$G^R(\omega) = \mathfrak{G}(-i\omega). \tag{17}$$

The converse problem of finding $\mathfrak{G}(\omega_n)$ from a given $G^R(\omega)$ is solved by Eq. (16) and the relation

$$\mathfrak{G}(\omega_n < 0) = \mathfrak{G}^*(|\omega_n|), \tag{16a}$$

which follows from Eq. (15).

Relations of the types (16) and (17) can in principle be obtained also for the vertex parts Γ and $\mathfrak{T}$. The general relations will, however, be more complicated. For physical applications one ordinarily needs not Γ and $\mathfrak{T}$ themselves, but the correlation functions, i.e., certain integrals of these quantities, and the practical procedure is to establish the connections between them separately for each concrete case.

In conclusion we express our gratitude to Academician L. D. Landau and L. P. Pitaevskiĭ for helpful discussions of the results of this work.

[1] A. Salam, Prog. Theor. Phys. **9**, 550, 1953.

[2] A. B. Migdal, J. Exptl. Theoret. Phys. (U.S.S.R.) **34**, 1438 (1958), Soviet Phys. JETP **7**, 996 (1958).

[3] V. M. Galitskiĭ and A. B. Migdal, J. Exptl. Theoret. Phys. (U.S.S.R.) **34**, 139 (1958), Soviet Phys. JETP **7**, 96 (1958).

[4] T. Matsubara, Prog. Theor. Phys. **14**, 351 (1955).

[5] C. T. Belyaev, J. Exptl. Theoret. Phys. (U.S.S.R.) **34**, 517 (1958), Soviet Phys. JETP **7**, 289 (1958).

[6] L. P. Gor'kov, J. Exptl. Theoret. Phys. (U.S.S.R.) **34**, 735 (1958), Soviet Phys. JETP **7**, 505 (1958).

[7] L. D. Landau, J. Exptl. Theoret. Phys. (U.S.S.R.) **34**, 262 (1958), Soviet Phys. JETP **7**, 182 (1958).

[8] H. Lehmann, Nuovo cimento **11**, 342 (1954).

[9] Ezawa, Tomozawa, and Umezawa, Nuovo cimento **5**, 810 (1957).

Translated by W. H. Furry
162

SOVIET PHYSICS JETP VOLUME 35 (8), NUMBER 1 JANUARY, 1959

A SUPERCONDUCTOR IN A HIGH FREQUENCY FIELD

A. A. ABRIKOSOV, L. P. GOR' KOV, and I. M. KHALATNIKOV

Institute for Physical Problems, Academy of Sciences, U.S.S.R.

Submitted to JETP editor March 4, 1958

J. Exptl. Theoret. Phys. (U.S.S.R.) 35, 265-275 (July, 1958)

We derive an equation describing the behavior of superconductors in a high frequency field. With the aid of this equation the frequency and temperature dependence of the impedance of a bulk superconductor have been evaluated.

As is well known Bardeen, Cooper, and Schrieffer[1] have recently constructed a microscopic theory of superconductivity in which they succeeded in explaining a whole number of properties of superconductors. In particular, in that paper they considered the behavior of superconductors in a constant weak field and obtained a new equation expressing the connection of the current with the field, replacing the equation of the phenomenological theory of F. and H. London. This equation turned out to be non-local; with its help the problem of the penetration of a weak static field into a bulk superconductor was solved in reference 1 and the dependence of the penetration depth on the temperature was found.

In the present paper we consider the behavior of a superconductor in a high frequency field. The calculation given below shows that in an alternating field the character of the connection between the current and the vector potential changes, and depends in an essential way on the frequency. For non-zero temperatures or a sufficiently high frequency this equation, written down for the Fourier components of the current, contains imaginary components, which are connected with the absorption of radiation in the superconductor. The impedance of the bulk superconductor is determined with the aid of the equation obtained.

1. THE EQUATION FOR THE CURRENT IN THE SUPERCONDUCTOR

In the non-stationary problem under consideration, it is most convenient to start directly from the expression for the current operator in second quantization:

$$\hat{j}(x) = -\frac{ie}{2m}\{\tilde{\psi}^+(x)\nabla\tilde{\psi}(x) - \nabla\tilde{\psi}^+(x)\tilde{\psi}(x)\} - \frac{e^2}{mc}\tilde{\psi}^+(x)\tilde{\psi}(x)\mathbf{A}(x) \tag{1}$$

The operators $\tilde{\psi}^+$ and $\tilde{\psi}$ are here written down in the Heisenberg representation and depend on the field **A**. This dependence has the usual form

$$\tilde{\psi}(x) = \hat{S}^{-1}(t)\psi(x)\hat{S}(t), \tag{2}$$

where $\psi(x)$ is the second quantization operator in the absence of the field, and $\hat{S}(t)$ the S-matrix satisfying the equation

$$i\frac{\partial}{\partial t}\hat{S}(t) = -\frac{1}{c}\left(\int \hat{j}(x)\mathbf{A}(x)\,d^3x\right)\hat{S}(t). \tag{3}$$

The field $\mathbf{A}(x)$ is, as usual, assumed to be adiabatically switched on at $t = -\infty$. The value of the current in the superconductor at a given point and at a given instant can clearly be obtained by taking the average of the operator (1) over the Heisenberg state of the system coinciding with the state of the system in the absence of the field:

$$j(x) = -\frac{ie}{2m}\langle\tilde{\psi}^+\nabla\tilde{\psi} - \nabla\tilde{\psi}^+\tilde{\psi}\rangle - \frac{e^2}{mc}\langle\tilde{\psi}^+\tilde{\psi}\rangle\mathbf{A}(x).$$

In a weak field it is sufficient to perform the calculations up to terms linear in **A**. We can therefore in the second term at once put $\tilde{\psi} = \psi$; this is the usual "London" term

$$-(e^2N/mc)\mathbf{A}(x). \tag{4}$$

In the first term it is necessary to expand the operators $\tilde{\psi}$ and $\tilde{\psi}^+$ according to (2) and (3) up to terms of the first order in **A**:

$$\tilde{\psi}(x) = \psi(x) - \frac{i}{c}\int_{-\infty}^{t}[\hat{j}_\alpha(y), \psi(y)]A_\alpha(y)\,d^4y \tag{5}$$

with a similar formula for $\tilde{\psi}^+$. It is sufficient in turn to retain in formula (5) for $\hat{j}(y)$ only the first part of expression (1)

$$\hat{j}_1(x) = -\frac{ie}{2m}(\nabla_x - \nabla_{x'})\psi^+(x')\psi(x)\Big|_{x'\to x}. \tag{6}$$

After substituting (5) into (6) we obtain the term in $\mathbf{j}_1(\mathbf{x}) \equiv \langle \hat{\mathbf{j}}_1(\mathbf{x}) \rangle$ which is linear in the field:

$$\mathbf{j}_1(x) = -\frac{ie^2}{4m^2c}(\nabla_x - \nabla_{x'}) \int_{-\infty}^{t} \left(\mathbf{A}(y)(\nabla_y - \nabla_{y'})\right) \langle\{\psi^+(y')\psi(y)\psi^+(x')\psi(x) - \psi^+(x')\psi(x)\psi^+(y')\psi(y)\}\rangle\, d^4y. \tag{7}$$

Here and henceforth the primed coordinates differ only in the space variables, which must be put equal after performing the differentiation.

The evaluation of the averages of the products of the four ψ and ψ^+ operators in (7) can most conveniently be performed by using the method developed by one of the authors in reference 2. These averages are expressed in terms of pair averages:

$$G_{\alpha\beta}(x - x') = -i\langle T(\psi_\alpha(x)\psi_\beta^+(x'))\rangle; \quad F_{\alpha\beta}^+(x - x') = \langle T(\psi_\alpha^+(x)\psi_\beta^+(x'))\rangle; \quad F_{\alpha\beta}(x - x') = \langle T(\psi_\alpha(x)\psi_\beta(x'))\rangle.$$

The dependence of these quantities on the spinor indices is as follows:

$$G_{\alpha\beta}(x - x') = \delta_{\alpha\beta}G(x - x'); \qquad F_{\alpha\beta}^+(x - x') = -F_{\alpha\beta}(x - x') = F(x - x')\begin{pmatrix} 0 & 1 \\ -1 & 0 \end{pmatrix}_{\alpha\beta}.$$

Performing the averaging in (7) we get

$$\mathbf{j}_1(x) = \frac{ie^2}{2m^2c}(\hat{\mathbf{p}}_x - \hat{\mathbf{p}}_{x'}) \int_{-\infty}^{t} (\mathbf{A}(y)(\hat{\mathbf{p}}_y - \hat{\mathbf{p}}_{y'}))\,\{[G(y - x')G(x - y') - F(x' - y')F(x - y)] - \text{C.C.}\}\, d^4y. \tag{8}$$

Going over to Fourier components we find:

$$\mathbf{j}_1(\mathbf{k}, \omega) = \frac{2e^2}{m^2c}\int \frac{d^3p}{(2\pi)^3} \int_{-\infty}^{\infty} \frac{d\omega_1}{2\pi} \int_{-\infty}^{\infty} \frac{d\omega_2}{2\pi}\, \mathbf{p}(\mathbf{A}(\mathbf{k}, \omega)\mathbf{p})$$

$$\times \left[G\left(\mathbf{p} - \frac{\mathbf{k}}{2}, \omega_1\right)G\left(\mathbf{p} + \frac{\mathbf{k}}{2}, \omega_2\right) - F\left(\mathbf{p} - \frac{\mathbf{k}}{2}, \omega_1\right)F\left(\mathbf{p} + \frac{\mathbf{k}}{2}, \omega_2\right) - \text{C.C.}\right]\frac{1}{\omega_1 - \omega_2 - \omega - i\delta}.$$

The Fourier components of the functions $G(x - x')$ and $F(x - x')$ were found in reference 2. It is convenient to write them in the form

$$G(\mathbf{p}, \omega) = \frac{u_p^2(1 - n_p)}{\omega - \varepsilon_p + i\delta} + \frac{v_p^2 n_p}{\omega + \varepsilon_p + i\delta} + \frac{v_p^2(1 - n_p)}{\omega + \varepsilon_p - i\delta} + \frac{u_p^2 n_p}{\omega - \varepsilon_p - i\delta};$$

$$F(\mathbf{p}, \omega) = iu_p v_p\left[\frac{1 - n_p}{\omega - \varepsilon_p + i\delta} - \frac{n_p}{\omega + \varepsilon_p + i\delta} - \frac{1 - n_p}{\omega + \varepsilon_p - i\delta} + \frac{n_p}{\omega - \varepsilon_p - i\delta}\right], \tag{9}$$

where

$$u_p^2 = {}^1/_2(1 + \xi_p/\varepsilon_p); \quad v_p^2 = {}^1/_2(1 - \xi_p/\varepsilon_p); \quad \xi_p = v(p - p_0); \quad \varepsilon_p = \sqrt{\xi_p^2 + \Delta^2}; \quad n_p = (e^{\varepsilon_p/T} + 1)^{-1}.$$

The quantity Δ depends on the temperature in the way found in the paper by Bardeen et al.

Substituting equations (9) into the expression for the current and integrating we find for the Fourier components of the current:

$$\mathbf{j}(\mathbf{k}, \omega) = \frac{2e^2}{(2\pi)^3 m^2 c}\int d^3p\, \mathbf{p}(\mathbf{pA}(\mathbf{k}, \omega))\left[(v_1u_2 - u_1v_2)\left(\frac{v_1u_2}{\varepsilon_1 + \varepsilon_2 + \omega + i\delta} + \frac{u_1v_2}{\varepsilon_1 + \varepsilon_2 - \omega - i\delta}\right)\right.$$

$$\left.\times(1 - n_1 - n_2) + (u_1u_2 + v_1v_2)\left(\frac{u_1u_2}{\varepsilon_2 - \varepsilon_1 + \omega + i\delta} + \frac{v_1v_2}{\varepsilon_2 - \varepsilon_1 - \omega - i\delta}\right)(n_1 - n_2)\right] - \frac{e^2}{mc}N\mathbf{A}(\mathbf{k}, \omega), \tag{10}$$

when the index 1 corresponds to the momentum $\mathbf{p} + \frac{1}{2}\mathbf{k}$, and the index 2 to the momentum $\mathbf{p} - \frac{1}{2}\mathbf{k}$. Substituting into (7), assuming the vector $\mathbf{A}$ to lie in the surface plane of the specimen and performing the integration over φ we get

$$\mathbf{j}(\mathbf{k}, \omega) = \frac{3e^2N\mathbf{A}(\mathbf{k}, \omega)}{32mc}\int_{-1}^{1} d\cos\theta \sin^2\theta \int_{-\xi_0}^{\xi_0} d\xi \left[\left(1 - \frac{\xi_1\xi_2 + \Delta^2}{\varepsilon_1\varepsilon_2}\right)\left(\tanh\frac{\varepsilon_1}{2T} + \tanh\frac{\varepsilon_2}{2T}\right)\left(\frac{1}{\varepsilon_1 + \varepsilon_2 + \omega + i\delta} + \frac{1}{\varepsilon_1 + \varepsilon_2 - \omega - i\delta}\right)\right.$$

$$\left. + \left(1 + \frac{\xi_1\xi_2 + \Delta^2}{\varepsilon_1\varepsilon_2}\right)\left(\tanh\frac{\varepsilon_1}{2T} - \tanh\frac{\varepsilon_2}{2T}\right)\left(\frac{1}{\varepsilon_1 - \varepsilon_2 + \omega + i\delta} + \frac{1}{\varepsilon_1 - \varepsilon_2 - \omega - i\delta}\right)\right] - \frac{e^2}{mc}N\mathbf{A}(\mathbf{k}, \omega). \tag{11}$$

In this expression we used the fact that terms containing one ξ give zero on integration. Indeed, we have

$$\xi_{1,2} = \xi \pm {}^1/_2\, vk \cos\theta, \tag{12}$$

and the integrations over $\cos\theta$ and ξ are performed over a symmetric domain. The quantity ξ_0 determines the upper limit of integration over ξ.

2. PIPPARD'S LIMITING CASE

Further evaluations are impossible to perform without more concrete assumptions about the magnitude of the parameters important for the problem considered. Of the greatest importance for us is the fact that, apparently, the penetration depth of the field for the majority of superconductors is much less than the quantity v/Δ, i.e., $vk \gg \Delta$ (See reference 1).* This inequality is violated only in the neighborhood of the transition point T_c, where the region with $vk \ll \Delta$ occurs. As was shown in reference 1, the basic domain of electrodynamics is founded upon the equation proposed earlier by Pippard.[3] We shall call this region the Pippard region. In the second region the applicability of the London electrodynamics[4] is conserved (London region).†

In the case of a variable field the London region gets narrowed. This is clear, though, from the fact that a normal metal is a Pippard one, since $\Delta = 0$ for it, whence follows that the London region cannot continue up to the transition point itself. Leaving a detailed analysis of this problem until the next section, we shall consider here how Eq. (11) simplifies in the Pippard region. We shall at the same time also assume $vk \gg \omega$. As we shall see in the following, this is always correct in the range of frequencies of most interest.

Equation (11) can be transformed in the following way. In view of the fact that ω only enters into denominators such as $\epsilon_1 + \epsilon_2 + \omega + i\delta$, we can subtract from the coefficient for $\mathbf{A}(\mathbf{k}, \omega)$ which we denote as $-(3e^2N\Delta/4mcvk)\,Q(\omega)$ its value in the static case. The remainder will be the integral over ξ and $\cos\theta$ where the important region of integration is the region $\xi \ll vk$, $\cos\theta \ll 1$. We introduce now as new variables ξ_1 and ξ_2. If we put $vk \gg T$ (this will be shown below) we can assume that the integrations over ξ_1 and ξ_2 proceed independently with limits from $-\infty$ to $+\infty$. Terms with the product $\xi_1\xi_2$ will thus drop out of the integral. Apart from that, the factor $\sin^2\theta$ can be replaced by unity. Finally we go over to the variables ϵ_1/Δ and ϵ_2/Δ (denoting them again by ϵ_1 and ϵ_2). As a result we get:

$$Q(\omega) - Q(0) = \frac{1}{2}\int_1^\infty \frac{d\varepsilon_1}{\sqrt{\varepsilon_1^2 - 1}} \int_1^\infty \frac{d\varepsilon_2}{\sqrt{\varepsilon_2^2 - 1}} \Big[(1 - \varepsilon_1\varepsilon_2)\Big(\tanh\frac{\varepsilon_1\Delta}{2T} + \tanh\frac{\varepsilon_2\Delta}{2T}\Big)\Big(\frac{1}{\varepsilon_1 + \varepsilon_2 + \omega/\Delta + i\delta}$$
$$+ \frac{1}{\varepsilon_1 + \varepsilon_2 - \omega/\Delta - i\delta} - \frac{2}{\varepsilon_1 + \varepsilon_2}\Big) - (1 + \varepsilon_1\varepsilon_2)\Big(\tanh\frac{\varepsilon_1\Delta}{2T} - \tanh\frac{\varepsilon_2\Delta}{2T}\Big)\Big(\frac{1}{\varepsilon_1 - \varepsilon_2 + \omega/\Delta + i\delta} + \frac{1}{\varepsilon_1 - \varepsilon_2 - \omega/\Delta - i\delta} - \frac{2}{\varepsilon_1 - \varepsilon_2}\Big)\Big]. \tag{13}$$

To fix ideas we shall now assume $\omega > 0$ and isolate the imaginary part from the integral (13). By means of slight transformations one can convince oneself that the real part of expression (13) is always of the same form as at $T = 0$, but with Δ depending on the temperature. The final expression is of the form

$$Q(\omega) - Q(0) = \int_1^\infty \frac{d\varepsilon_1}{\sqrt{\varepsilon_1^2 - 1}} \int_1^\infty \frac{d\varepsilon_2}{\sqrt{\varepsilon_2^2 - 1}} \Big[(1 - \varepsilon_1\varepsilon_2)\Big(\frac{1}{\varepsilon_1 + \varepsilon_2 + \omega/\Delta} + \frac{1}{\varepsilon_1 + \varepsilon_2 - \omega/\Delta} - \frac{2}{\varepsilon_1 + \varepsilon_2}\Big)\Big]$$
$$- \frac{i\pi}{2}\theta\Big(\frac{\omega}{2\Delta} - 1\Big)\int_1^{\omega/\Delta - 1} d\varepsilon \frac{[\varepsilon(\omega/\Delta - \varepsilon) - 1][\tanh(\omega/2T - \varepsilon\Delta/2T) + \tanh(\varepsilon\Delta/2T)]}{\sqrt{\varepsilon^2 - 1}\sqrt{(\omega/\Delta - \varepsilon)^2 - 1}} \tag{14}$$
$$- i\pi\int_1^\infty d\varepsilon \frac{[\varepsilon(\omega/\Delta + \varepsilon) + 1][\tanh(\omega/2T + \varepsilon\Delta/2T) - \tanh(\varepsilon\Delta/2T)]}{\sqrt{\varepsilon^2 - 1}\sqrt{(\varepsilon + \omega/\Delta)^2 - 1}}, \qquad \text{where } \theta(x) = \begin{cases} 1 & x > 0 \\ 0 & x < 0. \end{cases}$$

*Strictly speaking, the strong inequality $vk \gg \Delta$ is not valid. But the applicability of the formulae obtained in this limiting case is provided by the insertion of significant numerical factors.

†Such a family of superconductors we shall call Pippard superconductors. The opposite case (which can be called the London case) when the superconductor for $\omega = 0$, is described by the equations of F and H. London in the whole temperature range is, apparently, less common, and we shall not consider it.

One integration in the real part of $Q(\omega)$ can be performed. After this there remains

$$\mathrm{Re}\,[Q(\omega)-Q(0)] = \mathcal{J}(\omega/\Delta) + \mathcal{J}(-\omega/\Delta) - 2\mathcal{J}(0), \tag{15}$$

where

$$\mathcal{J}\left(\frac{\omega}{\Delta}\right) = \int_1^\infty \frac{d\varepsilon\,[\varepsilon(\varepsilon+\omega/\Delta)+1]}{\sqrt{\varepsilon^2-1}\sqrt{(\varepsilon+\omega/\Delta)^2-1}} \cosh^{-1}\left(\varepsilon+\frac{\omega}{\Delta}\right). \tag{16}$$

Although each of the integrals in (15) diverges, together they give a finite expression. In the integral $\mathcal{J}(-\omega/\Delta)$ in the region where $1 > \epsilon - \omega/\Delta > 0$, we must replace

$$\cosh^{-1}\left(\varepsilon-\frac{\omega}{\Delta}\right)\Big/\sqrt{\left(\varepsilon-\frac{\omega}{\Delta}\right)^2-1} \to \cos^{-1}\left(\varepsilon-\frac{\omega}{\Delta}\right)\Big/\sqrt{1-\left(\varepsilon-\frac{\omega}{\Delta}\right)^2},$$

and in the region where $\epsilon - \omega/\Delta$ is negative we must take $-\arccos(\omega/\Delta - \epsilon)$ or, respectively, $\frac{1}{2}\pi + \sin^{-1}(\omega/\Delta - \epsilon)$. The limiting value of this expression in the region of small frequencies is of the form

$$\mathrm{Re}\,[Q(\omega)-Q(0)] \approx (5\pi^2/12)(\omega/\Delta)^2. \tag{17}$$

In the region $\omega/\Delta \gg 1$ we get

$$\mathrm{Re}\,[Q(\omega)-Q(0)] = 2\ln(2\omega/\Delta) - \pi^2. \tag{18}$$

An expression for $Q(0)$ was found in Bardeen, Cooper, and Schrieffer's paper.[1] It is of the form

$$Q(0) = \pi^2 \tanh(\Delta/2T). \tag{19}$$

The integration (16) can be carried out in the case $\omega = 2\Delta$. In view of the fact that according to (14) the imaginary part of $Q(\omega)$ is equal to zero for $\omega < 2\Delta$, we get

$$Q(2\Delta) - Q(0) = 2\pi + 2\sqrt{3} + 4\ln(\sqrt{3}+1) - 2\ln 2 - \pi^2 \approx 2.5.$$

Hence it is clear that for $T = 0$ the magnitude of the current, and consequently the penetration depth which is proportional to $Q(\omega)^{1/3}$ (see Sec. 4), changes little when the frequency is changed from 0 to 2Δ:

$$\delta(2\Delta)/\delta(0)|_{T=0} \approx 0.93.$$

The imaginary part of formula (13) depends in an essential way on the relations between Δ, ω, and T. We consider first of all the case $T = 0$. Then there remains only the first term of the imaginary part in equation (14) (we shall call it the "pair" term) which is different from zero only for $\omega > 2\Delta$. Physically this is completely clear. As is well known, the imaginary part of the current determines the absorption and in the absence of an excitation such an absorption can only take place thanks to the destruction of a pair. The quantity 2Δ determines the energy necessary for this, when the statistics are taken into account.

Performing some simple transformations of the integral we find

$$\mathrm{Im}\,Q(\omega) = -2\pi\theta\left(\frac{\omega}{2\Delta}-1\right) \times \int_0^{\omega/2\Delta-1} d\varepsilon \frac{[(\omega/2\Delta)^2-\varepsilon^2-1]}{\{[(\omega/2\Delta+\varepsilon)^2-1][(\omega/2\Delta-\varepsilon)^2-1]\}^{1/2}}. \tag{20}$$

This integral can be reduced to an elliptical one. Near the threshold, i.e., for $\omega/2\Delta - 1 \ll 1$ it has the form

$$\mathrm{Im}\,Q(\omega) = -\pi^2(\omega/2\Delta - 1). \tag{21}$$

In the limiting case of large frequencies $\omega \gg \Delta$ it becomes equal to

$$\mathrm{Im}\,Q(\omega) = -\pi\omega/\Delta. \tag{22}$$

In the case of temperatures different from zero, of most interest are some limiting cases, as we shall see in the next section.

(a) The case $T \sim \omega \ll \Delta$. Here, of course, only the second term of the imaginary part of Eq. (14) (which we shall call the "electron" term) will take part. Transforming it, we find

$$\mathrm{Im}\,Q(\omega) = -4\pi \sinh\frac{\omega}{2T} K_0\left(\frac{\omega}{2T}\right) e^{-\Delta/T}, \tag{23}$$

where $K_0(x)$ is a Hankel function of an imaginary argument. In particular, we have for $T \ll \omega \ll \Delta$

$$\mathrm{Im}\,Q(\omega) = -2\pi\sqrt{\pi T/\omega}\, e^{-\Delta/T}, \tag{24}$$

and for $\omega \ll T \ll \Delta$

$$\mathrm{Im}\,Q(\omega) = -2\pi\frac{\omega}{T}\ln\frac{4T}{\gamma\omega} e^{-\frac{\Delta}{T}}; \quad (\gamma = e^C \approx 1.78). \tag{25}$$

(b) The case $\omega \ll T \sim \Delta$. Again only the electron term takes part. By means of a series of transformations we find

$$\mathrm{Im}\,Q(\omega) = -\pi\left[\frac{\omega}{T}\cosh^{-2}\frac{\Delta}{2T}\ln 2\sqrt{\frac{2\Delta}{\omega}} + \frac{\omega}{\Delta}\left(1-\tanh\frac{\Delta}{2T}\right) - 2\frac{\omega}{T}P\left(\frac{\Delta}{T}\right)\right], \tag{26}$$

where the function $P(x)$ is the integral

$$P(x) = \int_1^\infty \frac{d\varepsilon}{\varepsilon^2-1}\,\frac{\cosh x\varepsilon - \cosh x}{(\cosh x\varepsilon + 1)(\cosh x + 1)}. \tag{27}$$

The limiting values of this function are as follows:

$$x \gg 1 \quad P(x) = e^{-x} \ln 2\gamma x, \qquad x \ll 1 \quad P(x) = 4x\pi^{-2}\zeta(3) \approx 0.485x. \tag{28}$$

From expression (26) we find that for $\omega \ll T \ll \Delta$ the result coincides with (25) while for $\omega \ll \Delta \ll T$

$$\operatorname{Im} Q(\omega) = -\pi\omega/\Delta. \tag{29}$$

(c) The case $\omega \sim \Delta \ll T$. Although the pair term in principle takes part in Im Q, its contribution turns out to be small. As to the electron term, we get from it again equation (29).

(d) The case $T \ll \omega \sim \Delta$. Here there are two possibilities. If $\omega < 2\Delta$, only the electron term will take part which gives

$$\operatorname{Im} Q(\omega) = -2\pi\sqrt{\pi}\sqrt{\frac{T}{2\Delta} + \frac{T}{\omega}}\, e^{-\frac{\Delta}{T}}. \tag{30}$$

In particular, Eq. (24) is obtained for $T \ll \omega \ll \Delta$. If $\omega > 2\Delta$, expression (30) is exponentially small compared to the contribution from the pair term (of course slightly away from the threshold). In view of the smallness of the temperature we can then use Eqs. (20) to (21).

(e) The case $\Delta \ll T \sim \omega$. Here both the pair term and the electron term take part to an equal degree. The calculation leads to the result

$$\operatorname{Im} Q(\omega) = -\pi\left(\frac{\omega}{\Delta} - 2\tanh\frac{\omega}{2T}\right). \tag{31}$$

Although in this formula the second term is much smaller than the first one, we have retained it for a reason which will be given below.

(f) The case $T \sim \Delta \ll \omega$. Here the most important contribution is given by the pair term, and it is sufficient to consider only the main expression which is the same as Eq. (29).

At the end of this section we must note that in all cases when $\omega \gg \Delta$ the real part of $Q(\omega)$ is small compared to the imaginary part which is, independently of the temperature, equal to $-\pi\omega/\Delta$. In this way the relation between the current and the vector potential is in the case $\omega \gg \Delta$ of the form

$$j(\mathbf{k}, \omega) = i\,(3\pi e^2 N\omega / 4mcvk)\, A(\mathbf{k}, \omega). \tag{32}$$

This relation does not contain Δ and is exactly the same as the one obtained from the theory of the anomalous skin effect in a normal metal. This result is natural.

However, in the cases $\omega \gg \Delta$ we can not restrict ourselves only to the main term. In order that the difference between the superconducting state and the normal state can be revealed, it is necessary to take into account also terms of the next order. In the case $\Delta \ll T \sim \omega$ these terms come both from the real and from the imaginary part of $Q(\omega)$. Just for that reason was the main expression in Eq. (31) supplemented by a small term. In the last case $T \sim \Delta \ll \omega$ the main correction gives only the real part of $Q(\omega)$.

3. THE LONDON REGION

We shall now elucidate under what conditions the London region $vk \ll \Delta$ occurs. We shall not consider, as already stated, the possible, but apparently very uncommon, case when under static conditions the superconductor is a London one for the whole temperature range.

As was already noted in the preceding section the London region can appear only near T_c. On the other hand near T_c the gap width Δ becomes very small and since for $\omega \gg \Delta$ the metal differs little from a normal conductor, the London region cannot extend to the critical temperature itself and in any case has an upper limit through the condition $\omega \ll \Delta$.

To find the more exact location of the London region we find first of all the connection between the current and the vector potential in the case $\Delta \gg vk$. To do this we take into account the earlier noted fact that for the whole range of frequencies of interest we can assume $\omega \ll vk$ and also that the London region occurs only in the immediate neighborhood of T_c, i.e., that we can assume that $\Delta \ll T$. The necessary equation is obtained from Eq. (14). Only the integral with the difference of the tangents is then essential and the last term in that equation.

We find in that way

$$j(\mathbf{k}, \omega) = -\frac{e^2 N A(\mathbf{k}, \omega)}{mc}\left\{\alpha\frac{\Delta^2}{T_c^2} - \frac{3i\pi\omega}{4vk}\right\}; \tag{33}$$

where

$$\alpha = \frac{1}{4}\int_0^\infty \frac{\sinh xdx}{x\cosh^3 x} = \frac{7}{4\pi^2}\zeta(3) = 0.21.$$

If we restrict ourselves to the real part of this relation, we obtain the equation of the Londons, where the expression within the braces plays the role of the ratio of the number of superconducting electrons N_s to the total number of electrons. The imaginary part corresponds to absorption. We note that this expression is obtained under the assumption $\omega \ll \Delta \ll T$ and $\omega \ll vk$ but the relation between vk and Δ and between vk and T can be arbitrary. This explains indeed also the fact that this expression coincides within a small correction term with the corresponding equation for the Pippard case.

As a criterion for the transition from the Pippard region to the London region one can take the equality of the coefficients of **A** in the expression for the current. For the Pippard case, assuming $\omega \ll \Delta \ll T$, we find from (17), (19), and (29)

$$\mathbf{j}(\mathbf{k},\omega) = -\frac{e^2N}{mc}\mathbf{A}(\mathbf{k},\omega)\left\{\frac{3\pi^2}{8}\frac{\Delta^2}{vkT_c} - i\,\frac{3}{4}\pi\frac{\omega}{vk}\right\}. \tag{34}$$

We shall compare this formula with Eq. (33). We assume that the imaginary parts in both equations are small compared to the real parts. In that case the criterion coincides with the criterion for the static case. The London region occurs for

$$vk \ll vk_1 = 3\pi^2T_c/8\alpha \approx 17.7T_c. \tag{35}$$

Assuming $k \sim 1/\delta$ where δ is the penetration depth, one can show, using the well-known London equation for the penetration depth $\delta_L = (mc^2/4\pi N_s e^2)^{1/2}$ and Eq. (33), that this corresponds to the condition

$$\Delta \ll \frac{3}{16}\left(\frac{\pi}{\alpha}\right)^{3/2}\left(\frac{mc^2}{Ne^2v^2}\right)^{1/2}T_c^2. \tag{36}$$

From this inequality it follows that for instance for aluminium the London region occurs for $(T_c - T)/T_c \sim 4\times10^{-4}$, for tin for $(T_c - T)/T_c \sim 3\times10^{-2}$.*

When the temperature is increased the real part in Eq. (33) becomes small compared to the imaginary part. But since in the latter case the superconductor differs little from a normal metal, we obtain necessarily the Pippard case (we remember that the imaginary part in (33) refers to that case). In that way we can assume that the London region is bounded by the condition

$$vk \gg vk_2 = (3\pi\omega/4\alpha)(T_c/\Delta)^2 \tag{37}$$

or, after substituting for the London penetration depth,

$$\Delta \gg \frac{3^{1/3}\pi^{1/3}}{(4\alpha)^{1/3}}\left(\frac{mc^2}{Ne^2v^2}\right)^{1/3}T_c\omega^{1/3}. \tag{38}$$

Comparing Eqs. (34) and (35) we get the condition for the occurrence of the London region

$$\omega \ll \omega_c = \frac{9\pi^4}{(8\alpha)^3}\left(\frac{mc^2}{Ne^2v^2}\right)T_c^3. \tag{39}$$

Such a limitation can mean completely different frequencies for different substances. For aluminium, for instance, this frequency corresponds to $2\times10^{-2}\,T_c$ or $\sim 3\times10^9\ \mathrm{sec}^{-1}$. For tin it is equal to $0.5\,T_c \approx 2\times10^{11}\ \mathrm{sec}^{-1}$.

In view of the fact that of the greatest interest is the range of frequencies not too far from $\Delta(0)$ which for a Pippard metal is practically always beyond the limits of the region bounded by the inequality (39) we shall in general not consider the London region.

It is appropriate to note here also that apart from a limitation as to frequencies, the applicability of F. and H. London's equations in the form (33) is also limited from the high temperature side by the condition that the dimensions of pairs, which is of the order of v/Δ, must be small compared to the mean free path. In the opposite case the constant coefficient occurring in F. and H. London's equation can no longer be described by the first term within the brackets of Eq. (33). Since the mean free path can on the average be assumed to be of the order of 10^{-3} cm, and $v/\Delta(0) \sim 10^{-4}$ cm, the region of applicability of Eq. (33) is generally speaking small. As to the case $v/\Delta \gg l$, the electrodynamics for it has not been formulated and we shall not discuss it.

To conclude this section, we note that, as can be easily seen, for frequencies larger than ω_c a direct estimate gives $vk \gg T_c$ for the whole range of temperatures, including the neighborhood of T_c where the penetration depth is maximum. Such a relation between vk and T was essential for the conclusion reached in Sec. 2.

4. THE IMPEDANCE

The relation between **j** and **A** can be substituted into the Maxwell equations and one can obtain an expression for the dependence of the vector potential on the coordinates.

In the Pippard case considered by us the corresponding calculations are not different from the static case 1. Assuming the reflection of the electrons from the surface to be diffuse, we get for the penetration depth:

$$\begin{aligned}\delta &= \frac{1}{H(0)}\,\mathrm{Re}\int_0^\infty H\,dz = \frac{\mathrm{Re}\,A(0)}{H(0)}\\ &= \frac{\sqrt{3}}{2\pi}\left(\frac{c^2mv}{3\pi e^2N\Delta}\right)^{1/3}\mathrm{Re}\,[Q(\omega)]^{-1/3}.\end{aligned} \tag{40}$$

The impedance is determined as follows:

*Tin is in fact not a genuine Pippard metal and is on the boundary between the London and the Pippard situations, since for tin $vk \approx vk_1$ already at $T = 0$. But since far from T_c the temperature dependence and the penetration depth are very weak, essential departures from the Pippard equations occur only in the neighborhood of T_c.

$$Z = R + iX = E(0) \Big/ \int_0^\infty j\,dz = -\frac{4\pi}{c}\frac{E(0)}{H(0)}$$
$$= -\frac{4\pi i\omega}{c^2}\frac{A(0)}{H(0)} = -\frac{2\sqrt{3}\pi i\omega}{c^2}\left(\frac{mc^2v}{3\pi e^2 N\Delta Q(\omega)}\right)^{1/3} \tag{41}$$

For complex $Q(\omega)$ the value of the root is determined as the analytical continuation of the real root for real $Q(\omega)$.

It is convenient to relate the magnitude of the impedance to the value of the active resistivity in the normal state which is equal to

$$R_n = \sqrt{3}\,(mvc^2\pi / 3Ne^2)^{1/3}\,\omega^{2/3} / c^2. \tag{42}$$

Such an expression is obtained by substituting $Q(\omega) = -i\pi\omega/\Delta$ into (41) (see also reference 5). The ratio Z/R_n is determined by the equation

$$Z(\omega) / R_n = -2i\,(\pi\omega / \Delta Q(\omega))^{1/3}. \tag{43}$$

For the case $T = 0$ the frequency dependence of the impedance is obtained from Eqs. (43), (15), (19), and (20).

In the case of non-zero temperatures we carry out an analysis of the temperature dependence of the impedance for different frequencies.

(A) The case $\omega \ll \Delta(0)$

(1) For the lowest temperatures there occurs the region $T \ll \omega \ll \Delta$. Then T becomes of the order of ω and finally we go over to the region $\omega \ll T \ll \Delta$. A description of that transition is given by Eqs. (19), (23). Substituting into (43) and taking into account that $\mathrm{Re}\,Q(\omega) \gg \mathrm{Im}\,Q(\omega)$ we get

$$\frac{Z(\omega)}{R_n} = 2\left(\frac{\omega}{\pi\Delta}\right)^{1/3}\left[\frac{4}{3\pi}\sinh\frac{\omega}{2T}K_0\left(\frac{\omega}{2T}\right)e^{-\Delta/T} - i\right]. \tag{44}$$

(2) When the temperature is raised further, we get into the range $\omega \ll \Delta \sim T$ which then goes over into the range $\omega \ll \Delta \ll T$. This transition is described by Eqs. (19) and (26). As long as $\Delta/T \gg \omega/\Delta$, the imaginary part of $Q(\omega)$ will be small compared to the real part, as before, and we get in that way

$$\frac{Z(\omega)}{R_n} = 2\left[\frac{\omega}{\pi\Delta\tanh(\Delta/2T)}\right]^{1/3}\left[\frac{2\omega}{3\pi T}\frac{1}{\sinh(\Delta/T)}\ln 2\sqrt{\frac{2\Delta}{\omega}}\right.$$
$$\left.+\frac{1}{3\pi}\frac{\omega}{\Delta}\left(\coth\frac{\Delta}{2T} - 1\right) - \frac{2}{3\pi}P\left(\frac{\Delta}{T}\right)\coth\frac{\Delta}{2T} - i\right], \tag{45}$$

where $P(\Delta/T)$ is given by expression (27).

(3) For still higher temperatures, Δ decreases so much that it becomes comparable with and then less than ω. This region is described by Eqs. (19), (15), and (29); for $\omega \sim \Delta$ the real and imaginary parts of $Q(\omega)$ are of the same order of magnitude, and for $\omega \gg \Delta$ the real part of $Q(\omega)$ turns out to be small compared to the imaginary part. In that limiting case the equation for the impedance is of the form

$$\frac{Z(\omega)}{R_n} = 1 + \frac{1}{\sqrt{3}\pi}\frac{\Delta}{\omega}\left(2\ln\frac{2\omega}{\Delta} + \pi^2\right)$$
$$- i\sqrt{3}\left[1 - \frac{1}{3\sqrt{3}\pi}\frac{\Delta}{\omega}\left(2\ln\frac{2\omega}{\Delta} + \pi^2\right)\right]. \tag{46}$$

(B) The case $\omega \sim \Delta(0)$

(1) At low temperatures we have $T \ll \omega \sim \Delta$. In this region one applies Eqs. (15), (19), and (30) in the case when $\omega < 2\Delta(0)$, or Eq. (20) if $\omega > 2\Delta(0)$. In the first case the imaginary part of $Q(\omega)$ will be small compared to the real part, and the expression for the impedance will be of the form

$$\frac{Z(\omega)}{R_n} = 2\sqrt{\frac{\omega\pi}{\Delta \mathrm{Re}\,Q(\omega)}}$$
$$\times\left[\frac{2\pi\sqrt{\pi}}{3\mathrm{Re}\,Q(\omega)}\sqrt{\frac{T}{2\Delta} + \frac{T}{\omega}}\;e^{-\Delta/T} - i\right], \tag{47}$$

where $\mathrm{Re}\,Q(\omega)$ is given by Eq. (15) with an additional π^2. In the case $\omega > 2\Delta(0)$ the equations for the case $T = 0$ can be used.

(2) When the temperature is raised we go over to the range $\omega \sim \Delta \sim T$ and finally we get into the range $\Delta \ll T \sim \omega$. Here we can use equations (18) and (31). Inasfar as the real part of $Q(\omega)$ is less than the imaginary part we can again use an expansion. As a result we find:

$$\frac{Z(\omega)}{R_n} = 1 + \frac{1}{\sqrt{3}\pi}\frac{\Delta}{\omega}\left(2\ln\frac{2\omega}{\Delta} + \pi^2 + \frac{2\pi}{\sqrt{3}}\tanh\frac{\omega}{2T}\right)$$
$$- i\sqrt{3}\left[1 - \frac{1}{3\sqrt{3}\pi}\frac{\Delta}{\omega}\left(2\ln\frac{2\omega}{\Delta} + \pi^2 - 2\pi\sqrt{3}\tanh\frac{\omega}{2T}\right)\right]. \tag{48}$$

It is interesting to note that when the temperature is lowered starting from T_c, the real part of the impedance initially does not decrease, but slightly increases.

(C) The case $\omega \gg \Delta(0)$

In this case only the relation between T and Δ changes, but ω is all the time large compared with them. Here we apply Eqs. (18), (19), and (29). Taking into account again that the real part is small, we find:

$$\frac{Z(\omega)}{R_n} = 1 + \frac{1}{\sqrt{3}\pi}\frac{\Delta}{\omega}\left(2\ln\frac{2\omega}{\Delta} + \pi^2\left(1 - \tanh\frac{\Delta}{2T}\right)\right)$$
$$- i\sqrt{3}\left[1 - \frac{1}{3\sqrt{3}\pi}\frac{\Delta}{\omega}\left(2\ln\frac{2\omega}{\Delta} + \pi^2\left(1 - \tanh\frac{\Delta}{2T}\right)\right)\right]. \tag{49}$$

In view of the fact that a detailed comparison of the theory with experimental data requires large numerical calculations (first of all a tabulation of the functions (15) and (20), which at this moment has not yet been concluded) such a comparison will be given in a later paper.

In conclusion the authors express their deep gratitude to academician L. D. Landau for his interest in this paper.

[1] Bardeen, Cooper, and Schrieffer, Phys. Rev. **108**, 1175 (1957).

[2] L. P. Gor'kov, J. Exptl. Theoret. Phys. (U.S.S.R.) **34**, 735 (1958), Soviet Phys. JETP **7**, 505 (1958).

[3] A. B. Pippard, Proc. Roy. Soc. (London) **A216**, 547 (1953).

[4] F. and H. London, Proc. Roy. Soc. (London) **A149**, 71 (1935); Physica **2**, 341 (1935).

[5] G. E. H. Reuter and E. H. Sondheimer, Proc. Roy. Soc. (London) **A195**, 336 (1948).

Translated by D. ter Haar
37

SOVIET PHYSICS JETP VOLUME 13, NUMBER 1 JULY, 1961

RAMAN SCATTERING OF LIGHT IN SUPERCONDUCTORS

A. A. ABRIKOSOV and L. A. FAL'KOVSKIĬ

Institute for Physics Problems, Academy of Sciences, U.S.S.R.

Submitted to JETP editor July 25, 1960

J. Exptl. Theoret. Phys. (U.S.S.R.) **40**, 262-270 (January, 1961)

A computation is carried out of Raman scattering upon reflection of light from the surface of a superconductor. The distribution with angle and frequency of the scattered light and the absolute magnitude of the effect are found.

THE most interesting and important properties of superconductors consist of the special aspects of their behavior in an electromagnetic field. In particular, high-frequency impedance measurements provide the possibility of explaining directly the characteristic features of the electronic energy spectrum of superconductors. The presence of an energy gap Δ in this spectrum leads to the absence of absorption (at $T = 0$) of radiation having a frequency less than the threshold frequency, equal to 2Δ.

In addition to these experiments, which involve measurements at various frequencies, Khaĭkin and Bykov[1] made an attempt to investigate the electron spectrum using the Raman scattering of light in superconductors. The distribution of frequencies of the satellites, given sufficient intensity, might be measured by spectroscopic methods. The fundamental difficulty with this experiment lies in the extremely small amount of scattering. To a significant degree, this is associated with the fact that the skin effect, enables the light to penetrate only into a very thin surface layer of the metal, of the order of 10^{-5} cm. In Khaĭkin and Bykov's experiments, no satellites were found.

The question naturally arises: what increase in sensitivity is required for the successful performance of such an experiment, and what overall pattern is to be expected for the phenomenon? The present calculation was undertaken with this as its object.

Our goal is to derive the distribution with frequency and angle of the light reflected from the plane surface of a superconductor filling the semi-infinite volume $z > 0$. For simplicity, we shall assume $T = 0$, and limit ourselves to the case in which the incident and reflected waves do not make too large angles with the normal to the surface ($\sin^2\theta$, $\sin^2\theta' \ll |\epsilon|$, where ϵ is the complex dielectric constant); the latter limitation is of small significance in practice, since $|\epsilon| \sim 10$ for the majority of metals in the frequency range of interest to us. We shall assume that the frequency of the radiation falls in the optical region.

1. Let the incident wave be characterized by a vector potential

$$\mathbf{A}(z, y) = \mathbf{A}_0 \exp(ik_{0z}z + ik_{0y}y - i\omega t) + \text{compl. conj.}$$

The field which this wave generates in the metal we shall designate by $\mathbf{A}_2$. Under the influence of this field, the system of electrons can undergo a transition with the emission of a quantum $\mathbf{A}_2'$ (outside the metal there arises as a result a field $\mathbf{A}_0'$, corresponding to a quantum having frequency ω', direction in the interval $d\Omega'$ and a given polarization).

Let us introduce a reflection coefficient $d\sigma$, which we shall define as the fraction of the energy falling onto the surface of the superconductor (the latter, it is clear, is equal to $(2\pi)^{-1} \mathfrak{S} t\, |\mathbf{A}_0|^2 \omega^2 \times \cos\theta$, where $\mathfrak{S}$ is the area of the surface and t is time), which is reflected into the range of angles of $d\Omega'$ and frequencies $d\omega'$.*

Rather than solve this problem, it is simpler to find the probability of a transition of our system with absorption of a quantum ω'. In accordance with the principle of detailed balancing,[2] if the quantity

$$dW = B^+(\omega', \Omega') V\omega'^2 d\omega' d\Omega' / (2\pi)^3$$

represents the desired probability for the transition of the system (with V the normalization volume), then the probability for absorption of the quantum is $B^-(\omega', \Omega') = B^+(\omega', \Omega')$. The energy of the emitted quantum must not exceed the excitation energy of the system; i.e., in the present case, $\omega - \omega' > 0$.

Normalizing the amplitude of the secondary field outside the metal with the aid of the relation $(2\pi)^{-1} |\mathbf{A}_0'|^2 \omega'^2 V = \omega'$, we obtain

*Here and in what follows $\hbar = c = 1$.

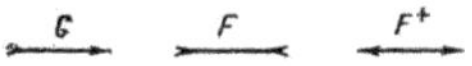

FIG. 1

$$d\sigma = \frac{\omega' dW}{(1/2\pi)\mathfrak{S}tA_0^2\omega^2\cos\theta} = \frac{(2\pi)^2\,[B^-(\omega',\Omega')/\mathfrak{S}t]}{|A_0|^2\,|A_0'|^2\cos\theta\,\omega^2}\,\frac{\omega'^2 d\Omega' d\omega'}{(2\pi)^3}. \quad (1)$$

In the quantum field theory technique the quantity $B^-(\omega', \Omega')$ can be determined from the relation

$$B^-(\omega'\Omega') = \sum_j |S_{j0}|^2,$$

where S_{j_0} is the scattering matrix element corresponding to a transition from the ground state to the state j, and the summation proceeds over all possible states attainable by absorption of a quantum. The problem is thus reduced to the calculation of the elements of the S-matrix for electrons in an external field $\mathbf{A}_2 + \mathbf{A}_2'$, bilinear relative to $\mathbf{A}_2$ and $\mathbf{A}_2'$.

In order to find the relation between $\mathbf{A}_2$ and $\mathbf{A}_2'$ and the corresponding fields outside the metal, one can make use of expressions obtained from the theory of the normal skin effect in a normal metal, since in accordance with our assumption the frequency of the light lies in the optical region. (Since $\omega \gg \Delta$, the distinction between normal and superconducting metals is not significant here, while in the optical frequency range the skin-effect in a normal metal is normal.)

2. We shall find the elements of the S-matrix in which we are interested by using a quantum field-theoretical technique developed for application to superconductors by Gor'kov.[3] The quantity Δ entering into the equations for the functions G and F clearly depends on the external field. To take proper account of the corresponding effects it will be more convenient for us not to write down the equations for the functions G and F, but rather to investigate the various Feynman diagrams for normal electrons having the interaction Hamiltonian:

$$\begin{aligned} H_i = {} & (ie/m)\,[\mathbf{A}_2(x) + \mathbf{A}_2'(x)]\,(\nabla_x - \nabla_{x'})\,\psi^+(x')\,\psi(x)\,|_{x'\to x} \\ & + (e^2/m)\,\mathbf{A}_2(x)\,\mathbf{A}_2'(x)\,\psi^+(x)\,\psi(x) \\ & + \tfrac{1}{2}g\psi^+(x)\,(\psi^+(x)\,\psi(x))\,\psi(x), \end{aligned}$$

i.e., incorporating the four-fermion interaction that leads to superconductivity.

We shall introduce, as was also done in the paper by Abrikosov and Gor'kov,[4] graphical representations for the functions G, F^+ and F (Fig. 1). As stated above, we are interested in S matrices bilinear in $\mathbf{A}_2$ and $\mathbf{A}_2'$. The simplest Feynman diagrams are represented in Fig. 2. The effect of the magnetic field upon Δ is manifested in the fact that along with the simple diagrams of Fig. 2 there appear the more complex diagrams represented in Fig. 3.

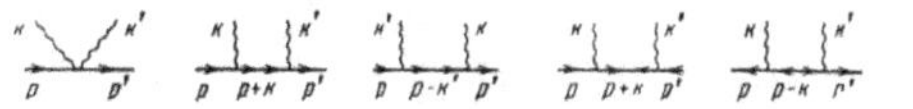

FIG. 2

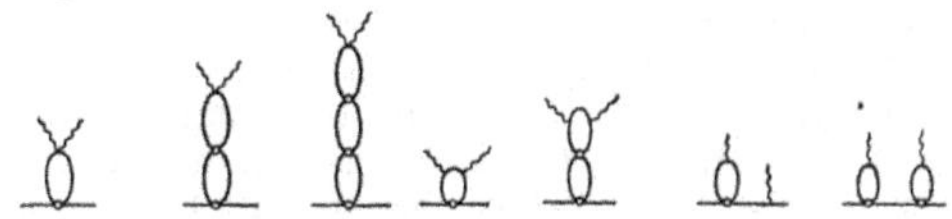

FIG. 3

Let us evaluate all diagrams of the types set forth. We shall demonstrate first that of all the diagrams in Fig. 2, only the diagram 2a is effective. The small contribution of diagrams of the types 2d and 2e follows directly from the form of the function $F(p)$:[3]

$$F(p) = i\Delta/(\varepsilon^2 - \xi^2 - \Delta^2 + i\delta).$$

Since the frequency of the light $\omega \gg \Delta$, we have $F(p+k) \sim \Delta/\omega^2$. Taking into account the fact that each peak with a single potential is proportional to

$$(e/m)\,\mathbf{pA} \sim (e/m)\,p_0 A,$$

it is not difficult to see that the ratio of these diagrams to the diagram 2a is of the order of $\Delta p_0^2/\omega^2 m \ll 1$.

It is possible in similar fashion to demonstrate that the summed contribution of the diagrams in Figs. 2b and 2c is also small. Actually, when summed these diagrams are proportional to

$$G(p+k) + G(p-k) \approx 1/\omega - 1/\omega'.$$

In what follows we shall be interested principally in frequency changes of order Δ, and, in any case, small by comparison with the fundamental frequency. In view of this, the summed contribution of b and c has the relative order of magnitude $(\omega - \omega')/\omega \ll 1$.

For the same reason, diagrams of the type 3b are small by comparison with diagrams 3a. Diagrams 3c can be eliminated if a transverse gauge is chosen for both fields; i.e., $\mathbf{k}\cdot\mathbf{A}_2 = \mathbf{k}'\cdot\mathbf{A}_2' = 0$. This develops quite analogously to what occurs in the derivation of the equations for the electrodynamics of superconductors, since the transverse gauge of the vector potential makes it possible to omit consideration of the variation of Δ with the field. Let us, indeed, consider a loop with a single electromagnetic vertex. This is proportional to

FIG. 4

$\int G(p)G(p+k)(\mathbf{p}\cdot\mathbf{A})d^4p$, or to a similar expression containing F functions instead of G's. From symmetry considerations it is clear that this integral will be proportional to $\mathbf{k}\cdot\mathbf{A} = 0$.

Let us now investigate in more detail diagrams of the type 3a. Each of these diagrams will be large only for the case in which the order of the small constant g will be compensated for by a large value for the integrals arising from the loop. It is not difficult to see that the largest integral will be that having two parallel G lines (Fig. 4), which will be of order $\ln(\omega_D/\eta)$, where $\eta = \max(\omega - \omega', \Delta, v|\mathbf{k} - \mathbf{k}'|)$, and ω_D is the Debye frequency. In view of the fact that $g \ln(\omega_D/\Delta) \sim 1$, it is clear that all of the diagrams in Fig. 3a will be equally effective.

At this point, however, it is necessary to make a reservation. In each of the diagrams 3a there must be at least one loop which does not yield a logarithmic integral. This is the loop with the electromagnetic vertex. In view of the fact that at this vertex the electron arrows must be directed to one side (the combination $\psi^+\psi$ enters into the Hamiltonian), the corresponding loop cannot have the same form as in Fig. 4. More precise analysis shows that diagrams of the type 3a are significant only in the most immediate vicinity of the "threshold" value $\omega - \omega' = 2\Delta$ (cf. below); over the whole range $\omega - \omega' - 2\Delta \gtrsim \Delta$, they may be neglected. Below, we shall consider this particular case. Changes which arise in the neighborhood of the threshold point will be indicated in the discussion of the final result.

3. We are thus left with only one diagram illustrated in Fig. 3a. The corresponding element of the S matrix has the form

$$S_{i0} = -i\int \langle f|(e^2/m)(\mathbf{A}_2(x)\mathbf{A}_2'(x))\,\psi_\alpha^+(x)\psi_\alpha(x)|0\rangle\, d^4x.$$

Inasmuch as we are not interested in the final state of the electronic system, we shall find the summed probability for all probable processes (for the given fields), which is proportional to

$$\sum_f |S_{i0}|^2 = \frac{e^4}{m^2}\int (\mathbf{A}_2^*(x')\mathbf{A}_2'(x'))(\mathbf{A}_2(x)\mathbf{A}_2'(x)) \times \langle \psi_\alpha^+(x')\psi_\alpha(x')\psi_\beta^+(x)\psi_\beta(x)\rangle\, d^4x d^4x'. \quad (2)$$

The mean over the ground state of the four ψ operators, which enters into Eq. (2), may be reduced to the sum of the products of pairs of ψ operators:

$$\langle \psi_\alpha^+(x')\psi_\alpha(x')\psi_\beta^+(x)\psi_\beta(x)\rangle = \langle \psi_\alpha^+(x')\psi_\beta(x)\rangle\langle \psi_\alpha(x')\psi_\beta^+(x)\rangle - \langle \psi_\alpha^+(x')\psi_\beta^+(x)\rangle\langle \psi_\alpha(x')\psi_\beta(x)\rangle$$

(means of the type $\langle\psi_\alpha^+(x')\psi_\alpha(x')\rangle = N$ can be neglected, since they do not depend upon the coordinate, and yield zeros when substituted into (2).

The expressions entering here may be found by the same method as the functions G and F in Gor'kov's theory.[3] For an infinite superconductor we find

$$\begin{aligned}
\langle \psi_\alpha^+(x')\psi_\beta(x)\rangle &= \delta_{\alpha\beta}\int v_p^2\delta(\varepsilon + \varepsilon_p)e^{ip(x-x')}\frac{d^4p}{(2\pi)^3},\\
\langle \psi_\alpha(x')\psi_\beta^+(x)\rangle &= \delta_{\alpha\beta}\int u_p^2\delta(\varepsilon - \varepsilon_p)e^{ip(x'-x)}\frac{d^4p}{(2\pi)^3},\\
\langle \psi_\alpha(x')\psi_\beta(x)\rangle &= -I_{\alpha\beta}\int u_p v_p\delta(\varepsilon - \varepsilon_p)e^{ip(x'-x)}\frac{d^4p}{(2\pi)^3},\\
\langle \psi_\alpha^+(x')\psi_\beta^+(x)\rangle &= I_{\alpha\beta}\int u_p v_p\delta(\varepsilon + \varepsilon_p)e^{ip(x-x')}\frac{d^4p}{(2\pi)^3}.
\end{aligned} \quad (3)$$

Here

$$p = (\varepsilon, \mathbf{p}), \qquad \varepsilon_p = \sqrt{\xi_p^2 + \Delta^2}, \qquad \xi_p = v(|\mathbf{p}| - p_0),$$

$$u_p^2 = \tfrac{1}{2}(1 + \xi_p/\varepsilon_p), \quad v_p^2 = \tfrac{1}{2}(1 - \xi_p/\varepsilon_p);$$

$I_{\alpha\beta}$ is a unitary antisymmetric two-row matrix.

In the present case we are dealing with a superconducting half-space. For this the expressions for the means of the operator pairs differ from the corresponding functions for an infinite space.* At the surface, the correlation functions must satisfy the appropriate boundary conditions. We shall limit ourselves here to the case of specular reflection of the electrons from the boundary. In this case the correlation functions fall to zero for $z = 0$ or $z' = 0$.

In place of the functions (3), it is sufficient in this case to take the differences $\Phi(z - z') - \Phi(z + z')$, which satisfy the same equations as the functions (3), as well as the necessary boundary conditions. Upon substitution into (2) we find that, with an accuracy up to insignificant corrections of order a/λ, where a is the inter-atomic distance and λ is the wavelength of the light, the problem reduces to that for an infinite superconductor. For this, the potentials $\mathbf{A}_2$, $\mathbf{A}_2'$ must be extended symmetrically into the region $z < 0$, while the factor $^1\!/_2$ appears before the integral in (2).

Substituting the expressions (3) into the integral in (2) we obtain

*Strictly speaking, this applies also to the evaluations of the various diagrams carried out above; it can, however, be shown that all results remain correct in the presence of a surface.

$$\sum_f |S_{f0}|^2 = \frac{e^4}{2m^2}\int (\mathbf{A}_2^*(x')\,\mathbf{A}_2^{*\prime}(x'))(\mathbf{A}_2(x)\,\mathbf{A}_2'(x))\,e^{i(p-p')(x-x')}$$
$$\times \frac{1}{2}\left(1-\frac{\xi_p\xi_{p'}}{\varepsilon_p\varepsilon_{p'}}+\frac{\Delta^2}{\varepsilon_p\varepsilon_{p'}}\right)\delta(\varepsilon+\varepsilon_p)\,\delta(\varepsilon'-\varepsilon_{p'})\frac{d^4p}{(2\pi)^3}$$
$$\times \frac{d^4p'}{(2\pi)^3}\,d^4x\,d^4x' = \frac{e^4}{2m^2}\,t\mathfrak{S}\int R(q_z)\,f(\mathbf{q})\,dq_z, \qquad (4)$$

where t represents time; the functions $f(\mathbf{q})$ and $R(q_z)$ have the form

$$f(\mathbf{q}) = \frac{1}{2}\int\frac{d^3p}{(2\pi)^3}\,\delta(q_0-\varepsilon_p-\varepsilon_{p+q})\left(1-\frac{\xi_p\xi_{p+q}}{\varepsilon_p\varepsilon_{p+q}}+\frac{\Delta^2}{\varepsilon_p\varepsilon_{p+q}}\right), \qquad (5)$$

$$R(q_z) = |[\mathbf{A}_2\mathbf{A}_2'(q_z)]|^2 \qquad (6)$$

The symbol $[\mathbf{A}_2\mathbf{A}_2'(q_z)]$ represents the Fourier component of the product $\mathbf{A}_2(x)\cdot\mathbf{A}_2'(x)$, i.e.,

$$\mathbf{A}_2(x)\,\mathbf{A}_2'(x) = \int [\mathbf{A}_2\mathbf{A}_2'(q_z)]\,e^{iqx}\,dq_z/2\pi.$$

The projections of the vector $\mathbf{q}$, with the exception of q_z, are equal to the corresponding differences in momentum and frequency between the incident and reflected quanta: $q_x = k_{0x} - k_{0x}'$, $q_y = k_{0y} - k_{0y}'$, $q_0 = \omega - \omega'$.

4. Let us compute first the function $f(\mathbf{q})$. In view of the fact that $\epsilon_p > 2\Delta$, the integral in (5) is different from zero only for $q_0 > 2\Delta$. This was to be expected, for $q_0 = \omega - \omega'$ is the energy transferred to the electronic system, which cannot be less than 2Δ. The integral (5) is fundamentally dependent upon the relation between $v\,|\mathbf{q}|$ and q_0. For the case $v\,|\mathbf{q}| \gg q_0$ the integral over $d^3p = (2\pi p_0^2/v)\,d\xi\,d\cos\theta$ may be replaced by

$$\frac{2\pi p_0^2}{v^2|\mathbf{q}|}\int_{-\infty}^{\infty}d\xi\int_{-\infty}^{\infty}d\xi' \qquad (\xi=\xi(q),\ \xi'=\xi(p+q)).$$

Terms in $\xi\xi'$ in the expression under the integral sign vanish. After this we proceed to integration over $d\epsilon$ and $d\epsilon'$.

As a result of all these simple operations we obtain

$$f(\mathbf{q}) = \frac{p_0^2}{\pi^2v^2|\mathbf{q}|}\left[\left(\frac{q_0}{2}+\Delta\right)\mathrm{E}\left(\frac{q_0-2\Delta}{q_0+2\Delta}\right)\right.$$
$$\left.-\frac{q_0\Delta}{q_0/2+\Delta}\mathrm{K}\left(\frac{q_0-2\Delta}{q_0+2\Delta}\right)\right], \qquad (7)$$

where E and K are complete elliptic integrals.

In accordance with (4) and (6), the function $f(\mathbf{q})$ is unique and depends on Δ. At the transition to the normal metal, therefore, only this function changes. As $\Delta \to 0$ we find

$$f_n(\mathbf{q}) = p_0^2q_0/2\pi^2v^2|\mathbf{q}|. \qquad (8)$$

In what follows we shall see that the presence of the factor $1/|\mathbf{q}|$ in $f(\mathbf{q})$ leads to the appearance of a large logarithmic integral over q_z. If $q_0 - 2\Delta \gtrsim \Delta$, then in the region $v\,|\mathbf{q}| \ll q_0$ the integral (5) does not depend upon $|\mathbf{q}|$. Therefore, the region $v\,|\mathbf{q}| \gg q_0$ is the principal region of integratiôn over q_z.

In the close vicinity of $q_0 = 2\Delta$ a special situation arises. Analysis of Eq. (5) shows that the range of values of $v\,|\mathbf{p}|$ for which the integral in (5) is proportional to $1/|\mathbf{q}|$ is determined by the inequality $v\,|\mathbf{q}| \geq \sqrt{q_0^2 - 4\Delta^2}$. At the same time, however, as has already been mentioned above, all the diagrams 3a assume significance in the vicinity of $q_0 = 2\Delta$.

Analysis of these diagrams, which, is too cumbersome to carry our here, shows that the lower limit of the logarithmic integration over q_z never falls below Δ, while in the immediate vicinity of $q_0 = 2\Delta$ this limit increases; this leads in the end to the elimination of the logarithmic integral, and then to the reduction of the whole effect to zero. In view of all that has been said, we can right up to the extremely close vicinity of the point $q_0 = 2\Delta$, make use of formula (7), and consider $v\,|\mathbf{q}| \gg q_0$.

5. Let us now go on to the calculation of the function $R(q_2)$. If we take yz as the plane of incidence, and if the field outside the metal is

$$\mathbf{A}(z, y) = \mathbf{A}_0\exp(ik_{0z}z + ik_{0y}y - i\omega t) + \text{compl. conj.},$$

where $k_{0z} = \omega\cos\theta$, and $k_{0y} = \omega\sin\theta$, with θ the angle of incidence, then within the metal the field will have the form

$$\mathbf{A}_2(z, y) = \mathbf{A}_2\exp(ik_{2z}z + ik_{2y}y - i\omega t) + \text{compl. conj.},$$

where $k_{2y} = k_{0y}$, and $k_{2z} = \omega\sqrt{\epsilon - \sin^2\theta}$ (ϵ is the complex dielectric constant), while that value of the radical is chosen for which $\mathrm{Im}\,k_{2z} > 0$. From Maxwell's equations and the boundary conditions at the surface it is possible to express $\mathbf{A}_2$ in terms of $\mathbf{A}_0$. It is necessary here to take account of the already-adopted condition $\mathbf{k}_2\cdot\mathbf{A}_2 = 0$.

a) If the incident wave is so polarized that the electric field is directed along the x axis, then the field will have the same orientation within the metal as well. In this case, the potentials are directed along the x axis, and the relation

$$A_{2x} = \frac{2\cos\theta}{\sqrt{\varepsilon-\sin^2\theta}+\cos\theta}A_{0(x)} \qquad (9a)$$

holds.

b) If in the incident wave the magnetic field is directed along the x axis, then the vector potentials $\mathbf{A}_0$ and $\mathbf{A}_2$ lie in the yz plane, and

$$A_{2z} = \frac{2\sin\theta\cos\theta}{\varepsilon\cos\theta+\sqrt{\varepsilon-\sin^2\theta}}A_{0(yz)},$$

$$A_{2y} = -\frac{2\cos\theta\sqrt{\varepsilon-\sin^2\theta}}{\varepsilon\cos\theta+\sqrt{\varepsilon-\sin^2\theta}}A_{0(yz)}. \qquad (9b)$$

The index (yz) indicates that the vector $\mathbf{A}_0$ lies in the yz plane. The same formulas connect $\mathbf{A}_2'$ with $\mathbf{A}_0'$.

In what follows, we shall investigate those angles θ (and θ') for which the condition $|\epsilon| \gg \sin^2\theta$ is fulfilled. In this event the expressions simplify; in particular, in case b) one may neglect A_{2z} by comparison with A_{2y}.

The Fourier component of the product of the vector potentials for the specular case $[\mathbf{A}_2(z) = \mathbf{A}_2(-z)]$ is

$$[\mathbf{A}_2\mathbf{A}_2'(q_z)] = \int_{-\infty}^{\infty} \mathbf{A}_2(z)\,\mathbf{A}_2'(z)\,e^{-q_z z}dz$$

$$= (\mathbf{A}_2\mathbf{A}_2'^{*})\frac{2i(k_{2z}-k_{2z}'^{*})}{(k_{2z}-k_{2z}'^{*})^2-q_z^2}.$$

We note that in carrying out the integration we shall select only those components of $\mathbf{A}_2(x)$ and $\mathbf{A}_2'(x)$ whose products have a time dependence $\exp\{-i(\omega-\omega')t\}$ (the remaining terms are of no interest, in view of the presence of the δ function in (5) and the condition $\omega-\omega'>0$). The difference $k_{2z}-k_{2z}'^{*}$ appearing in this expression is equal to $2i\omega\kappa$, $\kappa = \mathrm{Im}\sqrt{\epsilon}$, as a consequence of the inequalities $\sin^2\theta$, $\sin^2\theta' \ll |\epsilon|$.

Substituting all of the terms of the function $R(q_z)$ and carrying out a summation over the polarizations of the scattered, and averaging over those of the incident, light, we obtain

$$R(q_z) = 8|\mathbf{A}_0|^2|\mathbf{A}_0'|^2\cos^2\theta\cos^2\theta'$$
$$\times\left\{\frac{\cos^2\varphi}{[(n+\cos\theta)^2+\varkappa^2][(n+\cos\theta')^2+\varkappa^2]}\right.$$
$$+\frac{\sin^2\varphi}{[(n\cos\theta+1)^2+\varkappa^2\cos^2\theta][(n+\cos\theta')^2+\varkappa^2]}$$
$$+\frac{\sin^2\varphi}{[(n+\cos\theta)^2+\varkappa^2][(n\cos\theta'+1)^2+\varkappa^2\cos^2\theta']}$$
$$\left.+\frac{\cos^2\varphi}{[(n\cos\theta+1)^2+\varkappa^2\cos^2\theta][(n\cos\theta'+1)^2+\varkappa^2\cos^2\theta']}\right\}$$
$$\times\frac{16\varkappa^2\omega^2}{(q_z^2+4\varkappa^2\omega^2)^2}. \tag{10}$$

Here we have used the symbols: $\sqrt{\epsilon} = n + i\kappa$, and φ, the angle between the planes of incidence and reflection. Thus, the function $R(q_z)$ is equal to a constant for $q_z \ll 2\kappa\omega$, and falls off rapidly for $q_z \gg 2\kappa\omega$.

6. Let us now substitute (7) and (10) into (4). Taking $2\kappa\omega \gg q_0/v$, we arrive at the conclusion that the principal contribution to $\sum_j |S_{j0}|^2$ is provided by the logarithmic integral over q_z, in the region delimited by the inequalities $|\mathbf{q}| \gg q_0/v$ and $|q_z| \ll 2\kappa\omega$. In view of the fact that $|q_y| < \omega-\omega' \ll q_0/v$ we may, with logarithmic accuracy, write the result of the integration over q_z in the form

$$\int dq_z/|\mathbf{q}| = 2\ln(2\varkappa\omega v/q_0)$$

(the coefficient 2 results from integration over negative and positive q_z).

Substituting (7), (10), and (4) into (1), with attention to the fact that $\omega-\omega' \ll \omega$, we obtain the final expression for $d\sigma$ for the case $\omega-\omega' > 2\Delta$:

$$d\sigma = \frac{2e^4}{\pi^3}\frac{\cos\theta\cos^2\theta'}{\varkappa^2\omega^2}\{[(1+\cos^2\theta)(n^2+\varkappa^2+1)+4n\cos\theta]$$
$$\times[(1+\cos^2\theta')(n^2+\varkappa^2+1)+4n\cos\theta']$$
$$+\sin^2\theta\sin^2\theta'\cos2\varphi(n^2+\varkappa^2-1)^2\}$$
$$\times\{[(n\cos\theta+1)^2+\varkappa^2\cos^2\theta][(n\cos\theta'+1)^2$$
$$+\varkappa^2\cos^2\theta'][(n+\cos\theta)^2+\varkappa^2]$$
$$\times[(n+\cos\theta')^2+\varkappa^2]\}^{-1}\left[\left(\frac{\omega-\omega'}{2}+\Delta\right)\mathrm{E}\left(\frac{\omega-\omega'-2\Delta}{\omega-\omega'+2\Delta}\right)\right.$$
$$\left.-\frac{2(\omega-\omega')\Delta}{\omega-\omega'+2\Delta}\mathrm{K}\left(\frac{\omega-\omega'-2\Delta}{\omega-\omega'+2\Delta}\right)\right]\ln\frac{2\varkappa\omega v}{\omega-\omega'}d\omega' d\Omega. \tag{11}$$

The expression for $d\sigma$ for a normal metal is obtained by substituting $(\omega-\omega')/2$ for the square brackets containing the elliptic integrals E and K.

We note that the ratio of $d\sigma$ to $d\sigma_n$, for the same metal in the normal state, is a universal function of the ratio $(\omega-\omega')/2\Delta$, specifically:

$$(d\sigma - d\sigma_n)/d\sigma_n = \varphi((\omega-\omega')/2\Delta),$$
$$\varphi(x) = \left(1+\frac{1}{x}\right)\mathrm{E}\left(\frac{x-1}{x+1}\right)-\frac{2}{x+1}\mathrm{K}\left(\frac{x-1}{x+1}\right)-1. \tag{12}$$

A graph of the function $\varphi(x)$ is presented in Fig. 5. It is evident from this fugure that only in the region $\omega-\omega' < 4\Delta$ is there an intrinsic difference between the normal and superconducting metal.

The formula obtained is formally applicable over the whole range of frequencies for which $\ln(\kappa\omega v/|\omega-\omega'|)$ is sufficiently large; i.e., for $\omega-\omega' \ll \kappa\omega v/c$ (in the usual units). In reality, however, it is necessary to consider that treatment of the electrons as free is valid only for energies $\epsilon \ll \omega_D$ (ω_D is the Debye frequency). For $\epsilon \sim \omega_D$ there arises a strong attenuation of the electronic excitations, due to emission of phonons. This imposes the limitation $\omega-\omega' \ll \omega_D$, which more or less coincides with the preceding in the optical region. On the low-frequency side, as has already been stated, Eq. (11) becomes invalid in the region $\omega-\omega'-2\Delta \ll \Delta$. While this formula yields a finite

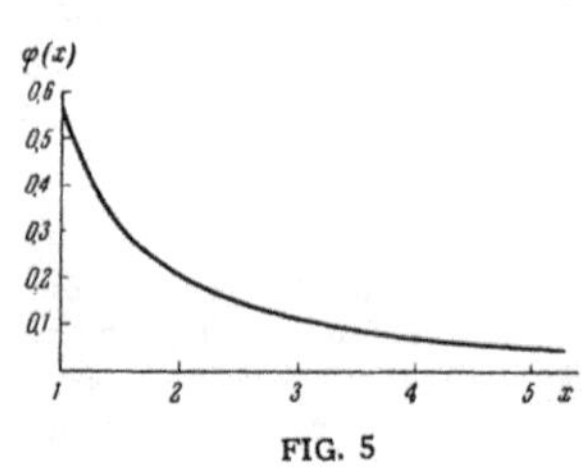

FIG. 5

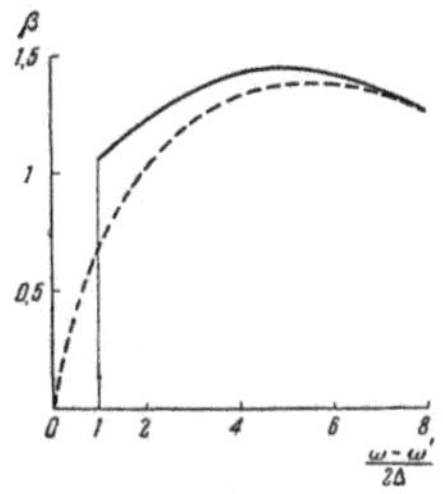

FIG. 6

value at $\omega = 2\Delta$, correct consideration of the graphs 3a gives at this point $d\sigma = 0$. In view of its narrow width, however, this region is not of great interest. For the case of a normal metal the applicability of the derived formula is limited only in the direction of large values of $\omega - \omega'$.

The absolute magnitude of the effect depends to a large degree upon the quantities κ and n. The data available on κ and n for certain metals[5] at $\lambda = 5800$ A and T = 300° K is presented below:

	V	Nb	Ta	Sn	Pb
n	3.03	1.80	2.05	1.48	2,01
$\varkappa$	3.51	2.11	2.31	5.25	3,48

At low temperatures κ changes little, while n decreases slightly. For Nb at $\lambda \approx 5800$ A ($\omega = 3.2 \times 10^{15}$ sec^{-1}), and $\theta = \theta' = 0$:

$$d\sigma = 0.6 \cdot 10^{-12} \beta d\Omega d\omega' / 2\Delta. \qquad (13)$$

The quantity β in (13) is of order unity; its dependence upon $(\omega - \omega')/2\Delta$ is illustrated in Fig. 6. The dashed line represents this same quantity for the normal state.

It follows from evaluation of (13) that for detection of the effect the sensitivity must be at least 10^5 times as high as in the experiment of Khaĭkin and Bykov.[1]

In conclusion, the authors express their gratitude to Academician L. D. Landau and to M. S. Khaĭkin for their consideration of the results of this work.

[1] M. S. Khaĭkin and V. P. Bykov, JETP **30**, 191 (1956), Soviet Phys. JETP **3**, 119 (1956).

[2] L. D. Landau and E. M. Lifshitz, Квантовая механика (Quantum Mechanics), Gostekhizdat, 1948; p. 451.

[3] L. P. Gor'kov, JETP **34**, 735 (1958), Soviet Phys. JETP **34**, 505 (1958).

[4] A. A. Abrikosov and L. P. Gor'kov, JETP **35**, 1558 (1959), Soviet Phys. JETP **35**, 1090 (1959).

[5] Smithsonian Physical Tables, 1954.

Translated by S. D. Elliott
40

SOVIET PHYSICS JETP VOLUME 12, NUMBER 6 JUNE, 1961

CONTRIBUTION TO THE THEORY OF SUPERCONDUCTING ALLOYS WITH PARAMAGNETIC IMPURITIES

A. A. ABRIKOSOV and L. P. GOR'KOV

Institute for Physics Problems, Academy of Sciences, U.S.S.R.

Submitted to JETP editor July 25, 1960

J. Exptl. Theoret. Phys. (U.S.S.R.) **39**, 1781-1796 (December, 1960)

We show in the framework of superconductivity theory that the transition temperature T_c is lowered in comparison with that of a pure metal when paramagnetic impurities are introduced into a superconductor. We have determined the dependence of T_c on the impurity concentration. Superconductivity vanishes at a certain critical concentration. We have made a detailed study of the thermodynamic and magnetic properties of superconducting alloys in the concentration range close to critical. We have found that in that range there is no exponential term in the expression for the heat capacity as $T \to 0$. The concentration dependence of the gap in the energy spectrum of such a system is determined. It is shown that the gap disappears at a concentration $n' \approx 0.9\ n_{cr}$, while at higher concentrations the energy spectrum of the system is continuous.

A large number of recent papers are devoted to a study of the influence of paramagnetic impurities on the transition of a metal into a superconducting state. It was observed in experiments by Matthias et al.[1] that rare earth impurities in lanthanum in the range of concentrations of about 1% appreciably lower the superconducting transition temperature. Herring[2] and also Suhl and Matthias[3] have in this connection expressed the hypothesis that the occurrence of this effect might be explained on the basis of an exchange interaction between the conduction electrons and the impurity atoms. A little later Matthias et al.[4] found also that a small amount of ferromagnetic impurities in titanium can lead to an increase of its superconducting transition temperature.

The present paper is devoted to a study of this problem on the basis of a microscopic superconductivity theory. It is well known that the mechanism responsible for superconductivity is the formation of bound electron pairs in a singlet state. The exchange interaction between electrons and spinning impurity atoms leads to nonconservation of the electron spin, which may affect the formation of Cooper pairs. It is most probable that the spin of the impurity should inhibit the appearance of the superconducting correlation, i.e., cause a decrease in T_c. The rigorous calculation given in the following confirms this supposition.

It is well known that the transition temperature of superconductors with nonmagnetic impurities remains practically unchanged in the region of low concentrations, and this was also shown in a paper by the present authors.[5] The significant change in T_c is in such cases caused by the influence of the foreign atoms on the basic lattice. One can neglect this effect in the low-concentration region in comparison with the change in the electromagnetic properties of the alloy, which are completely due to the scattering of the electrons by impurity atoms. After the metal has changed into the superconducting state, there arises a specific interelectron correlation which acts at distances of the order of $\xi_0 = \hbar v / 2\pi T_c$; these distances have the physical meaning of the average dimensions of a Cooper pair. The scattering of the electrons by the impurities decreases this correlation, restricting it at sufficiently high concentrations to distances of the order of the mean free path.

In the present paper we shall assume that there is a term due to exchange in the interaction between an electron and an impurity atom

$$\hat{v}(\mathbf{r}) = u_1(\mathbf{r}) + u_2(\mathbf{r})(\mathbf{S}\hat{\boldsymbol{\sigma}}), \tag{1}$$

where $\mathbf{S}$ is the moment of the impurity atom and $\hat{\boldsymbol{\sigma}}$ are the electron spin matrices. The Hamiltonian of the interaction between the electrons and the impurity atoms is in the second quantization representation of the form

$$H_{int} = (\psi^+ \hat{V} \psi) = (\psi^+ \sum_a \hat{v}(\mathbf{r} - \mathbf{r}_a)\psi).$$

1. THE DEPENDENCE OF THE TRANSITION TEMPERATURE ON THE IMPURITY CONCENTRATION

We consider first the problem of the dependence of the transition temperature on the impurity concentration. To do this we note that one can describe a superconductor by means of two Green's functions $\mathfrak{G}(x, x')$ and $\mathfrak{F}^+(x, x')$ as was shown by one of the authors.[6] In the thermodynamic approach the equations for the Fourier components of these functions are of the form

$$(i\omega_n + \nabla^2/2m + \mu)\hat{\mathfrak{G}}_\omega(\mathbf{r}, \mathbf{r}') - \hat{V}\hat{\mathfrak{G}}_\omega(\mathbf{r}, \mathbf{r}') + \hat{\Delta}(\mathbf{r})\hat{\mathfrak{F}}^+_\omega(\mathbf{r}, \mathbf{r}') = \delta(\mathbf{r} - \mathbf{r}'),$$

$$(i\omega_n - \nabla^2/2m - \mu)\hat{\mathfrak{F}}^+_\omega(\mathbf{r}, \mathbf{r}') + \hat{V}^t\hat{\mathfrak{F}}^+_\omega(\mathbf{r}, \mathbf{r}') + \hat{\Delta}^*(\mathbf{r})\hat{\mathfrak{G}}_\omega(\mathbf{r}, \mathbf{r}') = 0,$$

$$V^t_{\alpha\beta} = \hat{V}_{\beta\alpha}, \qquad \omega_n = (2n+1)\pi T. \tag{2}$$

We used the fact that in the constant field of the impurities the functions $\mathfrak{G}$ and $\mathfrak{F}^+$ will be functions of the difference between the "time" variables $\tau = \tau_1 - \tau_2$ only, and we have performed a Fourier transformation with respect to this difference. In particular,

$$\hat{\Delta}(\mathbf{r}) = |g| T \sum_\omega \hat{\mathfrak{F}}_\omega(\mathbf{r}, \mathbf{r}). \tag{3}$$

Unlike reference 7, we have written Eqs. (2) in matrix form as far as the spin variables are concerned, and the product $\hat{V}\hat{\mathfrak{G}}$, for instance, indicates a matrix product. Since $\hat{\Delta}(\mathbf{r})$ and $\hat{\Delta}^*(\mathbf{r})$ are expressed in terms of averages over $\psi\psi$ or $\psi^+\psi^+$ in one point,

$$\hat{\Delta}_{\alpha\beta}(\mathbf{r}) = \langle\psi_\alpha(x)\psi_\beta(x)\rangle, \quad \Delta^*_{\alpha\beta}(\mathbf{r}) = \langle\psi^+_\alpha(x)\psi^+_\beta(x)\rangle, \tag{4}$$

it is clear that the relation

$$\Delta_{\alpha\beta}(\mathbf{r}) = -\Delta_{\beta\alpha}(\mathbf{r}), \tag{5}$$

and also

$$\Delta^*_{\alpha\beta}(\mathbf{r}) = -\Delta^*_{\beta\alpha}(\mathbf{r}) \tag{6}$$

are always valid. When the metal is in its normal state with $T > T_c$, Eqs. (2) do not possess solutions with non-vanishing $\mathfrak{F}^+$ and Δ. The temperature T_c, at which the transition into the superconducting phase occurs, is determined by the fact that at $T = T_c$ there exist for the first time solutions of Eqs. (2) with non-vanishing, though arbitrary small, $\mathfrak{F}^+$ (and Δ). Retaining in (2) only the terms which are linear in $\mathfrak{F}^+$ (and Δ) we get

$$(i\omega_n + \nabla^2/2m + \mu)\hat{\mathfrak{G}}'(\mathbf{r}, \mathbf{r}') - \hat{V}(\mathbf{r})\hat{\mathfrak{G}}'_\omega(\mathbf{r}, \mathbf{r}') = \delta(\mathbf{r} - \mathbf{r}'),$$

$$(i\omega_n - \nabla^2/2m - \mu)\hat{\mathfrak{F}}^+_\omega(\mathbf{r}, \mathbf{r}') + \hat{V}^t(\mathbf{r})\hat{\mathfrak{F}}^+_\omega(\mathbf{r}, \mathbf{r}') + \hat{\Delta}^*(\mathbf{r})\hat{\mathfrak{G}}'_\omega(\mathbf{r}, \mathbf{r}') = 0. \tag{7}$$

The first of these equations determines the Green's function $\mathfrak{G}'_\omega(\mathbf{r}, \mathbf{r}')$ of the normal metal when there are impurity atoms present. One checks easily that the Green's function $\mathfrak{G}'_\omega(\mathbf{r}, \mathbf{r}')$, on the other hand, also satisfies the equation

$$(i\omega_n + \nabla'^2/2m + \mu)\hat{\mathfrak{G}}'_\omega(\mathbf{r}, \mathbf{r}') - \hat{\mathfrak{G}}'_\omega(\mathbf{r}, \mathbf{r}')\hat{V}(\mathbf{r}') = \delta(\mathbf{r} - \mathbf{r}'),$$

where the prime on the ∇^2 indicates that the differentiation is with respect to the coordinates $\mathbf{r}'$ in $\mathfrak{G}'_\omega(\mathbf{r}, \mathbf{r}')$. Comparing this last equation with the second of Eqs. (7) we find that $\mathfrak{F}^+_\omega(\mathbf{r}, \mathbf{r}')$ can be expressed in terms of Δ^* as follows

$$(\hat{\mathfrak{F}}^+_\omega(\mathbf{r}, \mathbf{r}'))_{\alpha\beta} = \int \mathfrak{G}'_{-\omega\gamma\alpha}(\mathbf{s}, \mathbf{r})\Delta^*_{\gamma\lambda}(\mathbf{s})\mathfrak{G}'_{\omega\lambda\beta}(\mathbf{s}, \mathbf{r}')d^3s. \tag{8}$$

Substituting this expression into (3) we get an equation for the function $\hat{\Delta}^*(\mathbf{r})$. The temperature at which this equation has for the first time a non-vanishing solution is just the transition temperature T_c which interests us:

$$\Delta^*_{\alpha\beta}(\mathbf{r}) = |g| T \sum_\omega \int \mathfrak{G}'_{-\omega\gamma\alpha}(\mathbf{s}, \mathbf{r})\Delta^*_{\gamma\lambda}(\mathbf{s})\mathfrak{G}'_{\omega\lambda\beta}(\mathbf{s}, \mathbf{r})d^3s. \tag{9}$$

We must average this equation over the positions of the impurity atoms. We developed in reference 5 a method which enabled us to perform such an averaging (see also reference 7). If on the diagrams for the Green's functions or for the other quantities we designate the scattering by an impurity atom by a cross, we see easily that to each such cross there corresponds a factor $v(\mathbf{q})e^{i\mathbf{q}\cdot\mathbf{r}_a}$ where $\mathbf{q}$ is the momentum transferred and $\mathbf{r}_a$ the position of the impurity atom.* The average over $\mathbf{r}_a$ vanishes when $\mathbf{q} \neq 0$ unless this diagram contains at least one scattering at $\mathbf{r}_a$ with $\mathbf{q}' = -\mathbf{q}$. In the latter case the result of the averaging will be

$$n|v(\mathbf{q})|^2, \tag{10}$$

where n is the number of impurity atoms per unit volume. It thus turns out that the averaging over the impurity-atom positions can be put in the form of a special diagram technique, by joining two identical atoms on the diagram with a dotted line, and this corresponds to a factor (10) in the matrix elements.

When taking the average of Eq. (9) one can independently take the average of $\Delta^*(\mathbf{s})$ and of the product of the two Green's functions on the right-hand side. One can directly check that this is correct. It can, however, also be seen from the fact that $\Delta^*(\mathbf{s})$, which is a constant when there

*We use here the Born approximation for the scattering of the electrons by an impurity atom. The physical results are independent of this approximation and we can therefore replace everywhere the Fourier component of the impurity potential by the corresponding scattering amplitude.

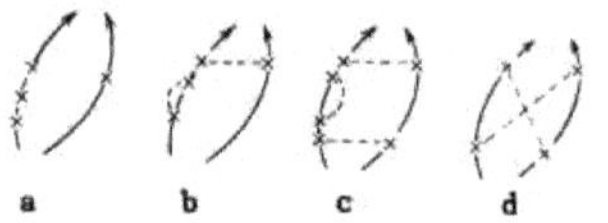

FIG. 1

are no impurities, is only changed near the impurity atoms while $\mathfrak{G}_\omega(\mathbf{r}, \mathbf{s})$ is a fast oscillating function. The result is

$$\overline{\Delta^{*}_{\alpha\beta}} = |g| T \sum_{\omega} \int \overline{\Delta^{*}_{\gamma\lambda}}\, \overline{\mathfrak{G}'_{-\omega\gamma\alpha}(\mathbf{s}, \mathbf{r})\, \mathfrak{G}'_{\omega\lambda\beta}(\mathbf{s}, \mathbf{r})}\, d^3 s. \quad (11)$$

The product of the two Green's functions $\mathfrak{G}'$ in (11) is depicted in Fig. 1a. The averaging over the impurity positions leads to the diagrams of Figs. 1b, c, and d. In this case, however, the dotted line no longer corresponds to a factor (10), but to a factor

$$n v_{\alpha\beta}(\mathbf{q})\, v_{\gamma\lambda}(-\mathbf{q}),$$

since the electron — impurity interaction contains, according to (1), an exchange part.

Apart from averaging over the impurity-atom coordinates, we must still average over the orientation of their spins. We assume that the spins are oriented arbitrarily relative to the lattice and that there is no correlation whatever between them. Therefore

$$\overline{S} = 0; \qquad \overline{S_i S_k} = \tfrac{1}{3} S(S+1)\,\delta_{ik}.$$

We shall discuss later on the basis of such an assumption. Assuming that the averaging sign in (11) includes also an average over the orientation of the spins of all impurity atoms, we shall let each dotted line in Fig. 1 correspond to a factor

$$n\left(|u_1(\mathbf{q})|^2 + \tfrac{1}{3} S(S+1)\, |u_2(\mathbf{q})|^2\, \sigma_{\alpha\beta}\sigma_{\gamma\lambda}\right). \quad (12)$$

It was shown in reference 5 that diagrams such as the ones of Fig. 1d, which contain intersections of dotted lines, and also any diagrams in which a dotted line includes a vertex with a momentum transfer of an electron line of the order of p_0, give a small contribution $\sim 1/p_0 l$, where l is the mean free path, and can therefore be discarded. Dotted lines such as the ones of Fig. 1b which join two crosses on the same electron line give a contribution to the averaged Green's function $\mathfrak{G}'_\omega(\mathbf{r}, \mathbf{r}')$ [we bear in mind that here $\mathfrak{G}'_\omega(\mathbf{r}, \mathbf{r}')$ refers to the normal metal]. One can easily check that in complete analogy with the result of reference 5 the sum of all such diagrams leads to the Green's function

$$\overline{\mathfrak{G}'_\omega(\mathbf{r}, \mathbf{r}')} = \widetilde{\mathfrak{G}}_\omega(\mathbf{r} - \mathbf{r}') = \mathfrak{G}_\omega(\mathbf{r} - \mathbf{r}') \exp(-|\mathbf{r} - \mathbf{r}'|/2l_1), \quad (13)$$

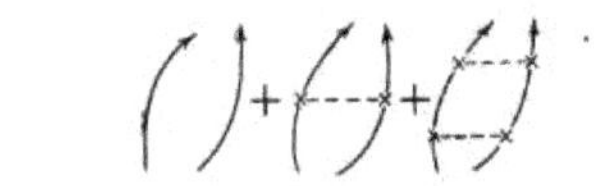

FIG. 2

where $\mathfrak{G}_\omega(\mathbf{r} - \mathbf{r}')$ is the Green's function of the normal metal when there are no impurity atoms. The Fourier components of the function

$$\widetilde{\mathfrak{G}}_\omega(\mathbf{r} - \mathbf{r}') = (2\pi)^{-3} \int \widetilde{\mathfrak{G}}_\omega(\mathbf{p})\, e^{i\mathbf{p}(\mathbf{r} - \mathbf{r}')}\, d^3 p$$

are equal to

$$\widetilde{\mathfrak{G}}_\omega(\mathbf{p}) = [i\omega\eta_1 - \xi]^{-1}, \quad (14)$$

where

$$\eta_1 = 1 + 1/2\tau_1|\omega|, \qquad \xi = v(p - p_0), \qquad v = p_0/m,$$
$$\frac{1}{\tau_1} = \frac{n m p_0}{(2\pi)^2} \int \left\{ |u_1(\mathbf{q})|^2 + |u_2(\mathbf{q})|^2 \frac{1}{3} S(S+1) \right\} do, \quad l_1 = v\tau_1.$$

The evaluation of the average of the product of the two Green's functions in (11) can thus be reduced to a summation of the "ladder" diagrams depicted in Fig. 2. In those diagrams the averaged electron Green's functions (13) and (14) occur as electron lines.

It follows from (5) and (6) that $\Delta^{*}_{\alpha\beta}$ is of the form

$$\Delta^{*}_{\alpha\beta} = \Delta^{*} g_{\alpha\beta}, \qquad g_{\alpha\beta} = \begin{pmatrix} 0 & 1 \\ -1 & 0 \end{pmatrix}_{\alpha\beta}, \quad (15)$$

where $g_{\alpha\beta}$ is the covariant metric tensor in spinor algebra.[8] We can now cancel the constant Δ^* in (11). We introduce the symbol $K_{\omega\alpha\beta}(\mathbf{p}_1, \mathbf{p}_2)$ for the Fourier component of the expression

$$\overline{\mathfrak{G}_{-\omega\gamma\alpha}(\mathbf{s}, \mathbf{r})\, g_{\gamma\lambda}\, \mathfrak{G}_{\omega\lambda\beta}(\mathbf{s}, \mathbf{r}')}$$
$$= (2\pi)^{-6} \int K_{\omega\alpha\beta}(\mathbf{p}_1, \mathbf{p}_2) \exp\{i\mathbf{p}_1(\mathbf{s} - \mathbf{r})$$
$$+ i\mathbf{p}_2(\mathbf{s} - \mathbf{r}')\}\, d^3 p_1 d^3 p_2. \quad (16)$$

The sum of all diagrams of Fig. 2 can be obtained by solving the following obvious integral equation

$$K_{\omega\alpha\beta}(\mathbf{p}_1, \mathbf{p}_2) = \widetilde{\mathfrak{G}}_\omega(\mathbf{p}_1)\, \widetilde{\mathfrak{G}}_{-\omega}(\mathbf{p}_2) \Big[g_{\alpha\beta} + \frac{n}{(2\pi)^3} \int \Big[|u_1|^2$$
$$+ \frac{1}{3} S(S+1)\, |u_2|^2 (\sigma_{\lambda\alpha}\sigma_{\rho\beta}) \Big] K_{\lambda\rho\omega}(\mathbf{p}_1', \mathbf{p}_2')\, d^3 p' \Big],$$

$$\mathbf{p}_1 - \mathbf{p}_1' = \mathbf{p}_2' - \mathbf{p}_2. \quad (17)$$

It is clear from this that $K_{\omega\alpha\beta}$ is proportional to $g_{\alpha\beta}$. One verifies easily by using spinor algebra that

$$\hat{g}^{-1} \hat{\sigma}^{t} \hat{g} = -\hat{\sigma}. \quad (18)$$

If $K_{\omega\alpha\beta}(\mathbf{p}_1, \mathbf{p}_2) = K_\omega(\mathbf{p}_1, \mathbf{p}_2)\, g_{\alpha\beta}$, we have thus

$$K_\omega(\mathbf{p}_1, \mathbf{p}_2) = \widetilde{\mathfrak{G}}_\omega(\mathbf{p}_1)\, \widetilde{\mathfrak{G}}_{-\omega}(\mathbf{p}_2) \Big[1 + \frac{n}{(2\pi)^3} \int \Big[|u_1|^2$$
$$- \frac{S(S+1)}{4} |u_2|^2 \Big] K_\omega(\mathbf{p}_1', \mathbf{p}_2')\, d^3 p' \Big]. \quad (19)$$

We are, according to (11), interested in the values for $p_1 = -p_2$. Substituting Eq. (14) for $\tilde{\mathfrak{G}}_\omega(p)$ into (19) and denoting the integral on the right-hand side of this expression by L_ω we find

$$K_\omega(\mathbf{p}, \mathbf{q}-\mathbf{p}) = \tilde{\mathfrak{G}}_\omega(\mathbf{p})\, \tilde{\mathfrak{G}}_{-\omega}(\mathbf{q}-\mathbf{p})\,[1 + L_\omega]. \qquad (20)$$

For L_ω we get the following expression

$$L_\omega = \frac{1}{2\tau_2\omega\eta_1}\left[1 - \frac{1}{2\tau_2\omega\eta_1}\right]^{-1} = \frac{1}{2\tau_2\omega\eta_s}, \qquad (21)$$

where

$$\frac{1}{\tau_2} = \frac{nmp_0}{(2\pi)^2}\int\left[|u_1|^2 - \frac{1}{4}S(S+1)|u_2|^2\right]do,$$

$$\eta_s = 1 + 1/\tau_s|\omega|; \qquad 1/\tau_s = 1/2\tau_1 - 1/2\tau_2 > 0.$$

Substituting the expression for L_ω into (20) and then substituting $K_\omega(p_1, p_2)$ into the kernel of Eq. (11) we get, if we use (21), the following relation

$$1 = \frac{|g|T_c}{(2\pi)^3}\sum_\omega\int\frac{\eta_1/\eta_s}{\omega^2\eta_1^2 + \xi^2}\,d^3\mathbf{p}; \qquad \omega_n = (2n+1)\pi T_c.$$

The sum and the integral over d^3p in this expression diverge for large ω and ξ. In the papers by Bardeen, Cooper, and Schrieffer[9] and Bogolyubov[10] this divergence was removed by cutting off the electron energies at $|\xi| = \omega_D$. It is convenient to subtract from the right-hand side of the relation which we have obtained the term that does not contain τ. The remaining difference can be integrated over ξ. Putting $1/|g| = (mp_0/2\pi^2)\ln(2\omega_D\gamma/\pi T_{c0})$, where T_{c0} is the transition temperature of the pure superconductor, we get

$$\frac{1}{2}\ln\frac{T_{c0}}{T_c} = \frac{1}{\pi T_c\tau_s}\sum_{n>0}\frac{1}{(2n+1)(2n+1+1/\tau_s\pi T_c)}.$$

We can express the expression on the right-hand side in terms of $\psi(x)$, the derivative of the logarithm of the Γ-function

$$\ln\frac{T_{c0}}{T_c} = \psi\left(\frac{1}{2} + \frac{\rho}{2}\right) - \psi\left(\frac{1}{2}\right), \qquad (22)$$

where $\rho = 1/\pi\tau_s T_c$. The shift in T_c is thus connected only with the exchange part of the interaction and depends on the parameter $1/\tau_s > 0$. Introducing f_+ and f_-, the scattering amplitudes of an electron with total angular momentum $S + 1/2$ and $S - 1/2$, we can express $1/\tau_s$ in the form

$$\frac{1}{\tau_s} = \frac{nmp_0}{2\pi^2}\frac{S(S+1)}{(2S+1)^2}\int|f_+ - f_-|^2\,do.$$

If $\rho \ll 1$, i.e., if the impurity concentration is low, we have

$$T_c = T_{c0} - \pi/4\tau_s. \qquad (23)$$

At low impurity concentrations the transition temperature decreases proportionally to the concentration.

If $\rho \gg 1$, i.e., if $T_c/T_{c0} \ll 1$, we have

$$T_c^2 = (6/\pi^2\tau_s^2)\ln(\pi T_{c0}\tau_s/2\gamma). \qquad (24)$$

At a certain critical concentration which is determined by the condition

$$\tau_{s\,cr} = 2\gamma/\pi T_{c0}, \qquad (25)$$

superconductivity vanishes at all temperatures in a system with paramagnetic impurities. The "critical mean free path" $l_{s\,cr} = v\tau_{s\,cr} \sim 10^{-4}$ cm. One should, however, bear in mind that the magnitude of the exchange interaction u_2 in (1) is always less than the non-exchange part so that the value of l_s is roughly two orders of magnitude longer than the mean free path which is determined, say, by the conductivity of the normal metal. Thus, $l_s \sim 10^{-4}$ cm corresponds to $l \sim 10^{-6}$ cm, or, an impurity concentration of about 1%.

The experimental data of Matthias, Suhl, and Corenzwit[1] indicate apparently the linear dependence of the change in the critical temperature of a solution of gadolinium in lanthanum. If we use the slope of their straight line to construct the theoretical dependence according to Eqs. (23) and (22), the disagreement with the experimental data is nowhere more than 10%. One must regard such an agreement as good. At the same time there exists another paper by the same authors[4] in which they report that for titanium with several ferromagnetic elements as impurities the transition temperature increases with increasing number of impurity atoms at comparatively low concentrations (1 - 5%). This result is in direct contradiction with the results obtained in the foregoing. The reason for this is not clear at the present time. Of course, one must note the well-known difficulties of experiments with alloys particularly in connection with the poor solubility of elements of the iron type in titanium.* Moreover, the concentrations used in reference 4 were of the order of several percent.

Generally speaking, the theory given in the foregoing contained an expansion in terms of the order a/l, where a is the interatomic distance and l the mean free path, and it should therefore be valid at sufficiently low concentrations when the presence of impurity atoms should lead solely to the scattering of the conduction electrons. It is possible that the increase in T_c observed in reference 4 is a secondary effect caused by different interactions between the impurity atoms which are important at the concentrations studied. In any

*The authors are grateful to A. I. Shal'nikov for drawing their attention to this fact.

case, the actual nature of the impurity atoms was inessential for our consideration.

Let us now consider in more detail the possibility of averaging over the orientation of the spin of a separate atom. There have appeared recently a number of papers[1,11] in which the existence of "impurity" ferromagnetism was detected in a number of alloys and, in particular, it was asserted in reference 11 that in some cases it was possible that there existed simultaneously superconductivity and ferromagnetism caused by paramagnetic impurities. If, indeed, some ferromagnetic ordering of spins were possible, the averaging over the spin orientation which was used by us should be completely unfounded. In that case it would, in general, not be possible to consider the spins on separate impurity atoms to be independent. In the framework of the model considered here, however, we see no positive basis whatever for such an ordering at low impurity concentrations.

Ruderman and Kittel[12] have considered the interaction between two paramagnetic atoms caused in a metal by the exchange of conduction electrons. The fact that the long-range part of this interaction oscillates over distances of the order of atomic distances, is typical of this interaction. At low concentrations, where the distances between the impurity atoms are sufficiently large, this interaction will thus vanish on the average for all atoms and, even if it leads to a ferromagnetic transition, this will occur with an exponentially small Curie temperature.* As far as the direct dipole interaction between impurities is concerned, this is also too small.

On these grounds, our averaging over the spins is fully valid for low concentrations. Apparently, this is precisely the situation with the experiments by Matthias et al.[1] from which one can see that for low concentrations the dependence of the ferromagnetic transition temperature on the concentration is not at all linear. The crossing of the ferromagnetic and the superconducting transition temperatures of an alloy, which was observed in reference 11, occurred when the concentration of one of the components was about 6%. Such a concentration is no longer low and all that has been said in the foregoing is inapplicable to it. As far as the nature of impurity ferromagnetism is concerned for such a system, it is probable that this is caused by the above-mentioned exchange interaction between impurities, which is due to the conduction electrons.

We must draw attention to the fact that order-of-magnitude estimates of all the quantities lead, as was shown in the foregoing, to the conclusion that the whole phenomenon breaks up in the concentration range on the order of 1%. The magnitude of the interval in which T_c changes with the impurity concentration as determined from our theory depends thus strongly on the nature of the effect studied or, to be more precise, on the magnitude of the exchange interaction u_2 in (1). The larger u_2, the faster will the T_c corresponding to the superconducting transition decrease and the lower the concentration that corresponds to the critical value $\tau_{s\,cr}$ of (25). It follows in any case from (23) that for low concentrations T_c must decrease linearly with increasing concentration. As we have said already, the fact that this was not observed in reference 11 shows that it is necessary to study those alloys at even lower concentrations.

2. THE THERMODYNAMIC AND ELECTROMAGNETIC PROPERTIES OF ALLOYS NEAR THE CRITICAL CONCENTRATION

We now turn to a study of the properties of alloys with paramagnetic impurities. Of most interest will be the results obtained in the concentration range where the temperature of the superconducting transition is small compared with T_{c0} of a pure superconductor, i.e., when $\tau_s \approx \tau_{s\,cr}$. In that range the quantities Δ and $\mathfrak{F}^+$ are small ($\Delta\tau_{s\,cr} \ll 1$). We write Eqs. (2) in yet another form, including in them an additional magnetic field

$$\left\{i\omega_n + \frac{1}{2m}\left(\frac{\partial}{\partial \mathbf{r}} - ie\mathbf{A}\right)^2 + \mu\right\}\hat{\mathfrak{G}}_\omega(\mathbf{r},\mathbf{r}') - \hat{V}\hat{\mathfrak{G}}_\omega(\mathbf{r},\mathbf{r}') + \hat{\Delta}(\mathbf{r})\hat{\mathfrak{F}}_\omega^+(\mathbf{r},\mathbf{r}') = \delta(\mathbf{r}-\mathbf{r}'),$$

$$\left\{i\omega_n - \frac{1}{2m}\left(\frac{\partial}{\partial \mathbf{r}} + ie\mathbf{A}\right)^2 - \mu\right\}\hat{\mathfrak{F}}_\omega^+(\mathbf{r},\mathbf{r}') + \hat{V}'\hat{\mathfrak{F}}_\omega^+(\mathbf{r},\mathbf{r}') + \hat{\Delta}^*(\mathbf{r})\hat{\mathfrak{G}}_\omega(\mathbf{r},\mathbf{r}') = 0. \quad (26)$$

In the region under consideration we can expand Eqs. (26) in terms of the quantity Δ. The corresponding calculations are similar to the derivation, by one of the authors,[7] of the equations of the Ginzburg-Landau theory[14] near T_c for superconducting alloys.

Introducing the Green's function $\mathfrak{G}'_\omega(\mathbf{r},\mathbf{r}')$ of the normal metal, which satisfies the equation

$$\left\{i\omega_n + \frac{1}{2m}\left(\frac{\partial}{\partial \mathbf{r}} - ie\mathbf{A}\right)^2 + \mu\right\}\hat{\mathfrak{G}}'_\omega(\mathbf{r},\mathbf{r}') - \hat{V}\hat{\mathfrak{G}}'_\omega(\mathbf{r},\mathbf{r}') = \delta(\mathbf{r}-\mathbf{r}'),$$

*We feel in this connection that the attempt by Brout[13] to explain impurity ferromagnetism by a model where the Hamiltonian contains a constant term proportional to $\Sigma S_i S_k$, where S_i are the spins of the impurity atoms, is incorrect.

we can use it to write for $\mathfrak{F}^+_\omega$ and $\mathfrak{G}_\omega$ up to terms of the third and second order in $|\Delta|$ respectively

$$\mathfrak{F}^+_{\omega\alpha\beta}(\mathbf{r}, \mathbf{r}') = \int \mathfrak{G}'_{-\omega\lambda\alpha}(\mathbf{s}, \mathbf{r})\, \Delta^*_{\lambda\nu}(\mathbf{s})\, \mathfrak{G}'_{\omega\nu\beta}(\mathbf{s}, \mathbf{r}')\, d^3\mathbf{s}$$
$$-\iiint \mathfrak{G}'_{-\omega\lambda\alpha}(\mathbf{s}, \mathbf{r})\, \Delta^*_{\lambda\nu}(\mathbf{s})\, \mathfrak{G}'_{\omega\nu k}(\mathbf{s}, \mathbf{l})\, \Delta_{kl}(\mathbf{l})\, \mathfrak{G}'_{-\omega\varkappa l}(\mathbf{m}, \mathbf{l})$$
$$\times \Delta^*_{\varkappa\nu}(\mathbf{m})\, \mathfrak{G}'_{\omega\nu\beta}(\mathbf{m}, \mathbf{r}')\, d^3\mathbf{m}\, d^3\mathbf{l}\, d^3\mathbf{s},$$

$$\mathfrak{G}_{\omega\alpha\beta}(\mathbf{r}, \mathbf{r}') = \mathfrak{G}'_{\omega\alpha\beta}(\mathbf{r}, \mathbf{r}') - \iint \mathfrak{G}'_{\alpha k}(r, l)\, \Delta_{kl}(\mathbf{l})\, \mathfrak{G}'_{-\omega\varkappa l}$$
$$\times (\mathbf{m}, \mathbf{l})\, \Delta^*_{\varkappa\rho}(\mathbf{m})\, \mathfrak{G}'_{\omega\rho\beta}(\mathbf{m}, \mathbf{r}')\, d^3\mathbf{m}\, d^3\mathbf{l}. \qquad (27)$$

Substituting Eq. (27) for $\mathfrak{F}^+$ into (3) we get an equation to determine $\hat{\Delta}^*(\mathbf{r})$

$$\Delta^*_{\alpha\beta}(\mathbf{r}) = |g|\, T \sum_\omega \Big\{ \int \mathfrak{G}'_{-\omega\lambda\alpha}(\mathbf{s}, \mathbf{r})\, \Delta^*_{\lambda\nu}(\mathbf{s})\, \mathfrak{G}'_{\omega\nu\beta}(\mathbf{s}, \mathbf{r})\, d^3\mathbf{s}$$
$$-\iiint \mathfrak{G}'_{-\omega\lambda\alpha}(\mathbf{s}, \mathbf{r})\, \Delta^*_{\lambda\nu}(\mathbf{s})\, \mathfrak{G}'_{\omega\nu k}(\mathbf{s}, \mathbf{l})\, \Delta_{kl}(\mathbf{l})\, \mathfrak{G}'_{-\omega\varkappa l}(\mathbf{m}, \mathbf{l})$$
$$\times \Delta^*_{\varkappa\nu}(\mathbf{m})\, \mathfrak{G}'_{\omega\nu\beta}(\mathbf{m}, \mathbf{r})\, d^3\mathbf{m}\, d^3\mathbf{l}\, d^3\mathbf{s} \Big\}. \qquad (28)$$

As far as the dependence of the Green's function $\mathfrak{G}'_{\omega\alpha\beta}(\mathbf{s}, \mathbf{r})$ on the magnetic field is concerned, it was established in reference 7 that since the screening radius of the electron in a magnetic field is large compared with the mean free path, one can write it in the form

$$\mathfrak{G}'_{\omega\alpha\beta}(\mathbf{s}, \mathbf{r}) = \exp\{iec^{-1}(\mathbf{A}, \mathbf{s} - \mathbf{r})\}\, \mathfrak{G}^0_{\omega\alpha\beta}(\mathbf{s}, \mathbf{r}), \qquad (29)$$

where $\mathfrak{G}^0_{\omega\alpha\beta}(\mathbf{s}, \mathbf{r})$ is the Green's function of the normal electrons when there is no field. The concentration range $n \sim n_{cr}$ is characterized, as we have seen, by the quantities $1/\tau_s \sim T_{c0}$. As has been stated above, the magnitude of the exchange interaction is appreciably less than the non-exchange part in (1). The normal mean free path l is thus in that case appreciably less than the correlation parameter $\xi \sim \hbar v/2\pi T_{c0}$.

It is natural to assume that under those conditions a localized situation occurs, i.e., that the function $\Delta(\mathbf{r})$ and the field $A(\mathbf{r})$ change over distances which are larger than the characteristic distances in the integrals of Eq. (28). For this reason we can in the third-order term neglect the dependence of $\Delta(\mathbf{r})$ on the coordinates and that of the functions $\mathfrak{G}'$ on the field, as is confirmed by the final result, and in the second term we can as far as this dependence is concerned expand in a Taylor series about the point $\mathbf{r}$, limiting ourselves to terms of the second order in $|\mathbf{r} - \mathbf{s}|$. Using (29) we get as a result

$$\frac{1}{|g|}\Delta^*(\mathbf{r})\, g_{\alpha\beta} = \Delta^*(\mathbf{r})\, Q^{(2)}_{\alpha\beta}$$
$$+ \frac{1}{6}\Big(\frac{\partial}{\partial \mathbf{r}} + 2ie\mathbf{A}(\mathbf{r})\Big)\Delta^*(\mathbf{r})\, Q^{(2)}_{\alpha\beta} + B_{\alpha\beta}|\Delta(\mathbf{r})|^2 \Delta^*(\mathbf{r});$$
$$Q^{(1)}_{\alpha\beta} = T \sum_\omega \int \overline{\mathfrak{G}^0_{-\omega\lambda\alpha}(\mathbf{s}, \mathbf{r})\, \hat{g}_{\lambda\nu}\, \mathfrak{G}^0_{\omega\nu\beta}(\mathbf{s}, \mathbf{r})}\, d^3\mathbf{s}, \qquad (30)$$

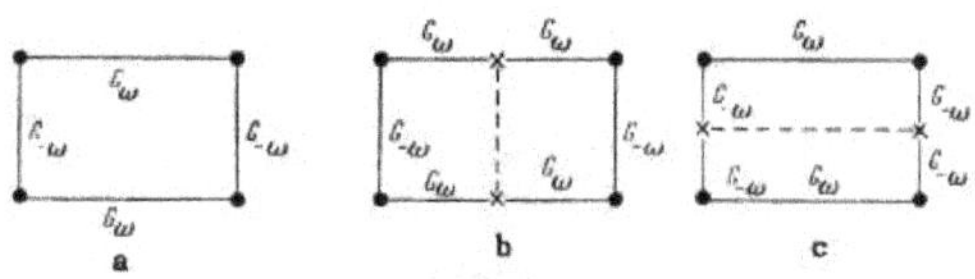

FIG. 3

$$Q^{(2)}_{\alpha\beta} = T \sum_\omega \int |\mathbf{r} - \mathbf{s}|^2\, \overline{\mathfrak{G}^0_{-\omega\lambda\alpha}(\mathbf{s}, \mathbf{r})\, g_{\lambda\nu}\, \mathfrak{G}^0_{\omega\nu\beta}(\mathbf{s}, \mathbf{r})}\, d^3\mathbf{s}, \quad B_{\alpha\beta} = T$$
$$\times \sum_\omega \iiint \overline{\mathfrak{G}^0_{-\omega\lambda\alpha}(\mathbf{s}, \mathbf{r})\, g_{\lambda\nu}\, \mathfrak{G}^0_{\omega\nu k}(\mathbf{s}, \mathbf{l})\, g_{kl}\, \mathfrak{G}^0_{-\omega\varkappa l}(\mathbf{m}, \mathbf{l})\, g_{k\nu}\, \mathfrak{G}^0_{\omega\nu\beta}(\mathbf{m}, \mathbf{r})}$$
$$\times d^3\mathbf{m}\, d^3\mathbf{l}\, d^3\mathbf{s}. \qquad (31)$$

As before, the bar indicates averaging over the impurity atom positions and over the orientation of their spins. The first two coefficients $Q^{(1)}_{\alpha\beta}$ and $Q^{(2)}_{\alpha\beta}$ can easily be connected with the kernel $K_{\omega\alpha\beta}(\mathbf{p}_1, \mathbf{p}_2)$ of (16). Indeed, we have

$$Q^{(1)}_{\alpha\beta} = \frac{T}{(2\pi)^3} \sum_\omega \int K_{\omega\alpha\beta}(\mathbf{p}, -\mathbf{p})\, d^3\mathbf{p},$$

$$Q^{(2)}_{\alpha\beta} = -\frac{T}{(2\pi)^3} \sum_\omega \int \Big[\frac{\partial^2}{\partial q_i^2} K_{\omega\alpha\beta}(\mathbf{p}, \mathbf{q} - \mathbf{p})\Big]_{q=0} d^3\mathbf{p}.$$

To find them we must thus solve Eqs. (19) and (20) up to terms of second order in q.

The product of four $\mathfrak{G}$-functions in B is schematically depicted in Fig. 3a. It follows from (20) and (21) that we average over the impurity atom positions first by replacing the lines in Fig. 3a by averaged Green's functions $\tilde{\mathfrak{G}}_\omega(\mathbf{p})$, and by replacing unity in the vertex by the ratio η_1/η_s. Moreover, the averaging gives rise to the two new diagrams of Figs. 3b and c, in which a dotted line joins two Green's functions with the same frequency. One verifies easily that all other diagrams are either equal to zero, (because all the poles lie in one half-plane) when integrating over $\xi = v(p - p_0)$ or contain intersecting dotted lines and their contribution can be neglected for the reasons given at the start of this paper. The actual determination of the coefficients $Q^{(1)}$, $Q^{(2)}$, and B requires elementary, but somewhat cumbersome algebraic calculations which we shall omit here.

As in (18), it is extremely essential to maintain the correct order of the spinor indices in the calculations. If we assume that in (1) the exchange part of the interaction is small compared with the non-exchange part, we get the final form of Eq. (30) in the temperature range $T \ll T_{c0}$ and range of concentrations $|n - n_{cr}| \ll n_{cr}$

$$\left\{\pi^2(T^2-T_c^2)+\frac{|\Delta|^2}{2}\right\}\Delta^*+2\left(\frac{\partial}{\partial \mathbf{r}}+2ie\mathbf{A}\right)^2\Delta^* v^2\frac{\tau_{tr}}{\tau_s}=0, \tag{32}$$

where τ_{tr} is the "transport" collision time, which enters in the conductivity of the normal metal

$$\sigma = Ne^2\tau_{tr}/m.$$

In the approximation considered we have

$$\frac{1}{\tau_{tr}}=\frac{nmp_0}{(2\pi)^2}\int |u_1|^2(1-\cos\theta)\,do.$$

When there is no field

$$\Delta^2=2\pi^2(T_c^2-T^2). \tag{33}$$

To determine the thermodynamic characteristics of the system we use the well-known formula for the variation of the thermodynamic potential

$$\delta\Omega/\delta g = g^{-1}\langle H_{int}\rangle.$$

We get from this[15]

$$\Omega_s-\Omega_n=\int_0^{|g|}\delta(1/|g|)\,\Delta^2. \tag{34}$$

Using the relation $\delta(1/|g|) = -(mp_0/2\pi^2)\delta T_{c0}/T_{c0}$ and Eq. (32) we get

$$F_s-F_n=-\frac{mp_0}{2\pi^2}\frac{\tau_s^2\Delta^4}{24}=-\frac{mp_0}{2\pi^2}\frac{\tau_s^2\pi^4}{6}(T_c^2-T^2)^2. \tag{35}$$

The critical field, defined in the usual way, is equal to

$$H_c=\frac{2\gamma}{T_{c0}}\sqrt{\frac{2mp_0\pi}{3}}\left(1-\frac{T^2}{T_c^2}\right)=H_{c0}\left(1-\frac{T^2}{T_c^2}\right).$$

We find from (35) the ratio of the heat capacity in the superconducting phase to the heat capacity in the normal phase at $T = T_c$

$$\begin{aligned}C_s(T)/C_n(T_c)&=T/T_c-(\pi^2\tau_s^2/T_c)(T_c^2T-3T^3)\\&=T/T_c-12.4(T_c/T_{c0})^2(T/T_c-3T^3/T_c^3).\end{aligned} \tag{36}$$

When there is a field one must add to Eq. (34) the equation of the electrodynamics of superconductors. To do this one must substitute in the definition of the electromagnetic current

$$\mathbf{j}(\mathbf{r})=\left\{\frac{ie}{m}(\nabla_r-\nabla_{r'})\mathfrak{G}(x,x')+\frac{2e^2\mathbf{A}(\mathbf{r})}{mc}\mathfrak{G}(x,x')\right\}_{\mathbf{r}=\mathbf{r}',\,t'=t+0}$$

the expression (27) for the Green's function. We shall not dwell on the calculations since they are, apart from the existence of spin at the impurities, similar to the calculations in reference 7. Performing the averaging and restricting ourselves to the first non-vanishing terms in **A** and Δ, we get

$$\mathbf{j}=\left\{\frac{ie}{m}\left(\Delta\frac{\partial\Delta^*}{\partial \mathbf{r}}-\Delta^*\frac{\partial\Delta}{\partial \mathbf{r}}\right)-\frac{4e^2}{mc}\mathbf{A}|\Delta|^2\right\}\frac{N\tau_s\tau_{tr}}{2\pi^2}. \tag{37}$$

Equations (32) and (37) together have the same form as the set of equations in the phenomenological Ginzburg-Landau theory.[14] The origin of the doubled elementary charge in these equations has a distinct physical meaning. It is related to the existence of pair correlations in present-day superconductivity theory.

In contrast to the case of pure superconductors, and also to the case of non-magnetic impurities,[7,15] these equations are applicable here in the whole temperature range of the superconducting phase. The temperature dependence enters into the different characteristics of the system through the factor $1 - T^2/T_c^2$ in Eq. (32). In particular, we get for the penetration depth of a weak magnetic field in superconductors

$$\begin{aligned}\delta&=\left[\frac{16\pi^3Ne^2\tau_s\tau_{tr}T_c^2}{mc^2}\left(1-\frac{T^2}{T_c^2}\right)\right]^{-1/2}\\&=\frac{c}{4\pi T_c}\left[\frac{2\sigma\gamma}{T_{c0}}\left(1-\frac{T^2}{T_c^2}\right)\right]^{-1/2}.\end{aligned} \tag{38}$$

For Eqs. (32) and (37) we can construct one dimensionless constant

$$\varkappa = 2\sqrt{2}eH\delta^2/\hbar c = ec(\pi\gamma)^{1/2}/4\sigma,$$

where γ is the coefficient in the linear law for the electronic part of the heat capacity of the normal metal. It is worth noting that as in reference 7 κ depends only on the parameters of the normal metal, while the dependence on τ_s disappears.

The most interesting result is that there is no exponential term in the heat capacity (36) as $T \to 0$. This indicates that there is no gap in the spectrum of such superconductors. This conclusion is confirmed also by the absence of a threshold in the absorption of electromagnetic radiation at $T = 0$. We can, namely, obtain by a method which we have used before[5] the following expression for the dielectric constant of paramagnetic alloys

$$\varepsilon(\omega)=-c^2/\omega^2\delta^2+4\pi i\sigma/\omega. \tag{39}$$

We study this problem now in more detail.

3. THE DEPENDENCE OF THE GAP IN THE SPECTRUM AT T = 0 ON THE IMPURITY CONCENTRATION AND THE HEAT CAPACITY AT LOW TEMPERATURES

Migdal and Galitskiĭ[16] have shown that when $\omega > 0$ the imaginary part of the Fourier component of the electron Green's function

$$G(x,x')=-i\langle T(\psi(x),\psi^+(x'))\rangle$$

can at absolute zero be written in the form

$$\operatorname{Im} G_\omega(\mathbf{p})=\sum_n \rho_n(\mathbf{p})\,\delta(\omega-\omega_{n0}),\quad (\omega>0),$$

where $\rho_n(\mathbf{p})$ is some real spectral function, and

$\omega_{n0} = E_n - E_0$ is the difference in energy between the n-th excited state and the ground state of the system. When there are impurities present the system has no single-particle spectrum, which is determined from the poles of the function $G_\omega(p)$. All the same, however, if the first excited state is separated from the ground state by a gap, the magnitude of the gap is determined by the value of the frequency ω at which the Green's function has a non-vanishing imaginary part for the first time.

One can show in the case of a superconductor that the same holds for the function $iF_\omega(p)$, where $F_\omega(p)$ is the Fourier component of $F^+(x, x') = \langle T(\psi^+(x), \psi^+(x'))\rangle$, or for the function $iF^+_{R\omega}(p)$, which is the Fourier component of the so-called retarded anticommutator of the two ψ^+ (the two functions are the same at $T = 0$ for $\omega > 0$). Bearing in mind the further calculation of the heat capacity, we consider the function $F^+_{R\omega}(p)$. Its characteristic property is its direct connection with the temperature-dependent function $\mathfrak{F}^+_{\omega_n}(p)$. Using methods similar to the ones used in reference 17, one can show that in the upper half-plane $iF^+_{R\omega}(p)$ is an analytic function that coincides with $\mathfrak{F}^+_{\omega_n}(p)$ in the points $\omega_n = i\pi T(2n + 1)$ (when $n > 0$). We have solved the equations for the temperature-dependent Green's functions in the limiting case where Δ and $\mathfrak{F}^+$ were small. These equations can be obtained in the general case by the same method as was used by us for the case of normal superconductors.[5] A new aspect is here the occurrence of two relaxation times τ_1 and τ_2 and the dependence of the quantity $\Delta^* = |g|\,\mathfrak{F}^+(x, x)$ on the concentration of the paramagnetic impurities. The functions $\mathfrak{G}_{\omega_n}$ and $\mathfrak{F}^+_{\omega_n}$ obtained in this way are of the form

$$\mathfrak{G}_{\omega_n}(p) = -\frac{\tilde{\omega}_n + \xi}{\tilde{\omega}_n^2 + \xi^2 + \tilde{\Delta}_n^2}, \qquad \mathfrak{F}^+_{\omega_n}(p) = \frac{\tilde{\Delta}_n}{\tilde{\omega}_n^2 + \xi^2 + \tilde{\Delta}_n^2}, \tag{40}$$

where

$$\tilde{\omega}_n = \omega_n + \frac{1}{2\tau_1}\frac{u_n}{\sqrt{u_n^2 + 1}}, \qquad \tilde{\Delta}_n = \Delta + \frac{1}{2\tau_2}\frac{1}{\sqrt{u_n^2 + 1}}, \tag{41}$$

$$u_n = \frac{\tilde{\omega}_n}{\tilde{\Delta}_n}, \qquad \frac{\omega_n}{\Delta} = u_n\left(1 - \frac{1}{\tau_s\Delta}\frac{1}{\sqrt{u_n^2 + 1}}\right). \tag{42}$$

One can obtain similar equations using the technique applicable at absolute zero. We can determine the function $F^+_{R\omega}(p)$ which interests us as follows. If we formally replace ω_n by $-i\omega$ and at the same time $\tilde{\omega}_n$ by $-i\tilde{\omega}$, u_n by $-iu$, and $\tilde{\Delta}_n$ by $\tilde{\Delta}$ we get for u the following equation

$$\frac{\omega}{\Delta} = u\left(1 - \frac{1}{\tau_s\Delta}\frac{1}{\sqrt{1 - u^2}}\right). \tag{42'}$$

The radical must be defined here as the analytical continuation of the positive square root $+\sqrt{1 - u^2}$ for real u and $|u| < 1$ in the upper half-plane of the variable u. (One can verify that this definition is correct by comparing these formulae with the results obtained by the usual technique at the absolute zero.)

In actual fact we need for the following not the whole of the function $F^+_{R\omega}(p)$ but only some of its properties. It follows from Eq. (42′) that when $\tau_s\Delta > 1$ and when ω is sufficiently small there exists a real solution for u, with $|u| < 1$. The connection with the limiting case $1/\tau_s \to 0$ shows that in our case we must also choose just that solution. Studying the right-hand side of (42′) one sees easily that it has a maximum for

$$u_0 = [1 - (1/\tau_s\Delta)^{2/3}]^{1/2},$$
$$\omega_0 = \Delta[1 - (1/\tau_s\Delta)^{2/3}]^{3/2}, \tag{43}$$

and that therefore all solutions of Eq. (42′) are complex when $\omega > \omega_0$.

Since we must take the real solution of Eq. (42′) when ω is small, it follows that u becomes complex and $F_{R\omega}(p)$ acquires thereby an imaginary part when $\omega > \omega_0$. The quantity ω_0 thus determines the size of the gap. When the concentration is increased, the value of ω_0 diminishes and tends to zero when $1/\tau_s\Delta = 1$. The magnitude of Δ is thus not at all equal to the gap in the energy spectrum for paramagnetic alloys.

We find now the dependence of Δ on the impurity concentration. The definition of $\Delta^* = |g|\,\mathfrak{F}^+(x, x)$ yields the equation

$$\Delta = \frac{|g|\,T}{(2\pi)^3}\sum_n \int \frac{d^3p\,\tilde{\Delta}_n}{\tilde{\omega}_n^2 + \xi^2 + \tilde{\Delta}_n^2}. \tag{44}$$

Since we are interested in the case $T = 0$, we can change from summation over n to integration, using the formula

$$T\sum_n \to (2\pi)^{-1}\int d\Omega.$$

After this it is convenient to add and subtract under the integral sign the expression $\Delta_0/(\Omega^2 + \xi^2 + \Delta_0^2)$ (Δ_0 is the value of Δ for $T = 0$). The result is

$$\Delta_0 \ln\frac{\Delta_0}{\Delta_{00}} = \int_0^\infty d\Omega\left\{\frac{1}{\sqrt{u^2 + 1}} - \frac{1}{\sqrt{\Omega^2/\Delta_0^2 + 1}}\right\}$$

(Δ_{00} is the value of Δ_0 for a pure superconductor). It is convenient to use Eq. (42) and change in this integral the integration over $d\Omega$ to one over du. The result is

$$\ln\frac{\Delta_0}{\Delta_{00}} = -\frac{\pi}{4\tau_s\Delta_{00}}, \quad \frac{1}{\tau_s\Delta_0} < 1;$$

$$\ln\frac{\Delta_0}{\Delta_{00}} = -\ln\left[\frac{1+\sqrt{1-(\tau_s\Delta_0)^2}}{\tau_s\Delta_0}\right] + \frac{1}{2}\sqrt{1-(\tau_s\Delta_0)^2} - \frac{1}{2\tau_s\Delta_0}\arcsin\tau_s\Delta_0,$$

$$\frac{1}{\tau_s\Delta_0} > 1. \tag{45}$$

We have $1/\tau_{s\,cr} = \pi T_{c0}/2\gamma = \Delta_{00}/2$. When $\tau_s\Delta_0 = 1$ we have $\Delta_0 = e^{-\pi/4}\Delta_{00}$. From this it is clear that there is some value of the concentration which is lower than the critical one for which the gap in the energy spectrum (43) vanishes

$$n' = 2e^{-\pi/4}n_{cr} \approx 0.91\, n_{cr}. \tag{46}$$

The fact that there exists a superconductor without a gap needs some explanations. We started from Eqs. (2) in our derivation. The possibility of describing a superconductor by such a set of equations for the functions $\mathfrak{G}$ and $\mathfrak{F}^+$ is based upon the hypothesis that several pair formations of electrons show a Bose condensation.[6] In a pure superconductor such formations are Cooper pairs, which are condensed in a state with total momentum equal to zero [$\mathfrak{F}^+(x, x')$ depends only on the difference between the coordinates]. Since the number of such pairs in an infinitely large system is arbitrarily large, the creation or annihilation of such a pair would not change the ground state of the system. This fact is indeed reflected in the existence of non-vanishing averages

$$\mathfrak{F}^+(x - x') = \langle T(\psi^+(x), \psi^+(x'))\rangle.$$

We have shown that when n is larger than the n′ given by (46) the functions $\mathfrak{F}^+$ are non-vanishing and finite, which confirms the validity of the assumption of a Bose condensation; the change in the gap in the spectrum (46) shows, however, that we are dealing here with a more complicated situation than in the case of a pure superconductor. There remains something obscure in the overall picture.

It is of interest to trace how the vanishing of the gap with increasing impurity concentration affects the temperature dependence of the electronic heat capacity in the range $T \ll T_c$. We first transform somewhat Eq. (34) for the thermodynamic potential. We subtract from (34) the analogous equation for a metal in which a different value for the coupling constant has been taken, $|g_1| > |g|$. This leads to

$$\Omega_s - \Omega_{s1} = -\int_{|g|}^{|g_1|} \delta\left(\frac{1}{|g|}\right)\Delta^2.$$

We use the equation for the gap, which we write in the form

$$\Delta = \frac{|g|\,mp_0}{2\pi^2} f_T(\Delta), \tag{47}$$

and we change from an integral over $|g|$ to one over Δ. Clearly

$$\Omega_s - \Omega_{s1} = -\frac{mp_0}{2\pi^2}\int_{\Delta}^{\Delta_1} [f_T'(y)\,y - f_T(y)]\,dy.$$

We now subtract from this relation its value for $T = 0$.

We assume that $\Delta_1(0) \gg \Delta(0)$. In that case $\Omega_{s1}(T) - \Omega_{s1}(0) \to 0$. We get thus

$$\Omega_s(T) - \Omega_s(0) = -\frac{mp_0}{2\pi^2}\left\{\int_{\Delta}^{\Delta_1}[f_T'(y)\,y - f_T(y)]\,dy - \int_{\Delta_0}^{\Delta_1}[f_0'(y)\,y - f_0(y)]\,dy\right\}_{\Delta_1\to\infty},$$

where $\Delta_0 \equiv \Delta(0)$ and the $f_0(y)$ are the values of the corresponding functions at $T = 0$. At low temperatures we can write

$$f_T(\Delta) = f_0(\Delta) + \delta f_T(\Delta), \qquad \Delta = \Delta_0 + \delta\Delta(T);$$

the additional temperature-dependent terms are small compared with the values at absolute zero. Using Eq. (47) we can express $\delta\Delta(T)$ in terms of $\delta f_T(\Delta)$:

$$\delta\Delta(T) = \Delta_0\delta f_T(\Delta)/(f_0 - f_0'\Delta_0).$$

Substituting this into the equation for the thermodynamic potential we find

$$\Omega_s(T) - \Omega_s(0) = \frac{mp_0}{\pi^2}\int_{\Delta_0}^{\infty}\delta f_T(y)\,dy. \tag{48}$$

It is thus sufficient for us to find merely the temperature-dependent addition to the right-hand side of Eq. (44) for Δ.

We start with the simplest case $\tau_s\Delta_0 < 1$. We note that the expression under the summation sign in (44) is symmetric with respect to a change in the sign of n. As before we go over from a summation to an integration but we now take into account a correction term using the well-known formula

$$T\sum_{n>0}\varphi(\omega_n) \to \frac{1}{2\pi}\left\{\int_0^\infty \varphi(\Omega)\,d\Omega - \frac{\pi^2T^2}{6}[\varphi'(\infty) - \varphi'(0)]\right\}.$$

The first term here is temperature-independent and the correction is proportional to T^2. After the necessary calculations we get

$$\delta f_T(\Delta) = -\pi^2T^2\tau_s/6\sqrt{(1/\tau_s\Delta)^2 - 1}.$$

When substituting this into (48) we must take into account (vide infra) the region where $\delta f \sim T^2$

extends only to $\Delta = 1/\tau_s$ and furthermore that this function depends exponentially on T, i.e., it tends to zero, even when we take terms in T^2 into account. The integral over Δ must thus only be taken up to that limit. As a result we get for the heat capacity

$$C_s = \tfrac{1}{3}\, m p_0 \sqrt{1-(\tau_s\Delta_0)^2}\, T. \tag{49}$$

When $\Delta_0\tau_s \ll 1$ this formula is the same as (36) in the limit $T \ll T_c$.

In the case $\Delta_0\tau_s > 1$ the method given here can not be applied, since it yields only an extra term which is quadratic in T in the thermodynamic potential and which is not present here. If we were to apply this method to the case we are now considering, we would only get the result that there is no such extra term. This is not surprising, since the presence of a gap in the energy spectrum compels us to assume that when $\Delta_0\tau_s > 1$ the extra term will be exponential. Because of this we shall apply another method.

If in Eq. (44) we change to a summation over positive n only, we can replace this sum by a contour integral. The right-hand side of Eq. (44) will be of the form

$$-\frac{1}{2\pi i}\int_C d\omega \tanh\frac{\omega}{2T}\int_{-\omega_D}^{\omega_D} d\xi\,\frac{mp_0}{2\pi^2}\,\mathfrak{F}(-i\omega,\xi).$$

The contour of integration is given in Fig. 4. We now distort the contour and take it along the real axis. This is possible since it was shown that the integral over ω can be considered to be the integral of the function $iF_{R\omega}(p)$ which is analytical in the upper half-plane. For real ω which are such that $|\omega| < \omega_0$ this function is real and symmetric with respect to replacing ω by $-\omega$. From this it follows that the real part of $iF_{R\omega}(p)$ on the real axis is symmetric with respect to a change in sign in ω while its imaginary part is antisymmetric. All that remains of this integral is therefore

$$\frac{1}{\pi}\,\mathrm{Im}\left\{\int_0^{\infty} d\omega \tanh\frac{\omega}{2T}\int_{-\omega_D}^{\omega_D}\frac{mp_0}{2\pi^2}\,d\xi\,\frac{\tilde{\Delta}}{-\tilde{\omega}^2+\xi^2+\tilde{\Delta}^2}\right\}.$$

From the foregoing it follows that the imaginary part of the function $iF_{R\omega}(p)$ appears only starting with $\omega = \omega_0$. If $T \ll \omega_0$ we can thus replace $\tanh \omega/2T$ by $1 - e^{-\omega/T}$. The term 1 gives a contribution to $f_T(\Delta)$ which is independent of the temperature. $\delta f_T(\Delta)$ is thus determined by substituting $-2e^{-\omega/T}$ for $\tanh \omega/2T$. The integral over ξ and ω is then formally convergent and we can first integrate over ξ. In this way we find

$$\delta f_T(\Delta) = -2\int_{\omega_0}^{\infty} e^{-\omega/T}\,\mathrm{Im}\left[\frac{1}{\sqrt{1-u^2}}\right] d\omega.$$

The $\sqrt{1-u^2}$ under the integral sign must be defined, as a function of ω, as the analytical continuation of its real positive value with a cut $\omega < \omega_0$ through the upper half-plane. Because of the presence of an exponential factor the most essential part of the integral will come from the neighborhood of the point ω_0. Expanding Eq. (42′) in a power series in $u - u_0$ and taking the definition of $\sqrt{1-u^2}$ for $\omega > \omega_0$ into account, we get

$$\mathrm{Im}\left[\frac{1}{\sqrt{1-u^2}}\right] = \sqrt{\omega-\omega_0}\left(\frac{3\Delta_0}{2}\right)^{-1/2}\left[1-\left(\frac{1}{\tau_s\Delta_0}\right)^{2/3}\right]^{1/4}(\Delta_0\tau_s)^{1/3}.$$

Substituting this into the integral for $\delta f_T(\Delta)$ we find

$$\delta f_T(\Delta) = -\sqrt{\frac{2\pi}{3}}\,T^{3/2}\tau_s^{1/2}\left[(\Delta_0\tau_s)^{2/3}-1\right]^{1/4} e^{-\omega_0/T}.$$

Substituting this expression into (48) and integrating over Δ we can find the temperature dependent extra term in the thermodynamic potential. We get for the heat capacity

$$C_s(T) = \left(\frac{mp_0}{\pi^2}\right)\sqrt{\frac{2\pi}{3}}\,\tau_s^{2/3}\Delta_0^{1/3}\left[\Delta_0^{2/3} - \tau_s^{-2/3}\right]^{11/4}\exp\left\{-\frac{1}{T}\left[\Delta_0^{2/3}-\tau_s^{-2/3}\right]^{3/2}\right\}. \tag{50}$$

One sees easily that this formula does not go over into the formula for the pure superconductor in the limit as $\tau_s \to \infty$. This is connected with the fact that the possibility of expanding Eq. (42′) in powers of $u - u_0$ imposes a restriction on the region of applicability of Eq. (50). The latter is obtained when one estimates the next term of such an expansion. This gives

$$T \ll \Delta_0^{1/3}\tau_s^{-2/3}.$$

At low concentrations this condition is stronger than the condition $T \ll \Delta_0$ which is necessary to separate off the exponential region of the heat capacity. The connection between Eq. (50) and the equation for a pure superconductor could be obtained by considering the case $\tau_s\Delta_0 \gg 1$, $T \ll \Delta_0$, but we shall not perform here the appropriate calculations as we are interested in the basic influence of the vanishing of the gap in the spectrum on the heat capacity.

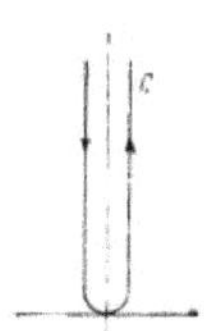

FIG. 4

It follows from Eq. (50) that when the impurity concentration increases the absolute magnitude of the exponent decreases while it does not vanish at $\Delta_0\tau_s = 1$. In the point $\Delta_0 = 1/\tau_s$ itself, where the exponential and the linear dependence come together one finds from calculating it that $C_s \sim T^{4/3}$. In view of the fact that this single point is not of great interest we shall not give the corresponding calculations or write out the coefficients.

In conclusion we express our gratitude to Academician L. D. Landau for many useful discussions of this paper.

[1] Matthias, Suhl, and Corenzwit, Phys. Rev. Letters **1**, 92 (1958).

[2] C. Herring, Physica **24**, S 184 (1958).

[3] H. Suhl and B. T. Matthias, Phys. Rev. **114**, 977 (1959).

[4] Matthias, Compton, Suhl, and Corenzwit, Phys. Rev. **115**, 1597 (1959).

[5] A. A. Abrikosov and L. P. Gor'kov, JETP **35**, 1558 (1958), Soviet Phys. JETP **8**, 1090 (1959); JETP **36**, 319 (1959), Soviet Phys. JETP **9**, 220 (1959).

[6] L. P. Gor'kov, JETP **34**, 735 (1958), Soviet Phys. JETP **7**, 505 (1958).

[7] L. P. Gor'kov, JETP **37**, 1407 (1959), Soviet Phys. JETP **10**, 998 (1960).

[8] L. D. Landau and E. M. Lifshitz, Quantum Mechanics, Pergamon Press, London, 1958, Sec. 54.

[9] Bardeen, Cooper, and Schrieffer, Phys. Rev. **108**, 1175 (1957).

[10] N. N. Bogolyubov, JETP **34**, 58 (1958), Soviet Phys. JETP **7**, 41 (1958).

[11] Matthias, Suhl, and Corenzwit, Phys. Rev. Letters **1**, 449 (1958).

[12] M. A. Ruderman and C. Kittel, Phys. Rev. **96**, 99 (1954).

[13] R. Brout, Phys. Rev. **115**, 824 (1959).

[14] V. L. Ginzburg and L. D. Landau, JETP **20**, 1064 (1950).

[15] L. P. Gor'kov, JETP **36**, 1918 (1959), Soviet Phys. JETP **9**, 1364 (1959).

[16] A. B. Migdal and V. M. Galitskiĭ, JETP **34**, 139 (1958), Soviet Phys. JETP **7**, 96 (1958).

[17] Abrikosov, Gor'kov, and Dzyaloshinskiĭ, JETP **36**, 900 (1959), Soviet Phys. JETP **9**, 636 (1959).

Translated by D. ter Haar
325

SOVIET PHYSICS JETP VOLUME 12, NUMBER 2 FEBRUARY, 1961

ON THE PROBLEM OF THE KNIGHT SHIFT IN SUPERCONDUCTORS

A. A. ABRIKOSOV and L. P. GOR'KOV

Institute for Physical Problems, Academy of Sciences, U.S.S.R.

Submitted to JETP editor March 23, 1960

J. Exptl. Theoret. Phys. (U.S.S.R.) **39**, 480-483 (August, 1960)

The problem of the Knight shift in superconductors of small size is discussed. It is shown that the derivation of a finite value of the shift at absolute zero obtained in the papers of Schrieffer[3] and of Anderson[4] is incorrect.

RECENTLY a number of papers[1-4] has been devoted to the theory of the so-called Knight shift in superconductors, i.e., the change in the nuclear resonance frequency as compared to dielectrics. The aim of these papers is to explain the experimental data obtained by Reif for superconducting emulsions.[5]

As is well known, the cause of the Knight shift in metals is the paramagnetism of the conduction electrons. Because the electron wave function is anomalously large in the neighborhood of the nucleus, the magnetization of the electrons changes the magnetic field acting on the nucleus. The difference between the effective and the external field has the following form

$$\Delta H = (8\pi/3N_{at})\,|\psi(0)|^2\chi H, \qquad (1)$$

where $|\psi(0)|^2$ is the probability density of finding the electron at the position of the nucleus, N_{at} is the number of atoms per unit volume, χ is the paramagnetic susceptibility of the electrons, and H is the external field.

Since the conduction electron states are altered in superconductors, one should expect a change in the magnitude of the Knight shift. Formula (1) is still valid in this case, with the susceptibility evidently undergoing the principal change in comparison with the normal metal. The observation of this effect is made difficult by the fact that the field does not penetrate deeply into a bulky superconductor. A homogeneous field can be produced only in superconductors of dimensions much smaller than δ, the penetration depth for a static field. It is just for this reason that an emulsion is utilized in Reif's experiments.

In Yosida's paper[1] a bulky superconductor in a homogeneous field was considered. Such a formulation of the problem does not correspond to actual experimental conditions. In order to be able to completely leave out of account the effect of the boundary conditions and of the inhomogeneity of the field it is necessary that the depth of penetration of the field and the dimensions of the sample should be large compared to the principal parameter of the theory $\xi_0 \sim \hbar v/kT_c$, the characteristic correlation radius. Only in such a case could Yoshida's results, obtained for an infinite superconductor, be used for the interpretation of Reif's experiments. However, it is well known that for the majority of superconductors δ is considerably smaller than ξ_0. From the formal point of view the calculations of the susceptibility in reference 1 for a bulk sample are of course correct, since in order to calculate the coefficient of proportionality between the magnetic moment and the field (assuming the latter to be homogeneous) it is not necessary to solve the problem of the penetration of the field into the sample. We are here simply discussing the fact that actually the field remains homogeneous at distances smaller than $\delta \ll \xi_0$. The attempt by Martin and Kadanoff[2] to take into account the inhomogeneity of the field in the sample cannot lead to any improvement, since such an inhomogeneity increases first of all the line width.

The problem can be formulated quite correctly for superconducting alloys. If the concentration of the impurities is sufficiently great then the role of the correlation parameter is played by the mean free path l. Under these conditions the situation described by London ($\delta \gg l$) arises in the superconductor, and it becomes possible to regard the field in the sample as homogeneous.

As has been pointed out earlier,[7,8] actual samples of small dimensions are conglomerates of crystallites with dimensions of the order of or smaller than the sample dimensions. Because of this, such samples have properties which are rather close to superconducting alloys with an effective mean free path of the same order of magnitude. The paramagnetic susceptibility of superconducting alloys has been recently investigated by Schrieffer[3] and by Anderson.[4] In particular, in these investigations the conclusion was reached that the presence of impurities leads to a finite Knight shift at T = 0 which is in agreement with the extrapolation of Reif's results to T = 0. In this note it will be shown that such a conclusion is erroneous.

The method for the study of superconductors containing impurities has been developed by us previously.[7,8] In the present case it is necessary to evaluate the spin magnetic moment of the system of electrons in a homogeneous magnetic field. It is given by

$$\mathbf{M}=\mu_0\,\mathrm{Sp}\left[\exp\left(\frac{\Omega+\hat{N}\mu-\hat{\mathcal{H}}}{T}\right)\psi_\alpha^+(x)\,\sigma_{\alpha\beta}\,\psi_\beta(x)\right], \qquad (2)$$

where μ_0 is the Bohr magneton and σ are the Pauli matrices. The Hamiltonian for the interaction between the electron spins and the magnetic field has the form

$$\hat{\mathcal{H}}_1=-\mu_0\int\psi_\alpha^+(\mathbf{x})\,(\sigma_{\alpha\beta}\mathbf{H})\,\psi_\beta(\mathbf{x})\,d^3\mathbf{x}. \qquad (3)$$

By utilizing the techniques used in quantum field theory at finite temperatures[8,9] and by restricting ourselves to terms of the first order in H we obtain

$$\mathbf{M}=-\mu_0^2\lim_{\substack{\tau'\to\tau+0\\ x'\to x}}\int_0^{1/T}d\tau_y\int d^3\mathbf{y}\,\mathrm{Sp}\Big[\exp\left(\frac{\Omega+\hat{N}\mu-\hat{\mathcal{H}}_0}{T}\right)$$
$$\times T_\tau\left(\psi_\beta(x)\,\psi_\gamma^+(y)\,\psi_\delta(y)\,\psi_\alpha^+(x')\right)\sigma_{\alpha\beta}(\sigma_{\gamma\delta}\mathbf{H})\Big] \qquad (4)$$

In this formula the operators ψ are not free, but include the interaction with the impurities:

$$\hat{\mathcal{H}}_2=\int\psi_\alpha^+(\mathbf{x})\,\psi_\alpha(\mathbf{x})\sum_a V(\mathbf{x}-\mathbf{x}_a)\,d^3\mathbf{x}, \qquad (5)$$

where $\mathbf{x}_a$ is the position vector of the impurity atom.

Further calculations have been carried out in complete analogy with the way this was done in references 7 and 8. It was shown there that averaging over the coordinates of the impurity atoms leads to the introduction of a special diagram technique. Each impurity line in a diagram arises from averaging two factors corresponding to scattering by one of the impurity atoms. It is sufficient to retain only the non-intersecting lines in order to obtain a result which differs from the correct one by small terms only. If we apply this procedure to formula (4) then the calculation of the magnetic moment obviously reduces to the evaluation of the sum of loops with two spinor vertices $\sigma_{\alpha\beta}$ and with all the nonintersecting impurity lines.

In averaging over the impurity coordinates we must take into account the fact that, for example, $\overline{G(x,y)\,G(y,x)}$ is not equal to $\overline{G(x,y)}\cdot\overline{G(y,x)}$. In the opposite case this expression would simply be given by

$$G_0(x-y)\,G_0(y-x)\exp(-|\mathbf{x}-\mathbf{y}|/l),$$

where l is the mean free path, while $G_0(x-y)$ is the Green's function for the pure superconductor. This incorrect assumption is the one made in Schrieffer's note[3] and leads to an erroneous result.

If we denote by $\Pi^1_{\beta\alpha}(\mathbf{p},\omega_n)$ the Fourier component of

$$\int_0^{1/T}d\tau_y\int d^3\mathbf{y}\,\mathrm{Sp}\Big[\exp\left(\frac{\Omega+\hat{N}\mu-\hat{\mathcal{H}}_0}{T}\right)$$
$$\times T_\tau\left(\psi_\beta(x)\,\psi_\gamma^+(y)\,\psi_\delta(y)\,\psi_\alpha^+(x')\right)\sigma_{\gamma\delta}\Big],$$

we can easily relate this quantity by means of appropriate equations to Fourier components of three other similar quantities:

$$\int d^4y\,\overline{[\psi_\beta(x)\,\psi_\gamma^+(y)\,\psi_\delta(y)\,(\psi(x')\,\hat{g})_\alpha]}\,\sigma_{\gamma\delta},$$

$$\int d^4y\,\overline{[(\hat{g}\psi^+(x))_\beta\,\psi_\gamma^+(y)\,\psi_\delta(y)\,\psi_\alpha(x')]}\,\sigma_{\gamma\delta},$$

$$\int d^4y\,\overline{[(\hat{g}\psi^+(x))_\beta\,\psi_\gamma^+(y)\,\psi_\delta^+(y)\,(\psi(x')\,\hat{g})_\alpha]}\,\sigma_{\gamma\delta},$$

where

$$\hat{g}=\begin{pmatrix}0&1\\-1&0\end{pmatrix}.$$

The magnetic moment is expressed in terms of $\Pi^1_{\beta\alpha}$ by means of the relation

$$\mathbf{M}=-\mu_0^2\,\sigma_{\alpha\beta}\,T\sum_n(2\pi)^{-3}\int\mathbf{H}\Pi^1_{\beta\alpha}(\omega_n,\mathbf{p})\,d^3\mathbf{p}, \qquad (6)$$

where $\omega_n=\pi T(2n+1)$.

Since all these calculations corespond completely to those carried out in reference 7, we shall not dwell on the solution of these equations, and shall simply state the answer:

$$\chi=2\mu_0^2(mp_0/2\pi^2)(N_n/N)=\chi_n N_n/N, \qquad (7)$$

where χ_n is the susceptibility of the normal metal, and N_n/N is the ratio of the number of "normal" electrons to the total number of electrons given by

$$\frac{N_n}{N_0}=\frac{1}{2T}\int_{\Delta(T)}^{\infty}\frac{\varepsilon\,d\varepsilon}{\sqrt{\varepsilon^2-\Delta^2(T)}\cosh^2(\varepsilon/2T)}. \qquad (8)$$

The function $N_s=N-N_n$ determines the penetration depth for a pure superconductor of the London type.

Formulas (7) and (8) agree with Yosida's result[1] obtained formally for the case of a bulk of pure superconductor in a homogeneous field. This is explained by the fact that these formulas do not contain the mean free path. We also note that in the case considered here $l \gg 1/p_0$ the size of the gap $\Delta(T)$ agrees with the same quantity for the case of a pure superconductor.

From our results it follows, in particular, that for $T = 0$ the susceptibility χ vanishes, i.e., that no Knight shift should be observed. This conclusion is perfectly natural. In its ground state the superconductor has no magnetic moment, and the excited state is separated from the ground state by an energy gap. This applies not only to pure superconductors, but also to alloys with nonmagnetic impurities. Nevertheless, this result contradicts Reif's experiments,[5] which, as has been noted by Yosida, give values of χ larger than those given by formula (7). The quantity χ does not vanish when we extrapolate to $T = 0$.

In this connection it is necessary to subject to some criticism the interpretation of Reif's experiment. As we have already mentioned, in these experiments emulsions were used, i.e., an aggregate of small spherical samples of different sizes. Apparently at the present time this is the only method for investigating nuclear resonance in superconductors, and it can give rise to no objections if the particles are sufficiently small. In this case they had a diameter d on the average equal to one third of the penetration depth. However, the distribution of particles with respect to size is fairly broad, and since the effectiveness of individual particles is proportional to their volume the most important role is played by the large particles, even though they are less numerous. In such particles the field is inhomogeneous, which is in complete agreement with the large width of the resonance line obtained by Reif at low temperatures.

One other essential circumstance should also be noted. As is well known, the transition from the superconducting to the normal state in particles of dimensions smaller than the penetration depth is a phase transition of the second kind for which the superconducting "gap" in the spectrum vanishes in a continuous manner. This means that particles for which the external magnetic field is only slightly lower than the critical value (we recall that $H_c \sim 1/d$) will be very close in their properties to the normal metal right down to very low temperatures. From this it follows that in the given external field (which, incidentally, was approximately equal to 1000 gauss) the particles cannot be separated into completely normal ones and totally superconducting ones with corresponding resonance frequencies, but give rise to various intermediate values of the resonance frequency.

In view of the foregoing, it appears to us that at the present time it is premature to speak of the existence of a contradiction between theory and experiment.

In conclusion we express our gratitude to Academician L. D. Landau for discussion of this work.

[1] K. Yosida, Phys. Rev. **110**, 769 (1958).

[2] P. Martin and L. Kadanoff, Phys. Rev. Letters **3**, 322 (1959).

[3] J. R. Schrieffer, Phys. Rev. Letters **3**, 323 (1959).

[4] P. W. Anderson, Phys. Rev. Letters **3**, 325 (1959).

[5] F. Reif, Phys. Rev. **106**, 208 (1957).

[6] Townes, Herring, and Knight, Phys. Rev. **77**, 851 (1950).

[7] A. A. Abrikosov and L. P. Gor'kov, JETP **35**, 1558 (1958), Soviet Phys. JETP **8**, 1090 (1959).

[8] A. A. Abrikosov and L. P. Gor'kov, JETP **36**, 319 (1959), Soviet Phys. JETP **9**, 220 (1959).

[9] Abrikosov, Gor'kov, and Dzyaloshinskiĭ, JETP **36**, 900 (1959), Soviet Phys. JETP **9**, 636 (1959).

Translated by G. Volkoff
93

SOVIET PHYSICS JETP VOLUME 15, NUMBER 4 OCTOBER, 1962

SPIN-ORBIT INTERACTION AND THE KNIGHT SHIFT IN SUPERCONDUCTORS

A. A. ABRIKOSOV and L. P. GOR'KOV

Institute for Physical Problems, Academy of Sciences, U.S.S.R.

Submitted to JETP editor November 4, 1961

J. Exptl. Theoret. Phys. (U.S.S.R.) **42**, 1088-1096 (April, 1962)

A theory of the Knight shift in superconductors is presented. It is shown that the effect can be completely explained if the spin-orbit part of the interaction in the scattering of electrons on the crystal boundaries is taken into account. The agreement between the theory and the experimental data [8] is found to be good.

1. INTRODUCTION

MANY recent papers [1-7] have been devoted to an explanation of the nonvanishing shift in the frequency of the nuclear magnetic resonance in a superconductor at zero absolute temperature. As is known, this so-called Knight shift is proportional to the paramagnetic susceptibility of the conduction electrons, χ:

$$\Delta\omega/\omega = (8\pi/3N_{at})\,|\psi(0)|^2\,\chi$$

where $|\psi(0)|^2$ is the probability density for finding an electron at the position of the nucleus and N_{at} is the number of atoms in a unit volume.

According to the microscopic theory of superconductivity, the paramagnetic susceptibility of the electrons in the superconducting state vanishes for $T = 0$. This is in contradiction with the nonvanishing Knight shift observed experimentally. The vanishing of χ for $T = 0$ is connected with the fact that the states of a superconductor are classified according to the eigenvalues of the total spin. In the ground state all electrons are paired in Cooper pairs, and the total spin of the system is zero. Excitation of the system requires a finite energy 2Δ, where Δ is the gap in the energy spectrum of the superconductor.

The most natural explanation of the finite Knight shift in superconductors is based on the assumption of Ferrell [5] and Anderson [6] that the spin-orbit interaction plays an important role for electrons in small specimens. Owing to this interaction the electron can change its spin direction when it is scattered at the boundary. The states of such a system can therefore no longer be characterized by the eigenvalues of the spin, and the above-mentioned conclusion is not correct.

A finite shift at $T = 0$ in tin has definitely been observed by Androes and Knight in recent work. [8] The earlier results of Reif [9] on the Knight shift in colloidal mercury are difficult to interpret, since, as noted by the authors, [4] the uncertainty in the size of the particles was so large that an appreciable number of large particles may have gone over into the normal state in fields of the order of a few kilooersted thus simulating a finite shift in the superconducting state.

The experiments of Androes and Knight [8] were performed with stacks of thin films with thickness of 40 Å. The fields used in this work (from 1 to 8 kilooersted) were weaker than the critical fields, which reach a strength of 25 kilooersted. The ratio χ_s/χ_n, extrapolated to $T = 0$, was equal to ~ 0.77.

In the present paper we shall show that the above-mentioned spin-orbit interaction provides a correct quantitative explanation of the phenomenon. For the mathematical formulation of the problem it is important to take into consideration that the thin films as well as all small-sized specimens, have a polycrystalline structure. At the boundaries of the crystallites the electrons are scattered, which leads to a destruction of the correlations between a pair of electrons which were introduced in the transition to the superconducting state. These correlations are thus restricted to distances of the order of the dimensions of the crystallites. The dimensions of the crystallites, therefore, play the role of a mean free path for the electrons. Hence we shall below consider the equivalent problem of the scattering of an electron by impurities in the superconductor, where spin flip can occur on account of the spin-orbit interaction. As will be shown below, the final result will involve only the mean free path of the electrons with respect to spin flip, $l_{s.o.}$.

The amplitude for the scattering of an electron by an atom of the impurity can be written in the form

$$\hat{f}(\mathbf{p}, \mathbf{p}') = a(\mathbf{p}, \mathbf{p}') + ib(\mathbf{p}, \mathbf{p}')\, p_0^{-2}([\mathbf{p}, \mathbf{p}']\, \hat{\sigma}), \qquad (1)^*$$

where $\mathbf{p}$ and $\mathbf{p}'$ are the momenta of the electron before and after the collision, $\hat{\sigma}$ are the Pauli spin matrices, and p_0 is the Fermi momentum. The second term in (1), corresponding to the spin-orbit interaction, is of order $b/a \sim (Ze^2/\hbar c)^2$ and may, in general, be appreciable for heavy elements. To keep the calculations simple, we shall, however, assume that $|b/a|^2 \ll 1$. In other words, we shall assume that the mean free path $l_{s.o.}$ is much larger than the mean free path between collisions of the electrons with the impurities, l, which is determined essentially by the amplitude $a(\mathbf{p}, \mathbf{p}')$ in (1). It was shown in [4] that the scattering does not change the paramagnetic susceptibility of the superconductor in the absence of the spin-orbit interaction, so that there will be a Knight shift of the order of unity at $T = 0$ if

$$l_{s.o.} \sim \xi_0 \sim \hbar v/T_c \sim 10^{-4} - 10^{-5} \text{ cm}.$$

2. CALCULATION OF THE PARAMAGNETIC SUSCEPTIBILITY

The spin magnetic moment of a system of electrons in a uniform magnetic field is equal to[4]

$$\mathbf{M} = -\mu_0^2 \lim_{\substack{\tau' \to \tau + 0 \\ r' \to r}} \int_0^{1/T} d\tau_y \int d^3\mathbf{y}\, \mathrm{Sp}\Big[\exp\Big(\frac{\Omega + \mu N - H_0}{T}\Big)$$

$$\times T_\tau\, (\psi_\alpha(x)\, \psi_\gamma^+(y)\, \psi_\delta(y)\, \psi_\beta^+(x'))\, \sigma_{\beta\alpha}(\sigma_{\gamma\delta}\mathbf{H})\Big]. \qquad (2)$$

Here μ_0 is the Bohr magneton. In the following we shall use the thermodynamic formulation[10,11] of perturbation theory, in which the field operators depend on the time parameter τ $(0 < \tau < 1/T)$.

Denoting the average over the Gibbs ensemble by $\langle \ldots \rangle$, we consider the average over the product of four operators appearing in (2):

$$\langle T\, (\psi_\alpha(x)\, \psi_\gamma^+(y) \psi_\delta(y)\, \psi_\beta^+(x')) \rangle.$$

In a superconductor this average can be written in the form

$$\langle T\, (\psi_\alpha(x)\, \psi_\gamma^+(y)\, \psi_\delta(y)\, \psi_\beta^+(x')) \rangle$$

$$= \mathfrak{G}_{\alpha\gamma}(x, y)\, \mathfrak{G}_{\delta\beta}(y, x') - \mathfrak{F}_{\alpha\delta}(x, y)\, \mathfrak{F}^+_{\gamma\beta}(y, x'), \qquad (3)$$

where

$$\mathfrak{G}_{\alpha\beta}(x, y) = -\langle T\, (\psi_\alpha(x)\, \psi_\beta^+(y)) \rangle;$$

$$\mathfrak{F}_{\alpha\beta}(x, y) = \langle T\, (\psi_\alpha(x)\, \psi_\beta(y)) \rangle;$$

$$\mathfrak{F}^+_{\alpha\beta}(x, y) = \langle T\, (\psi_\alpha^+(x)\, \psi_\beta^+(y)) \rangle.$$

The diagram corresponding to expression (3) is shown in Fig. 1. The line with both arrows pointing in the same direction represents the function $\mathfrak{G}$ and the lines with the arrows pointing both either in or out represent $\mathfrak{F}$ and $\mathfrak{F}^+$.

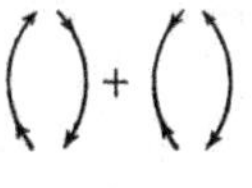

FIG. 1

In our case all quantities in (2) and (3) contain the interaction with the atoms of the impurities. In zeroth order in this interaction we must substitute the known expressions for the Green's functions of the pure superconductor in (3):

$$\mathfrak{G}_{\alpha\beta}(x - y) = \delta_{\alpha\beta}\, T\, (2\pi)^{-3} \sum_\omega \int d^3\mathbf{p}\, \mathfrak{G}(p) \exp\{i\mathbf{p}(\mathbf{x} - \mathbf{y})$$

$$- i\omega(\tau_x - \tau_y)\},$$

$$\mathfrak{F}^+_{\alpha\beta}(x - y) = -\mathfrak{F}_{\alpha\beta}(x - y) = \hat{g}_{\alpha\beta}\, T\, (2\pi)^{-3} \sum_\omega \int d^3\mathbf{p}\, \mathfrak{F}(p)$$

$$\times \exp\{i\mathbf{p}(\mathbf{x} - \mathbf{y}) - i\omega(\tau_x - \tau_y)\}, \qquad (4)$$

where

$$\mathfrak{G}(p) = -\frac{i\omega + \xi}{\omega^2 + \xi^2 + \Delta^2}, \qquad \mathfrak{F}(p) = \frac{\Delta}{\omega^2 + \xi^2 + \Delta^2}, \qquad (4')$$

g is the spin metric tensor ($g_{11} = g_{22} = 0$, $g_{12} = -g_{21} = 1$), Δ is the energy gap in the spectrum of the pure superconductor, and $\xi = v(p - p_0)$. The frequencies ω appearing in the Fourier series in τ take the values $\omega = (2n + 1)\pi T$ $(n = 0, \pm 1, \ldots)$, and $\sum_\omega \varphi(\omega)$ means $\sum_n \varphi(2n + 1)\pi T$.

Taking account of the interaction with the impurities changes expressions (4) and (4′). In Fig. 2 we show diagrams for expression (3) in which the crosses indicate the act of scattering of the electron with the separate atoms of the impurity. The scattering of the electron by an atom located at the point $\mathbf{r}_a$ gives rise to a factor $f_{\alpha\beta}(\mathbf{p}, \mathbf{p}') \times \exp\{i(\mathbf{p}' - \mathbf{p}) \cdot \mathbf{r}_a\}$ in the matrix element in the momentum representation.

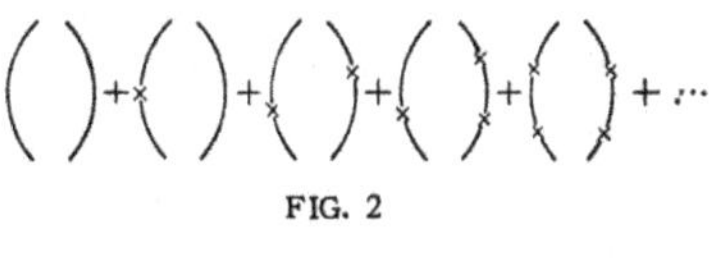

FIG. 2

$^*([\mathbf{p}, \mathbf{p}']\, \hat{\sigma}) = [(\mathbf{p} \times \mathbf{p}') \cdot \hat{\sigma}]$

In the following we shall be interested in the values of all quantities, including (2), averaged over the positions of all atoms of the impurity. The relevant momenta in our problem are all of the order $p \sim p_0 \sim 1/a$, where a is the interatomic distance. Therefore the average of the factor $\exp\{i(\mathbf{p}'-\mathbf{p})\cdot\mathbf{r}_a\}$ over $\mathbf{r}_a$ is zero and so is every matrix element unless it contains yet another factor $\exp\{-i(\mathbf{p}'-\mathbf{p})\cdot\mathbf{r}\}$ corresponding to the scattering on the same atom of the impurity with the initial and final states interchanged. In taking the average we must therefore sum over pairs of scatterings on the same atom.

The whole averaging procedure can be illustrated by a special diagram technique which has been developed by the authors.[12] The pairwise averaging is represented by a dotted line in the diagram, which joins two identical crosses (scattering by the same atoms). In the matrix element the dotted line corresponds to the factor

$$n(2\pi)^{-3} f_{\alpha\beta}(\mathbf{p}, \mathbf{p}') f_{\gamma\delta}(\mathbf{p}', \mathbf{p})$$

where n is the number density of the atoms of the impurity. With an accuracy up to small terms of order a/l it is sufficient to consider only nonintersecting dotted lines.

Let us introduce the notation

$$\Pi^{(1)}_{\alpha\beta}(x-y,\ y-x') = \overline{\langle T(\psi_\alpha(x)\psi^+_\delta(y)\sigma_{\delta\lambda}\psi_\lambda(y)\psi^+_\beta(x'))\rangle}. \tag{5}$$

(The bar denotes the average over the positions of the atoms of the impurity). Let us write

$$\Pi^{(1)}_{\alpha\beta}(x-y,\ y-x') = T^2(2\pi)^{-6}\sum_{\omega_+\omega_-}\iint d^3\mathbf{p}_+\, d^3\mathbf{p}_-\, \Pi^{(1)}_{\alpha\beta}(p_+,\ p_-)$$
$$\times \exp\{i\mathbf{p}_+(\mathbf{x}-\mathbf{y}) + i\mathbf{p}_-(\mathbf{y}-\mathbf{x}') - i\omega_+(\tau_x-\tau_y) - i\omega_-(\tau_y-\tau_{x'})\}. \tag{5'}$$

The magnetic moment is given in terms of $\Pi^{(1)}_{\alpha\beta}$ by the relation

$$\mathbf{M} = -\mu_0^2\sigma_{\alpha\beta}T(2\pi)^{-3}\sum_\omega\int(\mathbf{H}\Pi^{(1)}_{\alpha\beta}(p,\ p))\,d^3\mathbf{p}. \tag{6}$$

Some lowest order diagrams in the perturbation expansion of (5) are shown schematically (without indicating the direction of the arrows) in Fig. 3. In accordance with the general recipe we leave out all diagrams with intersecting dotted lines, as well as diagrams in which the dotted line goes into a vertex with a large ($\sim p_0$) momentum transfer. In these diagrams part of the integration is over momenta far away from the Fermi surface, which reduces the contribution from these diagrams by a factor of order $1/p_0 l \sim a/l \ll 1$ as compared to diagrams of the "ladder" type. In the remaining

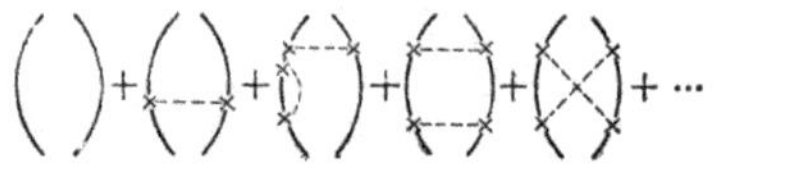

FIG. 3

diagrams the dotted line either joins two different electron lines or begins and ends on the same line (but without intersecting the "steps" of the ladder). In the latter case all such dotted lines evidently represent corrections from the interaction with the impurities to the Green's functions themselves. In order to derive an equation for $\Pi^{(1)}_{\alpha\beta}(p_+, p_-)$ it is therefore sufficient to consider only the "ladder" diagrams in which the dotted lines join different electron lines. The different sections of the electron lines in these diagrams represent the exact (averaged over the positions of the impurities) Green's functions $\mathfrak{G}$, $\mathfrak{F}$, and $\mathfrak{F}^+$ in the matrix element.

The equations for the Green's functions themselves in a superconductor with impurities coincide in this case exactly with the equations of [12]. Solving these equations, we find again that the averaged Green's functions for a superconductor with impurities can be obtained from (4) and (4') by the replacement

$$\{\omega,\ \Delta\} \to \{\omega\eta,\ \Delta\eta\}; \tag{7}$$

$$\eta = 1 + 1/2\tau\sqrt{\omega^2+\Delta^2}, \tag{8}$$

$$\frac{1}{\tau} = \frac{nmp_0}{4\pi^2}\int d\Omega\,(|a|^2 + |b|^2\sin^2\theta) = \frac{1}{\tau_0} + \frac{1}{\tau_1}. \tag{9}$$

Here we differ from [12] only in that $1/\tau$ now contains a contribution from the part of the scattering amplitude arising from the spin-orbit interaction:

$$\frac{1}{\tau_1} = \frac{nmp_0}{4\pi^2}\int|b|^2\sin^2\theta\, d\Omega. \tag{10}$$

As was shown in [12], formulas (4'), (7) to (9) imply that in the coordinate representation, the Green's function for a pure superconductor is to be multiplied by the factor

$$\exp\{-|r-r'|/2l\},$$

to account for the contamination by impurities, where $l = v\tau$ is the mean free path.* It follows from this that the introduction of impurities does not affect the thermodynamics of a superconductor, and, in particular, the transition temperature at which the system goes into the superconducting state remains unchanged. In the following all expressions will contain only the Green's functions for a superconductor with impurities, for which

*We note that $l_{s.o.} = v\tau_1$.

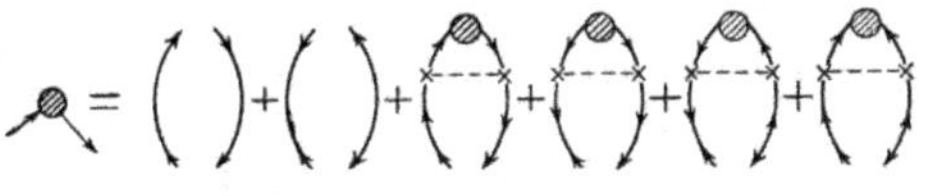

FIG. 4

we can, therefore, use the old symbols $\mathfrak{G}$, $\mathfrak{F}$, and $\mathfrak{F}^+$.

Figure 4 shows a graphical representation of the equation for $\Pi^{(1)}_{\alpha\beta}(p_+, p_-)$. We see that the determination of $\Pi^{(1)}_{\alpha\beta}(p_+, p_-)$ requires, in general, the knowledge of three new quantities which differ graphically from one another by the directions of the arrows. We define these quantities as Fourier transforms of the following expressions:

$$\Pi^{(2)}_{\alpha\beta}(x-y, y-x') = \langle \overline{T((\hat{g}\psi^+(x))_\alpha \psi^+_\delta(y)} \sigma_{\delta\lambda} \overline{\psi_\lambda(y)\psi^+_\beta(x'))} \rangle,$$

$$\Pi^{(3)}_{\alpha\beta}(x-y, y-x') = \langle \overline{T((\hat{g}\psi^+(x))_\alpha \psi^+_\delta(y)} \sigma_{\delta\lambda} \overline{\psi_\lambda(y)(\psi(x')\hat{g})_\beta)} \rangle,$$

$$\Pi^{(4)}_{\alpha\beta}(x-y, y-x') = \langle \overline{T(\psi_\alpha(x)\psi^-_\delta(y)} \sigma_{\delta\lambda} \overline{\psi_\lambda(y)(\psi^+(x')\hat{g})_\beta)} \rangle.$$

Expression (6) involves only the component $\Pi^{(1)}_{\alpha\beta}(p, p)$. Therefore we shall not write down the general equations for $\Pi^{(i)}_{\alpha\beta}(p_+, p_-)$, but remark only that for $p_+ = p_- = p$ the equations imply

$$\Pi^{(3)}_{\alpha\beta}(p, p) = -\Pi^{(1)}_{\alpha\beta}(p, p), \qquad \Pi^{(4)}_{\alpha\beta}(p, p) = \Pi^{(2)}_{\alpha\beta}(p, p).$$

Using these results, we can obtain from the equation Fig. 4 the following expression for $\Pi^{(1)}_{\alpha\beta}(p, p)$:

$$\Pi^{(1)}_{\alpha\beta}(p, p) = [\mathfrak{G}^2(p) + \mathfrak{F}^2(p)](\sigma_{\alpha\beta} + \Lambda^{(1)}_{\alpha\beta}(p)) + 2\mathfrak{F}(p)\mathfrak{G}(p)\Lambda^{(2)}_{\alpha\beta}(p), \tag{11}$$

where we have introduced the notation

$$\Lambda^{(1)}_{\alpha\beta}(p) = \frac{n}{(2\pi)^3}\int \hat{f}_{\alpha\delta}(\mathbf{p}, \mathbf{p}')\Pi^{(1)}_{\delta\lambda}(p', p')\hat{f}_{\lambda\beta}(\mathbf{p}', \mathbf{p})\,d^3\mathbf{p}',$$

$$\Lambda^{(2)}_{\alpha\beta}(p) = \frac{n}{(2\pi)^3}\int \hat{f}_{\alpha\delta}(\mathbf{p}, \mathbf{p}')\Pi^{(2)}_{\delta\lambda}(p', p')\hat{f}_{\lambda\beta}(\mathbf{p}', \mathbf{p})\,d^3\mathbf{p}'. \tag{12}$$

The analogous expression for $\Pi^{(2)}_{\alpha\beta}$ is

$$\Pi^{(2)}_{\alpha\beta}(p, p) = -[\mathfrak{G}(p)\mathfrak{F}(p) - \mathfrak{F}(p)\mathfrak{G}(-p)](\sigma_{\alpha\beta} + \Lambda^{(1)}_{\alpha\beta}(p)) - (\mathfrak{F}^2(p) - \mathfrak{G}(p)\mathfrak{G}(-p))\Lambda^{(2)}_{\alpha\beta}(p). \tag{13}$$

The vector $\Lambda^{(i)}$ can be directed either along $\boldsymbol{\sigma}$ or $\mathbf{p}$. Accordingly it is convenient to write $\Lambda^{(i)}$ as the sum of two parts:

$$\Lambda^{(i)}_{\alpha\beta}(p) = \sigma_{\alpha\beta}\Lambda^{(i)}_\sigma(p) + (\boldsymbol{\sigma}\mathbf{p})_{\alpha\beta}\mathbf{p}p_0^{-2}\Lambda^{(i)}_p(p). \tag{14}$$

Substituting (11) and (13) in (12), we obtain a system of four equations for the quantities $\Lambda^{(1,2)}_\sigma(p)$ and $\Lambda^{(1,2)}_p(p)$. Using the expressions (4′) and (7) for the Green's functions, we can easily show that all the quantities $\Lambda^{(1,2)}_{(\sigma,p)}(p)$ depend only on the fourth component of p, i.e., on the frequency ω. Owing to this circumstance the system of equations (11) to (14) can be reduced to algebraic form:

$$\left[(\omega^2+\Delta^2)^{3/2}\eta - \frac{\Delta^2}{2\tau_0}\right](1+\Lambda^{(1)}_\sigma) + \frac{i\Delta\omega}{2\tau_0} = (\omega^2+\Delta^2)^{3/2}\eta + \frac{1}{2\tau_0'}(\Delta^2\Lambda^{(1)}_p - i\Delta\omega\Lambda^{(2)}_p),$$

$$\left[(\omega^2+\Delta^2)^{3/2}\eta - \frac{\omega^2}{2\tau_0}\right]\Lambda^{(2)}_\sigma - \frac{i\Delta\omega}{2\tau_0}(1+\Lambda^{(1)}_\sigma) = \frac{1}{2\tau_0'}(i\Delta\omega\Lambda^{(1)}_p + \omega^2\Lambda^{(2)}_p),$$

$$\left[(\omega^2+\Delta^2)^{3/2}\eta - \frac{\Delta^2}{2\tau_0''}\right]\Lambda^{(1)}_p + \frac{i\Delta\omega}{2\tau_0''}\Lambda^{(2)}_p = -\frac{\Delta^2}{2\tau_1}(1+\Lambda^{(1)}_\sigma) + \frac{i\Delta\omega}{2\tau_1}\Lambda^{(2)}_\sigma,$$

$$\left[(\omega^2+\Delta^2)^{3/2}\eta - \frac{\omega^2}{2\tau_0''}\right]\Lambda^{(2)}_p - \frac{i\Delta\omega}{2\tau_0''}\Lambda^{(1)}_p = -\frac{i\Delta\omega}{2\tau_1}(1+\Lambda^{(1)}_\sigma) - \frac{\omega^2}{2\tau_1}\Lambda^{(2)}_\sigma. \tag{15}$$

Here

$$\frac{1}{\tau_0''} = \frac{1}{\tau_0} - \frac{3}{\tau_0'}; \qquad \frac{1}{\tau_0'} = \frac{nmp_0}{8\pi^2}\int |a|^2\sin^2\theta\,d\Omega.$$

The last two equations are convenient for expressing $\Lambda^{(1)}_p$ and $\Lambda^{(2)}_p$ in terms of $\Lambda^{(1)}_\sigma$ and $\Lambda^{(2)}_\sigma$:

$$\Lambda^{(1)}_p = -\frac{1}{3}\frac{\tau_0'}{\tau_1}\frac{\Delta^2(1+\Lambda^{(1)}_\sigma) - i\Delta\omega\Lambda^{(2)}_\sigma}{\omega^2+\Delta^2},$$

$$\Lambda^{(2)}_p = -\frac{1}{3}\frac{\tau_0'}{\tau_1}\frac{\omega^2(1+\Lambda^{(1)}_\sigma) + i\Delta\omega\Lambda^{(2)}_\sigma}{\omega^2+\Delta^2}. \tag{16}$$

Since $\tau_0'/\tau_1 \ll 1$, we have, in the derivation of the equations above, neglected interference terms and terms quadratic in the amplitude of the spin-orbit interaction in all expressions containing $\Lambda^{(1)}_p$ and $\Lambda^{(2)}_p$. The last-mentioned terms appear as a result of the multiplication of two scattering amplitudes (1) in the definition of the quantities $\Lambda^{(i)}$ [formula (12)].

With the help of (16) we find from the first two equations (15)

$$1 + \Lambda^{(1)}_\sigma = \frac{(\omega^2+\Delta^2)^{3/2}\eta - \omega^2(1/\tau_0 - 1/3\tau_1)/2}{(\omega^2+\Delta^2)[(\omega^2+\Delta^2)^{1/2}\eta - 1/2\tau_0 + 1/6\tau_1]},$$

$$\Lambda^{(2)}_\sigma = \frac{i\Delta\omega}{2}\frac{1/\tau_0 - 1/3\tau_1}{(\omega^2+\Delta^2)[(\omega^2+\Delta^2)^{1/2}\eta - 1/2\tau_0 + 1/6\tau_1]}. \tag{17}$$

We now have to calculate the susceptibility χ_s according to (6). Substituting (17) in (11), we obtain after integration finally

$$\frac{\chi_s}{\chi_n} = 1 - \Delta^2\pi T \sum_{\omega=-\infty}^{\infty} \frac{1}{(\omega^2+\Delta^2)\,[\sqrt{\omega^2+\Delta^2}+2/3\tau_1]}. \tag{18}$$

The contributions from $\Lambda_p^{(1)}$ and $\Lambda_p^{(2)}$ to χ can be neglected, since, according to (16), they are of the order $l/l_{s.o.}$, whereas the contributions from $\Lambda_\sigma^{(1)}$ and $\Lambda_\sigma^{(2)}$ are of the order ξ_0/l.*

In conclusion we write down the final formulas. If the spin-orbit interaction is very small ($l_{s.o.}$ large in comparison with ξ_0), then

$$\frac{\chi_s}{\chi_n} = \frac{N_n(T)}{N} - \frac{\pi\Delta}{6\tau_1}\frac{d}{d\Delta}\left(\frac{1}{\Delta}\,\text{th}\,\frac{\Delta}{2T}\right). \tag{19†}$$

The first term represents the ratio of the number of "normal" electrons over the total number of electrons and agrees with the results of Yosida and the earlier work of the authors.[1,4] For $T = 0$ we have, of course, $N_n(0) = 0$.

In the opposite limiting case of small mean free paths $l_{s.o.}$, we find that $\chi_s/\chi_n \to 1$:

$$\chi_s/\chi_n = 1 - \tfrac{3}{4}\pi\tau_1\Delta\,\text{th}\,(\Delta/2T). \tag{20}$$

The symbol Δ in formulas (18) to (20) denotes the equilibrium gap in the spectrum of the superconductor at the given temperature.

In the general case, when the mean free path $l_{s.o.}$ is comparable with the correlation length ξ_0, closed expressions can be written down for χ_s in the cases $T = 0$ and $|T_c - T| \ll T_c$. For $T = 0$ we have ($\rho_0 = \frac{2}{3}\tau_1\Delta_0$):

$$\frac{\chi_s}{\chi_n} = \begin{cases} 1 - \dfrac{1}{\rho_0}\left[\dfrac{\pi}{2} - \dfrac{\text{arch}\,\rho_0}{\sqrt{\rho_0^2-1}}\right], & \rho_0 > 1 \\[2ex] 1 - \dfrac{1}{\rho_0}\left[\dfrac{\pi}{2} - \dfrac{\text{arc cos}\,\rho_0}{\sqrt{1-\rho_0^2}}\right], & \rho_0 < 1 \end{cases} \tag{21‡}$$

This function is shown in Fig. 5.

For $|T - T_c| \ll T_c$ $(\Delta \to 0)$ we find

$$\frac{\chi_s}{\chi_n} = 1 - \frac{2\Delta^2}{(\pi T)^2\rho_c}\left\{\frac{\pi^2}{8} + \frac{1}{2\rho_c}\left[\psi\left(\frac{1}{2}\right) - \psi\left(\frac{1}{2}+\frac{\rho_c}{2}\right)\right]\right\}. \tag{22}$$

Here $\rho_c = \frac{2}{3}\tau_1\pi T_c = \rho_0/\gamma$, where γ is the Euler constant, $\gamma = 1.78$; $\psi(x)$ is the logarithmic derivative of the Γ function.

In the following section we shall compare our formulas with the experimental data.

3. DISCUSSION OF THE EXPERIMENTAL DATA

It has already been noted in [4] that such a comparison cannot be based on data which have been obtained with too large specimens and too strong fields. This limitation is due not only to the circumstance that the field is inhomogeneous in specimens with dimensions larger than the penetration depth. Another factor is important, too. It is known that if the dimensions of the specimen are smaller than the penetration depth the phase transition in the magnetic field is a second order transition and is connected with a gradual vanishing of the quantity Δ, the gap in the electron spectrum. The vanishing of the gap is, at the given temperature, accompanied by an increase in the number of "unpaired" electrons, i.e., an increase in the Knight shift. This last effect shows up, in particular, in the experiments of Androes and Knight (see Figs. 3 and 5 in [8]). In order to make this effect unimportant, the field must be weaker than the critical field of a specimen with the given dimensions d. Since $H_c \propto d^{-1}$, the effect of the change in the energy gap due to the external field is the smaller, the smaller the dimensions of the specimen.

For these reasons it is difficult to interpret the data of Reif[9] for colloidal mercury, for which the uncertainty in the size of the particles is large. This applies also to a considerable degree to the results of Androes and Knight[8] obtained with fields above 4 kilooersted. The critical field for the specimens used in [8] was 25 kilooersted at T = 0, and a field of 4 kilooersted could give a noticeable effect, in any case at temperatures which are close to the critical temperature. The attempt in [8] to interpret these data by taking account of the effect of the field through a change in the temperature scale does not seem convincing to us. A theory describing the effect of the field on the Knight shift does not exist as yet, and there are no indications whatever that a change in the temperature scale, and in particular, the one proposed in [8], has anything to do with the

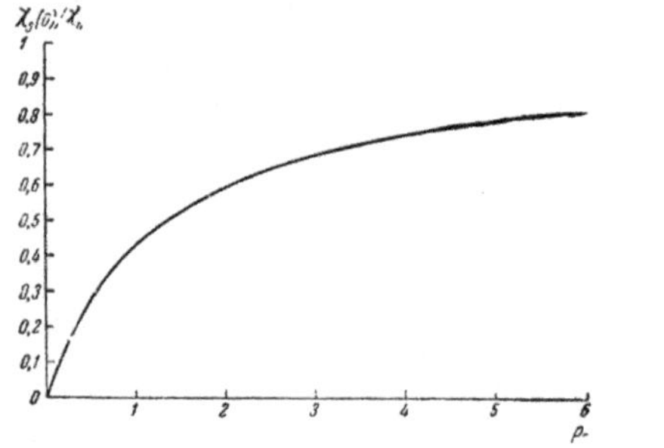

FIG. 5

*In the cases of interest to us the specimens have dimensions of the order 10^{-6} cm, i.e., $l \approx 10^{-6}$, so that $l \ll \xi_0 \approx 10^{-4}$ to 10^{-5} cm.

†th = tanh.

‡arch = $\cosh^{-1}$

problem. In particular, Androes and Knight [8] assume that near T_c one has $H_c \approx 2H(0)(1 - T/T_c)$, whereas it is known that for films $H_c \sim H_{cm}\delta/d$, where $H_{cm} \sim 1 - T/T_c$ is the field for a bulky specimen and $\delta \sim 1/\sqrt{1 - T/T_c}$ is the penetration depth, i.e., $H_c \sim \sqrt{1 - T/T_c}$.

Therefore, we have compared our results only with the data obtained in [8] with a field of 1.2 kilooersted (Fig. 6). The theoretical curves correspond to different values of ρ_0. The lower dashed curve refers to the case $\rho_0 = 0$, i.e., shows the dependence of the paramagnetic susceptibility of a superconductor in the absence of spin-orbit coupling. [1]

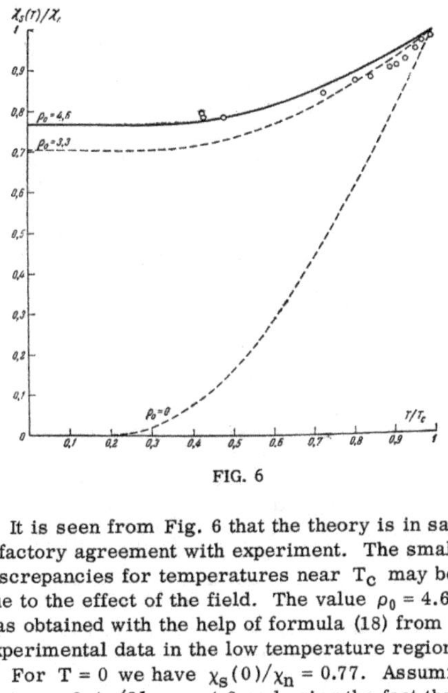

FIG. 6

It is seen from Fig. 6 that the theory is in satisfactory agreement with experiment. The small discrepancies for temperatures near T_c may be due to the effect of the field. The value $\rho_0 = 4.6$ was obtained with the help of formula (18) from experimental data in the low temperature region.

For $T = 0$ we have $\chi_s(0)/\chi_n = 0.77$. Assuming that $\rho_0 = 2\pi\xi_0/3l_{s.o.} = 4.6$ and using the fact that $\xi_0 = 2 \times 10^{-5}$ for tin, [7] we obtain $l_{s.o.} = 1.0 \times 10^{-5}$ cm. Assuming further that the mean free path l is of the order of the dimensions of the specimen, i.e., $l \sim 10^{-6}$, we find that $l \sim 0.1\, l_{s.o.}$, which implies that the spin-orbit interaction is somewhat weaker than usual.

In conclusion we make one further remark. Androes and Knight [8] have expressed doubt that the Knight shift can be explained with the help of the spin-orbit interaction, since they did not observe any appreciable dependence on the dimensions of the specimen in their experiments. Measurements on specimens with dimensions of ~ 1000 Å gave the same order of magnitude for the Knight shift as films with a thickness of 40 Å. In our opinion, these results are not sufficient to make the above-mentioned conclusion convincing. If the critical field for specimens with dimensions of 40 Å was 25 kilooersted, then specimens with dimensions of 1000 Å have a critical field of 1 kilooersted, and for this reason alone the Knight shift may have a magnitude of order unity.

[1] K. Yosida, Phys. Rev. **110**, 769 (1958).
[2] P. C. Martin and L. P. Kadanoff, Phys. Rev. Lett. **3**, 322 (1959).
[3] J. R. Schrieffer, Phys. Rev. Lett. **3**, 323 (1959).
[4] A. A. Abrikosov and L. P. Gor'kov, JETP **39**, 480 (1960), Soviet Phys. JETP **12**, 337 (1961).
[5] R. A. Ferrell, Phys. Rev. Lett. **3**, 262 (1959).
[6] P. W. Anderson, Phys. Rev. Lett. **3**, 325 (1959).
[7] J. Bardeen and J. R. Schrieffer, Progress in Low Temperature Physics **3**, Amsterdam (1961).
[8] G. M. Androes and W. D. Knight, Phys. Rev. **121**, 779 (1961).
[9] F. Reif, Phys. Rev. **106**, 208 (1957).
[10] T. Matsubara, Progr. Theor. Phys. **14**, 351 (1955).
[11] Abrikosov, Gor'kov, and Dzyaloshinskiĭ, JETP **36**, 900 (1959), Soviet Phys. JETP **9**, 636 (1959).
[12] A. A. Abrikosov and L. P. Gor'kov, JETP **35**, 1558 (1958) and **36**, 319 (1959), Soviet Phys. JETP **8**, 1090 (1959) and **9**, 220 (1959).

Translated by R. Lipperheide
179

SOVIET PHYSICS JETP VOLUME 16, NUMBER 3 MARCH, 1963

THEORY OF THE ELECTRON ENERGY SPECTRUM OF METALS WITH A BISMUTH TYPE LATTICE

A. A. ABRIKOSOV and L. A. FALKOVSKIĬ

Institute of Physics Problems, Academy of Sciences, U.S.S.R.

Submitted to JETP editor April 23, 1962

J. Exptl. Theoret. Phys. (U.S.S.R.) **43**, 1089-1101 (September, 1962)

The electron energy spectrum of metals of the fifth group of the periodic table is determined on the basis of some qualitative assumptions and symmetry theory.

1. As is well known, metals of the fifth group of the periodic system, i.e., arsenic, antimony, and bismuth, have several interesting distinguishing features. The crystal lattices of these metals have the same type. They are rhombohedral with two atoms per cell. A feature of these lattices is that they can be obtained from simple cubic lattices by slight displacement of the atoms. This is demonstrated in Fig. 1.

Let us separate two face-centered sublattices in a simple cubic lattice. The vertex of the cube of one of these sublattices is in the center of the cube of the other. Atoms of different sublattices are denoted by the filled and light circles. We now displace the sublattices relative to each other along the body diagonal of the cube. The lattice symmetry (class) then becomes immediately rhombohedral. In addition, it is necessary to carry out a slight rhombohedral deformation of the sublattices themselves. The resultant lattice is that of bismuth. The table lists the numbers characterizing the relative arrangement of the sublattices [the rhombohedral coordinates of two atoms belonging to two different sublattices are (s, s, s) and (−s, −s, −s)] and the angle between the main periods (sublattices in the case of a cube). These data show the extent to which the lattices of arsenic, antimony, and bismuth are close to a simple cubic lattice.

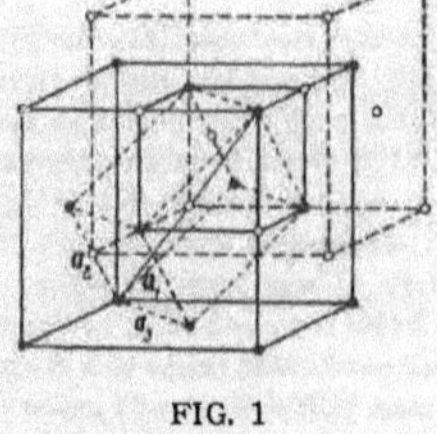

FIG. 1

	s	θ		s	θ
As	0.226	54°10′	Bi	0.237	54°14′
Sb	0.223	57°6′	cube	0.25	60°

Another feature of these elements is that they are poor metals, i.e., the number of their current carriers per atom is exceedingly small. Thus, in the case of bismuth, according to data obtained on the de Haas–van Alphen effect [1,2], the number of "free" electrons amounts to $\sim 10^{-5}$ per atom. According to Luttinger [3] the total electron density is $n = 2(2\pi)^{-3}(pV_0 + V_F)$, where V_0 is the volume of the reciprocal lattice cell, V_F the volume per cell bounded by all the Fermi surfaces (on the low-energy side), and p is an integer. This equation has the same form as for the model of interacting particles in a periodic field. Since $V_0 = (2\pi)^3/v_0$, where v_0 is the volume of the elementary cell, it follows from the Luttinger formula that if the number of electrons per elementary cell is odd, then the substance must be a metal, and if it is even it may be either a metal or a dielectric. In the latter case the number of carriers of the "electron" type should equal the number of "holes." It follows therefore that carriers of both types must exist in the metals under consideration.

In addition to determining the number of carriers, it is possible to obtain from the data on the de Haas–van Alphen effect rather detailed information on the electron spectrum. Apparently the Fermi surface for electrons has the form of three ellipsoids (or figures close to them) with centers on the binary axes [1]. It has been recently suggested on the basis of experimental data on the infrared absorption [4] that the form of these surfaces deviates from that of an exact ellipsoid [5]. The Fermi surface of holes is apparently not at all like an ellipsoid [2].

The present paper is devoted to the theory of the electron spectrum of metals of the fifth group. It is based on the form of the lattice of these metals and on certain qualitative assumptions. The results obtained furnish arguments in favor of the regularity in the occurrence of lattices of the bismuth type.

We note here that an attempt to relate the form of the bismuth lattice with the small number of carriers was already made by Jones,[6] who did not go, however, beyond qualitative discussions.

2. One of the authors[7] has shown that the spectrum of the excitations in a metal can be found by using the analogy with the model of noninteracting particles in a periodic field. We therefore base our arguments on such an analogy. In the metals we are considering the number of electrons is odd. Consequently, were they to have a simple cubic lattice, they would be good metals. As a result of the deformation, the lattice period changes, the number of electrons in the cell doubles, and the metals can in principle become dielectrics.

Such a transition is easiest to understand with the one-dimensional model of the metal as an example. Let us assume that by a slight change in potential we have doubled its period. This immediately gives rise to splitting of the energy bands. It is seen from Fig. 2 that the maximum $\epsilon(k)$ in the lower of the bands obtained and the minimum in the upper band occur at perfectly defined quasimomentum values, which are characterized by the following properties (we are considering one cell of the initial reciprocal lattice): a) in the initial reciprocal lattice these are points corresponding to the same value of the energy; b) in the new reciprocal lattice these are equivalent points, i.e., points differing by one period of the reciprocal lattice, whereas in the old lattice they were not equivalent. Even if the number of electrons per atom was odd in the initial lattice, the new substance will be a dielectric.

In the three-dimensional case, such a categoric statement is incorrect, and in the new lattice the minima of the upper band may turn out to be below the maxima of the lower band. Furthermore, if the deformation is infinitesimally small, the substance must remain a metal. This follows from the fact that a very small splitting cannot change the area of the Fermi surface (although of course it changes the number of carriers). However, if the splitting is finite, the substance may become a dielectric. We cannot offer an exhaustive proof that this should be the case in the situation which we are considering. We can, however, express a few ideas.

We note first that a mere change in the period can greatly decrease the number of carriers. For example, if we assume that in the initial lattice the Fermi surface is close to a sphere with a volume equal to half the volume of the reciprocal lattice cell, then the new faces of the Brillouin zone appearing as a result of the change in the period will cut away from the reciprocal lattice space volumes of the order of 0.01. It can be assumed that the increase in deformation will bring about a decrease in these volumes. On the other hand, a metal with a small number of carriers can turn into a dielectric.[1)] Obviously the ratio of the average distance between the quasiparticle to the screened radius of the Coulomb interaction is important here. If the screening radius is large compared with the average distance between quasiparticles, then the unscreened Coulomb interaction can give rise to bound pairs of excitations of the electron and hole type. Obviously the metal changes in this case into a dielectric. It is easy to see that the criterion for a changeover into a dielectric may be written in the form $v \ll 1/ma$, where v is the velocity on the Fermi boundary, m the mass of the free electron, and a the lattice period. In a spectrum of the type $\epsilon = p^2/2m^*$ we get $v \sim (N/V)^{1/3} m^{*-1}$, from which it follows that instability arises when the number of carriers is small.

The transition from a metal into a dielectric is a first-order phase transition[2)]. If we consider the

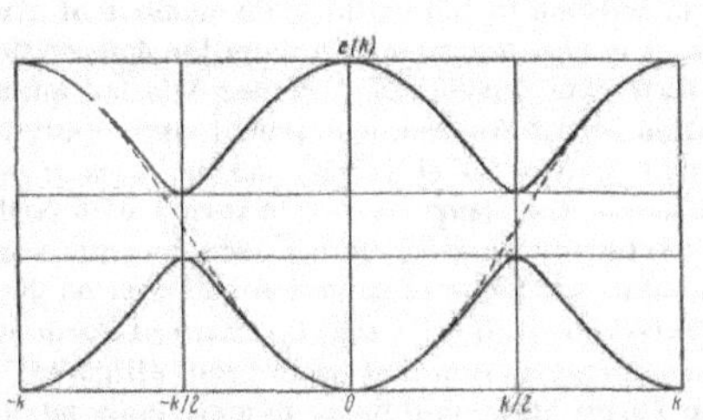

FIG. 2

[1)]The considerations advanced below are based on ideas of R. G. Arkhipov.

[2)]At first glance it may seem that the metal phase is not in equilibrium only because the number of current carriers is decreased, and that such non-equilibrium may occur even in the case of an infinitesimally small deformation. Actually this is not so, in view of the fact that in the case of infinitesimally small deformation the velocity on the Fermi surface has the same order of magnitude as before. A continuous transition with finite defomation is excluded, in view of the fact that the pair production leads to a decrease in screening and therefore, once initiated, should proceed to the end.

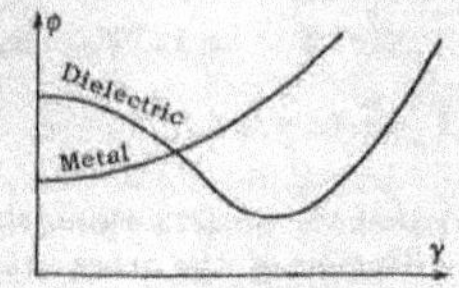

FIG. 3

thermodynamic potentials of both phases as functions of the deformation γ, then the corresponding curves should have the form shown in Fig. 3. If the minimum of the thermodynamic potential of the dielectric phase lies below the energy of the metal phase when $\gamma = 0$, such a deformation will occur spontaneously.

What is of importance to us for the time being is that both phases should exist above and below the points of intersection of the curves of Fig. 3, one in equilibrium and the other metastable. If we assume that in the case of bismuth this is precisely the situation, then we can conclude that the "dielectric" phase exists at as small a deformation as desired in the form of a metastable phase. At the point where the deformation vanishes, the "dielectric" phase becomes absolutely unstable.

We write "dielectric" in quotation marks, because in the lattice of the bismuth type, as will be shown below, this phase will actually be not a dielectric but a poor metal with few electrons. It turns out as well that the velocity on the Fermi boundary has the same order of magnitude as in good metals, so that the existence of such small groups does not contradict the ideas developed above.

To obtain the energy spectrum we first consider a lattice that is infinitesimally close to a cube, and then deform it and change it into the true lattice of bismuth. In order that the change in the spectrum be continuous in this case, without a phase transition, we consider only the "dielectric" phase, although it is not in equilibrium at small deformations.

If we separate two face-centered sublattices in the initial cubic lattice, then their periods will be

$$\mathbf{a}_1 = \mathbf{a}_2^0 + \mathbf{a}_3^0, \quad \mathbf{a}_2 = \mathbf{a}_3^0 + \mathbf{a}_1^0, \quad \mathbf{a}_3 = \mathbf{a}_1^0 + \mathbf{a}_2^0. \tag{1}$$

The periods of the new reciprocal lattice will be

$$\mathbf{K}_1 = 2\pi\,[\mathbf{a}_2\mathbf{a}_3] / (\mathbf{a}_1\,[\mathbf{a}_2\mathbf{a}_3]) = \tfrac{1}{2}(\mathbf{K}_2^0 + \mathbf{K}_3^0 - \mathbf{K}_1^0),$$

$$\mathbf{K}_2 = \tfrac{1}{2}(\mathbf{K}_3^0 + \mathbf{K}_1^0 - \mathbf{K}_2^0), \quad \mathbf{K}_3 = \tfrac{1}{2}(\mathbf{K}_1^0 + \mathbf{K}_2^0 - \mathbf{K}_3^0). \quad (2)^*$$

An arbitrary reciprocal lattice vector is equal to

$$\mathbf{K} = \tfrac{1}{2}(n_3 + n_2 - n_1)\,\mathbf{K}_1^0 + \tfrac{1}{2}(n_3 + n_1 - n_2)\,\mathbf{K}_2^0 + \tfrac{1}{2}(n_1 + n_2 - n_3)\,\mathbf{K}_3^0. \tag{3}$$

We consider points that lie on the body diagonals of the initial cube halfway between its vertex and the center. There are altogether eight such points, and all are nonequivalent in the old reciprocal lattice. In the new lattice they break up into four periods, each of which has two equivalent points $\mathbf{k}_i$ and $-\mathbf{k}_i$ (i.e., points differing by the new reciprocal lattice period). The coordinates of the points $\mathbf{k}_i$ can be written in terms of the coefficients in the formula $\mathbf{k} = \sum_s \alpha_s \mathbf{K}_s^0$:

$$\mathbf{k}_0 = \left(\tfrac{1}{4}\,\tfrac{1}{4}\,\tfrac{1}{4}\right), \quad \mathbf{k}_1 = \left(\tfrac{1}{4}\,\tfrac{\bar{1}}{4}\,\tfrac{\bar{1}}{4}\right), \quad \mathbf{k}_2 = \left(\tfrac{\bar{1}}{4}\,\tfrac{1}{4}\,\tfrac{\bar{1}}{4}\right),$$

$$\mathbf{k}_3 = \left(\tfrac{\bar{1}}{4}\,\tfrac{\bar{1}}{4}\,\tfrac{1}{4}\right). \tag{4}$$

The singular properties of the points $\mathbf{k}_i$ and $-\mathbf{k}_i$, namely the non-equivalence in the initial lattice and the equivalence in the new one, as well as the high symmetry, give grounds for expecting at these points maxima and minima of the energy as the function of the quasimomentum (see, for example, Fig. 2 for the one-dimensional model).

The location of the points $\mathbf{k}_i$ in the reciprocal lattice is illustrated in Fig. 4. After deformation, the point $\mathbf{k}_0$ is situated in the center of the rhombohedron. The little group of this point is D_{3d} and forms by itself a star. The points $\mathbf{k}_1$, $\mathbf{k}_2$, and $\mathbf{k}_3$ belong to one star. They are located on the binary axes of the rhombohedron. The little group of $\mathbf{k}_1$ is C_{2h}. We note that the location of these points corresponds exactly to the experimentally observed location of the centers of the Fermi surfaces.

To determine the spectrum it is necessary first to determine the basis functions and then construct $D_{ss'}$. Let us start with an initial cubic lattice. In the corresponding reciprocal lattice all the eight points $\mathbf{k}_i$ and $-\mathbf{k}_i$ belong to one star. The little group of each point is C_{3v}.

Since we are dealing with metals with large atomic numbers, it is necessary to take into ac-

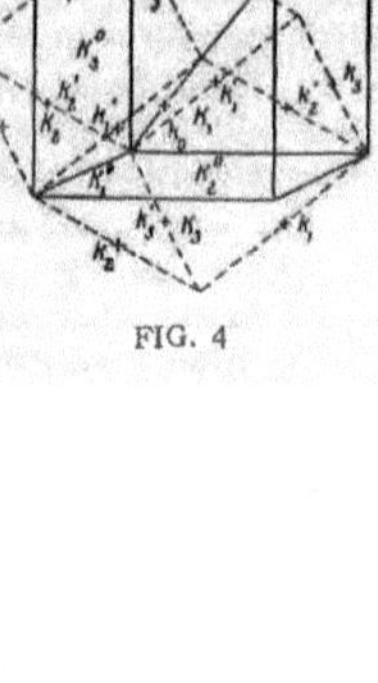

FIG. 4

*$[\mathbf{a}_2\,\mathbf{a}_3] = \mathbf{a}_2 \times \mathbf{a}_3$.

count the spin-orbit coupling, and consequently to consider doubly-valued representations of the group C_{3v}. We assume that the energy of the spin-orbit coupling is small compared with the ordinary electron energies. Without account of the spin orbit coupling, the C_{3v} group has three single-valued representations: two one-dimensional and one two-dimensional (see [8]). If the spin is taken into account, the corresponding representations can be multiplied by $D_{1/2} = E_2'$ (see [9]). In this case both one-dimensional representations go over into equivalent two-dimensional double-valued representations E_2', and the two-dimensional representation yields one two-dimensional representation E_2' and two complex conjugate one-dimensional representations E_1', combined into one.

In view of the fact that the lattice contains an inversion center, double degeneracy is maintained for all values of **k**. Consequently, for those representations which are obtained from the one-dimensional ones, account of the spin-orbit coupling does not lead to level splitting. The situation is different in the case of a two-dimensional single-valued representation, where the spin-orbit coupling splits the quadruple level into two double levels.

If we assume that the level considered corresponded to a one-dimensional (single-valued) representation of the C_{3v} group, then we can show (by a method analogous to that employed below) that no spectrum of the type observed in the case of bismuth can arise after the deformation. In view of this we start from the two-dimensional representation.

Now, however, we must recognize that we are considering a lattice with a modified period. In the new lattice the points $\mathbf{k}_i$ and $-\mathbf{k}_i$ become equivalent. The corresponding representations must be combined as well. As a result, the matrix $D_{ss'}$ in the vicinity of each point $\mathbf{k} = \mathbf{k}_i$ is of order 8×8.

To estimate the order of magnitude of the matrix coefficients it is convenient to introduce ψ_+ and ψ_-, which are basis functions of the two-dimensional (single-valued) representation of the group C_{3v}, having the properties $C_3\psi_\pm = e^{\pm 2\pi i/3}\psi_\pm$, $\sigma_{(1)}\psi_\pm = \psi_\mp$ ($\sigma_{(1)}$ is the reflection in the plane passing through the C_3 axis) and the spinors $\chi_\pm$, which transform in accordance with the rule $C_3\chi_\pm = e^{\pm i\pi/3}\chi_\pm$, $\sigma_{(1)}\chi_\pm = i\chi_\mp$. From these functions we can readily construct the bases of the double-valued representations of the C_{3v} group:

$$E_2': \quad \Psi_1 = \psi_+\chi_-, \ \Psi_2 = \psi_-\chi_+;$$

$$E_1': \quad \Psi_3 = 2^{-1/2}(\psi_+\chi_+ + \psi_-\chi_-), \ \Psi_4 = 2^{-1/2}(\psi_+\chi_+ - \psi_-\chi_-); \tag{5}$$

It is obvious that the matrix elements between the states with different χ can arise only as a result of spin-orbit coupling. The same pertains also to the difference in the diagonal matrix elements.

The functions (5) pertain to a definite $\mathbf{k}_i$. Corresponding to the point $-\mathbf{k}_i$ are the functions Ψ_1', Ψ_2', Ψ_3', and Ψ_4', obtained from (5) by inversion. It is obvious that the matrix elements of the transitions between these sets can arise only as a result of a change in the period, and will be proportional in first approximation to the relative shift of the sublattices.

3. The deformation by which a lattice of the bismuth type is obtained from the cubic lattice can be described by two quantities: the vector **u**, which is directed along the body diagonal of the cube and determines the displacement of the sublattices relative to the inversion center, and the component of the deformation tensor $u_{x'y'} = u_{y'z'} = u_{z'x'}$ (the coordinates x', y', and z' are defined relative to the cubic axes). We note that the vector **u** has a different sign for the two sublattices, and therefore, for example under the inversion transformation when a transition takes place from one sublattice to the other, this vector should remain unchanged. We disregard hydrostatic compression here, since it leads only to a displacement of the chemical potential.

To determine the dependence of the spectrum on the deformation, we should set up invariants of the basis functions, of the vector $\boldsymbol{\kappa} = \mathbf{k} - \mathbf{k}_i$, of the vector **u**, and of the deformation tensor u_{ik}. We shall assume that the components of $\boldsymbol{\kappa}$, the quantities defining the deformation, and also the quantity Δ which characterizes the spin-orbit coupling, are all quantities of the same order of magnitude (of course, with corresponding coefficients), and confine ourselves to first-order terms. We note that the components u_{ik} make up an invariant of the C_{3v} group: $\frac{1}{3}(u_{x'z'} + u_{y'z'} + u_{z'y'}) = w$. We take into consideration the fact that the functions Ψ_i' are obtained from the Ψ_i by inversion.

With these remarks taken into account, let us write down the matrix $D_{ss'}$. This is easiest to do by using a reference frame with the z axis in the $\mathbf{k}_i$ direction. Thereupon we obtain

$$D_{ss'}=\left\|\begin{matrix} f+a\varkappa_z+\Delta & 0 & \frac{b\varkappa_++\beta_+}{\sqrt{2}} & -\frac{b\varkappa_++\beta_+}{\sqrt{2}} & \gamma & 0 & \frac{\delta_+}{\sqrt{2}} & -\frac{\delta_+}{\sqrt{2}} \\ 0 & f+a\varkappa_z+\Delta & \frac{b\varkappa_-+\beta_-}{\sqrt{2}} & \frac{b\varkappa_-+\beta_-}{\sqrt{2}} & 0 & \gamma & \frac{\delta_-}{\sqrt{2}} & \frac{\delta_-}{\sqrt{2}} \\ \frac{b\varkappa_-+\beta_-}{\sqrt{2}} & \frac{b\varkappa_++\beta_+}{\sqrt{2}} & f+a\varkappa_z-\Delta & 0 & \frac{\delta_-}{\sqrt{2}} & \frac{\delta_+}{\sqrt{2}} & \gamma & 0 \\ -\frac{b\varkappa_-+\beta_-}{\sqrt{2}} & \frac{b\varkappa_++\beta_+}{\sqrt{2}} & 0 & f+a\varkappa_z-\Delta & -\frac{\delta_-}{\sqrt{2}} & \frac{\delta_+}{\sqrt{2}} & 0 & \gamma \\ \gamma & 0 & \frac{\delta_+}{\sqrt{2}} & -\frac{\delta_+}{\sqrt{2}} & f-a\varkappa_z+\Delta & 0 & \frac{-b\varkappa_++\beta_+}{\sqrt{2}} & \frac{b\varkappa_+-\beta_+}{\sqrt{2}} \\ 0 & \gamma & \frac{\delta_-}{\sqrt{2}} & \frac{\delta_-}{\sqrt{2}} & 0 & f-a\varkappa_z+\Delta & \frac{-b\varkappa_-+\beta_-}{\sqrt{2}} & \frac{-b\varkappa_-+\beta_-}{\sqrt{2}} \\ \frac{\delta_-}{\sqrt{2}} & \frac{\delta_+}{\sqrt{2}} & \gamma & 0 & \frac{-b\varkappa_-+\beta_-}{\sqrt{2}} & \frac{-b\varkappa_++\beta_+}{\sqrt{2}} & f-a\varkappa_z-\Delta & 0 \\ -\frac{\delta_-}{\sqrt{2}} & \frac{\delta_+}{\sqrt{2}} & 0 & \gamma & \frac{b\varkappa_--\beta_-}{\sqrt{2}} & \frac{-b\varkappa_++\beta_+}{\sqrt{2}} & 0 & f-a\varkappa_z-\Delta \end{matrix}\right\| \quad (6)$$

where

$$f=vu_z+\zeta w, \quad \gamma=cu_z,$$

$$\beta_\pm=eu_\pm=e(u_y\pm iu_x), \quad \delta_\pm=du_\pm. \quad (7)$$

The constants in the matrix $D_{ss'}$ do not depend on which of the points k_i is the one whose vicinity we are considering. This follows from the fact that the total quantity $Q(\omega, \mathbf{r}, \mathbf{r}')$ must be an invariant of the cubic group (see [7]).

If we are considering the vicinity of the point k_0, the z axis is directed along the rhombohedral axis. Therefore

$$f^{(0)}=vu+\zeta w, \quad \delta_\pm^{(0)}=\beta_\pm^{(0)}=0, \quad \gamma^{(0)}=cu. \quad (8)$$

On the other hand, if $k = k_1$, then the z axis is directed along the other C_3 axis in the cubic crystal. The transformation which changes the vector k_0 into k_1 is rotation through π around the x′ axis in the cubic reference frame. This rotation must be carried out relative to the axis passing through the cubic lattice site, for in this case no interchange of sublattices takes place. The projection of u on the new z axis is equal to $-u/3$. Upon rotation we obtain

$$\tfrac{1}{3}(u_{x'y'}+u_{x'z'}+u_{y'z'})\to\tfrac{1}{3}(u_{y'z'}-u_{x'y'}-u_{x'z'}) = -w/3.$$

The values of $u_\pm$ depend on the choice of the axes. It is most convenient to choose the x axis in the diagonal plane of the cube so that it is perpendicular to the cubic axis x′. After rotation, the new x axis will again remain perpendicular to the x′ axis and u_y will turn out to be $-2\sqrt{2}\,u/3$. We thus have

$$f^{(1)}=-vu/3-\zeta w/3, \quad \beta_+^{(1)}=\beta_-^{(1)}=-2\sqrt{2}eu/3,$$

$$\gamma^{(1)}=-cu/3, \quad \delta_+^{(1)}=\delta_-^{(1)}=-2\sqrt{2}du/3. \quad (9)$$

We use for simplicity the notation

$$f^{(0)}=f, \quad \gamma^{(0)}=\gamma, \quad \beta^{(1)}=\beta, \quad \delta^{(1)}=\delta.$$

In order to determine the spectrum it is necessary to find the eigenvalues of the matrix $D_{ss'}$. We assume that $\beta_+=\beta_-\neq 0$ and $\delta_+=\delta_-\neq 0$. This enables us to obtain a general formula for the vicinities of k_0 and k_1.

Expanding $\det(-\omega\delta_{ss'}+D_{ss'})$ we obtain an equation for the roots

$$\begin{aligned}&\Omega^4+\Delta^4+a^4\varkappa_z^4+b^4(\varkappa_x^2+\varkappa_y^2)^2+\gamma^4+\delta^4-2\Omega^2a^2\varkappa_z^2\\&-2\Omega^2\Delta^2-2a^2\varkappa_z^2\Delta^2-2\Omega^2\gamma^2+2\gamma^2a^2\varkappa_z^2-2\gamma^2\Delta^2+2\Delta^2\beta^2\\&+2\Delta^2b^2(\varkappa_x^2+\varkappa_y^2)-2\Omega^2\beta^2-2\Omega^2b^2(\varkappa_x^2+\varkappa_y^2)-2a^2\varkappa_z^2\beta^2\\&-2a^2b^2(\varkappa_x^2+\varkappa_y^2)\varkappa_z^2+2\delta^2a^2\varkappa_z^2+2\delta^2\Delta^2-2\delta^2\Omega^2-2\delta^2\beta^2\\&-2\gamma^2\beta^2+2\gamma^2b^2(\varkappa_x^2+\varkappa_y^2)-2\gamma^2\delta^2+8(\gamma\delta-\Omega\beta)ab\varkappa_z\varkappa_y\\&+2(\delta^2-\beta^2)b^2(\varkappa_y^2-\varkappa_x^2)+8\gamma\delta\beta\Omega=0,\end{aligned} \quad (10)$$

where $\Omega = f-\omega$; each root is double. Owing to the presence of terms that are linear in Ω, Eq. (10) can in general not be solved with respect to Ω. There are no such terms if we consider the vicinity of k_0, since $\beta^{(0)}=\delta^{(0)}=0$. We then obtain

$$-\Omega=\omega-f=\pm\left(\Delta^2+a^2\varkappa_z^2+\gamma^2+b^2\varkappa_\perp^2 \pm 2\sqrt{\gamma^2\Delta^2+a^2b^2\varkappa_z^2\varkappa_\perp^2+\Delta^2a^2\varkappa_z^2}\right)^{1/2}, \quad (11)$$

where $\kappa_\perp^2=\kappa_x^2+\kappa_y^2$.

Let us consider first the formula obtained in the limit of infinitesimally small deformation. Assuming f, γ, β, and δ to tend to zero, we obtain

$$\omega=\pm\left|\,|a\varkappa_z|\pm\sqrt{\Delta^2+b^2\varkappa_\perp^2}\,\right|. \quad (12)$$

It is easy to see that the equal-energy surface $\omega = 0$ degenerates into a double-cavity hyperboloid of revolution. This hyperboloid is a surface along which the two functions $\omega(k)$ are tangent. For one of these functions this surface corresponds to a minimum, while for the other it corresponds to a maximum. This takes place in the vicinity of any of the points k_i. It follows therefore that in the limit of infinitesimally small deformation we obtain a substance that is halfway between a metal and a dielectric, in which the conduction band is perfectly empty and the valence band is filled; furthermore, these bands touch each other along a certain surface. Such a degeneracy along an entire surface is possible only in the limiting case. Any finite deformation lifts this degeneracy.

We now turn to a finite deformation. According to (11), it is possible to find even in this case a value of ω corresponding to a minimum of one of the functions $\omega(\mathbf{k})$ in the vicinity of $\mathbf{k}_0$ and to the maximum of the other. To be sure, in this case the tangency is only in points, and not along the surface. The same holds true also for the vicinity of $\mathbf{k}_1$. However, the values of ω at which this occurs in the vicinity of $\mathbf{k}_0$ differ from those in the vicinity of $\mathbf{k}_1$, and this leads to a overlap of the bands. This means that, for any finite deformation, Fermi surfaces corresponding to electrons or holes appear in the vicinity of $\mathbf{k}_0$ and $\mathbf{k}_1$, $\mathbf{k}_2$, $\mathbf{k}_3$.

In principle formula (10) solves our problem. It remains only to choose the chemical potential correctly. For this purpose it is necessary to use the condition of the equality of the number of electrons to the number of holes. The remaining coefficients, namely a, b, Δ, γ, β, δ, and f are determined from the experimental data (say from the de Haas–van Alphen effect and cyclotron resonance). We note that these constants determine the entire electron spectrum, i.e., not only the Fermi surface, but any derivative $\partial^n\epsilon/\partial p_i \ldots \partial p_k$ on the boundary. The results obtained make it possible also to study the influence of the deformation on the electron spectrum. Here, of course, it is necessary to know not only the deformation tensor but also the displacement of the sublattices.

4. To facilitate the comparison of the results with experiment we shall analyze various cases. We start with the vicinity of $\mathbf{k} = \mathbf{k}_0$. According to the experimental data, the vicinity of $\mathbf{k} = \mathbf{k}_0$ contains holes while the vicinity of $\mathbf{k} = \mathbf{k}_1$ contains electrons (the sign of the carriers is chosen arbitrarily). In view of this, we must take the minus sign in front of the root in formula (11) for the vicinity of $\mathbf{k}_0$. To simplify the formulas we introduce the notation $p = a\kappa_z$ and $q = b\kappa_\perp$. In addition to formula (11), we will find it convenient to solve Eq. (10) with respect to p and with respect to q. We then obtain

$$p^2 = \Omega^2 + \Delta^2 - \gamma^2 + q^2 \pm 2\sqrt{\Omega^2\Delta^2 + (\Omega^2 - \gamma^2)\, q^2}, \quad (13a)$$

$$q^2 = \Omega^2 + p^2 - \gamma^2 - \Delta^2 \pm 2\sqrt{\gamma^2\Delta^2 + (\Omega^2 - \gamma^2)\, p^2}. \quad (13b)$$

Inasmuch as these equations contain only the squares of the coefficients γ and Δ, we shall simply assume them to be positive. The positiveness of Ω follows from (11).

From formula (13) follow immediately two principally different cases. If $\Omega > \gamma$, then p and q can assume infinite values, whereas if $\Omega < \gamma$, then p and q should remain finite. Actually, however, our entire analysis is valid so long as $\kappa_\perp$ and κ_z are noticeably smaller than the period of the reciprocal lattice. A more detailed analysis shows that several characteristic types of equal-energy surfaces are produced, depending on the ratios of the coefficients Ω, γ, and Δ. All these types are shown in Fig. 5.

Open surfaces arise in the case when $\Omega > \gamma$. The shown types of surfaces correspond to

$$a - \Omega > \Delta + \gamma;\ b - \gamma < \Omega < \gamma + \Delta,\ \Omega > \Delta - \gamma;$$

$$c - \gamma < \Omega < \Delta - \gamma.$$

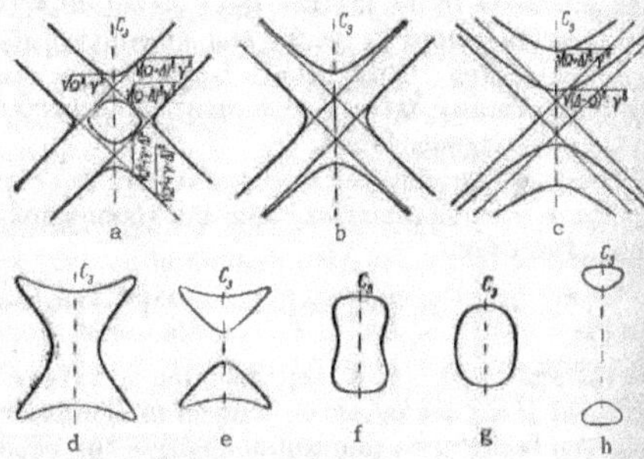

FIG. 5

We note that in case a there are two Fermi surfaces—closed and open (we have shown them together). Within each of these is located a region of large energies.

The closed Fermi surfaces arise in the cases

$$d - \gamma > \Omega > \sqrt{\Delta^2/4 + \gamma^2} - \Delta/2,\ \Omega > \Delta - \gamma;$$

$$e - \gamma > \Omega > \sqrt{\Delta^2/4 + \gamma^2} - \Delta/2,\ \Omega < \Delta - \gamma;$$

$$f - \sqrt{\Delta^2/4 + \gamma^2} - \Delta/2 > \Omega > \sqrt{\gamma(\gamma - \Delta)},\ \gamma > \Delta;$$

$$g - \sqrt{\gamma(\gamma - \Delta)} > \Omega > \gamma - \Delta,\ \gamma > \Delta;$$

$$h - \sqrt{\Delta^2/4 + \gamma^2} - \Delta/2 > \Omega > 0,\ \Omega < \Delta - \gamma.$$

We note that closed surfaces are obtained only when $\Delta \neq 0$, i.e., from spin-orbit coupling. A detailed comparison with the experimental data will be made in a future paper.

We consider now the situation with the electrons. In the vicinity of the points $\mathbf{k}_1$ the dependence $\omega(\mathbf{k})$ has a lower symmetry C_{2h} and depends on two additional parameters, β and δ. Formula (10) enables us to express in this case κ_x as a function of κ_y, κ_z, and Ω. Introducing the coordinates $q_x = b\kappa_x$ and $q_y = b\kappa_y$, we obtain

$$q_x^2 = p^2 - q_y^2 + \Omega^2 - \gamma^2 + \delta^2 - \beta^2 - \Delta^2 \pm 2\sqrt{(\beta q_y + \Omega p)^2 - (\delta q_y + \gamma p)^2 + (\Omega\delta - \gamma\beta)^2 + \Delta^2(\gamma^2 - \delta^2)}. \quad (14)$$

A complete analysis of all the possibilities afforded by this formula is a difficult matter. We determine only under what conditions will the surface ω = const be closed. For this purpose we consider the relation (14) when $\kappa \to \infty$. Discarding the constant terms, we obtain

$$q_x^2 \approx p^2 - q_y^2 \pm 2\sqrt{(\beta q_y + \Omega p)^2 - (\delta q_y + \gamma p)^2}. \quad (15)$$

The main terms are quadratic here. Obviously, in a direction corresponding to $q_y^2 > p^2$, the surface will be finite. In order for this to remain valid for all directions, it is necessary that the expression under the root be negative when $q_y^2 < p^2$, or in other words

$$(\beta s + \Omega)^2 - (\delta s + \gamma)^2 < 0, \; s^2 < 1, \quad (16)$$

where $s = q_y/p$. Writing this inequality in the form

$$[(\beta + \delta)s + (\gamma + \Omega)]\,[(\beta - \delta)s - (\gamma - \Omega)] < 0,$$

we immediately obtain the condition for the coefficients:

$$|\gamma + \Omega| > |\beta + \delta|, \; |\gamma - \Omega| > |\beta - \delta|, \; \gamma^2 - \Omega^2 > 0. \quad (17)$$

5. For a complete explanation of the singularities of bismuth, two problems remain to be solved. First, it is necessary to prove that the spectrum obtained is stable with respect to the Coulomb interaction of the quasiparticles. Second, it is necessary to demonstrate that the phase transition of the metal phase into the "dielectric" can actually take place.

The first of these problems is easy to solve. Inasmuch as we have used a linear approximation with respect to κ in the energy, the velocity on the Fermi boundary will be of the order of ordinary electronic velocities even when κ is small. This might be incorrect were $a\kappa_z$ or $b\kappa_\perp$ to be much smaller than γ, but the condition that the number of electrons and holes be equal and the fact that $f \sim \gamma$ excludes such a possibility. According to the foregoing, the screening radius has at the same time the order of magnitude of the distance between the quasiparticles.

The second question is much more complicated and we cannot resolve it completely. The main difficulty lies in the impossibility of determining the energy of the metallic phase. However, the energy of the dielectric phase can be determined. The change in energy on going over from a cubic lattice to the bismuth lattice is made up of the change in the ion energy, which is obviously proportional to the square of the deformation, and the change in the electron energy, including the electron-ion interaction. It is this energy that concerns us now. In view of the complexity of the spectrum in the general case, we simplify the problem by putting $\beta = \delta = 0$. In this case the spectrum in the vicinity of $\mathbf{k}_1$ will be the same as in the vicinity of $\mathbf{k}_0$, but with different coefficients.

We consider two cases: f, $\gamma \ll \Delta$ and f, $\gamma \gg \Delta$. We start with the case f, $\gamma \ll \Delta$. To find the electron energy, it is obviously necessary to calculate $\int_0^{N_0} \mu(N)\,dN$, where N_0 corresponds to the real number of electrons. This integral can be written in the form

$$\mu N_0 - \int_{\mu(0)}^{\mu} N(\mu_1)\,d\mu_1, \quad (18)$$

where $\mu = \mu(N_0)$ is the Fermi energy.

We first determine the value of μ in our model. For this purpose we find the value of ω at which the volumes of the "electron" and "hole" surfaces are equal. We assume that this occurs for closed Fermi surfaces, i.e., under our assumptions, in the case e or h. According to (13), the volume of the hole Fermi surface is in this case proportional to

$$\int_0^{q_{max}} q\,dq\,\{[\Omega^2 + \Delta^2 - \gamma^2 + q^2 + 2\sqrt{\Omega^2\Delta^2 + (\Omega^2 - \gamma^2)q^2}]^{1/2} - [\Omega^2 + \Delta^2 - \gamma^2 + q^2 - 2\sqrt{\Omega^2\Delta^2 + (\Omega^2 - \gamma^2)q^2}]^{1/2}\}$$
$$\approx 2\int_0^{q_{max}} y^2\,dy\left(\frac{\Delta^2\gamma^2}{\gamma^2 - \Omega^2} - y^2\right)^{-1/2}$$
$$= \Delta^2\left(-|\Omega| + \frac{\gamma^2}{\sqrt{\gamma^2 - \Omega^2}}\arcsin\frac{|\Omega|}{|\gamma|}\right), \quad (19)$$

where $q_{max} = |\Omega|\Delta/\sqrt{\gamma^2 - \Omega^2}$ [we neglect here terms of order $(\Omega/\Delta)^2$ and $(\gamma/\Delta)^2$].

For holes we must substitute $f - \mu$ in place of Ω; the former quantity is positive. For electrons we must take the positive root in the expression for

ω. Consequently, $\Omega = f^{(1)} - \mu < 0$, i.e., $|\Omega| = \mu - f^{(1)}$. According to formula (9), $\gamma^{(1)} = -\gamma/3$ and $f^{(1)} = -f/3$. In addition, it is necessary to take account of the fact that there are only three nonequivalent points $\mathbf{k}_1$, forming a star, i.e., the electron volume should be tripled. Taking these remarks into consideration, we can readily see that the condition that the volumes of the electron and hole surfaces be equal yields

$$\mu = 0.$$

We now proceed to calculate the electron energy. Since we are interested only in the change of energy upon deformation, we must subtract from (18) the corresponding expressions for the undeformed crystal. We then obtain

$$-\int_{\mu(0)}^{0} [N(f, \gamma, \mu) - N(0, 0, \mu)]\, d\mu.$$

As already noted, in the limit when the deformation decreases to zero the minimum of the upper energy band coincides with the maximum of the lower one. In this case the upper band will be empty and the lower one filled. It follows therefore that

$$N(f, \gamma, \mu) - N(0, 0, \mu) = N_e(f, \gamma, \mu) - N_h(f, \gamma, \mu) + N_h(0, 0, \mu),$$

where N_e and N_h are the electron and hole numbers, while $N_h(0, 0, 0) = 0$ (it is assumed here that $\mu < 0$).

In the vicinity of $\mathbf{k} = \mathbf{k}_0$ we always have holes when $\mu < 0$, while in the vicinity of $\mathbf{k} = \mathbf{k}_1$ we have electrons when $\mu > -f/3$ and holes when $\mu < -f/3$.

The sought integral is then

$$\int_{\mu(0)}^{0} N_{h0}(f, \gamma, \mu)\, d\mu + 3 \int_{\mu(0)}^{-f/3} N_{h1}(f, \gamma, \mu)\, d\mu - 3 \int_{-f/3}^{0} N_{e1}(f, \gamma, \mu)\, d\mu - 4 \int_{\mu(0)}^{0} N_h(0, 0, \mu)\, d\mu,$$

where N_{h0} corresponds to holes in the vicinity of $\mathbf{k}_0$ while N_{h1} pertains to the vicinity of $\mathbf{k}_1$. The coefficient 3 is the result of summation over all points of the star.

Since, as we have already seen, the volumes depend actually only on $|\Omega|$, we make a change of variable and obtain

$$E_e \int_{f}^{-\mu(0)+f} d|\Omega| N_h(\gamma, |\Omega|) + 3 \int_{f/3}^{-\mu(0)-f/3} d|\Omega| N_h\left(-\frac{\gamma}{3}, |\Omega|\right) - 4 \int_{0}^{-\mu(0)} N_h(0, |\Omega|)\, d|\Omega|. \tag{20}$$

The numbers of the holes are proportional to the corresponding volumes of the surface with specified Ω. It is obvious that the principal parts of the integrals will correspond to values $|\Omega| > \gamma$, where the surfaces are open. The volume bounded by any one of these surfaces is proportional in accordance to (13) to

$$N_h \infty \int_0^{\infty} [p_{max}(q) - p_{min}(q)]\, q\,dq = \int_0^{\infty} \{[\Omega^2 + \Delta^2 - \gamma^2 + q^2 + 2\sqrt{\Omega^2\Delta^2 + (\Omega^2 - \gamma^2)q^2}]^{1/2} - [\Omega^2 + \Delta^2 - \gamma^2 + q^2 - 2\sqrt{\Omega^2\Delta^2 + (\Omega^2 - \gamma^2)q^2}]^{1/2}\}\, q\,dq.$$

This formula is valid in this form, strictly speaking, only when $\Omega < \Delta + \gamma$. It will be presently shown, however, that it is the large values of q that are essentially important in the integral with respect to q, so that this formula turns out to be valid for arbitrary Ω.

Assuming q to be large and making a change of variable, we obtain

$$N_h \sim \int_{\frac{\Omega\Delta}{\sqrt{\Omega^2-\gamma^2}}}^{L} y^2 dy \sqrt{\Omega^2 - \gamma^2} \left(y^2 - \frac{\gamma^2\Delta^2}{\Omega^2 - \gamma^2} + \Omega^2 - \gamma^2\right)^{-1/2} \approx L^2 \sqrt{\Omega^2 - \gamma^2}. \tag{21}$$

We have chosen as the upper limit the value of the quasimomentum q in the region where the formulas obtained for the energy spectrum no longer correspond to reality. Obviously, $L \sim bK$, where K is a quantity on the order of the reciprocal lattice period. We note also that in the case when $f, \gamma \gg \Delta$, the electrons and holes can be equal in number only if the surfaces are open. The corresponding volumes are then expressed by formula (21). If we take one and the same L for the electrons and holes (which obviously follows from the fact that $L \gg \gamma, f$), this condition that the number of electrons and holes must be equal again yields $\mu = 0$. This follows from the fact that $|\gamma^{(1)}| = \gamma/3$ and $|f^{(1)}| > f/3$, while the number of electron surfaces is equal to three.

Substituting (21) into (20) we obtain

$$E_e \sim L^2 \left\{ \int_{\max(f,\gamma)}^{-\mu(0)} d|\Omega| \sqrt{\Omega^2 - \gamma^2} + 3 \int_{\max\left(\frac{f}{3}, \frac{\gamma}{3}\right)}^{-\mu(0)} d|\Omega| \sqrt{\Omega^2 - \frac{\gamma^2}{9}} - 4 \int_0^{-\mu(0)} \Omega\, d\Omega \approx -\frac{2}{3} L^2\gamma^2 \ln \frac{|\mu(0)|}{\max(f, \gamma)} \right\}. \tag{22}$$

A remarkable property of this energy is that at small deformations it can become larger than any term that is quadratic in γ.

It must be noted that this conclusion is not fully rigorous. In view of the fact that the large values of κ predominate in the integral with respect to κ, it would be necessary to take quadratic terms into account. However, in view of the fact that in the "dielectric" phase we have[3] $\Omega \ll |\mu(0)|$ and $N_h \sim \Omega$ as $\gamma \to 0$, we must obviously obtain $N_h = c_1\Omega + c_2\gamma^2/\Omega$ (c_1 and c_2 are constants) when $\gamma \ll \Omega \ll |\mu(0)|$, from which follows (22). The sign of the coefficient c_2 can generally speaking be arbitrary. Obviously, it must be regarded as negative, since the maximum values of the κ components are still several times smaller than the initial period of the reciprocal lattice, and therefore the conclusion drawn gives a correct indication of the sign of the energy.

Adding the ion energy, we determine the cubic lattice deformation energy for the dielectric phase (obviously $f \sim \gamma$):

$$E_{\text{diel}}(\gamma) = E_{\text{diel}}(0) + \sigma\gamma^2 - \chi\gamma^2 \ln \frac{P}{\gamma} \tag{23}$$

($P = \text{const} \approx 1$ eV; $\sigma, \chi > 0$). Minimization of this energy with respect to γ yields

$$\gamma_0 = Pe^{-(2\sigma+\chi)/2\chi}. \tag{24}$$

It seems likely that the constant σ, which determines the ion interaction energy $\sigma\gamma^2$, is somewhat larger than the constant χ, which corresponds to electron interaction with the modified potential. It follows as well from (24) that $\gamma \ll P$, i.e., the deformation is small. Substituting (24) in (23) we obtain

$$E_{\text{diel}} = E_{\text{diel}}(0) - \chi\gamma_0^2/2. \tag{25}$$

We were unable to calculate the energy for the metallic phase. If this energy increases with increasing deformation, and if furthermore

$$E_{\text{diel}}(0) - E_{\text{met}}(0) < \chi\gamma_0^2/2, \tag{26}$$

then the transition from the cubic lattice to the bismuth lattice is energetically favored. The smallness of the gap between the two phases denotes that the bismuth can be easily converted into the metallic state, something indeed observed in actual experiment [10].

In conclusion, the authors express their gratitude to I. M. Lifshitz, I. M. Khalatnikov, L. P. Gor'kov, and I. E. Dzyaloshinskiĭ for a discussion of the work.

[3]This is connected with the fact that when $\gamma \to 0$ the conduction band touches the valence band along the surface.

[1] D. Schoenberg, Proc. Roy. Soc. **A170**, 341 (1939); Phil. Trans. Roy. Soc. **A245**, 1 (1952). N. B. Brandt and M. V. Razumenko, JETP **39**, 276 (1960), Soviet Phys. JETP **12**, 198 (1961). N. B. Brandt and V. V. Shchekochikhina, JETP **41**, 1412 (1961), Soviet Phys. **14**, 1008 (1962).

[2] Brandt, Dubrovskaya, and Kytin, JETP **37**, 572 (1959), Soviet Phys. JETP **10**, 405 (1960). N. B. Brandt, PTE, No. 2, 1960. N. B. Brandt, JETP **38**, 1355 (1960), Soviet Phys. JETP **11**, 975 (1960).

[3] J. M. Luttinger, Phys. Rev. **119**, 1153 (1960).

[4] B. S. Boyle and K. F. Rodgers, Phys. Rev. Lett. **2**, 338 (1959).

[5] B. Lax, Bull. Am. Phys. Soc. **5**, 167 (1960).

[6] H. Johns, Proc. Roy. Soc. **A144**, 225 (1934); **A147**, 396 (1934).

[7] A. A. Abrikosov, JETP **43**, 1083 (1962), this issue p. 765.

[8] L. D. Landau and E. M. Lifshitz, Quantum Mechanics, Pergamon, 1958.

[9] R. Elliot, Phys. Rev. **96**, 280 (1954).

[10] N. B. Brandt and N. I. Ginzburg, JETP **39**, 1554 (1960), Soviet Phys. JETP **12**, 1082 (1961).

Translated by J. G. Adashko
184

SOVIET PHYSICS JETP VOLUME 20, NUMBER 2 FEBRUARY, 1965

CONCERNING SURFACE SUPERCONDUCTIVITY IN STRONG MAGNETIC FIELDS

A. A. ABRIKOSOV

Institute of Physics Problems, Academy of Sciences, U.S.S.R.

Submitted to JETP editor March 15, 1964

J. Exptl. Theoret. Phys. (U.S.S.R.) 47, 720-733 (August, 1964)

The question of surface superconductivity in fields above the critical field H_{c2} is examined. It is shown that the phenomenon exists at all temperatures below T_c. The critical field of thin films in a parallel field and the field dependence of the critical current for surface superconductivity of a bulk sample are determined.

IN a recently published article[1] Saint-James and de Gennes showed that the existence of a thin superconducting sheath near the surface of a superconductor is possible in fields above H_{c2}, at which the beginning of the superconducting phase occurs (the upper critical field in superconductors of the second kind[2] and the critical super-cooling field for superconductors of the first kind, see [3]). Superconductivity vanishes in this sheath only when the external field becomes equal to $H_{c3} = 1.7\,H_{c2}$.

To begin with it is necessary to briefly explain the origin of this effect. As is well-known,[2,3] the field H_{c2} is the field at which for the first time a small layer of the superconducting phase may appear in a normal metal. In view of this, in order to determine H_{c2} it is necessary to examine the linearized equations for the energy gap Δ (or what is the same, the Ψ function of Ginzburg and Landau[4]). The eigenvalue of the field, at which this equation has a solution different from zero well inside the sample, is H_{c2}. Studying the Ginzburg-Landau equation for Ψ, Saint-James and de Gennes[1] noticed that if one requires $\Psi \neq 0$ not inside but near the surface, then the corresponding eigenvalue of the field turns out to be higher.

In fact, the Ginzburg-Landau equation has the following boundary condition for Ψ:

$$\mathbf{n}(-i\hbar\nabla\Psi - 2ec^{-1}\mathbf{A}\Psi) = 0,$$

where $\mathbf{n}$ is the normal to the surface. If the problem deals with a field parallel to a plane boundary, then one can choose $\mathbf{A} \perp \mathbf{n}$ so that the condition reduces to $\partial\Psi/\partial\mathbf{n} = 0$ on the boundary. The function $\Psi_0 = \exp[-\kappa^2x^2/2\delta^2]$ satisfies the linearized Ginzburg-Landau equation and the boundary condition if the point $x = 0$ is far from the boundary (Fig. 1a). The corresponding eigenvalue of the field is $H_{c2} = \kappa\sqrt{2}\,H_{cm}$.[2] It is not difficult to see that if the point $x = 0$ is chosen on the boundary itself (Fig. 1b), then the function Ψ_0 will again satisfy both the equation and the boundary condition. If now the point $x = 0$ is chosen not on the surface itself but near it, then the function Ψ must be changed (Fig. 1c) and consequently the corresponding eigenvalue of the field must also change. It is easy to understand which way the change will go if it is remembered that the critical field is higher for thin films than for a bulk sample. It is clear from Fig. 1d how the Ψ function varies inside the sheath. Comparing with Fig. 1c, it is easy to see that in the case of Fig. 1c the critical field must turn out to be higher than in the cases of Figs. 1a and 1b. By selecting the position of the point $x = 0$, one can find the maximum field H_{c3}.

The presence of surface superconductivity must lead to a whole series of experimental consequences. Saint-James and de Gennes[1] mention such consequences as disagreement between the values of the critical field determined by the vanishing of the magnetic moment and the appearance of resistivity in superconductors of the second kind, an increase in the value of the supercooling field in superconductors of the first kind, etc.

Certain further aspects of this interesting phenomenon are considered in the present article.

1. SURFACE SUPERCONDUCTIVITY NEAR T_c AND AT LOW TEMPERATURES

In spite of the fact that the argument presented above makes the existence of surface superconductivity and the field H_{c3} quite certain at first glance, in reality a number of questions arise.

A. The boundary condition on the Ginzburg-Landau equation plays an extremely important role in this phenomenon. However, whereas the Ginzburg-Landau equation itself was obtained[5,6] from the microscopic theory, this did not occur with regard to the boundary condition.

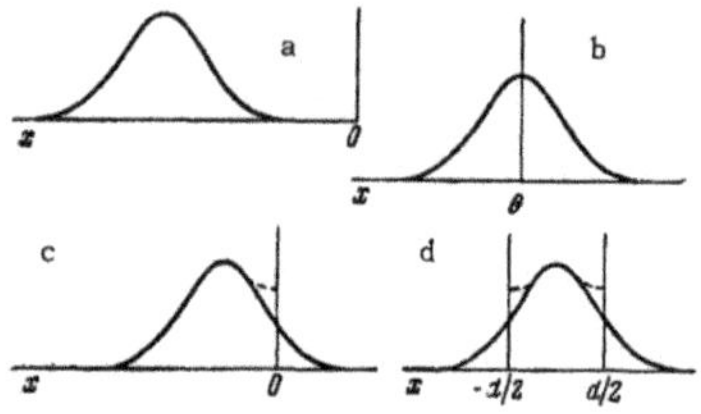

FIG. 1.

B. The region of applicability of the Ginzburg-Landau equations covers the temperatures at which all integral relations become local. This means that the penetration depth $\delta(T)$ and the characteristic length $\delta(T)/\kappa$ of superconducting structure must be considerably larger than the correlation parameter, which is of order $\delta(0)/\kappa$. In particular, it would be necessary that $(T_c - T)/T_c \ll 1$ in superconductors of the second kind with $\kappa > 1/\sqrt{2}$. If, however, $T_c - T \sim T_c$, then both the Ginzburg-Landau equation and the boundary condition on it become meaningless. The question arises of whether surface superconductivity is possible in this temperature range.

In order to answer these questions, we shall consider a general microscopic formulation of the problem where, for simplicity, we confine ourselves to an investigation of a pure unalloyed superconductor, and we assume specular reflection of electrons from the boundaries. The method which will be applied essentially follows the derivation of Gor'kov[7] for the critical supercooling field of pure superconductors. The difference is that we shall consider a semi-space instead of an infinite medium, and we require that

$$\mathfrak{G}_\omega(\mathbf{r}, 0) = \mathfrak{G}_\omega(0, \mathbf{r}') = 0, \quad \mathfrak{F}_\omega(\mathbf{r}, 0) = \mathfrak{F}_\omega(0, \mathbf{r}') = 0 \tag{1}$$

(we shall use the temperature technique).

It is easiest to solve this problem in the following manner. We shall consider the quasiclassical approximation. If we consider the vector potential A to be directed along the y axis and to depend on x (the field H_0 is along the z axis), then according to [7] the quasiclassical function $\mathfrak{G}_\omega(\mathbf{r}, \mathbf{r}')$ for the electrons in a normal metal occupying all space has the form

$$\mathfrak{G}_\omega^{(\infty)}(\mathbf{r}, \mathbf{r}') = \mathfrak{G}_{0\omega}(|\mathbf{r} - \mathbf{r}'|) \times \exp\{(ie/2\hbar c)[A(x) + A(x')](y - y')\}, \tag{2}$$

where $\mathfrak{G}_{0\omega}$ is the field-free function. In order to satisfy condition (1) we write $\mathfrak{G}_\omega(\mathbf{r}, \mathbf{r}')$ in the form

$$\mathfrak{G}_\omega(\mathbf{r}, \mathbf{r}') = [\mathfrak{G}_{0\omega}(|\mathbf{r} - \mathbf{r}'|) - \mathfrak{G}_{0\omega}(|\mathbf{r} - \bar{\mathbf{r}}'|)] \times \exp\{(ie/2\hbar c)[A(x) + A(x')](y - y')\}, \tag{3}$$

where $\bar{\mathbf{r}}' = (-x', y', z')$. In order for this function to satisfy the same equation as $\mathfrak{G}_\omega^{(\infty)}(\mathbf{r}, \mathbf{r}')$ it is necessary that $A(-x') = A(x')$, i.e., that the vector potential be symmetrically continued across the boundary. It is quite possible to assume this, since we are only interested in the region $x' > 0$.

The equation for the gap Δ is written in a form similar to Eq. (6) of the article by Gor'kov:[7]

$$\Delta^*(\mathbf{r}) = -\frac{gT}{\hbar}\sum_\omega \int_{x'>0} d\mathbf{r}'\, \mathfrak{G}_{-\omega}(\mathbf{r}', \mathbf{r})\, \mathfrak{G}_\omega(\mathbf{r}', \mathbf{r})\, \Delta^*(\mathbf{r}'). \tag{4}$$

In view of the fact that in the normal state the field inside a metal is homogeneous and equal to the external field, $A(x) = H_0(x - x_0)$. The constant x_0 corresponds to the possibility of movement of the center of the superconducting sheath relative to the boundary. The field H_{c3} should be defined as $\min_{x_0} H_c(x_0)$. Such an introduction of the constant x_0 is completely equivalent to the introduction of a phase factor e^{iky} into Δ, just as this done in [1].

Upon substitution of the function (3) into Eq. (4), one can immediately disregard the products $\mathfrak{G}_{0\omega}(|\mathbf{r} - \mathbf{r}'|)\mathfrak{G}_{0,-\omega}(|\mathbf{r} - \bar{\mathbf{r}}'|)$ since these terms are important only at distances from the boundary on the order of atomic distances and give a small contribution to the integral. Thus, we obtain

$$\Delta^*(\mathbf{r}) = -\frac{gT}{\hbar}\sum_\omega \int_{x'>0} d\mathbf{r}' [\mathfrak{G}_{0\omega}(|\mathbf{r}' - \mathbf{r}|)\, \mathfrak{G}_{0,-\omega}(|\mathbf{r}' - \mathbf{r}|) + \mathfrak{G}_{0\omega}(|\bar{\mathbf{r}}' - \mathbf{r}|)\, \mathfrak{G}_{0\omega}(|\bar{\mathbf{r}}' - \mathbf{r}|) \times \exp\left\{\frac{ieH_0}{\hbar c}(x + x' - 2x_0)(y - y')\right\}\Delta^*(\mathbf{r}'). \tag{5}$$

Carrying out transformations similar to those performed in [7], we obtain

$$\Delta^*(\mathbf{r}) \ln\frac{\hbar v}{\pi T_c \sigma} = \frac{T}{2v}\int_{x'>0,\, R>\sigma} d\mathbf{r}' \times \exp\left\{\frac{ieH_0}{\hbar c}(y - y')(x + x' - 2x_0)\right\} \times \left[R^{-2}\left(\operatorname{sh}\frac{2\pi T R}{\hbar v}\right)^{-1} + \bar{R}^{-2}\left(\operatorname{sh}\frac{2\pi T \bar{R}}{\hbar v}\right)^{-1}\right]\Delta^*(\mathbf{r}'), \tag{6*}$$

where $R = |\mathbf{r} - \mathbf{r}'|$, $\bar{R} = |\mathbf{r} - \bar{\mathbf{r}}'|$, and σ is an infinitesimal number which does not appear in the final result. In what follows we shall look for a solution $\Delta^*(\mathbf{r})$ which depends only on the coordinate x.

First let us consider the case of temperatures in the neighborhood of T_c. As will be clear from what follows, $\Delta(x)$ varies over those same distances as the argument of the exponential in Eq.

*sh = sinh.

(6), i.e., over distances on the order of $(\hbar c/eH_c)^{1/2}$, where H_c is the critical field, proportional to $T_c - T$. In the neighborhood of T_c, this distance may become much larger than $\hbar v/\pi T_c$ —the characteristic distance for the kernel of Eq. (6) inside the square brackets. As a result, $\Delta(x)$ and the exponential factor may be expanded in powers of $\mathbf{r}' - \mathbf{r}$. Here we obtain

$$\Delta(x) \ln \frac{\hbar v}{\pi T_c \sigma} = \frac{T}{2v} \int_{-\infty}^{\infty} dy'\, dz' \int_0^{\infty} dx' \left[\Delta(x) + \frac{d\Delta(x)}{dx}(x' - x) + \frac{1}{2}\frac{d^2\Delta}{dx^2}(x'-x)^2 - \frac{1}{2}\left(\frac{2eH_0}{\hbar c}\right)^2 \times (y'-y)^2 (x - x_0)^2 \Delta(x) \right] \times \left[R^{-2} \operatorname{sh}^{-1} \frac{2\pi T R}{\hbar v} + \overline{R}^{-2} \operatorname{sh}^{-1} \frac{2\pi T \overline{R}}{\hbar v} \right]$$

[since we are dealing everywhere with real solutions, we shall write $\Delta(x)$].

One can make the following substitution in the terms which do not contain derivatives of Δ:

$$\int_0^{\infty} dx' \left[R^{-2}\left(\operatorname{sh}\frac{2\pi T R}{\hbar v}\right)^{-1} + \overline{R}^{-2}\left(\operatorname{sh}\frac{2\pi T \overline{R}}{\hbar v}\right)^{-1} \right] = \int_{-\infty}^{\infty} dx'\, R^{-2} \operatorname{sh}^{-1} \frac{2\pi T R}{\hbar v}.$$

The integrals in such terms then become the same as for the case of infinite space. However, the terms with derivatives must be analyzed separately. Let us take the term with $d\Delta(x)/dx$ and change over from the variable x' to $x' - x$ in the term containing R and to $x' + x$ in the term containing $\overline{R}$. As a result we obtain

$$\frac{d\Delta(x)}{dx} \left[\int_{-x}^{\infty} \frac{X\,dX}{R^2 \operatorname{sh}(2\pi T R/v)} + \int_{x}^{\infty} \frac{X\,dX}{R^2 \operatorname{sh}(2\pi T R/\hbar v)} - 2x \int_x^{\infty} \frac{dX}{R^2 \operatorname{sh}(2\pi T R/\hbar v)} \right],$$

where $R^2 = X^2 + (y - y')^2 + (z - z')^2$. Combining the three integrals, we obtain

$$2\frac{d\Delta}{dx} \int_x^{\infty} \frac{(X - x)\,dX}{R^2 \operatorname{sh}(2\pi T R/\hbar v)}.$$

The coefficient of $d\Delta(x)/dx$ obviously differs from zero only at distances on the order of $\hbar v/\pi T_c$ from the boundary. Evaluating the integral of this coefficient with respect to x, we obtain

$$2\int_0^{\infty} dx \int_x^{\infty} \frac{(X - x)\,dX}{R^2 \operatorname{sh}(2\pi T R/\hbar v)} = \int_0^{\infty} \frac{X^2 dX}{R^2 \operatorname{sh}(2\pi T R/\hbar v)} = \frac{1}{2}\int_{-\infty}^{\infty} \frac{X^2 dX}{R^2 \operatorname{sh}(2\pi T R/\hbar v)}.$$

Hence it follows that the term containing $d\Delta(x)/dx$ can be approximately written in the form

$$\frac{7}{12}\zeta(3)\left(\frac{\hbar v}{2\pi T}\right)^2 \frac{d\Delta(x)}{dx}\delta(x - \sigma), \qquad \sigma \to +0.$$

Similarly transforming the term containing $d^2\Delta/dx^2$, we obtain

$$\frac{7}{12}\zeta(3)\left(\frac{\hbar v}{2\pi T}\right)^2 \frac{d^2\Delta}{dx^2}[1 - 2x\delta(x - \sigma)], \qquad \sigma \to +0.$$

It is obvious that the second term inside the square brackets is simply equal to zero.

Thus, the term containing $d\Delta/dx$ turns out to be the only term having a δ-function structure. It can be isolated by integrating over a small neighborhood of the point $x = 0$. As a result, we obtain the condition $d\Delta(x)/dx\,|_{x=0} = 0$, which coincides with the Ginzburg-Landau boundary condition. After this, the equation for Δ takes the usual form of the linear approximation to the Ginzburg-Landau equation:

$$\frac{\hbar^2}{4m}\left[\frac{d^2\Delta}{dx^2} - \left(\frac{2eH}{\hbar c}\right)^2 (x - x_0)^2 \Delta(x)\right] + \frac{1}{\lambda}\frac{T_c - T}{T_c}\Delta(x) = 0,$$

$$\lambda = 7\zeta(3)\varepsilon_F / 6(\pi T_c)^2.$$

Solving this equation, it is not difficult to see that Δ varies over distances on the order of $(\hbar c/eH_c)^{1/2}$, and H_c is proportional to $T_c - T$. Thus, in the neighborhood of T_c where Δ varies over distances which are large in comparison with $\hbar v/\pi T_c$, in fact not only the Ginzburg-Landau equation but also the boundary condition on it are valid.

The expansion in powers of $\mathbf{r} - \mathbf{r}'$ loses meaning at temperatures far from the critical temperature, because there Δ changes over distances on the order of $\hbar v/\pi T_c$; the same pertains to the argument of the exponential in Eq. (6) since the critical fields H_{c2} and H_{c3} are of order $(\hbar c/e)(T_c/\hbar v)^2$. Therefore, at such temperatures it is already impossible to use the Ginzburg-Landau equation, and it is necessary to solve the exact equation (6). This problem is very difficult, and therefore we shall only consider the case $T = 0$, and we confine ourselves to a demonstration that the surface superconductivity effect actually does exist in large fields.

At $T = 0$ Eq. (6) can be easily integrated with respect to $z - z'$ and $y - y'$. Also introducing a new dimension of length, namely $\xi = (2eH_0/\hbar c)^{1/2}x$, and putting $\eta = (2eH_0/\hbar c)^{1/2}x_0$, we obtain

$$\Delta(\xi)\ln\frac{v(2eH/\hbar c)^{1/2}}{\pi T_c\sigma_1} = \frac{1}{2}\int_{\sigma_1}^{\infty}\left[\frac{\exp\{-\frac{1}{2}|\xi-\xi'||\xi+\xi'-2\eta|\}}{|\xi-\xi'|} + \frac{\exp\{-\frac{1}{2}(\xi+\xi')|\xi+\xi'-2\eta|\}}{\xi+\xi'}\right]\Delta(\xi')d\xi'. \quad (7)$$

Let us assume that $\eta \gg 1$. In such a case it is convenient to introduce the integration variable $\xi-\eta$. If in addition we also suppose, and this is confirmed below, that $\Delta(\xi)$ will be large only for $\xi-\eta \sim 1$, then the lower limit with regard to the variable $\xi-\eta$, equal to $-\eta$, can be regarded as equal to $-\infty$. As a result we obtain

$$\Delta(\xi)\ln\frac{v(2eH_0/\hbar c)^{1/2}}{\pi T_c\sigma_1} = \frac{1}{2}\int_{\substack{-\infty\\|\xi-\xi'|>\sigma_1}}^{\infty}\left[\frac{\exp(-\frac{1}{2}|\xi^2-\xi'^2|)}{|\xi-\xi'|} + \frac{\exp(-\eta|\xi+\xi'|)}{2\eta}\right]\Delta(\xi')d\xi'. \quad (8)$$

In the second term of the kernel of the integral equation, we have neglected ξ in comparison with η. If this second term were not present, then Eq. (8) would coincide with Gor'kov's equation[7] for the field H_{c2} at $T = 0$. This term is exponentially small for $\eta \gg 1$. Values of $|\xi + \xi'| \sim \eta^{-1}$, in other words, $\xi \approx -\xi'$, are important in the evaluation of the integral containing the second term of the kernel. This gives the possibility of substituting $\Delta(\xi') \approx \Delta(\xi)$ in this integral and transferring this term to the left side.

As a result of evaluating the integral with respect to $\xi + \xi'$ we obtain

$$\Delta(\xi)\left[\ln\frac{v(2eH_0/\hbar c)^{1/2}}{\pi T_c\sigma_1} - \frac{1}{2\eta^2}\right] = \frac{1}{2}\int_{-\infty}^{\infty}\frac{\exp(-\frac{1}{2}|\xi^2-\xi'^2|)}{|\xi-\xi'|}\Delta(\xi')d\xi'. \quad (9)$$

This equation coincides exactly with the equation for H_{c2} if we make the replacement

$$H_0 = H_c(\eta) = H_{c2}(1+\eta^{-2}). \quad (10)$$

From here it follows that the critical field for nucleation increases as the nucleation center approaches the surface. This completely proves the existence of surface superconductivity beyond the region of applicability of the equations of Ginzburg and Landau.

It is unusually difficult to evaluate the field H_{c3} in the general case. However, on the basis of well-known properties of superconductors we can, it seems to us, draw the following conclusion. All experimental data indicates that the Ginzburg-Landau equations provide a good description of the magnetic properties of superconductors over the entire temperature interval, if it is assumed that the constants of this theory are slowly varying functions of the temperature. Of course, this is only an empirical fact. However, if we take such a viewpoint, then the ratio H_{c3}/H_{c2} must depend on temperature far more weakly than H_{c2}/H_{cm} does. Since the latter ratio only changes by 20% in the interval from T_c to absolute zero, we can assume that the relation

$$H_{c3}(T) = 1.7H_{c2}(T), \quad (11)$$

derived in [1] for the neighborhood of $T = T_c$, will be satisfied over this entire range of temperatures.

In concluding this section, we note the following circumstance. Saint-James and de Gennes already showed[1] that near the field H_{c3}, in spite of the presence of a surface superconducting layer, a total surface current shielding the inside of the sample from the external field is not created in the superconductor. This is not completely trivial, since the current density in the surface layer does not vanish; however, it changes sign upon going deeper inside the sample, so that on the whole no surface current is present. It is not difficult to see that this will be valid over the entire range of surface superconductivity, i.e., from the field H_{c2} up to H_{c3}. In fact, upon the appearance of a total surface current, the field well inside the sample is different from the external field. If the energy is measured from the normal state, then a positive increment to it appears, proportional to the volume of the sample:

$$\int\frac{(H-H_0)^2}{8\pi}dV.$$

At the same time, the change of the electron state occurs only near the surface, so the associated change of the energy is proportional to the surface area. Thus, one must necessarily have $H = H_0$ in bulk samples. Below (see Sec. 3) we shall demonstrate directly the absence of total current within the limits of the Ginzburg-Landau approximation (but far from H_{c3}).

2. CRITICAL FIELD OF THIN FILMS OF TYPE II SUPERCONDUCTORS

It follows from what has been said above that the magnetic moment of a bulk superconductor of the second kind must vanish at $H = H_{c2}$. However, if the critical field is defined by the appearance of electrical resistivity, then H_{c3} must be the critical field. The critical field of thin films[8] is measured in precisely this way, and in this connection the dependence of H_{c3} on the thickness of the film is of interest. A similar problem was considered by Saint-James and de Gennes [1] for a film located in a perpendicular field; however, having a comparison with the experimental data of Khukhareva [8] in mind, we shall investigate here the case of a parallel field.

In contrast our earlier determination of the critical field, [9] we must take into consideration here the effect of surface superconductivity, which is manifest in the fact that at sufficiently large thicknesses it may turn out that an asymmetric solution of the energy gap equation (with respect to the center of the film) can exist at larger fields than the symmetric solution (considered in [9]). In the same way as in [9], we confine ourselves to the neighborhood of T_c and use the Ginzburg-Landau equation.

Introducing the variable[1)] $\xi = \delta^{-1}(\kappa H_0/\sqrt{2}H_{cm})^{1/2}x$, where x is normal to the surface of the film, and $\beta = \kappa\sqrt{2}H_{cm}/H_0 = H_{c2}/H_0$, we obtain the equation

$$d^2\Psi/d\xi^2 + [\beta - (\xi - \eta)^2]\Psi = 0. \tag{12}$$

The boundary condition corresponds to $d\Psi/d\xi = 0$ for $\xi = \pm s/2$, where

$$s = \frac{1}{\delta}\left(\frac{\kappa H_0}{\sqrt{2}H_{cm}}\right)^{1/2} d = \frac{\kappa}{\delta}\left(\frac{H_0}{H_{c2}}\right)^{1/2} d = \frac{\kappa d}{\delta\sqrt{\beta}},$$

d is the thickness of the film. The quantity η in Eq. (12) corresponds, as in the preceding section, to a choice of the arbitrary constant in the vector potential such that Ψ is a function of only one coordinate. The problem consists in a determination of the minimum eigenvalue β of Eq. (12) as a function of s and η, and then minimization of β with respect to η in order to determine the final quantity $H_{c3}(d)$.

First of all we note that the final result must obviously have the form $H_{c3}/\kappa\sqrt{2}H_{cm} = f(\delta/\kappa d)$. Putting $\kappa = (2\sqrt{2}e/\hbar c)H_{cm}\delta^2$ and defining $\theta = (2e/\hbar c)^{1/2}H_{cm}\delta$, we obtain a dependence of the type

$$\frac{H_{c3}(d)}{2\theta^2} = f\left(\frac{1.29\cdot 10^{-4}}{\theta d}\right). \tag{13}$$

The same kind of dependence was obtained in [9], but in our case the function f is different (we also considered the doubling of the charge). It is characteristic that one parameter θ determines the complete curve $H_{c3}(d)$.

In the region of very small thicknesses, $s \ll 1$, the funtion Ψ will be close to constant. With increase of the thickness, it is obvious that the symmetric solution (i.e., $\eta = 0$) will correspond for some of the time to the largest field. This means that in the region of small thicknesses, the curve $H_{c3}(d)$ will coincide with $H_{c2}(d)$ found in [9], namely

$$H_{c3}(d)/H_{cm} \approx 2\sqrt{6}\,\delta/d, \quad d\kappa/\delta \ll 1. \tag{14}$$

However, at some value $s \sim 1$, η now becomes unequal to zero, and the curve $H_{c3}(d)$ will deviate from the curve found in [9]. An exact consideration of such thicknesses is unusually complicated. But we shall use the fact that in the region where values of $\eta \neq 0$ first appear, the formula for small thicknesses still gives a very good approximation for $H_{c3}(d)$. This will be clear from what follows.

If we change over from the variable ξ to $\xi - \eta$, then Eq. (12) coincides with the same equation which was solved in [9] and the boundary condition will be $d\Psi/d\xi = 0$ for $\xi = (s/2) - \eta$ and $\xi = -(s/2) - \eta$. Here the general solution of the equation will be

$$\Psi = \exp\left(-\frac{\xi^2}{2}\right)\left[C_1F\left(\frac{1-\beta}{4}, \frac{1}{2}, \xi^2\right) + C_2\xi F\left(\frac{3-\beta}{4}, \frac{3}{2}, \xi^2\right)\right] \tag{15}$$

(F is the confluent hypergeometric function). Expanding the function Ψ in powers of ξ and substituting into the boundary conditions, we obtain the following expression for β:

$$\beta = \frac{1}{3}\left(\frac{s}{2}\right)^2 - \frac{8}{35\cdot 27}\left(\frac{s}{2}\right)^6 + \eta^2\left[1 - \frac{8}{15}\left(\frac{s}{2}\right)^4\right] + \eta^4\frac{26}{63}\left(\frac{s}{2}\right)^6. \tag{16}$$

From here it follows that for $(s/2)^4 < 15/8$ the coefficient of the η^2 term is positive and the smallest value of β corresponds to $\eta = 0$. We note that on the boundary of this region, the second term in expression (16) amounts to only $1/21$ of the first. In reality, in the expression for H_{c3} the correction is smaller by another factor of two.

Using the expression for s and substituting $\beta \approx 1/3\,(s/2)^2$, we obtain the critical thickness

$$d_c = \sqrt{5/2}\,\delta/\kappa. \tag{17}$$

At larger thicknesses the coefficient of η^2 is negative. From the minimum of β with respect to η^2, with use of the term containing η^4, we obtain, substituting all the quantities,

[1)] This variable coincides with ξ in the preceding section. In fact, according to Gor'kov,[5] $\kappa = 2\sqrt{2}\,e\,H_{cm}\delta^2/\hbar c$, where H_{cm} is the thermodynamic critical field, δ is the penetration depth.

$$\frac{H_{c3}(d)}{H_{cm}}=\frac{2\sqrt{6}}{d}\delta\left\{1+\frac{d^2\varkappa^2}{105\delta^2}+\frac{84}{325}\left[\left(\frac{d}{d_c}\right)^2-1\right]^2\right\}. \quad (18)$$

The difference from the curve obtained in [9] lies in the last term.

One can obtain another limiting formula for large thicknesses. Since even for an infinite sample the calculation of H_{c3} is a difficult problem requiring numerical calculation, we employ the following method. The determination of the minimum value of β from Eq. (14) is equivalent to the variational problem of finding the minimum of the expression

$$\beta=\left[\int_0^s\left(\frac{d\Psi}{d\xi}\right)^2 d\xi+\int_0^s(\xi-\eta)^2\Psi^2 d\xi\right]\Big/\int_0^s\Psi^2 d\xi \quad (19)$$

(here we measure ξ and η from one of the boundaries). Let us assume that the solution of the variational problem is known as $s\to\infty$, where Ψ usually falls off with a characteristic distance on the order of unity. Then one can show that the difference between β of the variational problem (19) and β of the case $s\to\infty$ has the form

$$\beta=\beta_0+\left(\frac{d\Psi_0^2}{d\xi}\right)_{\xi=s}\Big/\int_0^\infty\Psi_0^2 d\xi, \quad (20)$$

where Ψ_0 and β_0 correspond to the problem with $s\to\infty$. The exact function Ψ_0 cannot be simply written. However, one can use the fact that the trial function $\Psi=\exp(-b\xi^2/2)$ gives, for $b=[1-(2/\pi)]^{1/2}$, a value

$$\beta_0=b=(1-2/\pi)^{1/2}, \quad (21)$$

differing from the true value by 2%. Substituting the approximate function Ψ_0 into (20), we obtain

$$H_{c3}(d)=H_{c3}(\infty)\left[1+\frac{4}{\sqrt{\pi}}\frac{\varkappa d}{\delta}\exp\left\{-\left(\frac{\varkappa d}{\delta}\right)^2\right\}\right]. \quad (22)$$

If the experimental data are reduced to the form (13), then it is useful to keep in mind that the new function f is very similar to the old function found in [9]. In the limiting cases it has the form $f(x)\to 1.7$ as $x\to 0$ (in contrast to unity in [9]) and $f(x)\approx 2\sqrt{3}x$ as $x\to\infty$ (the same as in [9]).

3. CRITICAL CURRENT IN THE CASE OF SURFACE SUPERCONDUCTIVITY

Since, as already mentioned, one can detect the critical field H_{c3} by the appearance of resistivity, the question arises: what limitation must be imposed on the magnitude of the measuring current? The critical current is also of interest from the point of view of the possibility of technological utilization of surface superconductivity. As in the preceding section, we shall confine ourselves to the approximation of Ginzburg and Landau, i.e., to the case when T is close to T_c.

First let us consider the neighborhood of $H=H_{c3}$. We assume that the current is parallel to the magnetic field. In this connection, the function Ψ acquires the factor e^{ikz}. If the current density is expressed in units of $cH_{cm}\sqrt{2}/4\pi\delta$, and the function Ψ, as usual, refers to the value for a bulk sample at the same temperature and without a magnetic field, then the expression for the current density has the form

$$j=q|\Psi|^2, \quad (23)$$

where $q=k\delta/\kappa$. The total current per unit length of contour of the transverse cross section, expressed in units of $cH_{cm}\sqrt{2}/4\pi$, is obviously equal to

$$J=\frac{\sqrt{\beta}}{\varkappa}\int_0^\infty j\,d\xi, \quad (24)$$

where ξ is the variable introduced earlier: $\xi=\kappa x/\delta\sqrt{\beta}$.

The equation in the linear approximation determines Ψ to within an arbitrary constant factor. In order to determine it, let us consider the complete equation for Ψ, where we take the factor e^{ikz} into consideration. Using the general equation for Ψ (see [4]) and changing over to the units introduced here, we obtain an equation for the correction of the first approximation with respect to Ψ^2, $\beta-\beta_0$, and q^2:

$$\frac{d^2\Psi_1}{d\xi^2}+[\beta_0-(\xi-\eta)^2]\Psi_1=\beta_0\left[\Psi_0^3+\frac{2}{\varkappa^2}(\xi-\eta)\Psi_0(\xi)\right.$$
$$\left.\times\int^{\xi}d\xi_1\int_0^{\xi_1}\Psi_0^2(\xi_2)(\xi_2-\eta)\,d\xi_2\right]-(\beta-\beta_0-\beta_0 q^2)\Psi_0. \quad (25)$$

Here the correction to the vector potential (in the usual units),

$$\frac{d^2A_1}{dx^2}=\frac{4\pi(2e)^2}{2mc^2}|\Psi|^2 A_0,$$

which is obtained from the equation for A, has been taken into consideration.

Ψ_0 is that solution of the homogeneous equation corresponding to (25) which satisfies the boundary conditions. The right side must be orthogonal to this solution. Hence we obtain the condition

$$(\beta-\beta_0-\beta_0 q^2)\int_0^\infty\Psi_0^2 d\xi=\beta_0\int_0^\infty\left(\Psi_0^4-\frac{2}{\varkappa^2}Q^2\right)d\xi,$$

$$Q(\xi)=\int_\xi^\infty(\xi-\eta)\Psi_0^2(\xi)\,d\xi, \quad (26)$$

which also determines the normalization of the function Ψ_0.

In order to establish the magnitude of the critical current, we substitute the found value of Ψ_0 into formula (26) and determine the unknown number q from the condition that J be a maximum. We have:

$$J = \frac{\sqrt{\beta_0}\,q}{\varkappa}\int_0^\infty \Psi_0^2 d\xi = \frac{\sqrt{\beta_0}\,q\,(\beta - \beta_0 - \beta_0 q^2)}{\varkappa\beta_0} \frac{\left[\int_0^\infty \Psi_0^2 d\xi\right]^2}{\int_\infty^0 \left(\Psi_0^4 - \frac{2}{\varkappa^2} Q^2\right) d\xi}. \quad (27)$$

The ratio of the integrals in (27) no longer contains normalization coefficients. Therefore, one can evaluate the maximum of only the preceding factor with respect to $\sqrt{\beta_0}\,q$. Performing this, we obtain

$$J_c = \frac{2}{3\sqrt{3}}\frac{\sqrt{\beta_0}}{\varkappa}\left(1 - \frac{\beta_0}{\beta}\right)^{3/2} \frac{\left[\int_0^\infty \Psi_0^2 d\xi\right]^2}{\int_0^\infty \left(\Psi_0^4 - \frac{2}{\varkappa^2} Q^2\right) d\xi}. \quad (28)$$

Now let us assume that the current flows along the surface but in a direction perpendicular to the field. This can occur only when η deviates from the value corresponding to the maximum field. As already mentioned in the preceding section, the determination of the smallest eigenvalue β_c is equivalent to the variational problem (19) (in the present case $s = \infty$). Since β_c is minimized with respect to η, the expansion of β_c in powers of $\eta - \eta_0$ begins with the quadratic term. Moreover, from (19) we obtain

$$\beta_c = \beta_0 + {}^1\!/_2 (\eta - \eta_0)^2 (\partial^2\beta / \partial\eta^2)_{\eta_0} = \beta_0 + (\eta - \eta_0)^2. \quad (29)$$

The current density in the usual units would be $-(2e^2/mc)\mathbf{A}|\Psi|^2$. In our units it equals

$$j = -(\delta / \sqrt{\beta})(\xi - \eta)|\Psi|^2. \quad (30)$$

But since the total current vanishes for $\eta = \eta_0$, then according to (24) we have

$$J = \frac{1}{\varkappa}(\eta - \eta_0)\int_0^\infty |\Psi|^2 d\xi. \quad (31)$$

Now let us consider the equation for Ψ. In the present case $q = 0$, but η differs from η_0. We transfer the term containing Ψ^3, and also the term containing the difference between β and the new critical value (30), to the right side. We have

$$\frac{d^2\Psi_1}{d\xi^2} + [\beta_c - (\xi - \eta)^2]\Psi_1 = \beta_0\Big[\Psi_{01}^3 + \frac{2}{\varkappa^2}(\xi - \eta)\Psi_{01}(\xi) \times \int^\xi d\xi_1 \int_0^{\xi_1} \Psi_0^2(\xi_2)(\xi_2 - \eta)\, d\xi_2\Big] - [\beta - \beta_0 - (\eta - \eta_0)^2]\Psi_{01}.$$

The solution Ψ_{01} of the homogeneous equation (for $\eta \neq \eta_0$) differs little from Ψ_0. The problem, as it is easy to see, reduces to the previous problem, except that $\eta - \eta_0$ appears instead of $\sqrt{\beta_0}\,q$. Thus, the critical current coincides in this case with (28). One can generalize this derivation without difficulty to the case of an arbitrary angle between field and current. Thus, the critical current is isotropic.

As already stated, the function $\Psi_0 = \exp(-\beta_0\xi^2/2)$ gives a value of β_0 in good agreement with the result of the exact calculation. Substituting this function into (28), we find

$$J_c = \frac{1}{3}\sqrt{\frac{2\pi}{3}}\frac{(1 - H/H_{c3})^{3/2}}{\varkappa(1 - 0.156/\varkappa^2)}. \quad (32)$$

The fact that a trial function of the type $\exp(-b\xi^2/2)$ satisfies the variational procedure well makes it realistic to attempt to obtain the critical current in the general case of fields ranging between H_{c2} and H_{c3}. We confine ourselves here to the case $\kappa \gg 1$. In this connection, we can neglect the deviation of the vector potential from $A_0 = H_0(x - x_0)$. It is necessary for us to find the exact solution of the equation for Ψ (the current is along the z axis):

$$d^2\Psi / d\xi^2 + [\beta - \beta q^2 - (\xi - \eta)^2]\Psi - \beta\Psi^3 = 0. \quad (33)$$

Instead of this, we can look for the minimum of the integral

$$F = \int_0^\infty \left(\frac{d\Psi}{d\xi}\right)^2 d\xi + \int_0^\infty [(\xi - \eta)^2 - \beta + \beta q^2]\Psi^2 d\xi + \frac{\beta}{2}\int_0^\infty \Psi^4 d\xi. \quad (34)$$

This expression in fact corresponds to the free energy, and it is necessary to interpret its minimization more broadly than simply as a formal way of obtaining Eq. (33). In particular, it is also necessary to minimize F with respect to η. From here and from expression (31) we immediately find that the transverse screening current is equal to zero, which is natural, as indicated earlier.

First let us consider the case when no current is present, i.e., $q = 0$. We choose a trial function in the form $\Psi = C\exp(-b\xi^2/2)$ and we determine the minimum of expression (34) with respect to η, C, and b. As a result we obtain

$$b = -{}^1\!/_3\beta + [{}^1\!/_9\beta^2 + {}^5\!/_3(1 - 2/\pi)]^{1/2},$$

$$C^2 = (4\sqrt{2}/5\beta)\{{}^4\!/_3\beta - [{}^1\!/_9\beta^2 + {}^5\!/_3(1 - 2/\pi)]^{1/2}\}. \quad (35)$$

As it should, $C^2 = 0$ for $\beta = \beta_0 = [1 - (2/\pi)]^{1/2}$. Here $b = \beta_0$. As $\beta \rightarrow \infty$ (a case which in fact is

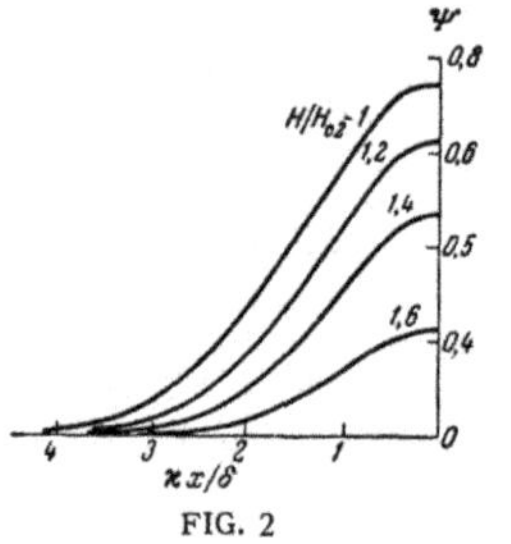
FIG. 2

not realized) $C^2 \to 4\sqrt{2}/5$ and $b \to 5\beta_0^2/2$. The form of Ψ for various values of the magnetic field H_0/H_{c2} is shown in Fig. 2.

In that case when a current exists in the z direction, it is necessary to make the substitution $\beta \to \beta(1-q^2)$ in the expression for b and in the numerator of the expression for C^2. According to (23) and (24) we obtain

$$J = \frac{1}{2}\sqrt{\frac{\pi}{b}}\frac{\sqrt{\beta}q}{\varkappa}C^2 = \frac{\sqrt{\pi}}{5\beta\varkappa b^2}(2b\beta - 3\alpha + 3b^2)^{1/2}(3\alpha - 5b^2), \tag{36}$$

where $\alpha = \frac{5}{3}[1-(2/\pi)] = \frac{5}{3}\beta_0$. Here for simplicity we have expressed q in terms of b. In this form it is more convenient to determine the maximum of J. The condition for a maximum gives the following equation:

$$t^4 + \tfrac{1}{3}\mu t^3 + t^2 + \mu t - \tfrac{10}{3} = 0, \tag{37}$$

where $t = b/\beta_0$, $\mu = \beta/\beta_0$. In turn, the expression for the current can be written in the form

$$J_c = (\sqrt{2\pi}/\varkappa)Q(H_{c3}/H);$$

$$Q(\mu) = (\mu t + \tfrac{3}{2}t^2 - \tfrac{5}{2})^{1/2}(1-t^2)/\mu t^2. \tag{38}$$

Here one should express t in terms of μ with the aid of Eq. (37). These formulas give a complete determination of the dependence of J_c on H. Near H_{c3} we obtain $1 - t \approx (\mu-1)/6$, $Q \approx (\mu-1)^{3/2}/3\sqrt{3}$, i.e., expression (32). For $H \to 0$, i.e., for $\mu \to \infty$, $t \approx 10/3\mu$, $Q(\mu) \approx (3/10)^{3/2}\mu/2$, i.e., $J_c \propto H^{-1}$. Of course, this case is actually not realized. The dependence of J_c on (H/H_{c2}) in the region $H_{c2} < H < H_{c3}$ is plotted in Fig. 3.

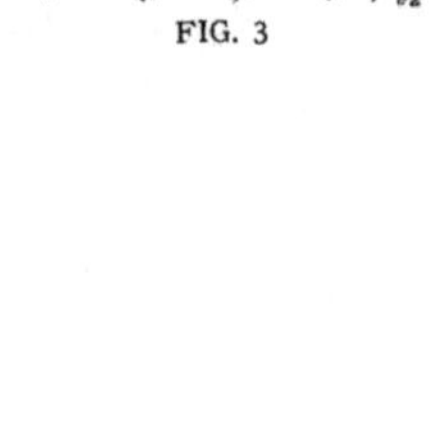
FIG. 3

Here we shall not carry out the derivation for current flowing at an angle to the field. This derivation is completely analogous to the preceding and leads to the conclusion that the critical current is isotropic.

4. EXPERIMENTAL DATA

Some of the experimental consequences of surface superconductivity were already discussed in the article by Saint-James and de Gennes.[1] The fundamental consequence, of course, is the fact that the vanishing of the magnetic moment of a bulk superconductor of the second kind and the appearance of resistivity in such superconductors occur at different values of the field. In experiments in which the moment is measured, surface superconductivity does not manifest itself in any way. However, in the extensive investigation of the critical fields of type II superconductors carried out by Berlincourt and Hake,[10] not the magnetic moment but the appearance of resistivity was observed. Therefore, one would think that they must have observed not H_{c2} but H_{c3}. At the same time, a comparison of data with the theoretical value of the field H_{c2} revealed agreement in this region of concentrations of alloys, where the mean free path was not overly small, and therefore the theory was applicable. In this same region, where the mean free path became too small, the theoretical field H_{c2} was above and not below the observed field.

It may be possible to attribute this clear contradiction to the large measuring current. In the experiments of Berlincourt and Hake, bars of length 1.3 cm, width from 0.02 to 0.08 cm, and thickness 0.005 cm were used. The current, relative to the cross sectional area, amounted to $10\,A/cm^2$. If we assume that surface superconductivity did exist, then the current must have flowed only on the surface. Recalculating, we obtain 2.5×10^{-2} A/cm. In order to estimate the critical current, we use formula (32). Changing over to the usual units, we obtain

$$J_c = \frac{5H_{cm}}{3\sqrt{3\pi}\varkappa}\left(1 - \frac{H}{H_{c3}}\right)^{3/2} \text{A/cm}. \tag{39}$$

If the most unfavorable numbers are substituted here, namely $H_{cm} = 100$ Oe and $\kappa = 100$, then we find that the current used in [10] should shift the field by only 16%. In fact, however, H_{cm} is closer to 1000 Oe and κ is hardly larger than 10 in the region where the quantitative theory is applicable. Therefore, we must admit that for some reason surface superconductivity was absent in the experiments of Berlincourt and Hake. It is possible that the reason was the inhomogeneity of the surface layers of the samples.

The experiments of Khukhareva[8] on the determination of the critical field of thin films condensed at low temperature, were more favorable for the observation of surface superconductivity. The method of preparation of these films would most likely give homogeneous samples. Khukhareva compared her data with theory[9] and observed quite good agreement. However, as we shall show, her results not only do not contradict, but perhaps are even in better agreement with the formulas of Sec. 2 of the present article.

In fact, the experimental data were reduced with the aid of a formula of type (13). Formula (13) exhibits two characteristic regions: the region of small relative thicknesses (neighborhood of T_c) where $H_{c3}(d)/2\theta^2 = (2\sqrt{3})1.29 \times 10^{-4}/\theta d$, and the region of large relative thicknesses where $H_{c3}/2\theta^2 \approx 1.7$. Since an incorrect relation $H_{c3}/2\theta_1^2 \approx 1$ was used by Khukhareva for large thicknesses, the value of θ_1 thus found was $\sqrt{1.7}$ times larger than the true value. Khukhareva recalculated from the value of θ, the coefficient δ_{00} in the temperature dependence of the penetration depth. $\delta_{00} = (22.6 \pm 2.3) \times 10^{-6}$ cm was obtained for mercury. A theoretical determination from the normal-state conductivity gives $\delta_{00} = 16 \times 10^{-6}$ cm. Since θ is proportional to δ_{00}, it follows that it is necessary to decrease the experimental value of δ_{00} by $\sqrt{1.7}$ times. This yields $\delta_{00} = (17.3 \pm 1.7) \times 10^{-6}$, in excellent agreement with the theoretical value.

To be sure, it is necessary to show that a decrease in θ leads, according to the relation in the region of small relative thicknesses, to the fact that one should regard the thickness of the films as 30% less than assumed earlier. This is apparently quite permissible. The point is that Khukhareva[8] determined the thickness of the films by two methods: a) from the amount of evaporated metal with assumption of a cosine law distribution on the surface, and b) by direct weighing of the collected evaporated substance (in one experiment). The first method gave a thickness 20% smaller than the second. A previous determination of the thickness of the film according to the curve of $H_c(d)$[9] agreed with the larger value obtained by the second method. Thus, a decrease of the thickness by 30% is in any case a change of the data in the right direction and by an amount of permissible order.

From everything that has been said, and also from the discussion of experimental data given by Saint-James and de Gennes,[1] it follows that at the present time there is still no clear experimental confirmation of surface superconductivity, and new more thorough experiments are necessary.

Remark (May 27, 1964). Recently a whole series of articles devoted to the measurement of the field H_{c3} have appeared in print. Surface superconductivity has been observed for a large number of different superconducting alloys.[11] The majority of these measurements well confirm the relationship $H_{c3} = 1.7 H_{c2}$ of Saint-James and de Gennes. It is of interest that surface superconductivity is observed even in superconductors of the first kind, for which $1/\sqrt{2} > \kappa > (1/\sqrt{2})/1.7$ (i.e., $H_{c2} < H_c < H_{c3}$ [12]).

[1] D. Saint-James and P. G. de Gennes, Phys. Letters 7, 306 (1963).

[2] A. A. Abrikosov, JETP 32, 1442 (1957), Soviet Phys. JETP 5, 1174 (1957).

[3] V. L. Ginzburg, JETP 31, 541 (1956), Soviet Phys. JETP 4, 594 (1957).

[4] V. L. Ginzburg and L. D. Landau, JETP 20, 1064 (1950).

[5] L. P. Gor'kov, JETP 36, 1918 (1959), Soviet Phys. JETP 9, 1364 (1959).

[6] L. P. Gor'kov, JETP 37, 1407 (1959), Soviet Phys. JETP 10, 998 (1960).

[7] L. P. Gor'kov, JETP 37, 833 (1959), Soviet Phys. JETP 10, 593 (1960).

[8] I. S. Khukhareva, JETP 41, 728 (1961), Soviet Phys. JETP 14, 526 (1962).

[9] A. A. Abrikosov, DAN SSSR 86, 489 (1952).

[10] T. G. Berlincourt and R. R. Hake, Phys. Rev. 131, 140 (1963).

[11] C. F. Hempstead and Y. B. Kim, Phys. Rev. Letters 12, 145 (1964); W. J. Tomasch and A. S. Joseph, Phys. Rev. Letters 12, 148 (1964); Bon Maridon, Goodman, and Lacaze, Phys. Letters 8, 15 (1964); Cardona, Fisher, and Rosenblum, Phys. Letters 8, 308 (1964).

[12] B. Rosenblum and M. Cardona, Phys. Letters 9, 220 (1964).

Translated by H. H. Nickle
102

SOVIET PHYSICS JETP VOLUME 18, NUMBER 5 MAY, 1964

THE CONDUCTIVITY OF STRONGLY COMPRESSED MATTER

A. A. ABRIKOSOV

Institute of Physics Problems, Academy of Sciences, U.S.S.R.

Submitted to JETP editor June 18, 1963

J. Exptl. Theoret. Phys. (U.S.S.R.) **45**, 2038-2047 (December, 1963)

Processes determining the conductivity of strongly compressed matter are considered for various temperatures and densities. It is shown that in the solid modification of strongly compressed hydrogen, which possesses a closed Fermi surface, the conductivity is much greater than that of solid modifications of other elements. The conductivity of various liquid modifications of strongly compressed matter is also considered, as well as the conductivity under conditions when the electrons are ultrarelativistic.

IT was shown earlier[1,2] that at high degrees of compression, when the interatomic distances are much smaller than the Bohr radius, matter should be in the solid state, provided that the temperature is not too high. An exception is hydrogen (and possibly helium), which at very high compressions can experience a transition into a quantum liquid reminiscent of superfluid helium[2]. The transition to the liquid phase occurs also at temperatures which exceed the lattice binding energy per atom (of the order of e^2p_0, where p_0 is the Fermi limiting momentum); with different cases may be realizable here, since both the electrons and the nuclei may be in degenerate and nondegenerate states.

If we recognize that the conductivity in the solid phase and in different temperature intervals is determined by different electron scattering processes, it becomes clear that the conductivity of strongly compressed matter can change quite appreciably with temperature and with density. The present article is devoted to a study of these processes. We confine ourselves to a determination of the order of magnitude of the conductivity, and consider for simplicity light elements ($Z \sim 1$).

1. SOLID PHASE

A. Scattering by impurities. As always, at the lowest temperature the most significant is the scattering of electrons by impurities and lattice defects. In the case considered here, such defects are charged. Consequently, the scattering probability has an order of magnitude

$$W_{\text{imp}} \sim N_{\text{imp}} \int \frac{e^4}{(\mathbf{p}-\mathbf{p}')^4} (1-\cos\theta)\,\delta(\varepsilon-\varepsilon')\,d^3\mathbf{p}',$$

where $\theta = (\widehat{\mathbf{p},\mathbf{p}'})$ and N_{imp} is the number of defects per unit volume.

The limits of the resultant logarithmic integral are the reciprocal Debye radius $\kappa \sim e p_0^{1/2} m^{1/2}$ and $p_0 = (3\pi^2 NZ)^{1/3}$, the Fermi limiting momentum [1)]. Taking the integral, we get

$$W_{\text{imp}} \sim \tau_{\text{imp}}^{-1} \sim N_{\text{imp}} e^4 m p_0^{-3} \ln (p_0/\kappa). \tag{1}$$

Substituting in the equation for the conductivity

$$\sigma \sim N e^2 \tau / m, \tag{2}$$

we obtain

$$\sigma_{\text{imp}} \sim N^2 / m^2 e^2 N_{\text{imp}} \ln (p_0/\kappa). \tag{3}$$

The lower limit σ_{imp} is attained when $N_{\text{imp}} \sim N$. We thus obtain

$$\sigma_{\text{imp}} > N / m^2 e^2 \ln (p_0/\kappa). \tag{4}$$

B. Scattering of electrons by electrons. As the temperature increases, other electron scattering mechanisms come into play, primarily scattering of electrons by electrons. Here, however, we must recognize that, as is well known (see [3]) contributions to scattering are made only by collisions with Umklapp. On the other hand, if we assume in the zeroth approximation a plane wave function for the electron, then the Coulomb interaction of the electrons will not lead to any Umklapp.

In view of this, we must consider the next approximation of the wave functions, namely

$$\psi_{\mathbf{p}} = \psi_{\mathbf{p}}^{(0)} + 4\pi e^2 Z N \sum_{\mathbf{K}\neq 0} \frac{1}{K^2} \frac{\psi_{\mathbf{p}-\mathbf{K}}^{(0)}}{\varepsilon(\mathbf{p}) - \varepsilon(\mathbf{p}-\mathbf{K})}, \tag{5}$$

where **K** is the reciprocal lattice vector. It is necessary that the second term of one of the wave functions (5) contribute to the matrix element of the electron Coulomb interaction energy.

1) Here and throughout we use units for which $\hbar = 1$.

Recognizing that processes with different Umklapps do not interfere, we have

$$W \sim \int \sum_{\mathbf{K}} \left[\frac{e^2}{(\mathbf{p}_1 - \mathbf{p}_1' - \mathbf{K})^2} \sum_{l} \frac{e^2 N}{\mathbf{K}^2 (\varepsilon(\mathbf{p}_l) - \varepsilon(\mathbf{p}_l \pm \mathbf{K}))^2} \right]^2 F(n) \times \delta(\varepsilon_1 + \varepsilon_2 - \varepsilon_1' - \varepsilon_2') \, d^3\mathbf{p}_1' d^3\mathbf{p}_2,$$

where $\mathbf{p}_i$ denotes all four momenta participating in the process, and the sign in front of $\mathbf{K}$ is taken with allowance for the conservation law $\mathbf{p}_1 + \mathbf{p}_2 - \mathbf{p}_1' - \mathbf{p}_2' - \mathbf{K} = 0$. The function $F(n)$ is a known form of Fermi equilibrium functions.

The presence of $F(n)$ together with $\delta(\epsilon_1 + \epsilon_2 - \epsilon_1' - \epsilon_2')$ causes all the integrations to be over vicinities of the Fermi surfaces. We can separate from the integrals with respect to $d^3\mathbf{p}_1'$ and $d^3\mathbf{p}_2$ the integrals with respect to the energies, which yield the factor $(T/\epsilon_F)^2$, where $\epsilon_F = p_0^2/2m$. The remaining three integrations pertain to the possible angles between the momenta $\mathbf{p}_i$ with allowance for the fact that each has an absolute value p_0 and that they satisfy the conservation laws $\mathbf{p}_1 + \mathbf{p}_2 - \mathbf{p}_1' - \mathbf{p}_2' = \mathbf{K}$.

In the case of hydrogen none of the denominators $\epsilon(\mathbf{p}_i) - \epsilon(\mathbf{p}_i - \mathbf{K})$ can vanish if the $\mathbf{p}_i$ lie on the Fermi surface (see [4]). In view of this we have

$$W_{e,\,\mathrm{H}} \sim \left(\frac{e^2}{p_0^2} \frac{e^2 N}{p_0^2 (p_0^2/m)} \right)^2 \left(\frac{T}{p_0^2/m} \right)^2 \frac{p_0^6}{p_0^2 m} \sim \frac{e^8 m^5 T^2}{p_0^6}. \tag{6}$$

Substituting (6) in (2) we get

$$\sigma_{e,\,\mathrm{H}} \sim N^3/e^6 m^6 T^2. \tag{7}$$

The lower limit of $\sigma_{e,\mathrm{H}}$ is obtained at a temperature corresponding to destruction of the lattice, $T \sim e^2/r \sim e^2 p_0$. We therefore have

$$\sigma_{e,\,\mathrm{H}} > N^{7/3}/e^{10} m^6. \tag{8}$$

Comparing with (3), we find that if the defect concentration in hydrogen is

$$N_{\mathrm{imp}}/N \gg (e^2 m/N^{1/3})^4 \sim (e^2/v)^4 \tag{9}$$

$(v = p_0/m)$, then the electron scattering does not play any role in the conductivity. We recall that the condition $e^2/v \ll 1$ is precisely the starting point of all the calculations (see [2]).

Starting with helium, the Fermi sphere can cross the boundaries of the Brillouin zone. Consequently, the differences $\epsilon(\mathbf{p}_i) - \epsilon(\mathbf{p}_i \pm \mathbf{K})$ can vanish. The most "dangerous" case is when for some $\mathbf{K}$ we have $\epsilon(\mathbf{p}_1') = \epsilon(\mathbf{p}_1' + \mathbf{K})$, in other words, $|\mathbf{p}_1' + \mathbf{K}| = p_0$. In the case of a common direction $\mathbf{p}_1$, the other denominator $\mathbf{p}_1 - \mathbf{p}_1' - \mathbf{K}$ cannot vanish here.

The scattering probability acquires an integral of the type

$$\int (|\mathbf{p}_1' + \mathbf{K}| - p_0)^{-2} d|\mathbf{p}_1' + \mathbf{K}|.$$

This integral diverges, and to estimate it correctly it is necessary to recognize that the form of the energy changes in the vicinity of $\epsilon(\mathbf{p}) = \epsilon(\mathbf{p} + \mathbf{K})$. This leads to cutoff of the integral at $\epsilon(\mathbf{p}) - \epsilon(\mathbf{p} + \mathbf{K}) \sim e^2 N/\mathbf{K}^2$. Taking these remarks into account we obtain for $Z \geq 2$

$$W_{e,\,Z \gg 2} \sim \left(\frac{e^2}{p_0^2} \frac{e^2 N}{p_0^2} \right)^2 \frac{m}{p_0^2} \frac{p_0^2}{e^2 N} \left(\frac{T}{p_0^2/m} \right)^2 \frac{p_0^6}{p_0^2/m} \sim \frac{e^6 m^4 T^2}{p_0^5}, \tag{10}$$

hence

$$\sigma_{e,\,Z \gg 2} \sim \frac{N^{8/3}}{e^4 m^5 T^2} \sim \frac{e^2}{v} \sigma_{e,\,\mathrm{H}} \ll \sigma_{e,\,\mathrm{H}}. \tag{11}$$

Substituting the upper temperature limit, we obtain

$$\sigma_{e,\,Z \gg 2} > \frac{N^2}{e^8 m^5}. \tag{12}$$

The electron scattering will be insignificant if the defect concentration is

$$\frac{N_{\mathrm{imp}}}{N} \gg \left(\frac{e^2 m}{N^{1/3}} \right)^3 \sim \left(\frac{e^2}{v} \right)^3.$$

<u>C. Scattering by phonons; high temperatures.</u> As is well known [12], the Debye frequency is of the order of

$$\omega_D \sim \omega_0 = \sqrt{\frac{4\pi N e^2 Z^2}{M}} \sim e N^{1/2}/M^{1/2}. \tag{13}$$

when $T \gg \omega_0$, the principal role is assumed by absorption and emission of phonons with momenta on the order of the reciprocal lattice period, and the corresponding Bose function yields a factor $[\exp(\omega_0/T) - 1]^{-1} \approx T/\omega_0$.

In an earlier paper [4], the author obtained the matrix element for the emission or absorption of a phonon. Its order of magnitude is

$$V \sim \frac{e^2 (\mathbf{k} + \mathbf{K})_\alpha}{(\mathbf{k} + \mathbf{K})^2 + \varkappa^2} \sqrt{\frac{N}{M\omega(\mathbf{k})}}.$$

Consequently, the probability of phonon emission or absorption is of the form

$$W_{\mathrm{ph},\,T \gg \omega_0} \sim \int V^2 \frac{T}{\omega_0} \delta(\varepsilon_1 - \varepsilon') \, d^3\mathbf{k} \sim \frac{e^4}{p_0^2} \frac{N}{M\omega_0} \frac{T}{\omega_0} \frac{p_0^3}{p_0^2/m} \sim \frac{e^2 m}{p_0} T, \tag{14}$$

hence

$$\sigma_{\mathrm{ph},\,T \gg \omega_0} \sim \frac{N^{4/3}}{m^2 T}. \tag{15}$$

Substituting the upper temperature limit $T \sim e^2 p_0$, we obtain

$$\sigma_{\mathrm{ph}} > \frac{N}{e^2 m^2}. \tag{16}$$

Comparing with (3), we see that when the defect concentration is $N_{\mathrm{imp}}/N \ll 1$ and the temperature is sufficiently high, the principal role is assumed by the scattering of electrons by phonons.

D. Scattering by phonons; low temperatures. At low temperatures $T \ll \omega_0$ the situation for hydrogen is again radically different from that for all other elements. As is well known (see [3]), to obtain a finite resistance in the case of scattering of electrons by phonons, it is necessary to take Umklapp into account. But at phonon momenta $k \sim T/u \ll p_0$, where u is the velocity of sound, the condition $\mathbf{p} + \mathbf{k} - \mathbf{p}' = \mathbf{K}$ cannot be satisfied if p and p′ lie on the Fermi surface and this surface is closed. This is precisely the situation in hydrogen. Thus it is necessary that **k** not be small, or else that either of the vectors **p** or **p′** not lie on the Fermi surface. This leads to the appearance of exponentially small factors. Since for a given momentum the phonon energy is smaller than the electron energy, phonons with large momenta are more likely and this gives an exponential factor in the form $\exp[-\alpha\omega_0/T]$ $(\alpha \sim 1)$.

The complete expression for $W_{ph,H,T\ll\omega_0}$ and $\sigma_{ph,H,T\ll\omega_0}$ is obtained from (12) and (13) by substituting $\exp(-\alpha\omega_0/T)$ for T/ω_0. As a result we have

$$W_{ph,H,T\ll\omega_0} \sim \omega_0 (e^2/v)\, e^{-\alpha\omega_0/T},$$

$$\sigma_{ph,H,T\ll\omega_0} \sim (M^{1/2} N^{4/3}/em^2)\, e^{\alpha\omega_0/T}. \tag{17}$$

Here, however, we must bear in mind the following. The calculation performed applies to the case when there are no processes other than Umklapp to disturb the momentum conservation law. Actually this is not so. The presence of lattice defects leads to phonon scattering in which the momentum is not conserved. This question deserves a more detailed analysis.

As is well known [3], the absence of resistance with momentum conservation is connected with the fact that there exists in this case for the additions to the distribution function a solution which causes both collision integrals to vanish, in the form

$$\delta n_e = n_e(1 - n_e) f_e,$$
$$\delta n_{ph} = n_{ph}(1 + n_{ph}) f_{ph},$$

where

$$f_e = cp_z, \quad f_{ph} = ck_z$$

(c is a constant).

We now take into account the scattering of the phonons by impurities. Then the equation $n_{ph} = 0$ can be arbitrarily written in the form

$$W_{ph,e}(f_{ph} + f_e - f_e') + W_{ph,imp} f_{ph} = 0, \tag{18}$$

where $W_{ph,e}$ and $W_{ph,imp}$ are the probability of the scattering of a phonon by the electrons and by the defects, and f_e and f_{ph} are unknown functions.

The collision integral in the electron equation is written in the form

$$W_{e,ph}(f_e - f_e' + f_{ph}).$$

Substituting f_{ph} from the preceding equation, we find that the collision integral of the electron equation assumes the form

$$(f_e - f_e')\, W_{e,ph}(W_{ph,e}/W_{ph,imp} + 1)^{-1}. \tag{19}$$

It follows therefore that as $W_{ph,imp}/W_{ph,e} \to 0$ this collision integral vanishes, while in the opposite limiting case it assumes the same form as for equilibrium phonon distribution.

Let us find the probabilities $W_{ph,e}$ and $W_{ph,imp}$. The former is obtained from the same matrix element that was used to derive formula (14) (in this case $K = 0$ and $k \ll \kappa$). We thus have

$$W_{ph,e} \sim \frac{e^4 k}{\varkappa^2} \frac{p_0^3}{p_0^2/m} \sim \frac{e^4}{e^2 p_0 m} \frac{T\sqrt{mM}}{p_0} \frac{p_0^3}{p_0^2 m} \sim \frac{e^2 T\sqrt{mM}}{p_0} \tag{20}$$

(we have substituted here $k \sim T/u$, $u \sim p_0/\sqrt{Mm}$).

To find $W_{ph,imp}$ we expand the interaction energy of the nuclei with charged defects in powers of the displacements of the nuclei, up to second order terms. The corresponding matrix element will take the form

$$\frac{\exp\{i(\mathbf{k}_1 - \mathbf{k}_2)\mathbf{R}_l\}(\mathbf{k}_1 - \mathbf{k}_2)_\alpha (\mathbf{k}_1 - \mathbf{k}_2)_\beta}{(\mathbf{k}_1 - \mathbf{k}_2)^2 + \varkappa^2} \frac{N}{[NM\omega(\mathbf{k}_1)]^{1/2}[NM\omega(\mathbf{k}_2)]^{1/2}}$$

($\mathbf{R}_l$ is the coordinate of the impurity atom). From this we obtain without difficulty the phonon-impurity scattering probability

$$W_{ph,imp} \sim \frac{e^4 k^4}{\varkappa^4 M^2 \omega^2(k)} \frac{k^3}{\omega(k)} N_{imp} \sim \frac{e^4 T^4}{u^7 \varkappa^4 M^2} N_{imp} \sim N_{imp} \frac{(mM)^{3/2}}{p_0^9} T^4. \tag{21}$$

We now find $W_{ph,e}/W_{ph,imp}$. According to (20) and (21) we have

$$W_{ph,e}/W_{ph,imp} \sim \frac{e^2 p_0^8}{MmT^3 N_{imp}}. \tag{22}$$

The smallest value of this ratio occurs when $N_{imp} \sim N$ and $T \sim \omega_0$. Therefore

$$W_{ph,e}/W_{ph,imp} > \frac{p_0^{1/2} M^{1/2}}{em} \sim \left(\frac{v}{e^2}\right)^{1/2} \left(\frac{M}{m}\right)^{1/2} \gg 1.$$

Thus, according to (19), we find that the role of the electron-phonon scattering probability can be assumed by the quantity $W_{e,ph}(W_{ph,imp}/W_{ph,e})$ provided it is larger than the probability of scattering with Umklapp (17).

The quantity $W_{e,ph}$ is the probability of the scattering of an electron by equilibrium phonons.

Using the matrix element V, we can write for the collision integral

$$\int |V|^2 (f_e - f_e') \delta \left(\frac{\mathbf{kp}}{m} - \frac{k^2}{2m} \right) d^3\mathbf{k}.$$

Choosing f_e in the form cp_z (z is along the field direction), we obtain

$$\int |V|^2 ck_z \delta \left(\frac{\mathbf{kp}}{m} - \frac{k^2}{2m} \right) d^3\mathbf{k} = \int |V|^2 cp_z \frac{(\mathbf{pk})}{p^2} \delta \left(\frac{\mathbf{pk}}{m} - \frac{k^2}{2m} \right) d^3\mathbf{k}$$
$$= cp_z \int |V|^2 \frac{k^2}{2p^2} \delta \left(\frac{\mathbf{kp}}{m} - \frac{k^2}{2m} \right) d^3\mathbf{k} \sim W_{\mathrm{e,ph}} cp_z.$$

Thus, recognizing that $k \ll \kappa$, we get

$$W_{\mathrm{e,ph}} \sim \frac{e^4 N k^2}{M\omega\varkappa^4} \frac{k^2}{p^2} \frac{k^3}{kp_0/m} \sim \frac{T^5 m^2 M^2}{p_0^8}. \tag{23}$$

The sought-for scattering probability takes the form

$$W_{\mathrm{e,ph}} \frac{W_{\mathrm{ph,imp}}}{W_{\mathrm{ph,e}}} \sim \frac{T^8 m^3 M^3 N_{\mathrm{imp}}}{p_0^{16} e^2}.$$

This value reaches its maximum when $T \sim \omega_0$, hence

$$W_{\mathrm{e,ph}} \frac{W_{\mathrm{ph,imp}}}{W_{\mathrm{ph,e}}} < \frac{e^6 m^3 N_{\mathrm{imp}}}{p_0^4 M}.$$

A comparison of this quantity with (1) shows that it amounts to $(e^2 m/vM) W_{\mathrm{imp}} \ll W_{\mathrm{imp}}$. This means that the account of relaxation of phonons on the impurities in hydrogen certainly does not make any noticeable contribution to the conductivity.

Let us make one more comparison, of (17) with (6) for $T \sim \omega_0$. We have

$$\frac{W_{\mathrm{ph,H},T\sim\omega_0}}{W_{\mathrm{e,H},T\sim\omega_0}} \sim \frac{\omega_0 e^2 m p_0^6}{p_0 e^8 m^5 \omega_0^2} \sim \frac{M^{1/2} p_0^{7/2}}{e^7 m^4} \sim \left(\frac{M}{m} \right)^{1/2} \left(\frac{v}{e^2} \right)^{7/2} \gg 1.$$

This means that in the temperature region $T \sim \omega_0$, the phonon scattering mechanism prevails over the electron mechanism[2)]. Thus, when $T \ll \omega_0$ the conductivity is determined by the smallest of the quantities (3), (7), or (17).

<u>E. Scattering by phonons; low temperatures; $Z \geq 2$.</u> The situation is different when $Z \geq 2$. Inasmuch as the Fermi surface crosses the borders of the Brillouin zone, Umklapp is no longer forbidden at low temperatures. To estimate the conductivity we assume that the additions to the distribution functions are of the form[3]

$$\delta n_{\mathrm{ph}} = n_{\mathrm{ph}}(1 + n_{\mathrm{ph}}) f_{\mathrm{ph}}, \quad \delta n_e = n_e(1 - n_e) f_e,$$

where $f_{\mathrm{ph}} = ck_z$ and $f_e = c_1 m v_{e,z}$ (c and c_1 are constants). The function f_e is chosen in this form because it must be periodic in the reciprocal lattice and continuous. If we were to use the phonon kinetic equation to express f_{ph} in terms of f_e, then this equation would be satisfied for $c = c_1$, were it not for the vicinity of the border of the Brillouin zone.

Two cases are possible. If the phonon momentum $k \sim T/u$ exceeds the interval of $e^2 p_0/v$ in which the form of the electron energy varies noticeably near the Brillouin zone, then we can neglect the variation of the energy. In such a case the electron velocity changes at the point of intersection of the Fermi surface with the Brillouin zone from $\mathbf{p}/m$ to $(\mathbf{p} - \mathbf{K})/m$, that is, by an amount of the order of p_0/m. Thus, if the electron crosses the border of the Brillouin zone on absorbing a phonon, then $f_e - f_e' \sim c_1 p_0$.

Since the intersection of the Fermi sphere with the border of the zone is a circle with a radius of the order of p_0, the effective section of the area at which the electron passes through the border has a relative dimension k/p_0.

In the case when $T/u \ll e^2 p_0/3$, it is necessary to take into account the change in the electron energy near the intersection. As is well known, in this case

$$\varepsilon_{1,2} = \tfrac{1}{2} [\varepsilon(\mathbf{p}) + \varepsilon(\mathbf{p} - \mathbf{k})]$$
$$\mp \{ \tfrac{1}{4} [\varepsilon(\mathbf{p}) - \varepsilon(\mathbf{p} - \mathbf{k})]^2 + U^2 \}^{1/2}, \tag{24}$$

where U is in this case equal to $4\pi Z^2 e^2 N/\mathbf{K}^2$. The normal component of the velocity obtained from this, in the momentum region about $e^2 p_0/3$ from the border of the Brillouin zone, is of the order of $sp_0^2/m^2 U$, where s is the distance to the border. Consequently, $f_e - f_e' \sim c_1 k/(e^2/v)$. But since relative order of the corresponding section of the surface is $\sim e^2/v$, we again find after integrating over the electron momenta that the vicinity of the intersection makes a contribution of the order of $c_1 k$.

Thus, the near vicinity of the intersection of the Fermi surface with the border of the Brillouin zone makes in both cases an addition to the function f_{ph} which is of the same order as the far part. It follows therefore that the constants c and c_1 in the functions f_{ph} and f_e are of the same order but not equal. As already noted, if we substitute these functions into the electron collision integral without account of Umklapp, then the integral vanishes if c and c_1 are equal.

On the other hand, if these constants are different, then a situation close to that for equilibrium phonons arises, that is, close to that for $f_{\mathrm{ph}} = 0$. In this case the collision probability is given by (23), and the corresponding value of the conductivity is

---

2)It is easy to see that this conclusion remains valid also when $Z \gg 2$.

$$\sigma_{\mathrm{ph},\, Z\geqslant 2,\, T\ll\omega_0} \sim e^2N^{11/3}/m^3M^2T^5. \quad (25)$$

It must be noted that as $T \to \omega_0$ formula (25) gives a limiting value

$$M^{1/3}p_0^{7/3}/e^3m^3,$$

whereas formula (15) gives as $T \to \omega_0$ the larger value

$$p_0^{8/3}M^{1/3}/m^2e.$$

This seemingly surprising circumstance is connected with the form of the spectrum of the longitudinal phonons, which reaches the limiting value ω_0 not for $k \sim p_0$ but for $k \sim \kappa$ (see [2]).

In the derivation of (25) we considered phonons with $\omega \ll \omega_0$, that is, $k \ll \kappa$. Thus, in the limit as $T \to \omega_0$ we obtain the contribution from scattering by phonons with $k \sim \kappa$. At the same time, (15) takes into account the contribution of phonons with $k \sim p_0$. It follows therefore that even below the Debye temperature the scattering of the electrons by the phonons with $k \sim p_0$ will be appreciable in some interval. The corresponding value of the conductivity is given by (18). Only at these temperatures, when the conductivity exceeds the value given by (15), do phonons with $k \ll \kappa$ begin to play the principal role.

F. Total resistivity. The complete expression for the temperature dependence of the resistivity is of the form

$$\rho = A_Z(N)N_{\mathrm{imp}} + B_Z(N)T, \quad T \gg \omega_0,$$

$$\rho = A_1(N)N_{\mathrm{imp}} + C_1(N)e^{-\alpha_1\omega_0/T} + E_1(N)T^2,$$

$$Z = 1, \quad T \ll \omega_0,$$

$$\rho = A_Z(N)N_{\mathrm{imp}} + C_Z(N)e^{-\alpha_Z\omega_0/T} + D_Z(N)T^5 + E_Z(N)T^2,$$

$$Z \geqslant 2, \quad T \ll \omega_0,$$

where $\omega_0 = \sqrt{Ne^2Z^2/M}$ and $\alpha_Z \sim 1$. The density dependence of the coefficients A, B, C, D, and E is given by formulas (3), (7), (11), (15), (18), and (25).

According to the foregoing, when $Z \sim 1$ all the coefficients A_Z, B_Z, C_Z, D_Z, and E_Z are of the same order when $Z \geq 2$, but for hydrogen the coefficient E_1 is of the order of $e^2E_{Z\geq 2}/v$, and the coefficient D_1 vanishes.

2. LIQUID PHASE

As noted in the beginning of the article, strongly compressed matter can be in the liquid state either at sufficiently high temperatures and large binding energies per atom, that is, when

$$T \gg e^2p_0, \quad (26)$$

or as a result of very large compressions, when the zero-point energy exceeds the binding energy, that is,

$$\omega_0 \gg e^2p_0. \quad (27)$$

Incidentally, as explained in [2], the latter occurs only for hydrogen and possibly for helium.

The liquid phase is a mixture of two liquids, electronic and nuclear, each of which can be degenerate. The degeneracy occurs at temperatures which are lower than p_0^2/m and p_0^2/M, respectively.

A. Electrons and nuclei degenerate. We assume that condition (26) is satisfied and at the same time the nuclear liquid is degenerate, that is,

$$p_0^2/M \gg T \gg e^2p_0.$$

From this it follows directly that condition (27) is satisfied, something which we shall assume can be realized only in hydrogen. Consequently, it is sufficient to consider only one case of degenerate nuclei, that of Fermi nuclei.

Let us find the probability of electron scattering in this case. The degeneracy of the nuclei leads to a limitation of the phase space for the nuclear recoil momenta and gives a factor $(TM/p_0^2)^2$. In all other respects the situation does not differ from the case considered above, that of scattering by defects (if we assume $N_{\mathrm{imp}} \sim N$). Thus, we have

$$W_{\mathrm{dd}} \sim \frac{Ne^4m}{p_0^3}\left(\frac{T}{p_0^2/M}\right)^2 \ln\frac{p_0}{\varkappa} \sim \frac{e^4mM^2}{p_0^4}T^2\ln\frac{p_0}{\varkappa}, \quad (28)$$

$$\sigma_{\mathrm{dd}} \sim \frac{N^{7/3}}{M^2m^2e^2T^2\ln(N^{1/3}/me^2)}. \quad (29)$$

B. Electrons degenerate, Boltzmann nuclei. At higher temperatures we get Boltzmann nuclei but the electron may remain degenerate. For this purpose we need

$$p_0^2/M \ll T \ll p_0^2/m.$$

In this case the scattering of the electrons does not differ at all from the scattering by impurities, and as a result we have

$$W_{\mathrm{db}} \sim e^4m\ln\frac{p_0}{\varkappa}, \quad (30)$$

$$\sigma_{\mathrm{db}} \sim \frac{N}{m^2e^2\ln(N^{1/3}/me^2)}. \quad (31)$$

C. Boltzmann electrons and nuclei. Finally, at temperatures $T \gg p_0^2/m$, the electron degeneracy is also lifted. For nondegenerate electrons formula (2) remains in force, as before. The scattering probability is in this case equal to

$$W_{bb} \sim \frac{e^4N}{p^4}\frac{p^3}{p^2/m}\ln\frac{p}{\varkappa} \sim \frac{e^4N}{T^{3/2}m^{1/2}}\ln\frac{\sqrt{mT}}{\varkappa}, \qquad (32)$$

$$\sigma_{bb} \sim \frac{T^{3/2}}{m^{1/2}e^2\ln(T^2m/e^2N)}. \qquad (33)$$

(we have substituted here $p \sim \sqrt{mT}$ and $\kappa \sim e\sqrt{N/T}$.) These results are well known and are cited here for completeness.

3. RELATIVISTIC ELECTRONS

In conclusion we note the following circumstance. In the entire calculation we have assumed that the electrons are nonrelativistic, that is, $p \gg mc$. This condition does not correspond very well to formula (27) [and consequently casts a shadow on (28) and (29)]. Indeed, according to (27),

$$p_0 \gg e^2M = (e^2M/cm)\,cm \gtrsim cm.$$

It is therefore of interest to ascertain the changes for the ultrarelativistic case, when momenta $p \gg mc$ are significant.

If the relativistic situation is due to large compression, then the corresponding momenta are obviously limited by the nuclear dimensions $p < 10^3\ \mathrm{cm}^{-1}$. This corresponds to energies $E/mc^2 \sim p/mc < 10^3$. But at these energies, processes like bremsstrahlung or pair production by electrons have relatively low probability. In view of this, the conductivity will be determined by the same collision processes which were considered previously. The only difference will be the replacement of the electron energy $p^2/2m$ by cp. Of course, all this may become incorrect at very high temperatures, but we shall not consider this case.

The general formula (2) for the conductivity remains the same, provided we write it in the form

$$\sigma \sim Ne^2 l/p, \qquad (2')$$

where $l \sim c\tau$ is the range and $p \sim p_0$ in the degenerate case or T/c in the Boltzmann case. In addition, we must take into consideration the fact that the velocity of sound is $u \sim \sqrt{cp_0/M}$ and $\kappa \sim p_0 e/\sqrt{c}$, so that $u\kappa \sim \omega_0$. As a result we find that the entire change in the formulas reduces to replacement of the electron mass by p/c.

In the solid phase which, as before, is realized when $T < e^2/p_0$, we have[3)]

$$\sigma'_{imp} \sim c^2N^{4/3}/e^2N_{imp}\ln(c/e^2), \qquad (3')$$

$$\sigma'_{e,H} \sim Nc^6/e^6T^2, \qquad (7')$$

$$\sigma'_{e,Z\geqslant 2} \sim Nc^5/e^4T^2, \qquad (11')$$

$$\sigma'_{ph,T\gg\omega_0} \sim N^{1/3}c^2/T, \qquad (15')$$

$$\sigma'_{ph,H,T\ll\omega_0} \sim (N^{1/6}M^{1/2}c^2/e)e^{\alpha\omega_0/T}, \qquad (18')$$

$$\sigma'_{ph,Z\geqslant 2,T\ll\omega_0} \sim N^{8/3}c^3e^2/M^2T^5. \qquad (25')$$

for the liquid phase we get

$$\sigma'_{dd} \sim N^{5/3}c^2/M^2e^2T^2\ln(c/e^2), \quad T \ll p_0^2/M, \qquad (29')$$

$$\sigma'_{db} \sim N^{1/3}c^2/e^2\ln(c/e^2), \quad p_0^2/M \ll T \ll cp_0, \qquad (31')$$

$$\sigma'_{bb} \sim cT/e^2\ln(T^3/e^2c^2N), \quad T \gg cp_0. \qquad (33')$$

---

3) As in the nonrelativistic case, (25') is valid only at the lowest temperatures. The complete formula for the conductivity in the case when $Z \geqslant 2$ and $T \ll \omega_0$ is of the form $\sigma = (\sigma'^{-1}_{(18')} + \sigma'^{-1}_{(25')})^{-1}$.

[1] A. A. Abrikosov, Astron. zhur. **31**, 112 (1954).

[2] A. A. Abrikosov, JETP **39**, 1797 (1960), Soviet Phys. JETP **12**, 1254 (1961).

[3] R. E. Peierls, Quantum Theory of Solids, Oxford, 1955.

[4] A. A. Abrikosov, JETP **41**, 569 (1961), Soviet Phys. JETP **14**, 408 (1962).

Translated by J. G. Adashko

Physics Vol. 2, No. 1, pp. 5-20, 1965. Physics Publishing Co. Printed in Great Britain.

ELECTRON SCATTERING ON MAGNETIC IMPURITIES IN METALS AND ANOMALOUS RESISTIVITY EFFECTS

A. A. ABRIKOSOV

Institute for Physical Problems of the Academy of Sciences of the USSR, Moscow USSR

(*Received* 17 *February* 1965)

Abstract

A special quantum field theory technique for a system of spins was used to evaluate the resistivity of metals containing paramagnetic impurities, assuming $J/\varepsilon_F \ll 1$, for an arbitrary value of $(J/\varepsilon_F)\ln(\varepsilon_F/T)$, where J is the exchange scattering amplitude and ε_F the Fermi energy. The first term of this series has been found earlier by Kondo [1]. It is shown that exchange and ordinary interactions give independent contributions to the resistivity. For a ferromagnetic sign of the exchange interaction between the electron and the impurity ($J > 0$), the exchange component of resistivity decreases with temperature and disappears at $T = 0$. In the reverse case ($J < 0$), the resistivity starts increasing when the temperature decreases. After going through a maximum (for $T = T_{max}$), where the exchange resistivity, due to a local impurity atom, is of the same order of magnitude as the usual resistivity, the exchange resistivity, even in this case, goes to zero at $T = 0$. Such a behavior is related to the resonant nature of the scattering amplitude for $J < 0$. The calculation assumes that the impurity spins are completely disordered, i.e., the temperature is higher than the Curie temperature of the impurity ferromagnetism. Since the latter is proportional to the concentration, while T_{max} does not depend on concentration, the results obtained are reasonable for sufficiently small concentration.

THE observation of a minimum in the resistivity of some metals as a function of temperature has attracted attention for a long time. A study of this effect has shown that it is probably related to the presence of impurity atoms with unfilled inner shells in the original non-magnetic metal. However, until recently there had been no satisfactory explanation of the interaction of electrons with such atoms and for the occurrence of a minimum. Such an explanation has been advanced by Kondo [1] who calculated the second order approximation (compared to Born's approximation) for the scattering probability of electrons by an impurity, caused by an exchange interaction J. It then turned out that in the scattering probability there is a correction of the order of $(J/\varepsilon_F)\ln(\varepsilon_F/\xi)$, where ξ is the electron energy measured from ε_F. This gives a correction to the resistivity of the order of $-(J/\varepsilon_F)\ln(\varepsilon_F/T)$, which results in an increase in resistivity with decreasing temperature for a negative sign of J (antiferromagnetic interaction between electrons and impurities).

Kondo's calculation is reasonable within two restrictions. First, the correction must be small. Second, the spins of the impurity atoms must not be correlated. In other words, the temperature must be above the Curie point for impurity ferromagnetism (or antiferromagnetism). It is well known (see for example Ref. 2) that the Curie point is proportional to the impurity concentration. Therefore if the concentration is small enough, the relative correction in resistivity, which is proportional to $(J/\varepsilon_F)\ \ln\ (\varepsilon_F/T)$ becomes of the order of or greater than unity even before the onset of spin ordering. In this case the perturbation theory is not usable and one must sum up the whole series. This calculation is the subject of this paper.

I. Techniques for Spins

The main difficulty in the calculation comes from the non-commutative nature of the spin operators for the impurities. Let us note that it is this very point which, according to Ref. 1, leads to logarithmic terms. This means that the exchange interaction Hamiltonian

$$H_{int} = -(J/N)\sum_n \psi_\alpha^+(\vec{r}_n)\vec{\sigma}_{\alpha\alpha'}\hat{\vec{S}}_n\psi_{\alpha'}(\vec{r}_n) \tag{1}$$

(we consider the interaction to be a δ-function) is in fact not equivalent to the interaction of the electron spin with an external magnetic field. On the other hand, the operator $\hat{\vec{S}}_n$ is not an ordinary field operator [3]. The fact is that every average of the type $\langle T\,[\hat{S}^i(t_1)\hat{S}^k(t_2), \ldots, \hat{S}^l(t_n)]\rangle$, where all the $\hat{S}^i$ refer to one atom, (i is the direction index x, y, z), $\langle \ldots \rangle = S_p \langle \ldots \rangle/(2S+1)$ cannot at all be represented by a sum of products of the type $\langle T\,[\hat{S}^i(t_1)\hat{S}^k(t_2)]\rangle\langle T\,[\hat{S}^l(t_3)\hat{S}^m(t_4)]\rangle \ldots + \ldots$. Because of this, in a calculation of effects where the non-commutation of the components of $\hat{\vec{S}}$ is essential a special technique is needed.

A number of such techniques were derived for spin $S = 1/2$ a few years ago by I.E. Dzyaloshinsky and the author (unpublished). Here, we use a variation of these which we were able to generalize to the case of an arbitrary spin.

Let us introduce operators for a virtual Fermi field $a_{n\beta}$, $a_{n\beta}^+$ corresponding to the spin operator $\hat{\vec{S}}_n$ according to the formula:

$$a^+_{n\beta}\vec{S}_{\beta\beta'}a_{n\beta'} = \hat{\vec{S}}_n \tag{2}$$

where the $S^i_{\beta\beta'}$ are spin matrices. Each index β and β' assumes $2S+1$ values: $-S, \ldots, S$. The operators $\hat{S}^i$, introduced according to (2), have the proper commutation relations when we assign the usual commutation relations to the operators $a_{n\beta}$:

$$\begin{aligned}\left\{a^+_{n\beta}a_{m\beta'}\right\} &= \delta_{\beta\beta'}\delta_{mn}\\ \left\{a_{n\beta}a_{m\beta'}\right\} &= 0\end{aligned} \tag{3}$$

However, they have one important inadequacy: The operators a_β introduce fictuous states. In fact, let us assume that spin $\vec{S}$ has a projection M on a chosen axis. There is a corresponding state φ_M for which $a_M\varphi_M \neq 0$, $a_{\beta\neq M}\varphi_M = 0$, i.e., there is only one "particle" corresponding to operator a_M and there are no "particles" corresponding to operators $a_{\beta \neq M}$. However, it is quite clear that by introducing operators a_β we have also brought into consideration non-physical states in which either all the occupation numbers are zero or more than one of them is equal to one. These states must be excluded.

In the case of spin $S = 1/2$, this happens automatically in the following way. All the physical quantities contain only the averages of products of operators $\hat{S}^i$. Other combinations of a_β are not involved. But the operators $\hat{S}^i$ themselves, for $S = 1/2$, have the property of giving zero when acting on the non-physical states (0, 0) or (1, 1). Thus, one can average over all the states. However, one must introduce a normalization. Since in a complete average, the result is divided by 4^n, while physically only 2^n states are relevant, where n is the number of different atoms participating in the average, the result must be multiplied by 2^n.

This procedure can be easily generalized to the case $S > 1/2$. The operators (2), in this case, give zero when acting on the state (0, 0 ... 0), but the result may be different from zero when they operate on the states where more than one occupation number is equal to 1. Because of this, we shall consider that each "particle" has an energy $\lambda \gg T$. Therefore, the greatest contribution will come from those states where $\sum_\beta N_\beta$ is minimum (i.e. 1). This procedure also requires a correct normalization so that the average of the form $(2S + 1)^{-1}S_p(\hat{S}^i\hat{S}^k, \ldots)$ (for one atom) has the correct value. The value calculated by our method must be multiplied by $e^{\lambda/T}/(2S + 1)$ and then we must let $\lambda \to \infty$. In the general case, the result must be multiplied by $[e^{\lambda/T}/(2S + 1)]^n$ where n is the number of participating impurity atoms.

But this normalization leads, in fact, to an essential inconvenience when the above technique is used. The fact is that, in contrast to the usual methods of field theory, the averages of the operators pertaining to a single atom and of the operators for different atoms do not have the same weight. This is very important in problems dealing with spin ordering. However, in the case considered here, each impurity atom scatters the electrons independently (this will be demonstrated later), and therefore one can consider a single atom. Then, the normalization causes no difficulty.

We use the finite temperature technique of field theory. The "free" Green's functions for the operators a_β (for one atom) have then the form:

$$\mathcal{G}_{\beta\beta'}(\tau - \tau') = -\langle T\,[a_\beta(\tau)a^+_{\beta'}(\tau')]\rangle$$

$$\mathcal{G}_{\beta\beta'}(\tau) = T\sum_\omega G_{\beta\beta'}(\omega)e^{-i\omega\tau} \tag{4}$$

$$\mathcal{G}_{\beta\beta'}(\omega) = \delta_{\beta\beta'}/(i\omega - \lambda), \qquad \omega = (2n + 1)\pi T$$

(here, the average is over all the states). On the diagrams, the $\mathcal{G}$ functions will be represented by dotted lines.

By way of example, let us calculate the simplest self energy diagram for the electron. Beside its illustrative value, this calculation will permit us to anticipate the most important portions of the complete analysis. Let us consider first the case of one impurity atom at point $\vec{r}$. The interaction Hamiltonian is given by formula (1). The diagram of the first non-vanishing approximation is shown in Fig. 1. It represents the expression:

$$\sum_{\alpha\alpha'}^{(1)}\left(\vec{p},\ \vec{p}',\ \omega\right) = -\frac{e^{\lambda/T}}{2S+1}\left(\frac{J}{N}\right)^2 Sp\left(S_i S_k\right)\left(\sigma_i\sigma_k\right)_{\alpha\alpha'}\ e^{i(\vec{p}-\vec{p}')\vec{r}}\ \times$$

$$\times\int d\vec{p}''/(2\pi)^3\ T^2\sum_{\omega_1\omega_2}\frac{1}{i(\omega+\omega_1-\omega_2)-\xi(\vec{p}'')}\times\frac{1}{i\omega_1-\lambda}\times\frac{1}{i\omega_2-\lambda},$$

where ξ is the electron energy with respect to the Fermi level.

FIGURE 1

Having summed over ω_1 and ω_2 we get in place of $T^2\sum\limits_{\omega_1\omega_2}$, the expression:

$$-\frac{1}{4}\frac{1}{i\omega_1-\xi(\vec{p}'')}\left(\tanh\xi/2T+\tanh\lambda/2T\right)\left(\coth\frac{\lambda+\xi}{2T}-\tanh\lambda/2T\right)$$

Assuming $\lambda \gg T$, we get simply $-e^{-\lambda/T}[i\omega-\xi(\vec{p}'')]^{-1}$. Averaging over the positions $\vec{r}$ of the impurity atom, and summing over all the impurity atoms, integrating over $\vec{p}''$ and discarding the constant part of Σ which gives only a small renormalization of the chemical potential, we get:

$$\sum_{\alpha\alpha'}\left(\vec{p},\ \vec{p}',\ \omega\right) = \delta_{\alpha\alpha'}\delta\left(\vec{p}-\vec{p}'\right)\left(2\pi\right)^3\sum\left(\omega\right)$$

$$\sum(\omega) = -(i/2\tau_0)\ \text{sign}\ \omega$$

$$\frac{1}{\tau_0} = (J/N)^2 S(S+1)\ N_i\ \frac{p_0 m}{\pi} \tag{5}$$

where N_i is the number of impurity atoms per unit volume.

Among the terms of further approximations to the self-energy there may be terms depending on only one impurity atom and terms depending on a great number of such atoms. Let us consider first of all the second order term, which depends only on a single impurity atom. There are two diagrams corresponding to this case (Fig. 2a and b). For diagram 2a, after averaging over the positions and summing over all the atoms, we have:

$$\Sigma^{(2a)}_{\alpha\alpha'}(\vec{p},\ \omega) = -\frac{e^{\lambda/T}}{2S+1}(J/N)^3 Sp(S_i S_k S_l)(\sigma_i \sigma_k \sigma_l)_{\alpha\alpha'} \times$$

$$\times T^3 \sum_{\omega_1\omega_2\omega_3} \int \frac{d\vec{p}\,'}{(2\pi)^3}\ \frac{d\vec{p}\,''}{(2\pi)^3}\ \frac{1}{i(\omega+\omega_3-\omega_1)-\xi(\vec{p}\,')} \times \frac{1}{i(\omega+\omega_3-\omega_2)-\xi(\vec{p}\,'')} \times$$

$$\times \frac{1}{i\omega_1-\lambda} \times \frac{1}{i\omega_2-\lambda} \times \frac{1}{i\omega_3-\lambda}$$

In this expression, one can easily see the rules for writing the formula corresponding to a diagram. (a) Each dotted line has its own frequency. The electron frequencies are determined by conservation rules. (b) The electron momentum on every internal line is independent, and one

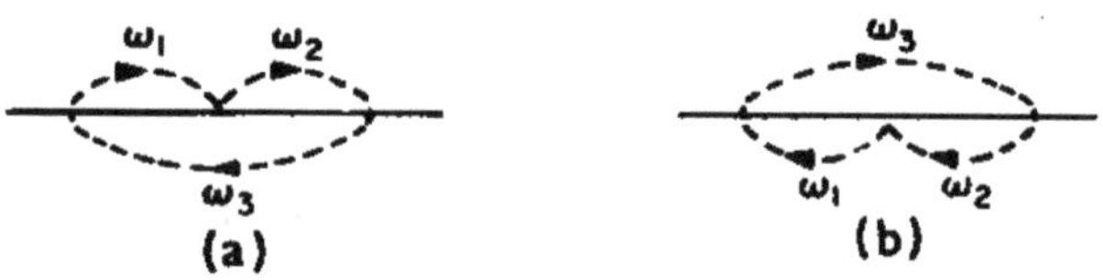

FIGURE 2.

should integrate over the momenta. (c) One takes the $\alpha\alpha'$ component of the product of electron spin operators in the order of their position on the electron line. (d) All the dotted lines corresponding to one atom form together a closed loop. A trace is taken of the product of the impurity spin operators in the order of their position along the loop.

Summing over ω_1, ω_2, and ω_3 and assuming $\lambda \gg T$, we get:

$$e^{-\lambda/T}\left[\frac{1}{e^{-\xi(\vec{p}\,'')/T}+1} \times \frac{1}{\xi(\vec{p}\,')-\xi(\vec{p}\,'')} \times \frac{1}{i\omega-\xi(\vec{p}\,')} + \right.$$

$$\left. + \frac{1}{e^{-\xi(\vec{p}\,')/T}+1} \times \frac{1}{\xi(\vec{p}\,'')-\xi(\vec{p}\,')} \times \frac{1}{i\omega-\xi(\vec{p}\,'')}\right]$$

These are two completely symmetrical expressions. One can readily see that the first term gives a logarithmic integral, for positive values of $\xi(\vec{p}\,'')$, in the region $T,\ \xi(\vec{p}\,') \ll \xi(\vec{p}\,'') \ll \epsilon_F$. In principle, the integral over $\vec{p}\,''$ extends over a larger range. In fact, its upper limit is a momentum of the order of the reciprocal radius of interaction; the latter can be considerably lower than $1/p_0$ for d or f orbitals. However, as will be shown in Section 3, this inconvenience disappears in a renormalization of the scattering amplitude J. As to the remaining integral, in that we can take $\xi(\vec{p}\,') \sim \omega$. Thus, in diagram 2$a$ one of the integrals over $\xi(\vec{p})$, can be logarithmic, but then the other one must necessarily be non-logarithmic. Since either integral may be logarithmic, two identical terms appear.

As a whole, we get, with logarithmic accuracy:

$$\sum^{(2a)} = \frac{i}{2\tau_0} \text{ sign } \omega \frac{2J}{N} \frac{p_0 m}{2\pi^2} \ln \frac{\varepsilon_F}{|\omega|}$$

In the same way we find that the Σ^{2b} corresponding to diagram (2*b*) is equal to Σ^{2a}. The result can be interpreted as a modification of the quantity $1/\tau_0$:

$$\frac{1}{\tau_0} \rightarrow \frac{1}{\tau_0}\left(1 - \frac{4J}{N} \frac{p_0 m}{2\pi^2} \ln \frac{\varepsilon_F}{|\omega|}\right) = \frac{1}{\tau_0}\left(1 - \frac{3zJ}{\varepsilon_F} \ln \frac{\varepsilon_F}{|\omega|}\right) \tag{6}$$

(z is the number of electrons per atom). This is in fact the result obtained by Kondo [1] by a different method.

Having analysed the above calculations, one can easily see that they can be described as follows: In each of the diagrams 2*a* and 2*b* we can draw two vertical sections through one electron line and two spin lines. Let us make such a section. In two summations and one integral corresponding to such a section, the essential values of the variables are of the order of ω. The remaining sum and integral are of logarithmic type, in which the essential values of the variables are:

$$|\omega| \ll |\omega_1| \sim |\xi_1| \ll \varepsilon_F.$$

In each diagram one must draw all the possible sections across three lines. The diagrams may be analysed in the following way. If one retains those terms which have the highest power of the log for a given value of J [these are the terms $J^{n+2}(\ln \varepsilon_F/|\omega|)^n$], then we must conclude that these are diagrams in which the integral along the section is taken for $|\xi|$, $|\omega_i| \sim |\omega|$, while all the other integrals are logarithmic and the important values of the variables are $|\xi_i|$, $|\omega_i| \gg |\omega|$. This situation strongly reminds one of the study of electron damping in a Fermi liquid by G.M. Eliashberg [4], and this is why we use his method.

However, before doing so, let us consider the diagrams of the next order in which not one, but two impurity atoms participate. Elementary diagrams are shown in Figs. 3*a* and 3*b*. If one

FIGURE 3

averages over impurity positions, one can readily see that the contribution of diagram 3*a* compared to diagram 1 is $1/\tau\varepsilon_F$), i.e. very small.

This pertains to all other diagrams with intersecting lines from various atoms. As to diagrams with internal inclusions (Type 3*b*), one can reason as follows. In accord with the foregoing, the role of the internal inclusion (see Ref. 5) consists in replacing $G = (i\omega - \xi)^{-1}$ by $G' = (i\omega + i/2\tau \text{ sign } \omega - \xi)^{-1}$. Let us introduce in diagram 3*b* the substitution $G' = G + (G' - G)$

and in the term with $G' - G$ we shall integrate again over ξ. This yields zero. This means that only the integral with G remains. This happens in all the diagrams with internal inclusions. Therefore, it is sufficient to consider only diagrams for a single impurity atom.

2. The Self-Energy Part

As already stated, we shall use the method of Reference 4. In any of the diagrams which may be of interest to us, we can make a section across three lines. The sums and integrals over these variables must be calculate accurately since the values of the arguments are quite small. As to the other portions of the diagram, their variables have quite large values, and, with logarithmic accuracy, we can replace all the sums over frequencies by integrals. Thus, we must calculate the full sum of diagrams having the form shown in Fig. 4, where the cross-hatched

FIGURE 4.

squares represent the sum of all "four tails" calculated with logarithmic accuracy. We shall designate this by the symbol:

$$\Gamma_{\alpha\beta,\alpha'\beta'}\left(\omega,\ \vec{p},\ \omega_1;\ \omega+\omega_1-\omega_2,\ \vec{q},\ \omega_2\right)$$

The following expression corresonds to the diagram of Fig. 4:

$$-\frac{e^{\lambda/T}}{2S+1} N_i T^2 \sum_{\omega_1\omega_2} \int \frac{d\vec{q}}{(2\pi)^3} \Gamma_{\alpha\beta,\alpha''\beta'}\left(\omega,\ \vec{p},\ \omega_1;\ \omega+\omega_1-\omega_2,\ \vec{q},\ \omega_2\right) \times \tag{7}$$

$$\times \Gamma_{\alpha''\beta',\alpha'\beta}\left(\omega+\omega_1-\omega_2,\ \vec{q},\ \omega_2;\ \omega,\ \vec{p},\ \omega_1\right) G\left(\omega+\omega_1-\omega_2,\ \vec{q}\right) \mathcal{G}(\omega_1)\ \mathcal{G}(\omega_2)$$

where $G(\omega,\ \vec{p}) = [i\omega - \xi(\vec{p})]^{-1}$, $\mathcal{G}(\omega) = (i\omega - \lambda)^{-1}$. We shall see later that Γ will depend only on $\ln \epsilon_F/|\omega|$. Therefore, we can make the transition from a summation over imaginary frequencies to an integration over real values and place ω on the real axis. From symmetry considerations, it follows that the spin dependence of Γ has the form:

$$\Gamma = \Gamma^{(0)} + \Gamma^{(\sigma)}(\vec{\sigma},\ \vec{S})$$

As in Reference (4) we have:

$$\frac{1}{2S+1} \Gamma_{\alpha\beta,\alpha''\beta'}(1,2;\ 3,4)\ \Gamma_{\alpha''\beta',\alpha'\beta}(3,4;\ 1,2) =$$

$$= \left[|\Gamma^{(0)}|^2 + S(S+1)|\ \Gamma^{(\sigma)}|^2\right] \delta_{\alpha\alpha'},$$

Having transformed th sum over the frequencies ω_1 and ω_2 as in reference (4) we get:

$$T^2 \sum_{\omega_1\omega_2} G(\omega + \omega_1 - \omega_2, \vec{q}) \, \mathscr{G}(\omega_1) \, \mathscr{G}(\omega_2) = \tag{8}$$

$$= \frac{i}{(2\pi)^2}\int Im\, \mathscr{G}_R(\omega_1)\, Im\, \mathscr{G}_R(\omega_2)\, Im\, G_R(\omega + \omega_1 - \omega_2) \times$$

$$\times \left(\tanh\frac{\omega_2}{2T} - \coth\frac{\omega_2 - \omega}{2T}\right)\left(\tanh\frac{\omega_1}{2T} - \tanh\frac{\omega + \omega_1 - \omega_2}{2T}\right) d\omega_1 d\omega_2$$

Here, we note that in the integrals small values of ω_1, ω_2 are essential (considering that $T \ll \lambda \ll \varepsilon_F$). Setting $Im\, G_R(\omega) = -\pi\delta(\omega - \xi)$, $Im\, \mathscr{G}_R(\omega) = -\pi\delta(\omega - \lambda)$ and taking into account the condition $\lambda \gg T$, we get:

$$\sum = -\, i \text{ sign } \omega\, N_i \frac{p_0 m}{2\pi}\left[\left|\Gamma^{(0)}(\omega, \lambda; \omega, \lambda)\right|^2 + S(S+1)\left|\Gamma^{(\sigma)}(\omega, \lambda; \omega, \lambda)\right|^2\right] \tag{9}$$

which differs from the first approximation (see eq. 5) by the substitution of the square bracket for $(J/N)^2 S(S+1)$.

Now let us examine Γ. As already mentioned, Γ can be calculated as if it were at $T = 0$, replacing all the summations over frequency by integrals. But these are still integrals along the imaginary axis. The author of this paper has demonstrated (unpublished) that all these integrals can be transformed to the real axis, together with all the external arguments. Thereupon, all the temperature Green functions may be replaced by time functions at $T = 0$ via the transformation $i\omega \to \omega + i\delta$ sign ω. Then we shall do the following. The integrals over ω_i involve only the frequencies on the spin lines. Since all these integrals are taken from $-\infty$ to $+\infty$, we can substitute $\omega - \lambda \to \omega$. Since at every vertex there is a spin line entering and one leaving, conservation laws are not violated. We must only make a transformation from $\omega - \lambda$ to ω in the external frequencies of the lines.

Since we must find $\Gamma(\omega, \lambda; \omega, \lambda)$, after all the transformations the task is reduced to a calculation of $\Gamma(\omega, 0; \omega, 0)$ in the usual method of time diagrams at $T = 0$, whereupon the G function becomes $(\omega - \xi + i\delta \text{ sign } \xi^{-1}$ and the $\mathscr{G}$ function $(\omega + i\delta)^{-1}$ since $\lambda > 0)$.

The first diagrams for the vertex part are shown in Fig. 5. We can express them as follows (corrections to J/N):

$$-\, i(J/N)^2(\vec{\sigma}\ \vec{S})^2 \frac{p_0 m}{2\pi^2}\int\frac{d\omega_1}{2\pi}\int d\xi_1 \frac{1}{\omega_1 + i\delta} \times \frac{1}{\omega - \omega_1 - \xi_1 + i\delta \text{ sign } \xi_1} =$$

$$= (J/N)^2 \frac{p_0 m}{2\pi^2}\left[S(S+1) - \vec{\sigma}\ \vec{S}\right]\left[\ln\frac{\varepsilon_F}{|\omega|} + i\pi\theta(\omega)\right] - \tag{10a}$$

$$-\, i(J/N)^2\ \sigma^i\sigma^k S^k S^i \frac{p_0 m}{2\pi^2}\int\frac{d\omega_1}{2\pi}\int d\xi_1 \frac{1}{\omega_1 + i\delta} \times \frac{1}{\omega + \omega_1 - \xi_1 + i\delta \text{ sign } \xi_1} =$$

$$= -\ (J/N)^2\ \frac{p_0 m}{2\pi^2}\left[S(S+1) + \vec{\sigma}\,\vec{S}\right]\left[\ln\frac{\varepsilon_F}{|\omega|} + i\pi\theta(-\omega)\right] \tag{10b}$$

where $\theta(x) = \begin{Bmatrix} 1, & x > 0 \\ 0, & x < 0 \end{Bmatrix}$ We left the non-logarithmic imaginary term here because it will be needed for further reasoning.

FIGURE 5

By examining higher order diagrams (Fig. 6) one can easily see that the diagrams with the highest power of the logarithm for each order are only those diagrams which can subsequently be reduced by two lines, an electron line and a spin line, in such a way that the result is a

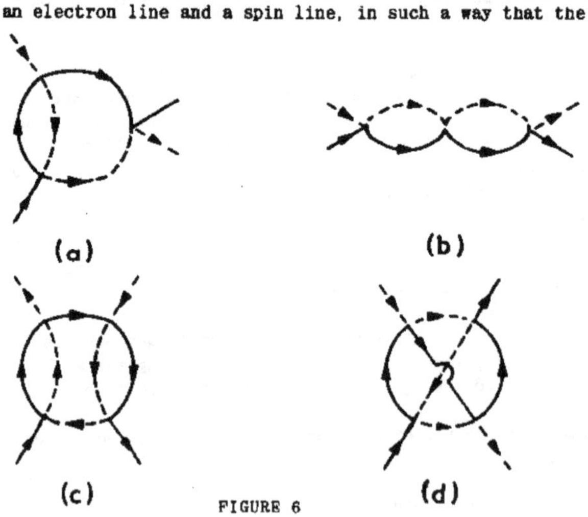

FIGURE 6

simple vertex. For example, by simplifying Fig. 6*c*, one can first obtain a diagram of the 6*a* type, then 6*b*, then a simple vertex. These properties are exhibited by so-called "parquet" diagrams 5*a* and *b*, 6*a*, *b*, *c* and are not exhibited by, for example, diagram 6*d*.

We make a summation of the "parquet" by the method of V.V. Sudakov [5]. Let us point out that there are two different elements of a "parquet", as shown in Fig. 5*a* and 5*b*. We can introduce two different summations of the diagrams as shown in Fig. 7*a* and 7*b*. Let us designate as Λ_1 the diagram which can be cut through two parallel lines (Fig. 7*a*) and Λ_2 the diagram which can be cut through two antiparallel lines (Fig. 7*b*). It is obvious that $\Gamma = \Gamma_0 + \Lambda_1 + \Lambda_2$, where Γ_0 is the original interaction $J/N(\vec{\sigma}\,\vec{S})$. Let us point out that in the logarithmic integration inside the internal Λ_1, Λ_2 of Fig. 7*a* and 7*b*, the essential values of arguments are larger than in the external lines, i.e. than in corresponding pairs of two parallel or two antiparallel lines.

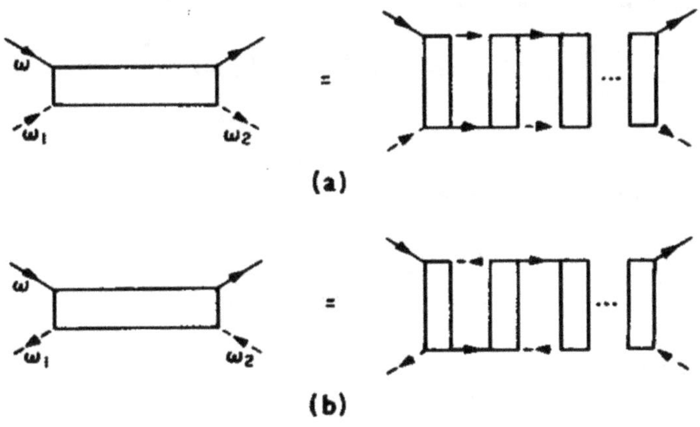

FIGURE 7

Let us examine now the diagram 7*a* in order to determine Λ_1. It is obvious that the lower limit for the logarithmic integrals in this diagram is ω (if the external $\omega_1 = \omega_2 = 0$). We shall select that internal dotted line in which ω_i is minimum. To the left and to the right of this element there can be any element of Γ, i.e. in other words, in the sum there is a full Γ both on the left and on the right sides. The quantity ω_i plays the role of ω for these Γ. Thus one can write (with logarithmic accuracy) the equation:

$$\Lambda_{2\;\alpha\beta,\alpha'\beta'}(\omega) = \frac{p_0 m}{2\pi^2}\int_{|\omega|}^{\epsilon_F}\frac{d\omega_1}{\omega_1}\,\Gamma_{\alpha\beta',\alpha''\beta''}(\omega_1)\,\Gamma_{\alpha''\beta'',\alpha'\beta}(\omega_1) \tag{11a}$$

and similarly:

$$\Lambda_{1\;\alpha\beta,\alpha'\beta'}(\omega) = \frac{p_0 m}{2\pi^2}\int_{|\omega|}^{\epsilon_F}\frac{d\omega_1}{\omega_1}\,\Gamma_{\alpha\beta'',\alpha''\beta'}(\omega_1)\,\Gamma_{\alpha''\beta,\alpha'\beta''}(\omega_1) \tag{11b}$$

Adding these expression we get:

$$\Gamma_{\alpha\beta,\alpha'\beta'}(\omega) = \frac{J}{N}(\vec{\sigma}\,\vec{S}) + \frac{p_0 m}{2\pi^2}\int_{|\omega|}^{\epsilon_F}\frac{d\omega_1}{\omega_1}\Big[\Gamma_{\alpha\beta,\alpha''\beta''}(\omega_1)\,\Gamma_{\alpha''\beta'',\alpha'\beta}(\omega_1) -$$

$$-\Gamma_{\alpha\beta'',\alpha''\beta'}(\omega_1)\,\Gamma_{\alpha''\beta,\alpha'\beta''}(\omega_1)\Big] \tag{12}$$

Let us introduce the transformation $\ln \epsilon_F/\omega = x$ and let us look for a solution in the form $\Gamma = \Gamma^{(0)} + (\vec{\sigma}\,\vec{S})\Gamma^{(\sigma)}$ Then we get:

$$\Gamma^{(\sigma)}(x) = \frac{J}{N} - \frac{p_0 m}{\pi^2}\int_0^x dy\left[\Gamma^{(\sigma)}(y)\right]^2$$

$$\Gamma^{(0)} = 0 \tag{13}$$

Solving the integral equation, we obtain:

$$\Gamma = \frac{(J/N)\,\vec{\sigma}\,\vec{S}}{1 + \frac{J}{N}\frac{p_0 m}{\pi^2}\ln\frac{\epsilon_F}{|\omega|}} \tag{14}$$

Substituting into (9), we get:

$$\sum = -\frac{i}{2\tau_0}\,\mathrm{sign}\,\omega\left[1 + \frac{3Jz}{2\epsilon_F}\ln\frac{\epsilon_F}{|\omega|}\right]^{-2} \tag{15}$$

Expressions (14) and (15) have an important property. If $J < 0$, for some value of $|\omega|$ these expressions have a pole. The presence of such a pole indicates that there is a resonance in the scattering. When taking into account the final resonance width one must add to the logarithmic term in the denominator of formula (14) an imaginary component proportional in $i\frac{J}{N}\frac{p_0 m}{\pi^2}$. This can already be seen from formula (10). Our approximation is not applicable for finding such terms. Hence, we shall use expressions (14) and (15) far from the pole; but in the vicinity of the pole we shall evaluate only orders of magnitude, on the basis that a term proportional to $i\pi$ is added to $\ln \epsilon_F/|\omega|$ in the denominator.

3. Ordinary Interaction. Transition to the Scattering Amplitude

Until now we have limited ourselves to the study of the exchange interaction. However, the interaction of the electron with the impurity atom involves necessarily the usual non-exchange term which is of considerably larger magnitude. Naturally, this raises the question of possible interference effects.

The usual interaction, in the absence of exchange interaction, can be considered as an external field, and it is designated by a cross on the electron line.

First of all let us mention that, for the same reasons as earlier, it is sufficient to examine the interaction with only one atom. Now let us assume that in diagram 5*a* we have placed on the electron line a number of crosses (Fig. 8). Considering for simplicity's sake that the interaction is isotropic, we get

$$-\frac{i}{2\pi}\left(\frac{J}{N}\right)^2\left[S(S+1)-\vec{\sigma}\,\vec{S}\right]\left(\frac{p_0 m}{2\pi^2}\right)^2\varphi\int d\omega_1 d\xi_1 d\xi_2\,\frac{1}{\omega_1+i\delta}\,\frac{1}{\omega-\omega_1-\xi_1+i\delta\,\mathrm{sign}\,\xi_1}\times$$

$$\times\frac{1}{\omega-\omega_1-\xi_2+i\delta\,\mathrm{sign}\,\xi_2}$$

where φ designates the whole insert from the first to the last cross. An integration will convince us that the logarithmic expression is not obtained in this case. Taking into account the fact that the sum of all the possible φ has a magnitude ε_F/N, we find that the diagram having the insert is $(\ln\varepsilon_F/\omega)^{-1}$ times smaller than the diagram without insert.

However, a cross can be placed also on that electron line which does not participate in logarithmic integrals. The simplest is to place crosses in the very first diagram 1*a*. Then, one can be convinced that the diagram with the insert has a magnitude ω/ε_F. From this it follows that the usual interaction gives a separate component in S and cannot interfere with the exchange part. The usual part has been studied in a whole series of papers (see for example, ref. 3, § 39.2). It has the form $-i/2\tau_{ord}$ sign ω, where $\dfrac{1}{\tau_{ord}} = N_i\,\dfrac{mp_0}{\pi}\,|a|^2$; a is the scattering amplitude corresponding to the non-exchange interaction.

Now, let us examine the second relevant problem. In all of the foregoing calculations, the integrals over ξ were taken in the range $-\varepsilon_F<\xi<\varepsilon_F$. As a matter of fact, these integrals are limited, for positive values of ξ only by the condition $|\vec{p}|<1/r_{int}$ where r_{int} is the interaction radius, which may be much smaller than $1/p_0$. Therefore a renormalization of the interactions must be made.

Instead of the Born scattering amplitude, we shall introduce the true amplitude of scattering of an electron by an atom. If $\hat{f}$ is the Born amplitude and $\hat{f}_1$ is the actual amplitude, then by analogy of ref. 3 § 25.4 we have:

$$\hat{f}=\hat{f}_1-\hat{f}\hat{f}_1\frac{p_0 m}{2\pi^1}\int\frac{d\xi_1}{\xi-\xi_1+i\delta}\tag{16}$$

Until now, the role of $\hat{f}$ was played by the quantity $-\frac{J}{N}\vec{\sigma}\,\vec{S}$. But now we shall take $\hat{f}=-\frac{J}{N}\vec{\sigma}\,\vec{S}+f_0$ where f_0 does not depend on spins, and we shall choose f_0 in such a way that equation (16) will be satisfied for $\hat{f}_1=-\frac{J_1}{N}(\vec{\sigma}\,\vec{S})$ This means that in the interaction Hamiltonian we have included a part of the usual interaction term. The diagram-method does not suffer from this since the operator $f_0 a_\beta^{+}a_\beta$ also gives zero when it operates on the state with zero occupation numbers, while the other improper states are excluded by our procedure. This situation, generally speaking, permits a study of the usual interaction together with the exchange interaction. However, since the usual term, except for the renormalization, makes no contribution to the logarithmic integral (inserting the usual interaction in integrals 10*a* and 10*b* gives a mutual cancellation), we have considered it separately. This has the advantage that the usual interaction U is expected to be stronger than the exchange interaction, and therefore a simultaneous study would have imposed an unnecessary limitation: $J/N \ln \varepsilon_F/T \gg U$. The inserted term f_0, as we shall soon see, has a magnitude of $(J/N)(J/\varepsilon_F)$ and does not impose new limitations.

FIGURE 8

Solving equation (16) to second order accuracy we have:

$$\hat{f} = \frac{J_1}{N}\vec{\sigma}\vec{S} - \left(\frac{J_1}{N}\right)^2 \frac{p_0 m}{2\pi^2}\left[S(S+1) - \vec{\sigma}\vec{S}\right]\int \frac{d\xi_1}{\xi - \xi_1 + i\delta} \tag{17}$$

When we find second order terms (formula 10), we renormalize in the following way. We replace J by J_1 in (10); furthermore, in first order terms, at the pole in Γ we change to J_1/N using (17). If the integration over ξ is not performed, we get in the sum:

$$\left(\frac{J_1}{N}\right)^2 \left[S(S+1) - \vec{\sigma}\vec{S}\right]\frac{p_0 m}{2\pi^2}\int d\xi_1\left[\frac{\theta(\xi_1)}{\xi_1 - \omega - i\delta} - \frac{1}{\xi_1 - \xi + i\delta}\right] \tag{10a'}$$

Later on we shall be convinced that to find the conductivity, we shall need the case $\xi \ll \varepsilon_F$. In this case we obtain with logarithmic accuracy:

$$-\left(\frac{J_1}{N}\right)^2 \left[S(S+1) - \vec{\sigma}\vec{S}\right]\frac{p_0 m}{2\pi^2}\int d\xi_1 \frac{\theta(-\xi_1)}{\xi_1 - \omega} = \left(\frac{J_1}{N}\right)^2 \frac{p_0 m}{2\pi^2}\left[S(S+1) - \vec{\sigma}\vec{S}\right]\ln\frac{\varepsilon_F}{|\omega|}$$

Since now the integration is performed over negative $\xi_1(|p| < p_0)$, there is no question about the range of integration.

This is the situation for the lower approximation. We shall show that such a renormalization can also be performed for higher orders of approximation. It is easiest to do so by going from the n-th order to the order $n+1$. This transition can be accomplished by replacing in turn each simple vertex by the sum of the two vertices shown in Fig. 5a and 5b.

If in the n-th order all the vertices contained J_1/N, then in the new diagrams of order $n+1$ there will be a J_1/N everywhere after this transformation. Furthermore, we must perfect the n-th order diagram by substituting in turn instead of J_1/N the expression (17). In other words, such a transition from n to $n+1$ order of approximation changes the integrals of type (10a) into those of type (10a). If in addition, we take into account that in all the logarithmic integrals $\omega \sim \xi$, then it practically amounts to having a J_1/N in all vertices and all the integrals are taken over $\xi < 0$, i.e., one can take ε_F as the range for integrals over ξ.

4. Resistivity

As we know (see for example ref. (3), § 39.3), the relation between current and vector potential, which is frequency dependent, has the following form in the absence of spatial dispersion:

$$J_i(\omega) = Q_{ik}(\omega)A_k(\omega) \tag{18}$$

'here, in the case of isotropic scattering, $Q(\omega)$ represents an analytic continuation to real ω of the expression:

$$\frac{Ne^2}{mc}\delta_{ik} + \frac{2e^2}{(2\pi)^3m^2c}T\sum_{\omega'}\int d\vec{p}\ p_ip_kG(\vec{p},\ \omega')G(\vec{p},\ \omega' - \omega) \tag{19}$$

where G are temperature Green functions. Let us change from a summation over ω' to an integration over real frequencies (see Ref. (3), § 21.2); then, considering $\omega > 0$ in (19), we can continue this expression analytically to real values. We obtain then:

$$\frac{Ne^2}{mc}\delta_{ik} + \frac{e^2}{\pi(2\pi)^3m^2c}\int_{-\infty}^{\infty}d\omega'\int d\vec{p}\ p_ip_k\ Im\ G_R(\vec{p},\ \omega')\ \times$$
$$\times\left[G_A(\vec{p},\ \omega' - \omega) + G_R(\vec{p},\ \omega' + \omega)\right]\tanh\frac{\omega'}{2T} \tag{20}$$

In view of the fact that the G functions do not depend on the direction of $\vec{p}$ and that the integrals over $\vec{p}$ are taken in the neighborhood of the Fermi surface, we can make the substitution:

$$\int\frac{d\vec{p}}{(2\pi)^3}p_ip_k \rightarrow \frac{1}{3}\delta_{ik}\frac{p_0{}^3m}{(2\pi)^2}\int d\xi$$

The integration must first be performed over ω and then over ξ. However, as always [3], we shall use the following approach. We shall add and subtract from the expression under the integral the "free" term, i.e. the term corresponding to no interaction with impurities. Then, in the integral with the "free" expression, we shall first integrate over the frequency but, in the convergent term with the difference, we shall integrate first over ξ.

Instead of G_R we shall insert the expression ($G_A = G_R{}^*$):

$$\left(\omega - \xi + \frac{i}{2\tau(\omega)}\right)^{-1} \tag{21}$$

where:

$$\frac{1}{\tau(\omega)} = \frac{1}{\tau_{ord}} + \frac{1}{\tau_{ex}(\omega)} \tag{22}$$

where τ_{ord} is the non-exchange collision time, and $1/\tau_{ex}(\omega)$ corresponds to the expression (15). After all the operations, assuming $\omega \to 0$, we get:

$$\vec{j} = \frac{i\sigma\omega}{c}\vec{A}$$

where

$$\sigma = \frac{Ne^2}{m}\int_0^\infty \frac{d\omega\tau(\omega)}{2T\cosh^2(\omega/2T)}$$

With logarithmic accuracy, we obtain the resistivity:

$$\rho = \rho_{ord} + \rho_{ex}$$

where

$$\rho_{ex} = \rho_{0,ex}\left[1 + \frac{3J_1 z}{2\varepsilon_F}\ln\frac{\varepsilon_F}{T}\right]^{-2} \tag{23}$$

and where

$$\rho_{0,ex} = \frac{3\pi m J_1{}^2 S(S+1)c}{2N\varepsilon_F e^2 h} \tag{24}$$

and c is the atomic concentration.

If $J_1 > 0$, ρ decreases with decreasing temperatures and becomes zero when $T \to 0$. In the case of $J_1 < 0$, expression (23) becomes infinite when

$$T_{max} = q\,\varepsilon_F \exp\left[-\frac{2\,\varepsilon_F}{3|J_1|z}\right] \tag{25}$$

($q \sim 1$). According to the considerations of Section 2, in the vicinity of such a point the resistivity goes through a maximum. The maximum value is

$$\rho_{ex} \sim \rho_{0,ex}\left(\frac{\varepsilon_F}{J_1}\right)^2$$

According to (24), this expression is of the same order of magnitude as the ordinary resistivity due to the same impurities. The linewidth of the temperature dependent peak is $\Delta T \sim T_0$. With further decrease in temperature $\rho_{ex} \to 0$.

We have not cleared up the effect of ferromagnetic ordering. However, it probably liquidates the effect we have analysed. Since the ferromagnetic transition temperature $T_c \sim cJ_1{}^2/\varepsilon_F$ (see Ref. 2), where c is the impurity concentration, in principle the peak can be obtained with any combination of metals with $J_1 < 0$ and for a sufficiently small impurity concentration.

In a future publication we shall consider the effect of impurity ferromagnetism and of external magnetic field.

I am grateful to I.E. Dzyaloshinski for numerous useful discussions.

REFERENCES

1. J. KONDO, *Prog. Theor. Phys.* 32, 37 (1964).

2. A.A. ABRIKOSOV and L.P. GORKOV, *JETP* 43, 2230 (1962).

3. A.A. ABRIKOSOV, L.P. GORKOV and I.E. DZYALOSHINSKI, *Quantum Field Theory in Statistical Physics*, Prentice Hall (1963).

4. G.M. ELIASHBERG, *JETP* 42, 1658 (1962).

5. V.V. SUDAKOV, *Dokl. Akad. Nauk SSSR* 111, 338 (1956).

SOVIET PHYSICS JETP VOLUME 29, NUMBER 4 OCTOBER, 1969

GALVANOMAGNETIC PHENOMENA IN METALS IN THE QUANTUM LIMIT

A. A. ABRIKOSOV

Institute of Technical Physics, U.S.S.R. Academy of Sciences

Submitted November 20, 1968

Zh. Eksp. Teor. Fiz. 56, 1391–1406 (April, 1969)

The conductivity tensor in a strong magnetic field is determined for the case when magnetic splitting is of the order or more than the Fermi energy. The case of T = 0 and neutral point impurities is considered. Two models are analyzed, the first with a "good" metal with a quadratic isotropic dispersion law, and the second with a semimetal with isotropic and quadratic spectra of electrons and holes separated by a large momentum in momentum space. The calculation method, which makes use of Green's function expanded in the eigenfunctions of the problem, considerably simplifies the calculations for the first model, which has already been considered in previous papers, and refines the results. The conductivity tensor for the second model is found for the limiting quantum case in which, owing to level overlapping, the number of electrons and holes becomes much greater than in the absence of a field. Other possible applications of the calculation method are discussed.

1. INTRODUCTION

As is well known, the galvanomagnetic properties of metals, even in very strong fields, are well described by the quasiclassical approximation, in which the electrons can be regarded as moving in their classical orbits. This is connected with the fact that usually the distance between the Landau levels μ^*H is much smaller than the Fermi energy ϵ_F. The fields required to violate this relation at $\epsilon_F \sim 1$ eV and $\mu^* \sim \mu_{free}$ are of the order of 10^8 Oe, which at any rate are presently not attainable under laboratory conditions. When $\mu^*H \ll \epsilon_F$, allowance for the quantum effects yields all the small oscillatory additions to the conductivity tensor.

The relation $\mu^*H \ll \epsilon_F$ can be violated in semimetals, where, first, the Fermi energy is of the order of several hundred eV, and on the other hand the small effective masses give large values of μ^*. As a result, fields on the order of 10^5 Oe may suffice. In principle, this is the case also for semiconductors. However, inasmuch as the presence of free carriers is ensured in semiconductors by the presence of impurities, the number of carriers is approximately equal to the number of impurities, the condition $\hbar/\tau \ll \epsilon_F$ is violated in semiconductors (τ—free path time), leading to additional difficulties. Such a case will not be considered here.

To study the main features of the phenomenon, we confine ourselves to two simple models. The first is a metal with a dispersion law $p^2/2m_1$, with m_1 of the order of the electron mass and with the usual electron density ($n_e \sim 10^{22}$ cm^{-3}). This model will help us develop a calculation method in simple form, and to investigate the role of various contributions to the effect.

The magnetic field is assumed arbitrary. In principle, this case was investigated earlier.[1,2] However, the method which is employed here is much simpler and is more general. This has made it possible to refine the results obtained in [1,2].

The second model is a simplified variant of a semimetal, which, insofar as we know, was never investigated in the quantum limit.

The calculation method is based on expanding the Green's function of the electron in terms of the eigenfunctions in the magnetic field. This method was used by Bychkov[3] to investigate the role of collisions in the de Haas–van Alphen effect. We shall show here that it is also very convenient for the calculation of kinetic phenomena.

2. ISOTROPIC MODEL OF METAL

Let us consider a metal with a dispersion law $\epsilon = p^2/2m_1$, an electron density n_e, and an impurity density N_i. If the magnetic field is directed along the z axis, then the electron energy is described by the formula

$$\varepsilon_{n\pm}(p_z) = \frac{eH}{m_1c}\left(n+\frac{1}{2}\right) \mp \frac{eH}{2mc} + \frac{p_z^2}{2m_1} \tag{1}$$

(we put here $\hbar = 1$). For each value of n and for each index "plus" or "minus" there corresponds a separate Fermi momentum $p_{0n\pm}$, determined from the conditions

$$p_{0n\pm} = \left[\left(\zeta - \frac{eH}{m_1c}\left(n+\frac{1}{2}\right) \pm \frac{eH}{2mc}\right)2m_1\right]^{1/2}, \tag{2}$$

$$n_e = \sum p_{0n\pm}\frac{eH}{2\pi^2c}, \tag{3}$$

where ζ is the chemical potential. It is defined by condition (3), in which account is taken of the degeneracy of the Landau levels. The sum is over all the real $p_{0n\pm}$.

If we assume the interaction of the electrons with the impurity to be independent of the electron spin, then the projection of the electron spin does not change upon scattering, and we can consider the corresponding currents produced by electrons with spin projections $\pm 1/2$ independently. We therefore consider, say, electrons with spin projection $+1/2$. If a gauge $A_y = Hx$ is chosen, the eigenfunctions take the form

$$\psi_{p_zp_yn} = \exp\{ip_zz + ip_yy\}\psi_n\left(x - \frac{cp_y}{eH}\right), \tag{4}$$

where

$$\psi_n(x) = \left(\frac{eH}{\pi c}\right)^{1/4}\frac{1}{2^{n/2}\sqrt{n!}}\exp\left\{-\frac{eH}{2c}x^2\right\}H_n\left(\sqrt{\frac{eH}{c}}x\right)$$

(H_n is a Hermite polynomial).

With the aid of the functions $\psi_{p_z p_y n}$ we can easily write down the Green's function of the free electrons (without interaction with the impurity) at T = 0

$$G^{(0)}(p_z, p_y, x, x' \omega) = \sum_n \frac{\psi_n(x - cp_y/eH)\psi_n(x' - cp_y/eH)}{\omega - \xi_n(p_z) + i\delta \operatorname{sign}\omega}, \qquad (5)$$

where

$$\xi_n(p_z) = \varepsilon_{n+}(p_z) - \zeta.$$

We now consider electron scattering, assuming the interaction to be pointlike and the scattering amplitude to be isotropic (U_0).[1)] Averaging over the random distribution of the impurities, we have in first order in U_0

$$\delta^{(1)}G = \int G^{(0)}(x, x_1) G^{(0)}(x_1, x') dx_1 U_0 N_i$$
$$= U_0 N_i \sum_n \frac{\psi_n(x - cp_y/eH)\psi_n(x' - cp_y/eH)}{[\omega - \xi_n(p_z) + i\delta \operatorname{sign}\omega]^2}$$

(we use here the fact that the ψ_n are orthonormal). This correction corresponds to diagram a of Fig. 1.

We now consider the next-order correction, diagram b of Fig. 1, and assume that both crosses pertain to one atom. We then obtain after averaging over the impurity positions

$$\delta^{(2)}G = N_i U_0^2 \int G^{(0)}(p_z, p_y, x, x_1) G^{(0)}(p_{z1}, p_{y1}, x_1, x_1)$$
$$\cdot G^{(0)}(p_z, p_y, x_1, x') \frac{dp_{y1}}{2\pi}\frac{dp_{z1}}{2\pi} dx_1.$$

Since the ψ_n in the second G-function contain identical arguments, we obtain the factor $eH/2\pi c$ by integrating over p_{y1} and using the normalization condition. We are then left with the expression

$$\sum_n [\omega - \xi_n(p_{z1}) + i\delta \operatorname{sign}\omega]^{-1}.$$

Integration with respect to p_{z1} under the condition $\omega \ll p_{n_0}^2/2m_1$ yields $-im_1 \operatorname{sign}\omega \sum_n p_{0n}^{-1}$. Then, integrating with respect to x_1, we obtain

$$\delta^{(2)}G = -i \operatorname{sign}\omega N_i U_0^2 \frac{eHm_1}{2\pi c} \sum_{n_1} \frac{1}{p_{0n_1}}$$
$$\times \sum_n \frac{\psi_n(x - cp_y/eH)\psi_n(x' - cp_y/eH)}{[\omega - \xi_n(p_z) + i\delta \operatorname{sign}\omega]^2}$$

Thus, the correction $\delta^{(2)}G$, relative to $\delta^{(1)}G$, is of the order of $U_0 eHm_1/cp_0$. If we recognize that $U_0 \sim (p_0^2/m) \times (1/p_0^3)$ (U_0 is the Born scattering amplitude with dimension erg-cm^3), then the ratio $\delta^{(2)}G/\delta^{(1)}G$ is of the order of $(eH/mc)(p_0^2/m)$. We consider the case when this ratio is of the order of or larger than unity. Therefore all the corrections of this kind should be taken into account.

The sum of all the diagrams of type c in Fig. 1 yields

$$\delta G = \Sigma \sum_n \frac{\psi_n(x - cp_y/eH)\psi_n(x' - cp_y/eH)}{[\omega - \xi_n(p_z) + i\delta \operatorname{sign}\omega]^2},$$

where

$$\Sigma = N_i U_0 \left[1 + i \operatorname{sign}\omega\, U_0 \frac{eHm_1}{2\pi c} \sum_n \frac{1}{p_{0n}}\right]^{-1}. \qquad (6)$$

[1)]Such a description of the interaction of the electrons with the impurity in a superstrong magnetic field is in fact far from obvious, and should be regarded as a very simple model assumption.

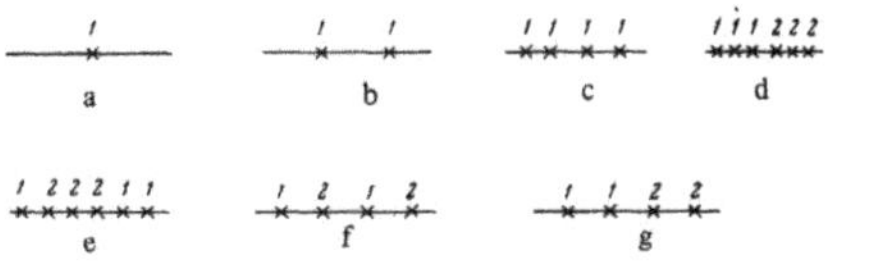

FIG. 1

If we now take into account the diagrams of type d in Fig. 1, then it is easy to see that we obtain as the sum

$$G = \sum_n \frac{\psi_n(x - cp_y/eH)\psi_n(x' - cp_y/eH)}{\omega - \xi_n(p_z) - \Sigma}. \qquad (7)$$

We can choose for Σ only its imaginary part, since the real part is a renormalization of the chemical potential. If $\operatorname{Im}\Sigma \ll p_0^2/2m$, then the integral with respect to p_z of the denominator, including Σ, does not differ from the case when we had $i\delta \operatorname{sign}\omega$ in lieu of Σ. In view of this, expression (6) actually takes into account also diagrams of type E in Fig. 1, where all the crosses from atom 2 fall between the two nearest crosses of atom 1.

We have not yet taken into account the diagrams with meshing of the crosses from different atoms. Let us consider the simplest of these diagrams, f of Fig. 1. It contains five Green's functions and integration with respect to two coordinates x, two coordinates p_y, and two coordinates p_z. Since all the $\psi_n(x - cp_y/eH)$ are exponentially decreasing functions, it follows that, as a result of the normalization condition, the integration with respect to x_1 and x_2 annihilates four of these functions, and the integration with respect to p_{y1} and p_{y2} annihilates four more, with a factor $(eH/c)^2$ appearing as a result. We are left with the integrals of the denominators with respect to p_z. These integrals can be readily evaluated. As a result we find that this diagram is in order $\omega/(p_0^2/m)$ relative to diagram g of Fig. 1 with non-meshing crosses. Since we shall henceforth be interested throughout in ω of the order of τ^{-1}, where $1/\tau = -2\operatorname{Im}\Sigma$, and Σ has been defined earlier, we can neglect diagrams with meshing if $\tau p_0^2/m \gg 1$. This will serve as a condition for the applicability of the present results (with respect to limitation on the impurity density).

We now proceed to calculate the conductivity tensor. We describe the external electric field by a vector potential $\mathbf{A}_1(\mathbf{r}, t)$, and we denote the vector potential of the large static magnetic field by $\mathbf{A}_0(\mathbf{r})$. The current operator is of the form

$$\tilde{\mathbf{j}} = -\frac{ie}{2m_1}(\nabla_r - \nabla_{r'})\tilde{\psi}^+(r', t)\tilde{\psi}(r, t)\Big|_{r' \to r} - \frac{e^2}{m_1 c}\tilde{\psi}^+\tilde{\psi}(\mathbf{A}_0 + \mathbf{A}_1). \qquad (8)$$

This expression must be averaged and, since there is no current under equilibrium, it is necessary to separate the terms linear in $\mathbf{A}_1$. We shall use the formula

$$\delta\mathscr{H}_A = -\frac{1}{c}\int \mathbf{j}\delta\mathbf{A}\, dV.$$

In this case, the role of $\delta\mathbf{A}$ is played by $\mathbf{A}_1$. Expressing

$$\tilde{\mathbf{j}} = \exp\left(i\int_{-\infty}^{t} \delta H_A dt\right)\mathbf{j}\exp\left(-i\int_{-\infty}^{t} \delta H_A dt\right)$$

and separating terms linear in $\mathbf{A}_1$, we obtain

$$j_i(\mathbf{r}, t) = \frac{ie^2}{m_1^2 c} \int dt_1\, dr_1 \left[\frac{i}{2}\left(\frac{\partial}{\partial x_i} - \frac{\partial}{\partial x_i'}\right)_{\mathbf{r}'\to\mathbf{r}} + \frac{e}{c} A_0^i(\mathbf{r})\right]$$
$$\times\left[\frac{i}{2}\left(\frac{\partial}{\partial x_{1k}} - \frac{\partial}{\partial x_{1k}'}\right)_{\mathbf{r}_1'\to\mathbf{r}_1} + \frac{e}{c} A_0^k(\mathbf{r}_1)\right] \langle[\psi^+(\mathbf{r}'t)\psi(\mathbf{r}t), \psi^+(\mathbf{r}_1't_1)\psi(\mathbf{r}_1t_1)]$$
$$\times\Theta(t - t_1)\rangle A_1^k(\mathbf{r}_1t_1) - \frac{e^2}{m_1 c} A_1^i n_e \qquad (9)$$

(where $\Theta(x) = 1$ when $x > 0$ and $\Theta(x) = 0$ when $x < 0$).

As is well known, the mean value of a retarded commutator can be expressed in terms of the time-ordered mean value, by simply reversing the sign of the imaginary part of the Fourier component with respect to $t - t_1$ when $\omega < 0$. In view of this we define in lieu of (9) the j_i^t with time-ordered mean value. It is easy to go over in it to Green's function, and in place of $\langle \ldots \rangle$ in (9) we obtain $G(\mathbf{r}, \mathbf{r}_1', t - t_1)G(\mathbf{r}, \mathbf{r}', t_1 - t)$. As usual, this product must be averaged over the positions of the impurity centers.

We begin with the component σ_{zz}. We consider first diagrams in which different impurity atoms enter in both G-functions. Then the mean value of the product is equal to the product of the mean values, and changing over to Fourier components we obtain

$$j_{0z}^t(k_y, k_z, x, \omega_0) = \frac{ie^2}{m_1^2 c} \int p_z^2 G\left(p_y - \frac{k_y}{2}, p_z - \frac{k_z}{2}, x, x_1, \omega\right)$$
$$\times G\left(p_y + \frac{k_y}{2}, p_z + \frac{k_z}{2}, x_1, x, \omega + \omega_0\right) dx_1\, dp_y\, dp_z\, d\omega/(2\pi)^3$$
$$\times A_1^z(k_y, k_z, x_1, \omega_0) - \frac{n_e e^2}{m_1 c} A_1^i(k_y, k_z, x, \omega_0). \qquad (10)$$

This corresponds to a diagram a of Fig. 2. This is the general expression, and we are interested in the static tensor σ_{ik}. We therefore assume the potential $\mathbf{A}_1$ to be independent of x, and put k = 0, assuming ω_0 to be small.

We substitute in the integral (10) the expression (7). Integration with respect to x_1 and p_y yields, as a result of the normalization of the functions ψ_n,

$$j_{0z} = \frac{ie^2}{m_1^2 c} \frac{eH}{2\pi c} \int dp_z\, d\omega/(2\pi)^2 p_z^2 \left[\sum_n (\omega - \xi_n - \Sigma(\omega))^{-1}\right.$$
$$\left.\times(\omega + \omega_0 - \xi_n + \Sigma(\omega + \omega_0))^{-1}\right] A_1^z(\omega_0) - \frac{n_e e^2}{m_1 c} A_1^z(\omega_0). \qquad (11)$$

As already mentioned, it is sufficient to take into account only the imaginary part of Σ. Putting $\operatorname{Im}\Sigma = -(i/2\tau)\operatorname{sign}\omega$, and assuming that $\omega_0 > 0$, we write the integral with respect to ω in the form

$$\int_{-\infty}^{-\omega_0}\left[\left(\omega - \xi_n - \frac{i}{2\tau}\right)\left(\omega + \omega_0 - \xi_n - \frac{i}{2\tau}\right)\right]^{-1} d\omega$$
$$+ \int_0^{\infty}\left[\left(\omega - \xi_n + \frac{i}{2\tau}\right)\left(\omega + \omega_0 - \xi_n + \frac{i}{2\tau}\right)\right]^{-1} d\omega$$
$$+ \int_{-\omega_0}^{0}\left[\left(\omega - \xi_n + \frac{i}{2\tau}\right)\left(\omega + \omega_0 - \xi_n + \frac{i}{2\tau}\right)\right]^{-1} d\omega.$$

This integral is equal to

$$\frac{i/\tau}{\omega_0(\omega_0 + i/\tau)} \ln\frac{(\xi_n + \omega_0 + i/2\tau)(\xi_n - \omega_0 - i/2\tau)}{(\xi_n + i/2\tau)(\xi_n - i/2\tau)}.$$

We then integrate with respect to p_z. In view of the fact that only the vicinity of $p_z = \pm p_{n_0}$ is important, we can replace p_z^2 by $p_{n_0}^2$ and integrate with respect to ξ, with

$$dp_z \to \frac{2m_1}{p_{n0}} \int d\xi$$

(the factor 2 is due to the two possible signs of p_0). We then obtain (we omit the symbol t, since j^t coincides with j when $\omega_0 > 0$)

$$j_{0z} = \frac{e^2}{m_1 c}\left[\frac{eH}{2\pi^2 c} \sum_n p_{0n} \frac{i/\tau}{\omega_0 + i/\tau} - n_e\right] A_z(\omega_0).$$

To determine the static conductivity, it is sufficient to put $\omega_0 \ll 1/\tau$. Then the zeroth term in ω_0 vanishes by virtue of (3). Since $E_k = i\omega_0 A_1^k/c$, we obtain from the first approximation, adding the currents from the electrons with spin projections $\pm {}^1/_2$,

$$\sigma_{zz} = (n_+\tau_+ + n_-\tau_-)e^2/m_1, \qquad (12)$$

where

$$\tau_\pm = -(2\operatorname{Im}\Sigma)^{-1} = \left\{1 + \left(U_0 \frac{eHm_1}{2\pi c} \sum \frac{1}{p_{0n}}\right)^2\right\} \Big/ N_i U_0^2 \frac{eHm_1}{\pi c} \sum \frac{1}{p_{0n}} \qquad (13)$$

(the summation for τ_+ and τ_- is over the corresponding p_{0n}, see (2)).

We have omitted here the zero index of σ_{zz}, since we shall show presently that other diagrams make no appreciable contribution. Indeed, let us consider the simplest of the unaccounted for diagrams. It is shown in Fig. 2b. Integrations with respect to p_{z1} and p_{z2} are independent. Since the Green's functions are even in p_z, and the vertices are odd, the diagram b of Fig. 2 vanishes. For the same reason, all the diagrams with parallel dashed lines for σ_{zz}, and in general all diagrams for the conductivity tensor components σ_{zx} and σ_{zy} vanish (in one vertex p_z, and in the other p_y or $\partial/\partial x$). To be sure, this argument cannot be used to eliminate diagrams for σ_{zz} with meshing lines. The simplest diagrams of this type are shown in Figs. 2c, d, and e. Although each of them is of the same order as that of Fig. 2a, the main term cancels out in the sum, and the result is of order $m/p_0^2\tau$. We do not present here the corresponding derivations. They are not complicated but quite long.

We now consider the component σ_{yy}. For the diagram of type a on Fig. 2 we obtain

$$j_{0y} = \frac{ie^2}{m_1^2 c} \int \left(p_y - \frac{eH}{c} x\right)\left(p_y - \frac{eH}{c} x_1\right)$$
$$\times \sum_n \frac{\psi_n(x - cp_y/eH)\psi_n(x_1 - cp_y/eH)}{\omega - \xi_n + (i/2\tau)\operatorname{sign}\omega}$$
$$\times \sum_{n_1} \frac{\psi_{n_1}(x_1 - cp_y/eH)\psi_{n_1}(x - cp_y/eH)}{\omega + \omega_0 - \xi_{n_1} + (i/2\tau)\operatorname{sign}(\omega + \omega_0)}$$
$$\times A_1^y(\omega_0)\, dx_1\, dp_y\, dp_z\, d\omega/(2\pi)^3 - \frac{e^2 n_e}{m_1 c} A_1^y(\omega_0).$$

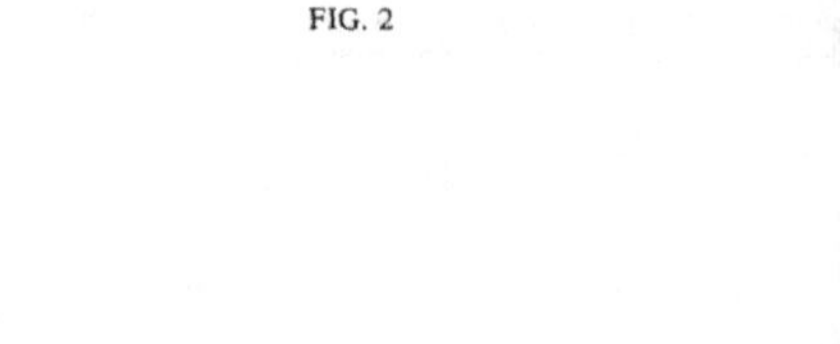

FIG. 2

The integral with respect to x_1 is in fact the matrix element of the oscillator coordinate. After obtaining this integral, a similar role is played by the integral with respect to p_y. In this case

$$x_{n,\,n+1} = x_{n+1,\,n} = \sqrt{(n+1)c/eH}.$$

We then obtain

$$j_{0y} = \frac{ie^2}{m_1{}^2c}\frac{1}{4\pi}\left(\frac{eH}{c}\right)^2\int\frac{dp_z\,d\omega}{2\pi\;2\pi}\sum_n(n+1)\Big\{\Big[\Big(\omega-\xi_{n+1}+\frac{i}{2\tau}\operatorname{sign}\omega\Big)$$
$$\times\Big(\omega+\omega_0-\xi_n+\frac{i}{2\tau}\operatorname{sign}(\omega+\omega_0)\Big)\Big]^{-1}$$
$$+\Big[\Big(\omega-\xi_n+\frac{i}{2\tau}\operatorname{sign}\omega\Big)\Big(\omega+\omega_0-\xi_{n+1}+\frac{i}{2\tau}\operatorname{sign}(\omega+\omega_0)\Big)\Big]^{-1}\Big\}$$
$$\times A_1{}^y(\omega_0)-\frac{e^2n_e}{m_1c}A_1{}^y(\omega_0). \tag{14}$$

This integral is easiest to calculate by subtracting and adding an analogous expression with $\tau\to\infty$. In the "free" term we can replace all the sign ω by sign ξ, where ξ corresponds to the given denominator. Then, when integrating with respect to ω, we should have $\xi_{n+1}>0$ and $\xi_n<0$ (the opposite case is impossible). We then obtain for the free term

$$\frac{e^2}{m_1{}^2c}\frac{1}{4\pi}\left(\frac{eH}{c}\right)^2\sum_n(n+1)\int\limits_{\xi_{n+1}>0,\,\xi_n<0}\frac{dp_z}{2\pi}$$
$$\times\left(\frac{1}{\omega_0+\xi_{n+1}-\xi_n}-\frac{1}{\omega_0-\xi_{n+1}+\xi_n}\right)=\frac{e^2}{m_1{}^2c}\frac{1}{2\pi^2}\left(\frac{eH}{c}\right)^2$$
$$\times\frac{eH/m_1c}{(eH/m_1c)^2-\omega_0{}^2}\sum_n(n+1)(p_{0n}-p_{0n+1})A_1{}^y(\omega_0)$$
$$\approx\frac{e^2}{m_1c}\frac{eH}{2\pi^2c}\sum_n p_{0n}A_1{}^y(\omega_0)=\frac{e^2n_e}{m_1c}A_1{}^y(\omega_0).$$

(We have neglected additions of order $\omega_0^2/(eH/m_1c)^2$.) Thus, this term cancels the last term in the current.

It remains to calculate the difference between the integral with τ and with the free term. Since the integral converges, we shall first integrate with respect to p_z. Assuming $\omega_0\ll 1/\tau\ll eH/m_1c$, we obtain

$$\sigma_{yy}=\frac{e^2m_1}{2\pi^2}\left(\frac{1}{\tau_+}\sum_n\frac{n+{}^1\!/_2}{p_{0n+}}+\frac{1}{\tau_-}\sum_n\frac{n+{}^1\!/_2}{p_{0n-}}\right). \tag{15}$$

We have again omitted the index zero, since the next diagrams make no contribution. Indeed, let us consider the diagram of type b on Fig. 2. In each vertex there is contained in lieu of p_z an operator having matrix elements with $n' = n\pm 1$. Integrating with respect to the x corresponding to the vertex, we obtain the product $\psi_n(x_1-cp_y/eH)\,\psi_{n\pm1}(x_1-cp_y/eH)$, which then vanishes upon integration with respect to p_y. But it is necessary to be cautious with the next diagrams, for according to (15) the main expression has been decreased compared with σ_{zz} by a factor $1/(\Omega\tau)^2$, where $\Omega = eH/m_1c$. Calculation of the diagrams of types c, d, and e of Fig. 2 show that terms of the required order vanish as a result of integration with respect to p_y. We succeeded in proving this by direct calculation only in the limiting case of a very strong field, when all the levels with n = 0 and 1 take part. However, in all probability, this is valid also in the general case.

The calculation of σ_{xx} and σ_{yy} is analogous. We use here the matrix elements

$$\left(\frac{\partial}{\partial x}\right)_{n,\,n+1}=-\left(\frac{\partial}{\partial x}\right)_{n+1,\,n}=\sqrt{\frac{(n+1)eH}{2c}}.$$

From the symmetry of the problem it is obvious that $\sigma_{xx}=\sigma_{yy}$. To calculate σ_{xy} we determine j_{0x} under the influence of the potential A_1^y. We then obtain in lieu of (14)

$$j_{x0}=-\frac{e^2}{m_1{}^2c}\frac{1}{4\pi}\left(\frac{eH}{c}\right)^2\int\frac{dp_z}{2\pi}\frac{d\omega}{2\pi}\sum_n(1+n)$$
$$\times\{[(\omega-\xi_n+(i/2\tau)\operatorname{sign}\omega)(\omega+\omega_0-\xi_{n+1}+(i/2\tau)\operatorname{sign}(\omega+\omega_0))]^{-1}$$
$$-[(\omega-\xi_{n+1}+(i/2\tau)\operatorname{sign}\omega)(\omega+\omega_0-\xi_n$$
$$+(i/2\tau)\operatorname{sign}(\omega+\omega_0))]^{-1}\}A_1{}^y(\omega_0). \tag{16}$$

In this case the answer does not depend on τ (since the terms of zeroth order in $1/\tau$ do not cancel out). This means that we can put $1/\tau\to 0$ and replace sign ω by sign ξ of the corresponding ξ in the denominator. The subsequent integration is then analogous to the preceding one, and we obtain

$$\sigma_{xy}=n_e ec/H, \tag{17}$$

i.e., the usual Hall conductivity. Obviously, $\sigma_{yx}=-\sigma_{xy}$.

We now analyze the derived formulas. We start with the limiting quantum case $eH/mc\gg p_0^2/m$, when all the electrons are concentrated on one Landau level. In this case

$$p_0=\frac{2\pi^2c}{eH}n_e, \tag{18}$$

with $n_e\ll(eH/c)^{3/2}$. Since $n\sim p_{00}^3$, where p_{00} is the Fermi momentum in the absence of a field, it follows that $p_0\ll p_{00}$. If we recognize that the interaction with the impurity atom is of the order of $p_{00}^2/2m$, we get $U_0\sim p_{00}^2/mn$. It is then easily seen that

$$U_0\frac{eHm_1}{cp_0}\sim\frac{p_{00}{}^2}{p_0^2}\gg 1.$$

According to (13), we obtain

$$\tau=\frac{eHm_1}{4\pi cp_0N_i}=\frac{1}{(2\pi)^3}\left(\frac{eH}{c}\right)^2\frac{m_1}{nN_i}, \tag{19}$$

i.e., τ depends only on the impurity density, but not on the concrete form of the interaction. The condition $1/\tau\ll p_0^2/m$ denotes in this case that $N_i\ll n$, i.e., a low impurity density. From (12) and (15), we get

$$\sigma_{zz}=\frac{1}{(2\pi)^3}\left(\frac{e^2H}{c}\right)^2\frac{1}{N_i}, \tag{20}$$

$$\sigma_{xx}=\sigma_{yy}=\frac{ecN_i}{\pi H}. \tag{21}$$

The qualitative form of these formulas has a natural physical interpretation. Since each electron moves on a helix with axis along z, the impurities interfere with the conductivity along this axis, and conversely contributes to conductivity in the perpendicular direction. The decrease of the radius of the helix with increasing field decreases also the collision probability, thus leading to an increase of σ_{zz} and to a decrease of σ_{xx} and σ_{yy}. From (17), (20), and (21) we obtain the resistivity tensor $\rho_{ik} = \sigma_{ik}^{-1}$:

$$\rho_{zz} = (2\pi)^3 N_i \left(\frac{c}{e^2 H}\right)^2,$$

$$\rho_{xx} = \rho_{yy} = \frac{N_i H}{\pi n^2 ec}, \quad \rho_{xy} = \frac{H}{nec}. \tag{22}$$

We now consider the opposite limit $eH/mc \ll p_0^2/2m$. This is the quasiclassical case, in which it is possible to change from summation with respect to n to integration. The small addition gives the so-called "quantum oscillations."

Using the Poisson summation formula

$$\sum_0^\infty \varphi(n) = \int_{-1/2}^\infty \varphi(x)\,dx + 2\,\mathrm{Re}\sum_{k=1}^\infty \int_{-1/2}^\infty \varphi(x) e^{2\pi i k x} dx,$$

we obtain

$$\sum_n \frac{1}{p_{0n\pm}} = \frac{1}{\sqrt{2m_1}} \int_0^{\zeta_\pm/\Omega} \frac{dx}{\sqrt{\zeta_\pm - \Omega x}} + \frac{2}{\sqrt{2m_1}}\,\mathrm{Re}\sum_{k=1}^\infty \int_0^{\zeta_\pm/\Omega} \frac{e^{2\pi i k x} dx}{\sqrt{\zeta_\pm - \Omega x}}$$

where

$$\zeta_\pm = \zeta \pm eH/2mc, \quad \Omega = eH/m_1 c.$$

Changing over to the variable $y = \zeta_\pm - \Omega x$ and continuing in the second integral the limit of integration with respect to y to ∞, we obtain

$$\sum_n \frac{1}{p_{0n\pm}} \approx \frac{2}{\Omega}\sqrt{\frac{\zeta_\pm}{2m_1}} + \frac{1}{\sqrt{m_1\Omega}} \sum_k \frac{1}{\sqrt{k}} \cos\left(2\pi k \frac{\zeta_\pm}{\Omega} - \frac{\pi}{4}\right).$$

In the first term we can put $\zeta_\pm \approx p_{00}^2/2m_1$. This cannot be done in the second term, in view of the fact that this term oscillates.[2)] After substitution, we obtain

$$\sum_n \frac{1}{p_{0n\pm}} = \frac{cp_{00}}{eH}\left[1 + \sqrt{\frac{eH}{cp_{00}^2}} \sum_k \frac{1}{\sqrt{k}} \cos\left(2\pi k \frac{\zeta_\pm}{\Omega} - \frac{\pi}{4}\right)\right]. \tag{23}$$

The second term is of relative order $(\Omega/\zeta)^{1/2}$. Therefore, upon substitution in (12) for τ we can expand with respect to this term. It is convenient here to go over from the Born amplitude to the true amplitude in the absence of the field. Considering Fig. 1c for an individual electron interacting with an impurity, we obtain

$$A = \frac{U_0}{1 + iU_0 p_{00} m_1/2\pi}.$$

Expressing U_0 in terms of A_1 we get

$$\tau_\pm = (N_i |A|^2 p_{00} m/\pi)^{-1} \Big\{1 + [2|A|^2 (p_{00} m/2\pi)^2 - 1] \cdot \sqrt{\frac{eH}{cp_{00}^2}} \sum_k \frac{1}{\sqrt{k}} \cos\left(\frac{2\pi k \zeta_\pm}{\Omega} - \frac{\pi}{4}\right)\Big\}. \tag{24}$$

When this expression is substituted in (12), we can assume that $n_+ \approx n_-$. After this, addition of terms with indices $\pm$ denotes

$$\cos(2\pi k\zeta_+/\Omega - \pi/4) + \cos(2\pi k\zeta_-/\Omega - \pi/4) \approx \cos(2\pi k\zeta/\Omega - \pi/4)\cos(\pi k m/m_1).$$

Thus, as a result we get

$$\sigma_{zz} = \sigma_0 \Big\{1 + [2|A|^2 (p_{00} m_1/2\pi)^2 - 1] \sqrt{\frac{eH}{cp_{00}^2}} \times \sum_k \frac{1}{\sqrt{k}} \cos\left(\frac{\pi k c p_{00}^2}{eH} - \frac{\pi}{4}\right) \cos\left(\pi k \frac{m}{m_1}\right)\Big\}, \tag{25}$$

where

$$\sigma_0 = ne^2\tau_0/m_1, \quad 1/\tau_0 = N_i |A|^2 p_{00} m_1/\pi.$$

In analogy with (23), we get

$$\sum_n \frac{n + 1/2}{p_{0n\pm}} \approx \frac{p_{00}^3 c^2}{3(eH)^2}\left[1 + \frac{3}{2}\sqrt{\frac{eH}{cp_{00}^2}} \sum_k \frac{1}{\sqrt{k}} \cos\left(2\pi k \frac{\zeta_\pm}{\Omega} - \frac{\pi}{4}\right)\right]. \tag{26}$$

Substituting in (15), we get

$$\sigma_{yy} = \sigma_{xx} = \frac{ne^2 m_1}{\tau_0}\left(\frac{c}{eH}\right)^2 \Big\{1 + \left[\frac{5}{2} - 2|A|^2 \left(\frac{p_{00} m_1}{2\pi}\right)^2\right] \sqrt{\frac{eH}{cp_{00}^2}} \times \sum_k \frac{1}{\sqrt{k}} \cos\left(\frac{\pi k c p_{00}^2}{eH} - \frac{\pi}{4}\right) \cos\left(\pi k \frac{m}{m_1}\right)\Big\}. \tag{27}$$

We recall, incidentally, that we have assumed throughout that $\Omega \gg 1/\tau_0$.

Finally, let us consider the case $\Omega \sim \zeta$ and ascertain what occurs when the chemical potential passes through one of the Landau levels, say n_0. We assume that this is a level with spin orientation along the field. We then can easily obtain from (23) the corresponding value of the field

$$\left(\frac{eH_{n_0}}{c}\right)^{3/2} = \sqrt{2}\,\pi^2 n_e \Big[\sum_{n=1}^{n_0} \sqrt{n} + \sum_n \sqrt{n_0 - n - \frac{m_1}{m}}\Big]^{-1}.$$

Let us examine τ in the vicinity of this point. So long as the field is somewhat larger than H_{n_0}, according to (13), τ behaves like a smooth function of the field and is of the order of

$$\tau \sim \frac{eH_{n_0} m_1}{cN_i p_{0n}} \sim \sqrt{\frac{eH_{n_0}}{c}} \frac{m_1}{N_i} \sim n_e^{1/3} \frac{m_1}{N_i}.$$

Accordingly, the components of the conductivity tensor are of the order of

$$\sigma_{zz}(H_{n_0} + 0) \sim e^2 n_e^{1/3} \frac{n_e}{N_i},$$

$$\sigma_{xx} = \sigma_{yy}(H_{n_0} + 0) \sim e^2 n_e^{1/3} \frac{N_i}{n_e}. \tag{28}$$

However, if we approach this point from below, then the momentum p_{n_0} becomes small near H_{n_0}, and this immediately affects the conductivity. From (3) we have

$$p_{0n_0} = 2\pi^2 n_e \frac{c}{H_{n_0} e} \frac{H_{n_0} - H}{H_{n_0}}.$$

This yields directly

$$\tau \sim n_e^{1/3} \frac{m_1}{N_i} \frac{H_{n_0}}{H_{n_0} - H},$$

$$\sigma_{zz}(H \to H_{n_0} - 0) \sim e^2 n_e^{1/3} \frac{n_e}{N_i} \frac{H_{n_0}}{H_{n_0} - H},$$

$$\sigma_{xx} = \sigma_{yy}(H \to H_{n_0} - 0) \sim e^2 n_e^{1/3} \frac{N_i}{n_e}. \tag{29}$$

Thus, the component σ_{zz} increases when H approaches H_{n_0} from below. As to σ_{xx} and σ_{yy}, they are of the same order as when $H = H_{n_0} + 0$, but it can be shown that the coefficient will be different in this case. Consequently, these components experience a discontinuity.

---

2)Strictly speaking, ζ also has an oscillating addition. It can be obtained with the aid of formulas (2) and (3). However, when we sum Σp_{0n} in analogy with the summation in (23), we find that the relative order of the addition is $(\Omega/\zeta)^{3/2}$, whereas other sums make additions of order $\sqrt{\Omega/\zeta}$ to the conductivity. For this reason, we can assume that $n_+ = n_-$.

Actually these singularities become smoothed out somewhat. The point is that the condition for the applicability of the formulas obtained above is, in particular, $p_{0n_0}^2/m_1 \gg 1/\tau$. This imposes the limitation $(H_{n_0} - H)/H_{n_0} \gg N_i/n_e$. Thus, the singularity of σ_{ik} should be smoothed out over this region. In particular, it follows therefore that the maximum reached by σ_{zz} is of the order of

$$\sigma_{zz\,max} \sim e^2 n_e^{1/3} (n_e/N_i)^2. \qquad (30)$$

2. MODEL OF A SEMIMETAL

We now consider a model closer to the real situation. We assume that near $\mathbf{p} = 0$ we have isotropic holes with small effective mass, and near $\mathbf{p} = \mathbf{P}$ we have "electrons" that are isotropic with respect to $\mathbf{p} - \mathbf{P}$ and also have a small mass (Fig. 3a). Then, in principle, three different cases are possible.

1. The valence and the conduction bands intersect at H = 0, and this intersection remains for all fields (Fig. 3b).

2. The bands do not intersect at H = 0, but the intersection occurs at some field value (Fig. 3c).

The fields have intersected at H = 0, but the intersection was lifted under the influence of the field (Fig. 3d).

We have considered only the limits of the strongest fields, when only one of the Landau levels is filled. The energies of the electrons and holes are given by

$$\varepsilon_e(H, p_z) = \varepsilon_e + \frac{eH}{2m_{e1}c} - \frac{eH}{2m_e c} + \frac{(\mathbf{p}-\mathbf{P})_z^2}{2m_{e1}},$$

$$\varepsilon_h(H, p_z) = \varepsilon_h - \frac{eH}{2m_{h1}c} + \frac{eH}{2m_h c} - \frac{p_z^2}{2m_{h1}}. \qquad (31)$$

The effective masses m_e and m_h for the spin splitting are also assumed to be small, of the order of m_{e1} and m_{h1}. This corresponds to the real situation in semimetals where, as a result of the spin-orbit interaction, the spin splitting of the levels is of the order of the orbital splitting. We put

$$\Delta = \varepsilon_h - \varepsilon_e, \quad \frac{1}{m^*} = \frac{1}{m_h} - \frac{1}{m_{h1}} + \frac{1}{m_e} - \frac{1}{m_{e1}}. \qquad (32)$$

Then, obviously, the aforementioned cases correspond to:

1) $\Delta > 0,\ m^* > 0,$
2) $\Delta < 0,\ m^* > 0,$
3) $\Delta > 0,\ m^* < 0.$

The number of electrons (which equals the number of holes) we have

$$n_h = n_e = \frac{eH}{2\pi^2 c} p_0, \qquad (33)$$

where p_0 is determined from the condition that the energies of the electrons and holes be equal to the chemical potential, i.e.,

$$p_0 = \sqrt{2m^{**}\left(\Delta + \frac{eH}{2m^* c}\right)}, \qquad (34)$$

where $(m^{**})^{-1} = m_{h1}^{-1} + m_{e1}^{-1}$. With the aid of (34) we easily obtain the conditions under which only the lower level takes part in the spectrum. To this end, obviously, the following inequality should be satisfied:

$$\frac{cp_0^2}{2eH} < 1, \quad \frac{m_{e1}}{m_e}, \quad \frac{m_{h1}}{m_h}.$$

We shall not write out explicitly the relations obtained in this case for the masses and the field H. It is important that these conditions can exist.

We note first that the possible existence of a Fermi spectrum in a semimetal is not obvious in the extreme quantum case. The motion becomes one-dimensional. As is well known, in the one-dimensional case the attraction of the particles must lead to the formation of bound states.[3)]

The situation becomes more complicated by the fact that this question is solved differently for two isolated particles and for two Fermi distributions of the electrons and the holes. It is more or less obvious that the transition from one of these variants to the other occurs when the Fermi energy becomes comparable with the binding energy of the pair (exciton). The latter can be easily estimated. It turns out to be of the order of $m_1 e^4$ if $e^2 m_1/\sqrt{eH/c} \lesssim 1$.

For semimetals without a magnetic field, the velocity on the Fermi surface is $p_{00}/m_1 \sim 10^8$ cm/sec, i.e., $e^2 m_1/p_{00} \sim 1$. This means that in strong fields, where $p_0 \sim \sqrt{eH/c} \gg p_{00}$,[4)] the ratio of the exciton binding energy to the Fermi energy is of the order of $(e^2 m_1/p_0)^2 \ll 1$. It follows therefore that in this case a consideration of individual electrons and holes will not do.

We do not present the details of the estimate in which allowance is made for the Fermi distributions. It is obtained in the same manner as for superconductivity. The binding energy has in this case the form $a(p_0^2/m_1) \times \exp(-bp_0/m_1 e^2)$, where $a \sim 1$ and $b \sim 1$. This energy is much smaller than p_0^2/m_1. Consequently, there exists an interval of temperatures much lower than the Fermi energy and much higher than the binding energy, in which temperature effects can be neglected and at the same time it is possible to disregard exciton formation. All the following will pertain precisely to this case (see also footnote 4).

Let us consider Fig. 1a. Averaging over the random arrangement of the impurities gave us in the preceding model the product $\psi_n(x - cp_y/eH)\psi_n(x' - cp_y/eH)$. This was due to the fact that the functions ψ_n are orthonormal. In the present case we may encounter a situation wherein the Green's function on one side of the cross corresponds to holes, and on the other side to electrons. In this case we have the integral

$$\int \exp[i(p_z - p_z' + P_z)z_1 + i(p_y - p_y' + P_y)y_1 + iP_x x_1] \times \psi_n\left(x_1 - \frac{cp_y}{eH}\right)\psi_{n1}\left(x_1 - \frac{cp_y'}{eH}\right) d^3\mathbf{r}_1$$

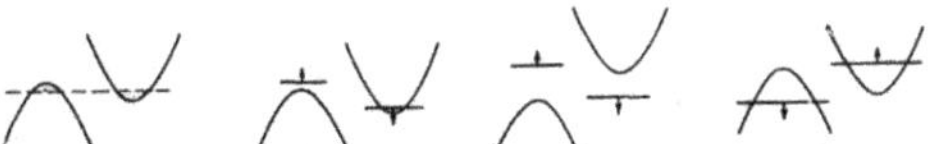

FIG. 3

[3)]The author is grateful to L. P. Gor'kov for suggesting this problem.

[4)]Of course, according to (34), it is necessary to have for this purpose $m^* > 0$, i.e., all this pertains only to cases 1) and 2).

$$\times[\omega - \xi_{nh}(p_z) + i\delta \operatorname{sign} \omega]^{-1}[\omega - \xi_{ne}(p_z') + i\delta \operatorname{sign} \omega]^{-1},$$

where

$$\xi_{nh}(p_z) = (p_0^2 - p_z^2)/2m_{h1}, \quad \xi_{ne} = (p_z^2 - p_0^2)/2m_{e1}.$$

Although the integral written out above does not vanish exactly, it is very small. Indeed, by definition, at least one of the nonvanishing components of the vector **P** is of the order of $1/a$, where a is an atomic dimension. The functions $\psi_n(x - cp_y/eH)$ vanish at a distance $\sqrt{c/eH}$ along x or $\sqrt{eh/c}$ along p_y. We are considering a case in which $\sqrt{eH/c} \sim p_0 \ll P$. Consequently, if $P_y \neq 0$, then the product $\psi_n(x_1 - cp_y/eH)\psi_{n_1}(x - c(p_y - P_y)/eH))$, which appears after integration with respect to y_1, should be small. If $P_x \neq 0$, then the smallness appears in the result of integration with respect to x_1 owing to the rapidly oscillating factor $\exp[iP_x x_1]$ in the integral. Finally, if $P_z \neq 0$, then P_z appears in one of the denominators, making it large.

Thus, the "diagonality" of G in averaging over the positions of the impurities is retained in this case, too. This, of course, does not exclude summations over either the electron or the hole states in the intervals between the crosses in the diagrams of types b and c on Fig. 1. It is easy to verify that both contributions have the same form, i.e., $m_{e1} + m_{h1}$ appears in (6) in lieu of m_1.

We now estimate the order of magnitude of the correction to the Born approximation in our case. E_0 is of the order of $P^2/m^{(0)}P^3 \sim 1/m^{(0)}P$, where $m^{(0)}$ is the mass of the free electron and P is of the order of the reciprocal atomic distances. Thus, the relative order of the correction is

$$(eHm_1/cp_0)/m^{(0)}P.$$

In the quantum region $p_0 \sim \sqrt{eH/c}$, and the correction is of the order of $p_0 m_1/m^{(0)}P$. In this case p_0 is of the order of the limiting Fermi momentum without a field, or larger. But in semimetals the velocity on the Fermi boundary is of the same order as in ordinary metals, i.e., without a field we have $p_0/m_1 \sim P/m^{(0)}$ (p_0—Fermi momentum of the semimetal without a magnetic field). It follows therefore that the correction at not too large fields is of the order of $(p_{00}/P)^2 \sim 10^{-3}$. This means that it can be neglected. Of course, this smallness can be overcome by increasing the magnetic field. But whereas on the limit of the ultraquantum region the field should be of the order of 10^5 Oe, in order to make the correction large the field must be increased by six orders of magnitude, which is not realistic. Thus, we can always use the value of τ^{-1} in the Born approximation, i.e.,

$$\frac{1}{\tau} = N_i U_0^2 \frac{eH}{\pi c}(m_{1e} + m_{1h})\frac{1}{p_0}. \tag{35}$$

In general, the condition $1/\tau \ll p_0^2/m_1$ does not impose any limitations on the concentration of the impurity in the ultraquantum region.

We proceed to the conductivity tensor. It is easy to see, first, that the entire loop on Fig. 2 should pertain either to electrons or to holes, i.e., the current consists of two independent parts. For the diagonal components we obtain (only the diagrams of type a in Fig. 2 are significant)

$$\sigma_{zz} = e^2 n\tau\left(\frac{1}{m_{e1}} + \frac{1}{m_{h1}}\right), \tag{36}$$

$$\sigma_{yy} = \sigma_{xx} = \frac{e^2}{4\pi^2 p_0 \tau}(m_{e1} + m_{h1}). \tag{37}$$

The Hall component σ_{xy} in the approximation of formula (17) vanishes in this case. In the next approximation, contributions are made by diagrams of types c, d, and e in Fig. 2. We can therefore determine σ_{xy} only in order of magnitude:

$$\sigma_{xy} = -\sigma_{yx} \sim \frac{e^2 m_1^3}{p_0 \tau^2}\frac{c}{eH}. \tag{38}$$

As already mentioned, the obtained formulas, strictly speaking, are valid only for the case of strong fields $eH/m_1 c \gg \Delta$, $\sqrt{eH/c} \gg p_{00}$. Then, according to (33)–(38)

$$p_0 = \sqrt{\frac{m^{**}}{m^*}\frac{eH}{c}}, \tag{39}$$

$$n_h = n_e = \frac{1}{2\pi^2}\sqrt{\frac{m^{**}}{m^*}}\left(\frac{eH}{c}\right)^{3/2}, \tag{40}$$

$$\frac{1}{\tau} = \frac{1}{\pi} N_i U_0^2 (m_{e1} + m_{h1})\sqrt{\frac{m^*}{m^{**}}}\sqrt{\frac{eH}{c}}, \tag{41}$$

$$\sigma_{zz} = \frac{e^2 n\tau}{m^{**}} = \frac{1}{2\pi}\frac{e^2}{m^*(m_{e1} + m_{h1})}\frac{eH}{c}\frac{1}{N_i U_0^2}, \tag{42}$$

$$\sigma_{xx} = \sigma_{yy} = \frac{e^2}{4\pi^3}\frac{m^*}{m^{**}} N_i U_0^2 (m_{e1} + m_{h1})^2, \tag{43}$$

$$\sigma_{xy} = -\sigma_{yx} \sim e^2 m_1^4 N_i^2 U_0^2 \sqrt{c/eH}. \tag{44}$$

In view of the fact that in this case $\sigma_{xy} \ll \sigma_{xx}$, the diagonal components of the tensor ρ_{ik} are simply the reciprocals of the corresponding σ_{ik}. As to the component ρ_{xy}, we obtain for it

$$\rho_{xy} = -\rho_{yx} \propto \frac{1}{e^2}\sqrt{\frac{c}{eH}}. \tag{45}$$

Thus, the resistance along the field decreases with increasing H, the transverse resistance reaches saturation, and ρ_{yx} does not depend on the impurity density. The behavior of ρ_{xx} differs from the non-quantum case $p_{00}^2/m_1 \gg \Omega \gg 1/\tau$, where $\rho_{xx} \propto H^2$.

4. CONCLUSION

The described method can be generalized and applied to a large group of various quantum phenomena in metals.

a) Quasiclassical limit. Quantum oscillations. The obtained formulas can be usually generalized to the case of a finite temperature. Then the expressions of type (11), (14), and (16) can be used for an arbitrary energy spectrum. After summation over ω, there remains summation over n and p_z, which is carried out by the same method as in the paper of I. Lifshitz and A. Kosevich.[4] Unlike the previously employed complicated and inconvenient methods for calculating the quantum oscillations in kinetic phenomena[1,2,5,6], this method has the advantage of being highly automatic and providing for the elimination of errors. Of course, in order to consider high-frequency phenomena it is necessary to introduce further modifications, in which account is taken of the inhomogeneity of the field at the surface, as well as boundary effects.

b) Quantum limit. In addition to calculating other characteristics of the two models considered in the present paper, we can study also more concrete spectra in the quantum limits. To this end it is necessary to know the entire function $\epsilon(\mathbf{p})$, and not only the vicinity of the Fermi surface. The problem is complicated by the fact that usually, the different branches of the spectrum are close to each other for a large number of values of **p**, and cannot be regarded separately upon quantization. In the most important case, however, that of real semimetals, the spectrum is known[7] and in principle we know how to solve the quantum problem, although this is a complicated matter and was completed only for holes in a field parallel to the principal axis.[8] 5)

Recently a new phenomenon was discovered,[10, 11] namely, the quantization of the electron motion over jumping trajectories along the surface. In this case the lower quantum levels are important. Although this problem has its own specific features, its solution can apparently likewise be simplified and refined by using Green's functions expanded in terms of the eigenfunctions of the problem.

The author is grateful to L. P. Gor'kov, Yu. A. Bychkov, and M. Ya. Azbel' for a discussion of the work.

5)The energy spectrum in very strong fields may be altered by striction effects (the author thanks M. Ya. Azbel' for pointing out this circumstance). Obviously, this occurs when the splitting of the magnetic levels becomes of the same order as the energies responsible for the structure of the spectrum. For semimetals (see [9]), this energy (γ) is of the order of several tenths of an electron volt, whereas the Fermi energy (reckoned from the bottom of the band) is of the order of several hundreths of an electron volt. Consequently, there is an interval of strong fields where the change of the parameters of the spectrum (such as the effective mass) can still be disregarded.

[1] V. G. Skobov, Zh. Eksp. Teor. Fiz. **38**, 1304 (1960) [Sov. Phys.-JETP **11**, 941 (1960)].

[2] E. N. Adams and T. D. Holstein, J. Phys. Chem. Sol. **10**, 254 (1959).

[3] Yu. A. Bychkov, Zh. Eksp. Teor. Fiz. **39**, 1401 (1960) [Sov. Phys.-JETP **12**, 971 (1961)].

[4] I. M. Lifshitz and A. M. Kosevich, Zh. Eksp. Teor. Fiz. **29**, 730 (1955) [Sov. Phys.-JETP **2**, 636 (1956)].

[5] I. M. Lifshitz, Zh. Eksp. Teor. Fiz. **30**, 814 (1956) [Sov. Phys.-JETP **3**, 774 (1956); **32**, 1509 (1957) [Sov. Phys.-JETP **5**, 1227 (1957)].

[6] A. M. Kosevich and V. V. Andreev, Zh. Eksp. Teor. Fiz. **38**, 882 (1960) [Sov. Phys.-JETP **11**, 637 (1960)].

[7] A. A. Abrikosov and L. A. Fal'kovskiĭ, Zh. Eksp. Teor. Fiz. **43**, 1089 (1963) [Sov. Phys.-JETP **16**, 769 (1963)].

[8] L. A. Fal'kovskiĭ, Zh. Eksp. Teor. Fiz. **44**, 1935 (1963) [Sov. Phys.-JETP **17**, 1302 (1963)].

[9] L. A. Fal'kovskiĭ and G. S. Razina, Zh. Eksp. Teor. Fiz. **49**, 265 (1965) [Sov. Phys.-JETP **22**, 187 (1966)].

[10] M. S. Khaĭkin, Zh. Eksp. Teor. Fiz. **39**, 212 (1960) [Sov. Phys.-JETP **12**, 152 (1961)].

[11] T. W. Nee and R. E. Prange, Phys. Lett. **25A**, 583 (1967).

Translated by J. G. Adashko
159

Journal of Low Temperature Physics, Vol. 3, No. 5, 1970

On the Theory of the Kondo Effect

A. A. Abrikosov and A. A. Migdal

L. D. Landau Institute for Theoretical Physics
Academy of Sciences of the USSR, Vorobyevskoe Shosse 2, Moscow, USSR

(Received July 28, 1970)

The Kondo effect is investigated by methods developed in the theory of fields with strong coupling. Renormalizability relations analogous to the scaling laws in the theory of second-order phase transitions are found. The behavior of the scattering amplitude of an electron by an impurity is analyzed as a function of the electron energy. Formulas are derived for the resistance as a function of the temperature and for the magnetic moment as a function of the magnetic field and temperature. For the antiferromagnetic interaction sign the resistance tends to a constant and the magnetic moment to zero as $T \to 0$ *according to power laws with universal exponents that depend only on the spin of the impurity. For sufficiently large spin the susceptibility tends to infinity as* $T \to 0$.

1. INTRODUCTION

Since the publication of Kondo's paper[1] which explained the logarithmic increase of the resistance with decreasing temperature for metals containing magnetic impurities, a great many papers have been published on the various properties of these objects (see Ref. 2, which contains an extensive bibliography). In the perturbation series for the scattering amplitude of electrons by impurities the summation of the principal terms involving the maximum power of $\ln(\varepsilon_F/\omega)$ (where ω is the electron energy relative to the Fermi level) in each order on the interaction constant yielded the following result[3,4]:

$$A = -\frac{(J/N)\boldsymbol{\sigma}\mathbf{S}}{1 + 2(J/N)\nu \ln(\varepsilon_F/\omega)} \tag{1}$$

Here, J/N is the constant appearing in the Hamiltonian for the exchange interaction of an electron with an impurity:

$$H_{int} = -\frac{J}{N}\Psi_\alpha^+(0)\boldsymbol{\sigma}_{\alpha\alpha'}\Psi_{\alpha'}(0)\mathbf{S} \tag{2}$$

(the impurity atom is situated at the point $\mathbf{r} = 0$), and $\nu = p_0 m/2\pi^2$ is the density of electron states at the Fermi surface.

In the case of the antiferromagnetic interaction sign $J < 0$ the scattering amplitude (1) goes to infinity at the so-called "Kondo temperature":

$$T_K \propto \varepsilon_F \exp\left(-\tfrac{1}{2}|J/N|\nu\right) \tag{3}$$

This demonstrates the inadequacy of the assumed approximation when the temperature or electron energy is sufficiently low.

The difficulty in formulating a theory of the Kondo effect lies in the fact that for sufficiently low electron energy the interaction of the electron with an impurity becomes intense for any "bare" interaction constant, no matter how small. One can seek to avoid this difficulty by various methods.

The notion has been advanced in several papers that the electrons form a bound state with the impurity, and this state has been described by the onset of a certain "order parameter." This notion is not justifiable in the strict sense, because in the Kondo effect the impurities are uncorrelated, and one individual impurity cannot induce the order parameter. Nevertheless, it is conceivable that the "bound-state" hypothesis affords a better approximation at temperatures well below T_K.

However, other hypotheses are also possible. For example, attempts have been made to solve the problem by truncation of the infinite chains of equations. Another approach has been based on the analytic properties of the scattering amplitude and the unitarity relations neglecting the many-particle intermediate states.[4]

In the present article we set forth a more general approach to the problem. It exhibits the fact that methods based on truncation of the chains or neglect of the many-particle intermediate states cannot yield a correct result in the general case, because the higher-order terms do not play a lesser role than the low-order terms.

2. RENORMALIZABILITY AND ITS IMPLICATIONS

In Ref. 3 a diagram technique has been developed for the spin operators by means of the introduction of pseudofermions with particle creation and annihilation operators a_M^+ and a_M. The impurity spin operators are written in the form

$$S^i = a_M^+ S^i_{MM'} a_{M'} \tag{4}$$

where a_M^+ and a_M obey the usual Fermi anticommutation rules, the index M takes all values from $-S$ to $+S$, and $S^i_{MM'}$ are the spin matrices. For the elimination of spurious states each pseudofermion is assigned a large positive energy λ. The zero-order pseudofermion Green function has the form

$$g^{(0)} = \frac{1}{i\omega - \lambda} \tag{5}$$

The renormalizability of the theory can be proved in general by the standard technique (see, e.g., Ref. 5). For the determination of the divergent irreducible diagrams it is required to bear in mind that every internal electron line is integrated near the Fermi surface as $d^3p \to p_0^2\, d\xi/v\, [\xi = v(p - p_0)]$. This leaves integration* over the frequencies, which obey conservation rules. Consequently, we obtain the degree of divergence:

$$K = 1 - \frac{N_e}{2} \tag{6}$$

where N_e is the number of external electron vertices. This number must by necessity be even. For $N_e = 0$ we have $K = 1$. This represents the divergence in the self energy of the pseudofermion and can be eliminated by renormalization of the pseudofermion "mass" λ. The remaining logarithmically divergent part is excluded by renormalization of the pseudofermion Green function. One more logarithmic divergence occurs for $N_e = 2$, i.e., in the vertex part Γ corresponding to the scattering of an electron by a pseudofermion. It is eliminated by renormalization of the interaction constant J/N in the vertex part.

Renormalizability implies that with the variation of the upper limit of integration, equal to the Fermi energy ε_F, and the corresponding variation of the interaction constant all the Green functions and vertices are multiplied by some constants. The invariant charge, i.e., the invariant quantity associated with this transformation, is the combination $\Gamma\mathcal{G}$, where Γ is the electron–pseudofermion vertex and $\mathcal{G}$ is the total Green function of the pseudofermion. The electron Green function enters into all the integrals in the form $\int G(\omega, \xi)\, d\xi$, where $\xi = v(p - p_0)$ (see Ref. 3). This integral is not significantly altered by interaction with impurities.

Now let us consider a certain physical quantity Q in the absence of a magnetic field. It is expressed as the frequency integral of some combination of $\mathcal{G}$ and Γ and depends on the ratio T/ε_F. With variation of ε_F Q is multiplied by a constant. Consequently, it must satisfy the condition (see Ref. 6)

$$\frac{d(\ln Q)}{dx} = \mathrm{Inv}$$

where $x = \ln(\varepsilon_F/T)$. Indeed, the left-hand side of the above relation is invariant under the renormalization transformation. The right-hand side must comprise an invariant function of the interaction constant and ε_F/T. If z is some invariant combination of the type indicated, we can write

$$\frac{d(\ln Q)}{dx} = \phi(z) \tag{7}$$

The quantity z may be determined as follows. We consider the quantity $v\omega_1\mathcal{G}(\omega_1)\Gamma(\omega, \omega_1; \omega, \omega_1)$ ($v = p_0 m/2\pi^2$ is the density of electron states and is included for normalization) and continue it analytically to the real axis for the electron frequency ω and pseudofermion frequency ω_1. Then put $\omega_1 = \lambda$, $\omega = 0$. It can be shown (by a method analogous to the one in Section 3) that this quantity is proportional to $\boldsymbol{\sigma}\mathbf{S}$. We call the coefficient of $\boldsymbol{\sigma}\mathbf{S}$ the charge z. In the

*Sums over imaginary frequencies can be converted to integration over real frequencies.

first logarithmic approximation z is given by

$$z = \frac{-(J/N)\nu}{1 + 2(J/N)\nu \ln(\varepsilon_F/T)}$$

and does not change as

$$\varepsilon_F \to \varepsilon_{F'}, J\nu/N \to J'\nu/N = [(J\nu/N)^{-1} + 2\ln(\varepsilon_F/\varepsilon_{F'})]^{-1}$$

Relation (7), in particular, can be applied to z itself:

$$\frac{d(\ln z)}{dx} = f(z) \tag{8}$$

Solving this equation, we have

$$x = q(z) - q(z_0) \tag{9}$$

where $q(z) = \int \frac{d\ln z}{f(z)}$ and z_0 is the value of z at $x = 0$, i.e., $-(J/N)\nu$. We define the "Kondo temperature" by the relation

$$T_K = \varepsilon_F \exp[q(z_0)] \tag{10}$$

Then we obtain from (9),

$$z = q^{-1}[\ln(T_K/T), S] \tag{11}$$

The dependence on S comes from the fact that the function $f(z)$, as we shall find out below, depends explicitly on the magnitude of the impurity spin.

We now investigate Eq. (7) for the physical quantity Q. From (8) we obtain

$$\frac{d(\ln Q)}{d(\ln z)} = \phi(z)/f(z) \tag{12}$$

Hence,

$$Q = Q_0 \exp\left[\int_{z_0}^{z} \phi(z)\, dz/f(z)\right] \tag{13}$$

From this we obtain, by virtue of Eq. (11),

$$Q = Q_0 \frac{\eta[\ln(T_K/T), S]}{\eta[\ln(T_K/\varepsilon_F), S]} \tag{14}$$

For $|J/N|\nu \ll 1$ any variable Q will differ only slightly from its zero value (i.e., the value at $T \sim \varepsilon_F$) down to temperatures of the order T_K. Therefore, taking the temperature T in the interval $T_K \ll T \ll \varepsilon_F$, we can verify that the factor associated with Q_0 in (14) remains close to unity down to T_K. This makes it clear that under suitable normalization the denominator in Eq. (14) may be assumed equal to unity, and in place of (14) we can write

$$Q = Q_0 \zeta[\ln(T_K/T), S] \tag{15}$$

However, when the small corrections of the order Jv/N are taken into account it is required to use Eq. (14).

In the case when $T = 0$ but a magnetic field is present the role of the lower limit of the frequency integrals is taken over by $g\mu H$, where g is the gyromagnetic ratio and μ is the Bohr magneton. This can be perceived by analysis of the first logarithmic corrections to the physical variables. As a result we obtain relations analogous to (15) but with $\ln(T_K/g\mu H)$ instead of $\ln(T_K/T)$. In general, we write

$$Q = Q_0\zeta[\ln(T_K/T), g\mu H/T, S] \tag{16}$$

It follows from all the preceding that for any physical quantity the ratio Q/Q_0 is a universal function of T_K/T and $g\mu H/T$, depending only on the spin S.

3. SCATTERING AMPLITUDE AT $T = 0$

The form of the universal functions in the above relations can be readily determined for $T \gg T_K$, where perturbation theory is applicable. In particular, for the scattering amplitude with $\omega \sim T$ we have from (1)

$$\frac{A}{A_0} = \frac{1}{x} \tag{17}$$

For $J > 0$, as we shall see later, perturbation theory is applicable at all temperatures. In the case $J < 0$, however, strong coupling appears for $T \lesssim T_K$, and the situation changes.

Two conceptually different possibilities can be considered. First, it may be assumed that in the strong coupling domain all Γ diagrams containing exact Γ, $\mathcal{G}$ internally are of the same order of magnitude. This hypothesis has been utilized successfully in various problems involving strong coupling (see, e.g., Ref. 7), in particular in the problem of second-order phase transitions, and it yields power laws. The other possibility is an increase of the diagrams with increasing order. We shall confine ourselves to the first of these two possibilities.

We begin with a calculation of the scattering amplitude of an electron by an impurity at $T = 0$ as a function of the electron energy ω. It has been studied in Ref. 3. All the integrations over the frequency were transferred to the real axis, and the pseudofermion frequencies were measured relative to λ. The free-pseudofermion Green function, in this case, had the form

$$\mathcal{G}^{(0)}_{MM'} = \frac{\delta_{MM'}}{\omega + i\delta} \tag{18}$$

where $\delta \to +0$.

In principle, at $T = 0$ one can manage without λ, by the introduction of a magnetic field (passing to the limit $H \to 0$ for problems without a magnetic field). In this case we can determine the ground state of the system of pseudofermions, namely, $a_S\Phi_0 \neq 0$, $a_{M<S}\phi_0 = 0$. Then the corresponding free $\mathcal{G}$

function has the form

$$\mathscr{G}^{(0)}_{MM'} = \frac{\delta_{MM'}}{\omega + g\mu H + i\delta \operatorname{sgn}(M - S)} \tag{19}$$

where g is the gyromagnetic ratio, μ is the Bohr magneton, and $\delta \to +0$. For $H = 0$ the result turns out to be the same for both approaches. We shall therefore use the simpler formula (18), leaving (19) for problems involving a magnetic field.

Consider the vertex γ at which two pseudofermion lines join (Fig. 1). It is readily perceived that the diagrams involved in $\gamma(\omega, \omega)$ can be obtained from the diagrams for the pseudofermion internal energy by differentiation with respect to the external frequency. This permits us to write the Ward identity

$$\gamma(\omega, \omega) = \frac{\partial \mathscr{G}^{-1}}{\partial \omega} \tag{20}$$

We now calculate the first correction to γ (shown in Fig. 1(b)). A simple calculation yields the following, to logarithmic accuracy:

$$\gamma^{(1)}(\omega, \omega) = 2\left(\frac{J}{N}\nu\right)^2 S(S + 1) \ln (\varepsilon_F/\omega) \tag{21}$$

Integrating over the frequency and recognizing that the constant of integration can be absorbed into λ, we obtain to the first approximation

$$\mathscr{G}(\omega) = \frac{1}{\omega + i\delta}\left[1 - 2\left(\frac{J}{N}\nu\right)^2 S(S + 1) \ln (\varepsilon_F/\omega)\right] \tag{22}$$

Thus, the pseudofermion Green function acquires logarithmic corrections.

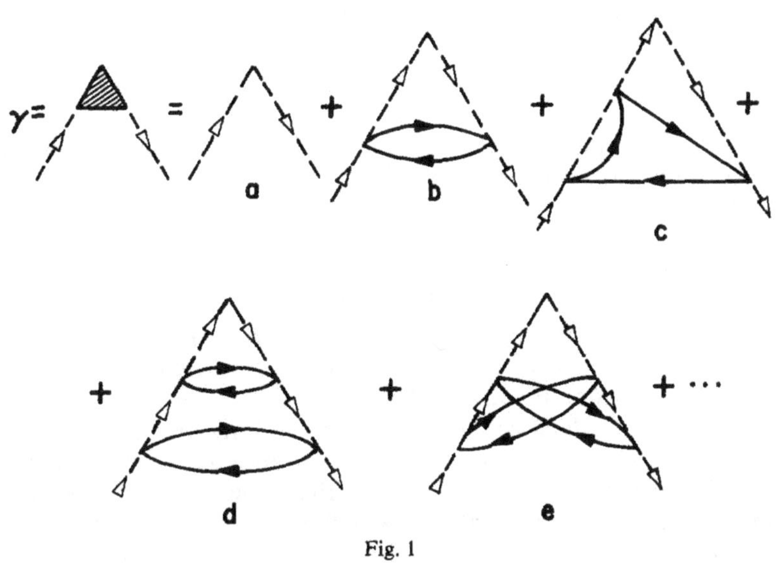

Fig. 1

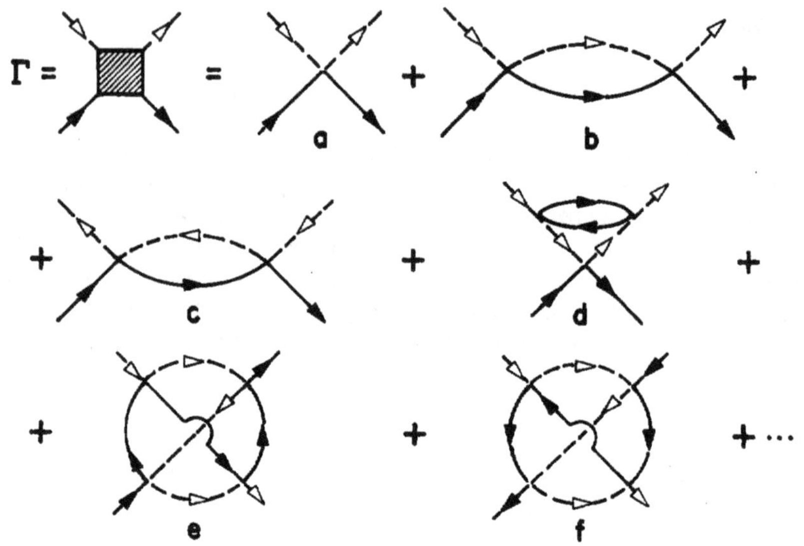

Fig. 2

Other logarithmic corrections enter into the scattering amplitude of an electron by an impurity. The first correction comes from Fig. 2(b and c). In the first approximation (provided all external frequencies are of the same order) we obtain the following, to logarithmic accuracy:

$$\Gamma^{(1)} = \left(-\frac{J}{N}\boldsymbol{\sigma}\mathbf{S}\right)\left[-2\frac{J}{N}\nu \ln(\varepsilon_F/\omega)\right] \tag{23}$$

Another type of logarithmic correction comes from Fig. 2(d). It is equal to

$$\Gamma^{(1')} = \left(-\frac{J}{N}\boldsymbol{\sigma}\mathbf{S}\right)\cdot 2[S(S+1)-1]\left(\frac{J}{N}\nu\right)^2 \ln(\varepsilon_F/\omega) \tag{24}$$

Consider now more complex diagrams. We start with the approximation which we call the "logarithmic" approximation. The first term for Γ is obtained by mutual insertion of the elements in Fig. 2(b and c). This is the "parquet" considered in Ref. 3, which can be written in terms of full vertices. The parquet term is sufficient as long as $(J\nu/N)\ln(\varepsilon_F/\omega) \lesssim 1$ but $\Gamma \ll 1$. As Γ approaches unity it is necessary to include the higher-order corrections.

The first such correction for Γ is shown in Fig. 2(d) and for γ in Fig. 1(b). We can incorporate all corrections of this type by substituting $\mathscr{G}$ for $\mathscr{G}^{(0)}$ in the parquet term and writing Fig. 2(d) in terms of the total functions Γ and $\mathscr{G}$. Next come Fig. 2(e and f), which contain the logarithm only two times less than the parquet diagram of the same order. We write it once again in terms of the full vertices and Green functions $\mathscr{G}$. A term of the same type is derived from the preceding diagrams if the principal term with the maximum number of logarithms is subtracted.

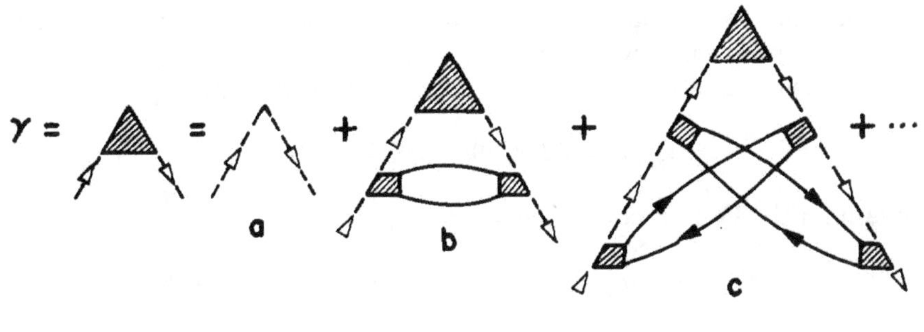

Fig. 3

Following this procedure, we get a sequence of terms containing full vertices and $\mathcal{G}$ functions, each of which represents a single logarithmic integral over a domain in which all internal frequencies are of the same order. The same procedure is applicable to γ (Fig. 3). In this approach we neglect the nonlogarithmic parts of the integrals, which originate from the residues in the integration over the internal frequencies and which contain Γ and $\mathcal{G}$ with arguments of the same order as the external frequency. We shall discuss below the extent to which this approximation is suitable for our purposes.

The given approximation is clearly renormalizable. The invariant charge, which is left unaltered by renormalization, is the combination $\Gamma\mathcal{G}$ or Γ/γ. The latter plays the part of the true scattering amplitude of an electron by an impurity. This ensues from the fact, demonstrated in Ref. 3, that the transport time of the electron is determined by the diagram in Fig. 4, in which the internal lines correspond to the imaginary parts of the appropriate Green functions.

As noted, if all the diagrams for Γ and γ (expressed in terms of full quantities $\Gamma, \mathcal{G}$) are limited to logarithmic integrals, then Γ with arguments of the same order will again be expressed in terms of the same type of Γ. Moreover, it can be shown in this case that Γ is necessarily proportional to $\boldsymbol{\sigma}\mathbf{S}$. This follows from the symmetry between electrons and holes.

The initial Hamiltonian

$$H = \int \Psi_\alpha^+ \varepsilon \Psi_\alpha \, dv - \frac{J}{N} \Psi_\alpha^+(0) \sigma_{\alpha\alpha'}^i \Psi_{\alpha'}(0) a_\beta^+ S_{\beta\beta'}^i a_{\beta'}$$

where $\varepsilon = p^2/2m - \mu$ is invariant under the transformation $\Psi \to g\Psi^+$, $\varepsilon \to -\varepsilon$, where $g = \begin{pmatrix} 0 & +1 \\ -1 & 0 \end{pmatrix}$ is the metric matrix for spinors. The change of the sign of ε is actually inconsequential, since the quantities with which we are concerned always

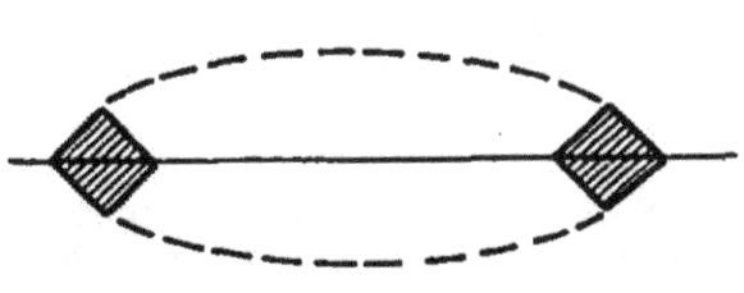

Fig. 4

involve integrals over ε with symmetric limits $((-\varepsilon_F \div \varepsilon_F)$; see Ref. 3). This is how the electron–hole symmetry enters into the picture.

Carrying out the transformation $\Psi \to g\Psi^+$ in the quantity

$$W_{\alpha\beta\alpha'\beta'} = \langle T\Psi_\alpha(0)a_\beta(t_1)\Psi^+_{\alpha'}(t_2)a^+_{\beta'}(t_3)\rangle$$

we obtain for the Fourier components

$$W(\omega_1, \omega_2, \omega'_1, \omega'_2) = -gW^+(-\omega'_1, \omega_2, -\omega_1, \omega'_2)g^{-1} \tag{25}$$

where the frequencies ω_1 and ω'_1 refer to the electron. From this it follows that

$$W(\omega, 0, \omega, 0) = W_1(\omega)\,\mathrm{sgn}\,(\omega) + W_2(\omega)\boldsymbol{\sigma}\mathbf{S} \tag{26}$$

where $W_{1,2}(\omega)$ are even functions. The same applies to Γ with frequencies of the same order to logarithmic accuracy. Inasmuch as sgn ω can only appear as a result of taking a residue in integration, then as long as the diagrams are limited to logarithmic integrals such terms will be absent.

We denote $\Gamma = \boldsymbol{\sigma}\mathbf{S}\tilde{\Gamma}/\nu$. Since all the quantities depend only on $\ln(\varepsilon_F/\omega)$, the renormalizability condition yields

$$\frac{d}{dx}(\ln \gamma) = F_1(z)$$

$$\frac{d}{dx}(\ln \tilde{\Gamma}) = F_2(z) \tag{27}$$

$$z = \tilde{\Gamma}/\gamma$$

where $x = \ln(\varepsilon_F/\omega)$. Combining these two equations, we have

$$\frac{d}{dx}(\ln z) = F_2(z) - F_1(z) = F(z) \tag{28}$$

These equations need to be augmented by the boundary conditions $\gamma(x = 0) = 1$, $\tilde{\Gamma}(x = 0) = -J\nu/N$.

The first terms of the expansion of the functions F_1 and F_2 on z are obtained from the first diagrams. Thus, we have*

$$F_1(z) = 2S(S + 1)z^2 + 0(z^4)$$

$$F_2(z) = 2z + 2[S(S + 1) - 1]z^2 - \left\{\frac{8\pi^2}{3}[S(S + 1) - \tfrac{1}{2}] - 4\right\}z^3 + 0(z^4)$$

$$F(z) = 2z - 2z^2 - \left\{\frac{8\pi^2}{3}[S(S + 1) - \tfrac{1}{2}] - 4\right\}z^3 + 0(z^4) \tag{29}$$

*Note that in order to obtain the terms $\sim z^3$ in F_2 and F it is necessary to take into account not only diagrams of the type 2*e* and 2*f*, but also the corresponding parts of the diagrams of the parquet class.

Now consider the behavior of the solution of Eq. (15) for small x. Inasmuch as $|\tilde{\Gamma}| = |J|\nu/N \ll 1$ for $x = 0$, for small x we can limit ourselves with the first few terms of (29). Assuming $F(z) = 2z$, we obtain the following result corresponding to Eq. (1):

$$z = \frac{\tilde{\Gamma}}{\gamma} = -\frac{J\nu/N}{1 + 2J\nu x/N}$$

This expression is valid for $z \ll 1$.

In the case $J > 0$, beginning with the value $-J\nu/N$ for $x = 0$, the absolute value of z decreases with increasing x, and, as a result, the subsequent terms in (29) never become significant. Consequently, with $J > 0$, $J\nu/N \ll 1$ Eq. (1) is valid for any ω.

The situation is altered for $J < 0$. In this case z increases with x, and it is therefore required to include higher-order terms in $F(z)$.

Two hypotheses present themselves with regard to the subsequent behavior of $z(x)$. One possibility is the case when $z \to$ const as $x \to \infty$. This case corresponds to the first of the two possibilities set forth in the introduction, i.e., z reaches a limit such that the contributions of all the diagrams become of the same order. According to Eq. (28), this requires that as z increases the function $F(z)$ begins to decrease, attaining a value $F(z_0) = 0$ for some z_0.

In the neighborhood of this point we can write

$$\frac{1}{z_0}\frac{dz}{dx} = F'(z_0)(z - z_0) \qquad [\text{where } F'(z_0) < 0]$$

Hence it follows that

$$z \to z_0 - \alpha e^{z_0 F_0' x} = z_0 - a(\omega/T_K)^{-z_0 F_0'} \tag{30}$$

where $a \sim z_0 \sim 1$. We find in this case, therefore, that as $x \to \infty$ the function $\tilde{\Gamma}/\gamma = z$ tends to a finite limit by a power law. The functions γ and $\tilde{\Gamma}$ themselves obey a power law with respect to ω as $\omega \to 0$:

$$\tilde{\Gamma} \sim \gamma \propto (\omega/T_K)^{-F_1(z_0)} \tag{31}$$

Another case is possible wherein the function $F(z)$ does not attain a zero value. According to Eq. (28), the role of the diagrams now grows as their order increases, and the situation corresponds to the second possibility indicated in the introduction. We shall not pursue this case further.

However, even in the case investigated here there is a flaw in our reasoning. So far we have confined ourselves exclusively to logarithmic integrals. This was permissible as long as z was small. However, it may be concluded from the first terms written for the function F that $z_0 \sim 1$. In this case the nonlogarithmic parts of the integrals become significant. We are at once confronted with a number of new difficulties. First, Γ acquires, in addition to the term proportional to $\boldsymbol{\sigma}\mathbf{S}$, a term that is independent of $\boldsymbol{\sigma}$. Then, an explicit dependence on ratios of the external frequency appears.

However, even in this situation the power laws are justifiable if we rely, not on the renormalizability equations, but on the hypothesis that all the diagrams become of the same order in the strong-coupling domain (see Ref. 7). This condition together with the Ward identity (20) is satisfied by the functions

$$\gamma(\omega_1, \omega_2) = \omega_1^{\alpha} f(\omega_2/\omega_1)$$

$$\Gamma(\omega_1, \omega_2, \omega_3) = \omega_1^{\alpha}[f_1(\omega_2/\omega_1, \omega_3/\omega_1) + \boldsymbol{\sigma}\mathbf{S} f_2(\omega_2/\omega_1, \omega_3/\omega_1)] \tag{32}$$

In order to prove this we need to be able to write down equations (at least in infinite-series form) involving only full lines and vertices. The procedure followed above is only applicable in the logarithmic approximation. This is because the diagrams entering into the "parquet" are expressed in terms of the full Γ only with logarithmic accuracy.

This difficulty can be overcome, however, if instead of the direct electron–impurity interaction we introduce particles transmitting that interaction and write

$$H'_{int} = \lambda_1 \Psi^+(0)\sigma^i \Psi(0)\eta_i + \lambda_2 a^+ S^i a \eta_i^+ + M\eta_i^+\eta_i \tag{33}$$

(where $M \gg \varepsilon_F$). The above Hamiltonian is non-Hermitian, but for $M \to \infty$ it yields correct diagrams and facilitates their classification. By comparing the diagrams in theories with H_{int}(2) and H'_{int} (33), one readily grasps their mutual correspondence, where

$$\frac{\lambda_1 \lambda_2}{M} = \frac{J}{N}$$

We can introduce the full vertex for the interaction of pseudofermions with η particles. It may be expressed as a sum of diagrams involving the full functions. The same is true of the function γ, from which the pseudofermion Green function is derivable by means of the Ward identity. In this form the power laws (19) ensue from the requirement that all the diagrams be of the same order. Sample diagrams for the total vertex are shown in Fig. 5.

Somewhat more knowledge can be acquired with regard to the case $S \to \infty$. The spin S enters into the coefficients of the expansion in Eq. (29). It is reasonable to assume that for this reason the value $F(z_0) = 0$ should be attained for some small z_0. Clearly the most interesting in this respect are terms containing the maximum power of S in a given order.

Let us look more thoroughly at the logarithmic series (29). We note, first of all, that the z^2 term does not contain S, so that as long as z is small this term is inconsequential. It was obtained from diagrams of the type in Fig. 1(b) and 2(d) where the electron loop encircles bare vertices. If we consider all diagrams of this type for the functions F_1 and F_2, we notice at once that they differ only to the extent that in F_1 the loops encircle unity, whereas in F_2 they encircle the operator **S**. Consequently in the difference of such diagrams the power of S is lowered by 2. As for the remaining diagrams, inasmuch as $\Gamma \sim \boldsymbol{\sigma}\mathbf{S}$, the diagrams in Fig. 2 containing even powers of Γ are necessarily combined so as to make the power of S in the function $F(z)$ lower than the order of the diagram by 2. These diagrams give $F(z)$ terms with odd powers of z. As for the diagrams containing odd powers of Γ, they also cannot give the maximum power of S. This is related to the fact that the logarithmic

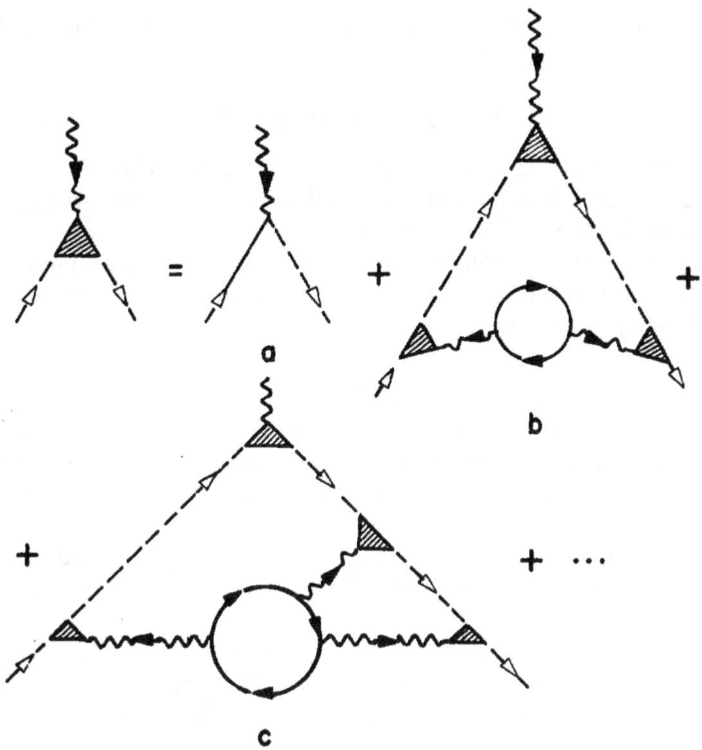

Fig. 5

integrals originate from the noncommutativity of the spin operators. In the case when the spin is regarded as classical the logarithmic integrals do not occur. Consequently, there occurs here a reduction in the power by 2.

Thus, we should expect F to acquire terms of the type S^2z^3, S^2z^4, S^4z^5, $S^4z^6, \ldots$. It is clear that as long as z is small the even terms will always be smaller than the odd terms in $F(z)$. The condition $F(z_0) = 0$ in this case gives $z_0 \sim 1/S$, which is in fact small. In order to find the coefficient it is necessary in principle to know the entire series of terms with odd powers of z.

Despite the smallness of z_0, we are not justified in neglecting the nonlogarithmic terms in this case. Indeed, the nonlogarithmic part in the parquet gives $S^2z^2 \sim 1$. On the other hand, the amplitude found in the logarithmic approximation, when squared, gives $S^2z^2 \sim 1$. Thus, both parts offer comparable contributions to the scattering cross section.

Since the actual calculation of the scattering amplitude presents enormous difficulties, the best we can hope for is an estimation of the behavior of the powers as a function of S. Considering that the nonlogarithmic terms are of the same order as the logarithmic terms for all S, it seems to us that the logarithmic approximation affords reasonably reliable information as to the order of the powers. We shall make use of this in the following discussion. For the time being we merely note that for $S \gg 1$ the power exponent in Eq. (30) is of the order $1/S$, while in (31) it is of unit order.

4. PHYSICAL IMPLICATIONS

Although the foregoing results contain some unknown functions or constants, they still permit us to draw a number of qualitative conclusions. We begin with the resistance as a function of the temperature.

Inasmuch as the scattering amplitude of an electron by an impurity is proportional to the invariant charge z, on the basis of the general rule implied by renormalizability we can write

$$\rho = \frac{mN_i}{e^2 v} f(S, T/T_K) \tag{34}$$

where N_i is the impurity concentration and f is a universal function. This equation must hold for all temperatures and any sign of J, regardless of whether or not the earlier assumption is true that all diagrams become of the same order in the strong-coupling domain.

In the preceding section we calculated the scattering amplitude for $T = 0$ as a function of ω. The resistance is proportional to the scattering amplitude squared for $\omega \sim T$.

In the case $J > 0$ the logarithmic approximation suffices. Then in the expression for the amplitude for $T = 0$ we can simply replace ω by T, whereupon we obtain the equation of Ref. 3 corresponding to (34), where

$$f = \frac{\pi}{4} S(S+1)[\ln(T_K/T)]^{-2}$$

$$T_K = \varepsilon_F \exp(N/2Jv) \gg \varepsilon_F$$

In the case $J < 0$ it is necessary to recognize that for $\omega \ll T$ the temperature begins to play the role of the lower limit of the integrals, rather than ω. It follows from the matching condition for $\omega \sim T$ that the exponents in the power laws deduced in the preceding section carry over to the temperature dependence for $\omega \lesssim T$. Hence, it follows that for $T \ll T_K$

$$\rho = \frac{mN_i}{e^2 v}[c(S) - a(S)(T/T_K)^{b(S)/S}] \tag{35}$$

where the coefficients $a(S)$, $b(S)$, and $c(S)$ are of order of unity and tend to finite limits as $S \to \infty$.

Two things are worth noting. Equation (35) implies that the power exponent in the temperature term depends on the impurity spin and it is by no means necessarily true that the value of $b(S)/S = 2$ with which the experimental data are usually compared. Moreover, for sufficiently large spin this power is less than unity, i.e., for $T = 0$ the curve for $\rho(T)$ will have a vertical, rather than a horizontal, tangent.

The second consideration is the fact that for large spins the expansion in the coefficients of (29) is in powers of $[S(S+1)]^{-1}$. Consequently, the difference between the coefficients $a(S)$, $b(S)$, and $c(S)$, even for spins such as $\frac{3}{2}$ and 2, must be about 10% and vanish rapidly for large spins.

Next we consider the magnetic moment. According to (16) we have

$$M = g\mu S\phi[\ln(T_K/T), g\mu H/T, S] \tag{36}$$

This relation implies, in particular, that for $T = 0$

$$M(H) = g\mu S\phi[\ln(T_K/g\mu H), S] \tag{37}$$

and, conversely, for $g\mu H \ll T$

$$\frac{\partial M}{\partial H} = \chi = \frac{(g\mu)^2}{T}\phi[\ln(T_K/T), S] \tag{38}$$

These equations are of a general character.

We now seek more detailed results. We begin with the case $T = 0$. It will be more to our advantage to use the second type of diagram technique for $T = 0$, in which the large pseudofermion energy λ is not introduced and the free Green function is described by Eq. (19). The magnetic moment of the impurity is

$$\mathcal{M} = g\mu \sum_M M \int \mathcal{G}_{MM}(\omega)e^{i\omega\tau}\frac{d\omega}{2\pi i} \tag{39}$$

where $\tau \to +0$. The quantity on the right-hand side represents the sum of all loops closed on the spin $S_z = M$. The first diagrams are shown in Fig. 6. The diagram in Fig. 6(a) corresponds to

$$\mathcal{M}_0 = g\mu S \tag{40}$$

The calculation of the diagram in Fig. 6(b) yields the result

$$\mathcal{M}_1 = g\mu S[-2(J\nu/N)^2\ln(\varepsilon_F/g\mu H)] \tag{41}$$

In the integral for $\mathcal{M}_1$ the logarithm can be obtained as follows. One of the frequencies corresponding to the $\mathcal{G}$ lines is put $\omega \sim g\mu H$, while the integration over the other frequency is in the interval $g\mu H \ll \omega \ll \varepsilon_F$. It is readily seen that now we obtain either a logarithmic correction to the vertex of the $\mathcal{G}$ line or to the lower vertex with S_z. We denote the full vertex with S_z as $S_z\gamma_S$ and regard γ_S as a slowly varying (logarithmic) function. We can write the following to logarithmic accuracy:

$$\mathcal{M} = g\mu S\cdot\gamma_S(g\mu H)/\gamma(g\mu H) \tag{42}$$

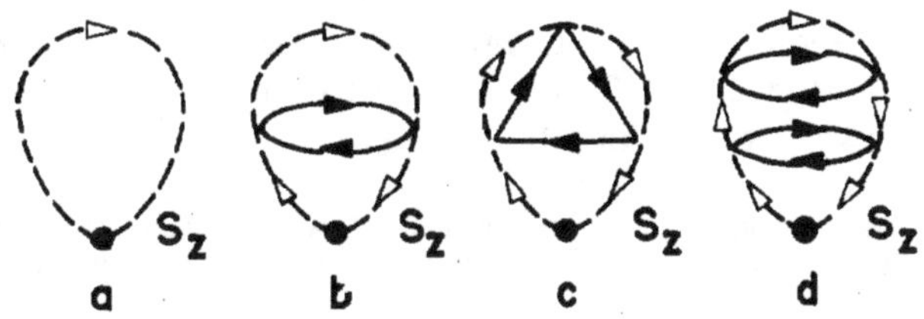

Fig. 6

The quantity γ_S is also renormalizable, and to Eq. (22) we can add

$$\frac{d}{dx}\ln\gamma_S = F_3(z)$$

Combining with (42), we obtain

$$\frac{d}{dx}\ln\mathcal{M} = F_3(z) - F_1(z) \tag{43}$$

In the first approximation $F_3(z)$ has the form

$$F_3(z) = 2[S(S+1) - 1]z^2 + 0(z^4) \tag{44}$$

For the difference $F_3 - F_1$ we obtain in this order $-2z^2$.

For $J > 0$ we can restrict everything to the first approximation. Solving Eq. (43), we obtain

$$\mathcal{M} = g\mu S\left[1 - \frac{2(J\nu/N)^2 x}{1 + 2(J\nu/N)x}\right] \tag{45}$$

where $x = \ln(\varepsilon_F/g\mu H)$. Consequently, in this case the impurity spin is screened as $H \to 0$ only by a small quantity of the order $J\nu/N$.*

The situation is quite different for $J < 0$. As $H \to 0$ we obtain a constant in Eq. (43), where, inasmuch as the impurity spin clearly cannot grow to infinity at $H = 0$, we must conclude that this constant is negative. Hence, the quantity $\mathcal{M}$ decreases according to a power law as $H \to 0$.

This brings us to an interesting observation with regard to the large spin case. We have already seen that in the first approximation the principal term proportional to S^2 cancels out of the difference $F_3 - F_1$. It is readily perceived that the same thing happens in the next orders. In fact, the diagrams for γ_S and γ have exactly the same appearance, and the entire difference between diagrams of the same type is associated with commutations of the spin operators. For this reason the power of S is lowered by 2 in the difference between two such diagrams. Therefore, the series for $F_3 - F_1$ contains terms $z^2, S^2z^4, \ldots$. When z reaches its limit $z_0 \sim 1/S$ all these terms become of the same order, namely, $1/S^2$.

Consequently, for $T = 0$ and $g\mu H \ll T_K$ we have

$$\mathcal{M}(H, T = 0) = g\mu S p(S)(g\mu H/T_K)^{q(S)/S^2} \tag{46}$$

where $p(S)$ and $q(S)$ are constants of the order of unity, which tend to finite limits as $S \to \infty$.

Thus, the foregoing arguments reveal that for $J < 0$ the impurity spin is *completely screened* for $H \to 0$, $T \to 0$; for not too small S the power-law dependence of $\mathcal{M}(H)$ is weaker than for $\rho(0) - \rho(T)$.

*Equation (45), in which the small corrections are included, satisfies relation (14). This is evident if we write the following, which is obtained directly from the solution of Eq. (43) before expansion in $J\nu/N$:

$$\mathcal{M} = \mathcal{M}_0 \exp\left[J\nu/(N + 2J\nu x) - J\nu/N\right]$$

We now consider the susceptibility for $T \gg g\mu H$. This requires that we use the temperature technique of Ref. 3 with the pseudofermion energy λ. Now, the pseudofermion magnetic moment is equal to

$$\mathcal{M}(H, T) = g\mu \frac{\sum_M M T \sum_\omega \mathcal{G}_{MM}(\omega) \exp(i\omega\tau)}{\sum_M T \sum_\omega \mathcal{G}_{MM}(\omega) \exp(i\omega\tau)}, \tau \to +0 \tag{47}$$

Again in order to exhibit the qualitative character of the effects we confine our discussion to the logarithmic approximation. The following is essential here. For $T = 0$, working with real frequencies and measuring ω relative to λ, we have $\mathcal{G}(\omega) = \mathcal{G}_0(\omega)/\gamma$. Since γ is a logarithmic function, it is permissible with logarithmic accuracy to replace the frequency ω in γ by $i\omega - \lambda$ (we return to the old origin). The significant values of the argument $i\omega - \lambda$ will be of the order $(g\mu H, T)$. The numerator will contain an expression of the same type as calculated in the preceding section. This yields the factor γ_S/γ for max$(T, g\mu H)$. In the denominator we also have a sum of diagrams of the type in Fig. 6, but without S_z in the vertex. Therefore, in place of γ_S/γ we get 1.

Thus, all of the foregoing implies that

$$\mathcal{M}(H, T) = \mathcal{M}_0(H, T)\cdot\frac{\gamma_S}{\gamma} \tag{48}$$

where $\mathcal{M}_0$ is the free-impurity magnetic moment described by the Brillouin function

$$\mathcal{M}_0 = g\mu S B_S(g\mu H S/T) \tag{49}$$

In particular, for $g\mu H \ll T$ we obtain the susceptibility for small fields:

$$\chi = \frac{(g\mu)^2 S(S+1)}{3T}\gamma_S(T)/\gamma(T) \tag{50}$$

Consequently, for $J > 0$ the susceptibility is practically the same as before. But if $J < 0$ the situation changes significantly, and a new power-law dependence occurs. We thus have for $g\mu H \ll T \ll T_K$

$$\chi = \frac{(g\mu)^2 S(S+1)}{3T} p_1(S)(T/T_K)^{q(S)/S^2} \tag{51}$$

where $p_1(S) \sim 1$ and is finite as $S \to \infty$. The power $q(S)$ must remain the same as in (46) according to the matching condition for $g\mu H \sim T$. In view of the fact that $S(S+1) = \frac{15}{4}$ already for $S = \frac{3}{2}$, it may be inferred that $\chi \to \infty$ as $T \to 0$, although by a weaker law than $1/T$.

The above inference contradicts most of the theoretical results in which it has been stated that $\chi \to$ const as $T \to 0$. However, there is experimental evidence that $\chi(T)$ does indeed grow.[8] Moreover, the behavior of the magnetic moment for T, $g\mu H \ll T_K$ is determined by the relation of $g\mu H$ and T. In particular, if one measures the susceptibility in a finite field, then for $T \ll g\mu H$ it ceases to depend on the temperature, and the constant value attained in this event is of the order

$$\frac{g\mu S}{H}(g\mu H/T_K)^{q(S)/S^2}$$

The total power of H is negative (for not too small S), so that this constant becomes greater the smaller the field.

Another curious fact comes to our attention. The limit of the logarithmic integrals is $\max(g\mu H, T)$, whereas the Brillouin function involves the ratio $g\mu HS/T$. Therefore, for $S \gg 1$ there is an interval $T \gg g\mu H \gg T/S$ in which

$$\mathcal{M} = g\mu Sp_1 \cdot (T/T_K)^{q/S^2}, p_1 = p_1(\infty), q = q(\infty) \tag{52}$$

Consequently, for the smallest fields the magnetic moment depends linearly on the field; it then reaches saturation, and with a further increase in the field it begins once again to increase.

All of these considerations, of course, apply to the case when $g\mu H \ll T_K$, $T \ll T_K$.

5. CONCLUSION

We have thus obtained two types of equations. One is deduced solely from the renormalizability of the theory and must be necessarily true if the assumed form of electron–impurity interaction (2) is valid.

The other equations, which pertain to the more detailed behavior of the variables for $J < 0$, $T \ll T_K$, must be fulfilled only in the case if the assumed hypothesis is valid with regard to the behavior of the diagrams in the strong-coupling domain. Of special significance in this respect are the predictions of the growth of $\chi(T)$ with decreasing temperature and of the change of the dependence $\rho(T)$ for sufficiently large spins (appearance of the vertical tangent at $T = 0$).

In the case the experimental data do not confirm these predictions, it must be supposed that the second of the indicated possibilities described in Section 3 is realized, i.e., that the diagrams grow with increasing order. Not to be overlooked is the possibility that the "bound-state" approach might afford a better approximation in this case.

ACKNOWLEDGMENTS

We would like to take this opportunity to thank I. E. Dzyaloshinskii, A. I. Larkin, and V. L. Pokrovskii for a discussion of the paper.

NOTE ADDED IN PROOF

After this article was submitted for publication, we received the preprint of M. Fowler and A. Zawadowski, where the ideas of renormalizability were applied to the Kondo problem. The equation analogous to ours (28) was obtained but only the first two terms in $F(z)$ were taken into account in this work. An arbitrary neglection of higher-order terms led to $b(S)/S = 2$ in (35) for which we do not see any theoretical reasons. Moreover, the best fit of the experimental results (W. A. Steyert, private communication) corresponds to $b(S)/S = 1.3 \div 1.6$ for $S = 3/2$. The authors would like to express their gratitude to Dr. W. A. Steyert for this remark.

REFERENCES

1. J. Kondo, *Progr. Theor. Phys.* **32**, 37 (1964).
2. A. A. Abrikosov, *Usp. Fiz. Nauk* **97**, 403 (1969).
3. A. A. Abrikosov, *Physics* **2**, 21 (1965).
4. H. Suhl, *Phys. Rev.* **138**, A515 (1965).
5. H. H. Bogolyubov and D. V. Shirkov, *Introduction to the Quantum Theory of Wave Fields* [in Russian], Fizmatgiz, Moscow, 1957.
6. V. V. Sudakov, *Zh. Eksp. Teor. Fiz.* **31**, 729 (1956).
7. V. N. Gribov and A. A. Migdal, *Zh. Eksp. Teor. Fiz.* **55**, 1498 (1968); A. A. Migdal, *Zh. Eksp. Teor. Fiz.* **55**, 1964 (1968); A. A. Migdal, *Zh. Eksp. Teor. Fiz.* **59**, 1015 (1970).
8. M. D. Daybell and W. A. Steyert, *Phys. Rev.* **167**, 536 (1968).

SOVIET PHYSICS JETP VOLUME 32, NUMBER 4 APRIL, 1971

POSSIBLE EXISTENCE OF SUBSTANCES INTERMEDIATE BETWEEN METALS AND DIELECTRICS

A. A. ABRIKOSOV and S. D. BENESLAVSKIĬ

L. D. Landau Institute of Theoretical Physics

Submitted April 13, 1970

Zh. Eksp. Teor. Fiz. 59, 1280–1298 (October, 1970)

The question of the possible existence of substances having an electron spectrum without any energy gap and, at the same time, not possessing a Fermi surface is investigated. First of all the question of the possibility of contact of the conduction band and the valence band at a single point is investigated within the framework of the one-electron problem. It is shown that the symmetry conditions for the crystal admit of such a possibility. A complete investigation is carried out for points in reciprocal lattice space with a little group which is equivalent to a point group, and an example of a more complicated little group is considered. It is shown that in the neighborhood of the point of contact the spectrum may be linear as well as quadratic.

The role of the Coulomb interaction is considered for both types of spectra. In the case of a linear dispersion law a slowly varying (logarithmic) factor appears in the spectrum. In the case of a quadratic spectrum the effective interaction becomes strong for small momenta, and the concept of the one-particle spectrum turns out to be inapplicable. The behavior of the Green's functions is determined by similarity laws analogous to those obtained in field theory with strong coupling and in the neighborhood of a phase transition point of the second kind (scaling). Hence follow power laws for the electronic heat capacity and for the momentum distribution of the electrons.

1. INTRODUCTION

ONE of the basic assumptions of the Landau[1] theory of a Fermi liquid is the relationship, according to which the limiting momentum of the excitations in an isotropic Fermi liquid is determined in the same way as in a gas, by the density of the atoms in the liquid, i.e., $p_0 = (3\pi^2 n)^{1/3}$ (where n is the density of particles). According to the work of Luttinger and Ward,[2] with a certain amount of alteration this theorem is also applicable to the electronic liquid in metals. Namely, it turns out that the total volume, bounded by all Fermi surfaces, determines the density of the electrons:

$$n = 2(2\pi)^{-3}(pV_0 + V_F),$$

where V_0 denotes the volume of an elementary cell of the reciprocal lattice, p is an integer, and V_F denotes the total volume inside all of the Fermi surfaces (from the side of smaller energies), referred to a single cell of the reciprocal lattice.

It is not difficult to see that the cited relation is exactly the same as for a noninteracting Fermi system in a periodic field. In view of the fact that $V_0 = (2\pi)^3/v$, where v is the volume of an elementary cell of the crystal, the substance can be dielectric only when the number of electrons per elementary cell of the crystal is even. If in this case the substance nevertheless is a metal, then the number of holes must necessarily be equal to the number of electrons. For metals with an odd number of electrons per cell, on the other hand, such compensation cannot occur.

In this connection the question arises whether substances can exist in nature which are neither genuine metals nor dielectrics in the sense that there are no free electrons in them, and at the same time no energy gap is present. One can represent such a substance as the limiting case of a metal with a point Fermi surface. In the language of the theory of noninteracting particles in a periodic field, this corresponds to the case when a completely filled valence band touches a completely empty conduction band. Indications of the possible existence of such substances have recently appeared in the literature (gray tin[3] and mercury telluride[4]).

In the present article the conditions under which a point Fermi surface may appear in a model of noninteracting electrons will be investigated, and the question of the influence of the Coulomb interaction of the electrons on the energy spectrum in the neighborhood of the Fermi point will be considered.

In the case of interest to us, the maximum of the valence band must coincide with the minimum of the conduction band. In general an increased symmetry of the appropriate point of the reciprocal lattice favors degeneracy of the energy levels, but it may also occur at accidental points. However, here somewhat more is required, namely, that this point simultaneously corresponds to extrema of the two bands. The occurrence of such a situation at an accidental point is an improbable special case. In view of this we shall consider only "regular" cases, i.e., points of high symmetry. A complete analysis is given in the case when the little group of the corresponding point is equivalent to the point group, and an example of a more complicated group is considered. It turns out that a "Fermi point" is possible in a whole series of different cases, where in the most typical cases the dispersion law in the vicinity of such a point will either be linear or quadratic.

However, in general a model of noninteracting electrons does not correspond to a real electronic liquid. In the case of an isotropic liquid with short-range interaction forces, Landau was able to show (see[5]) that the energy spectrum of the quasi-particles retains the same form as for a Fermi gas. This proof was then generalized to a Fermi liquid in a crystal[2,6] and to a Fermi liquid with long-range Coulomb forces.[7] In the last case the fact that the presence of "free electrons" leads to a screening of the Coulomb field and converts it into a short-range field is extremely important. Close-lying excited states play the most important role in the integral which determines the screening. Therefore, in any case the screening is absent if there is a gap in the spectrum. On the other hand, in a real metal the screening radius is of the order of interatomic distances.

From this point of view it is necessary to investigate what kind of influence the Coulomb field has in the case of a "Fermi point." The absence of any gap in the spectrum must necessarily lead to a weakening of the Coulomb potential at large distances. At the same time the replacement of the Fermi surface by a point leads to a change in the nature of the screening, and it is not obvious beforehand how this will affect the energy spectrum.

The usual way to prove the presence of a Fermi spectrum in the theory of a Fermi liquid (see[5]) consists in the following. It is assumed that the spectrum has the appropriate form and, starting from this, the Green's function is written down. After this it is proved that the constant coefficients appearing in it are actually constants which depend on the interaction of the particles. This method of proof is universal for that case when one can foresee the result beforehand. However, in the case under consideration this is by no means so. Therefore, let us proceed according to a different method. Let us consider the Coulomb interaction, which we shall assume to be weak and, using the Green's function of noninteracting particles as the zero-order approximation, let us find the Green's function and the other characteristic quantities associated with the presence of the interaction.

The supposition about the weakness of the interaction may be justified by the following consideration. In the cases under consideration there exists a large range of momenta for which the valence band is close to the conduction band. As we shall see below, in the very neighborhood of the point of contact between the bands a nontrivial situation arises which, in particular, leads to a change in the nature of the interaction at small values of the momenta. However, the nearness of the bands is preserved even at a certain distance from this neighborhood. If we shall consider the electronic transitions outside the "dangerous neighborhood", then they lead to the appearance of the ordinary dielectric constant ϵ_0 which, in view of the proximity of the bands, should be rather large (as, for example, in Ge and Si). In view of this the bare Coulomb interaction $e^2/\epsilon_0 r$ or, more precisely, the characteristic dimensionless constant $e^2/\epsilon_0 v$, where $v \sim 10^8$ cm/sec, must be small. Of course, in addition the case is possible when the dielectric constant is not large and consequently the interaction will not be weak which, in turn, leads to an increase of the "dangerous neighborhood" up to dimensions which are comparable with the period of the reciprocal lattice. However, to judge from the existing data on α-Sn and HgTe, this case scarcely occurs. But even if ϵ_0 is not overly large then, in any case, with the aid of a study of the weakened interaction one can formulate some idea about what happens in the neighborhood of the Fermi point.

We have considered the two most typical cases: linear and quadratic spectra in the neighborhood of the contact point. It turns out that both types of spectra lead to quite different consequences upon taking the interaction into account. One can assume that these two characteristic types of behavior exhaust all possibilities.

2. CONTACT IN THE GROUPS D_n

First of all let us consider a point $\mathbf{k}_0$ in reciprocal lattice space with a little group which is equivalent to one of the point groups. Since the question involves the contact of bands, then degeneracy of the levels must occur at the point $\mathbf{k}_0$, which is lifted (if only partially) upon a change of $\mathbf{k}$ in an arbitrary direction. This means that the little group of $\mathbf{k}_0$ must have a multidimensional representation, i.e., contain axes of the third, fourth, or sixth orders. In addition, it is obvious that an intersection of the axes and planes of symmetry must occur at the point $\mathbf{k}_0$. If the spin-orbit coupling is small enough so that one can neglect it, then it is necessary to consider single-valued irreducible representations. Upon taking account of the spin-orbit coupling it is necessary to consider double-valued representations.

Let us start with simple groups. Let us demonstrate that contact is possible in the groups D_3, D_4, and D_6. As is shown in article[5], in order to determine the dispersion law it is necessary to construct an invariant form $Q(\mathbf{k};\, \mathbf{r}, \mathbf{r}')$ containing the components of the vector $\mathbf{k}$ and which is bilinear in the basis functions $u_\alpha(\mathbf{r})$ and $u_\alpha^*(\mathbf{r}')$. Let us consider single-valued representations. In the groups under consideration, one invariant exists which is linear in k: $[u_+(\mathbf{r})u_+^*(\mathbf{r}') - u_-(\mathbf{r})u_-^*(\mathbf{r}')]k_z$. The presence of such an invariant is very important because it is precisely this property which guarantees the possibility of a "Fermi point", where the functions $u_\pm$ transform like $x \pm iy$. With the components k_x and k_y one can form only the quadratic invariants

$$u_+(\mathbf{r})u_-^*(\mathbf{r}')k_-^2 + u_-(\mathbf{r})u_+^*(\mathbf{r}')k_+^2, \quad k_\pm = k_x \pm ik_y$$

and

$$[u_+(\mathbf{r})u_+^*(\mathbf{r}') + u_-(\mathbf{r})u_-^*(\mathbf{r}')]k_\perp^2, \; k_\perp^2 = k_x^2 + k_y^2.$$

Since the latter combination of $u_\pm$ is an invariant by itself, then one can also multiply it by the invariant k_z^2. Finally the invariant

$$u_+(\mathbf{r})u_-^*(\mathbf{r}')k_+^2 + u_-(\mathbf{r})u_+^*(\mathbf{r}')k_-^2.$$

still exists in the group D_4. Collecting everything together one can write down the following invariant form:

$$\begin{aligned} Q(\mathbf{k};\, \mathbf{r}, \mathbf{r}') = {} & [u_+(\mathbf{r})u_+^*(\mathbf{r}') + u_-(\mathbf{r})u_-^*(\mathbf{r}')](\alpha k_\perp^2 + \beta k_z^2) \\ & + v[u_+(\mathbf{r})u_+^*(\mathbf{r}') - u_-(\mathbf{r})u_-^*(\mathbf{r}')]k_z + \\ & + \gamma[u_+(\mathbf{r})u_-^*(\mathbf{r}')k_-^2 + u_-(\mathbf{r})u_+^*(\mathbf{r}')k_+^2] \\ & + \delta[u_+(\mathbf{r})u_-^*(\mathbf{r}')k_+^2 + u_-(\mathbf{r})u_+^*(\mathbf{r}')k_-^2]. \end{aligned} \tag{1}$$

In the groups D_3 and D_6 the coefficient $\delta = 0$. Putting together the secular equation, we obtain the following result for the energy levels:

$$\varepsilon_{1,2} = \alpha k_\perp^2 + \beta k_z^2 \pm [v^2 k_z^2 + (\gamma + \delta)^2 k_\perp^4 - 8\delta\gamma k_x^2 k_y^2]^{1/2}. \tag{2}$$

It is not difficult to see that ranges of values of the

coefficients exist for which one of the branches is positive for all directions **k** and the other is negative, i.e., a point Fermi surface is possible. It is of interest to note that the behavior of the branches of the spectrum is strongly anisotropic. In a direction outside the basis plane $\epsilon_{1,2}$ changes basically according to a linear law, but in the basis plane it changes according to a quadratic law.

In the presence of strong spin-orbit coupling it is necessary to use double-valued representations. Two double-valued representations exist in the group D_3 (see[6]). With the aid of the basis functions for the representation E_2', which transform like the components of a spinor, one can set up invariants which are linear in k_z and $k_\pm$. The general invariant form has the form

$$Q(\mathbf{k};\ \mathbf{r},\ \mathbf{r}') = \alpha[u_1(\mathbf{r})u_1^*(\mathbf{r}') - u_2(\mathbf{r})u_2^*(\mathbf{r}')]k_z + \beta[u_1(\mathbf{r})u_2^*(\mathbf{r}')k_- + u_2(\mathbf{r})u_1^*(\mathbf{r}')k_+]. \tag{3}$$

The solution of the corresponding secular equation has the form

$$\varepsilon_{1,2} = \pm\sqrt{(\alpha k_z)^2 + (\beta k_\perp)^2}. \tag{4}$$

As to the other double-valued representation E_1' (which in the present case is a combination of two complex-conjugate one-dimensional representations), then from the corresponding basis functions one can construct only an invariant or a combination which gives an invariant together with k_z. Thus, for the energy spectrum we obtain

$$\varepsilon_{1,2} = \alpha k^2 + \beta k_z^2 \pm \gamma k_z. \tag{5}$$

In this connection a "Fermi point" cannot appear.

In the case of the group D_6 the representations E_1' and E_2' give a situation analogous to E_2' in D_3, i.e. to formula (4), and the representation E_3' is analogous to E_1' in D_3, i.e., it does not correspond to a "Fermi point." Finally, in the group D_4 both double-valued two-dimensional representations are analogous to E_2' in the group D_3 and give formula (4).

In all of the described cases of the occurrence of a "Fermi point" the fact that the basis functions permitted the construction of an invariant linear in k_z was very important. This does not hold in other point groups with one axis of high order. In these groups the component k_z may appear in the function $Q(\mathbf{k};\ \mathbf{r},\ \mathbf{r}')$ only in the form k_z^2. But k_z^2 is an invariant by itself, and therefore the energy spectrum will have the form

$$\varepsilon_{1,2} = \alpha k_z^2 + f_{1,2}(k_x, k_y). \tag{6}$$

Consequently the branches do not separate along the direction k_z, i.e., a "Fermi point" is impossible.

3. CUBIC GROUPS. SINGLE-VALUED REPRESENTATIONS

Let us go on to the cubic groups. We note that since the functions u_i enter by pairs, then there is no difference between even and odd representations, so that one cannot determine the parity of the representation.

The components of the vector **k** transform according to one of the three-dimensional representations of the cubic groups, and from the quadratic combinations one can set up the bases of three representations: the invariant k^2, the basis of a two-dimensional representation $(k_z^2 + \theta k_x^2 + \theta^2 k_y^2,\ k_z^2 + \theta^2 k_x^2 + \theta k_y^2)$ where $\theta = e^{2\pi i/3}$, and the basis of one of the three-dimensional representations $(k_x k_y,\ k_y k_z,\ k_z k_x)$.

Let the functions u_i correspond to the two-dimensional representation. By the usual method one can establish that from such functions one can make up only the invariant $u_1(\mathbf{r})u_1^*(\mathbf{r}') + u_2(\mathbf{r})u_2^*(\mathbf{r}')$ and the basis of the two-dimensional representation $u_1(\mathbf{r})u_2^*(\mathbf{r}')$, $u_2(\mathbf{r})u_1^*(\mathbf{r}')$. Owing to this we can easily determine the function $Q(\mathbf{k};\ \mathbf{r},\ \mathbf{r}')$ (see[5]) and find an expression for the energy levels

$$\varepsilon_{1,2} = \alpha k^2 \pm \beta[k^4 - 3(k_x^2 k_y^2 + k_y^2 k_z^2 + k_z^2 k_x^2)]^{1/2}. \tag{7}$$

Having taken **k** in the [111] direction we obtain $\epsilon_{1,2} = \alpha k^2$, from which it follows that although degeneracy occurs at the point $\mathbf{k} = 0$ it does not simultaneously correspond to a minimum of one of the functions $\epsilon_i(\mathbf{k})$ and a maximum of the other.

Let us go on to the three-dimensional representations. Here it is easiest of all to follow the method indicated in[6] (see also[8]). The bilinear combinations of interest to us are $u_i^*(\mathbf{r}')[f(\hat{J})]_{ik}u_k(\mathbf{r})$, where $f(\hat{J})$ is some kind of product of the spin projection operators (in the present case $J = 1$). These combinations transform in the same way as the operators $f(\hat{J})$ themselves, and therefore it is necessary for us in turn to construct invariants out of the components of **k** and of the operators $\hat{J}$. Since $J = 1$ the invariants may be either linear or quadratic in $\hat{J}$. From the requirement of invariance under time reversal it follows that the total power of **k** and $\hat{J}$ should be even.

If the little group of **k** is equivalent to the group T or O, then an invariant exists which is linear in **k**, namely, the product $\mathbf{k}\cdot\hat{\mathbf{J}}$. Confining our attention to this approximation we obtain

$$\varepsilon_{1,2,3} = +vk,\quad 0,\quad -vk. \tag{8}$$

For the final solution of the question about whether the point $\mathbf{k} = 0$ possesses the necessary properties, it is necessary to determine whether it corresponds to an extremum of the function $\epsilon_2(\mathbf{k})$. For this it is necessary to find the next approximation for $\epsilon_2(\mathbf{k})$. Among the invariants is αk^2, and one can assume that it is larger than all remaining invariants. In this connection, the function $\epsilon_2(\mathbf{k})$ will have the same sign for all values of **k**. Therefore, in the present case the point $\mathbf{k} = 0$ may correspond to contact of the conduction band with the valence band.

In the case when the little group is equivalent to T_h or O_h, i.e., when it contains a center of inversion, the invariants must necessarily contain an even number of **k** and an even number of $\hat{J}$. The same pertains in the present case to the group T_d, although it does not contain a center of inversion. This is related to the fact that the scalar product $(\mathbf{k}\cdot\hat{\mathbf{J}})$ is not an invariant for this group, but the invariants which are linear in **k** and cubic in $\hat{\mathbf{J}}$ are absent for $J = 1$. One can show that in all three groups there is one and the same common invariant:

$$\hat{D} = \alpha k^2 + \beta[(J_z^2 + \theta J_x^2 + \theta^2 J_y^2)(k_z^2 + \theta^2 k_x^2 + \theta k_y^2) + (J_z^2 + \theta^2 J_x^2 + \theta J_y^2)(k_z^2 + \theta k_x^2 + \theta^2 k_y^2)] + \gamma[k_x k_y\{J_x, J_y\} + k_y k_z\{J_y, J_z\} + k_z k_x\{J_z, J_x\}], \tag{9}$$

where $\{J_x, J_y\} = J_x J_y + J_y J_x$. The corresponding secular equation has the form

$$(\varepsilon-\alpha k^2)^3-(\varepsilon-\alpha k^2)[3\beta^2k^4+(\gamma^2-9\beta^2)(k_x^2k_z^2+k_y^2k_z^2+k_x^2k_y^2)]+2\beta^3k^6+\beta(\gamma^2-9\beta^2)k^2(k_x^2k_z^2+k_y^2k_z^2+k_y^2k_x^2)-(2\gamma^3-9\beta\gamma^2-9\beta^3)k_x^2k_y^2k_z^2=0. \quad (10)$$

Let us demonstrate that the coefficients α, β, and γ can be chosen in such a way that the eigenvalues $\epsilon_i(\mathbf{k})$ will have different signs, where each of them will retain its sign for all directions **k**. The determinant of the matrix D_{ik} given by Eq. (9) is the product of the three functions $\epsilon_1(\mathbf{k})\epsilon_2(\mathbf{k})\epsilon_3(\mathbf{k})$. If one of them changes sign for any value of **k**, then the determinant must vanish at this point. Hence one of the conditions is $\|D_{ik}(\mathbf{k})\| \neq 0$. For this it is obviously necessary that all extrema of $\|D_{ik}(\mathbf{k})\|$ as a function of the direction of **k** do not vanish and have one and the same sign. But the extrema of the determinant correspond to the most symmetric directions, namely [100], [110], and [111]. Therefore, it is sufficient to establish the conditions under which, let us say, $\|D_{ik}\| < 0$ in these three directions. In addition, it is necessary that in one of the indicated directions the functions $\epsilon_i(\mathbf{k})$ have different values, i.e., for example, for $\|D_{ik}\| < 0$ the values $\epsilon_1 < 0$ and $\epsilon_2, \epsilon_3 > 0$. The condition that there be one negative function and two positive functions $\epsilon_i(\mathbf{k})$ for all three symmetric directions of **k** is obviously equivalent to these requirements.

For such directions the functions $\epsilon_i(\mathbf{k})$ can easily be found:

$$\begin{aligned}
&[100]: \quad \varepsilon_1=(\alpha-2\beta)k^2, \quad \varepsilon_2=\varepsilon_3=(\alpha+\beta)k^2,\\
&[110]: \quad \varepsilon_1=[\alpha-{}^1\!/_2(\beta+\gamma)]k^2, \quad \varepsilon_2=[\alpha-{}^1\!/_2(\beta-\gamma)]k^2, \ \varepsilon_3=(\alpha+\beta)k^2\\
&[111]: \quad \varepsilon_1=[\alpha-{}^2\!/_3\gamma]k^2, \quad \varepsilon_2=\varepsilon_3=(\alpha+{}^1\!/_3\gamma)k^2.
\end{aligned} \quad (11)$$

It is not difficult to show that ranges of the values of α, β, and γ exist in which $\epsilon_1 < 0$ and $\epsilon_2, \epsilon_3 > 0$. Thus, contact of the conduction band with the valence band is possible in the groups T_d, T_h, and O_h.

4. CUBIC GROUPS. DOUBLE-VALUED REPRESENTATIONS

Let us go on to double-valued representations. There are two kinds of such representations: two-dimensional corresponding to J = 1/2 and four-dimensional corresponding to J = 3/2. If the little group k = 0 is equivalent to T or O then again the invariant $\mathbf{k}\cdot\hat{\mathbf{J}}$ exists. For two-dimensional representations we obtain

$$\varepsilon_1=vk, \quad \varepsilon_2=-vk. \quad (12)$$

One cannot consider two-dimensional representations of the remaining groups. In fact, in the presence of a center of inversion in the lattice, the two-fold degeneracy is retained for arbitrary **k**, which corresponds to the fact that the energy does not depend on the direction of the spin. As for the group T_d, then here, just as for the single-valued three-dimensional representations, the necessary invariant which is linear in **k** is absent. Therefore, in the present case the fundamental invariant will simply be αk^2, i.e., the two functions $\epsilon_{1,2}(\mathbf{k})$ coincide in this approximation. This degeneracy is lifted only by small terms which are cubic in **k**, from which it is clear that in the present case the point k = 0 does not possess the required properties.

Let us go on to four-dimensional representations. In the absence of a center of inversion (T, T_d, or O) the admissible invariants can be linear in **k** and cubic in **J**. For the group T the general form of the invariant is given by

$$\hat{D}=\alpha(\mathbf{k}\hat{\mathbf{J}})+\beta(k_xJ_x^3+k_yJ_y^3+k_zJ_z^3)+\gamma[k_x\{J_x,(J_y^2-J_z^2)\}+k_y\{J_y(J_z^2-J_x^2)\}+k_z\{J_z,(J_x^2-J_y^2)\}], \quad (13)$$

where $\{A, B\} = AB + BA$.

In the group O the last invariant is absent, i.e., $\gamma = 0$, and in the group T_d, $\alpha = \beta = 0$. The possible existence of a "Fermi point" in the groups T and O is determined by the following simple consideration. Let $\alpha \gg \beta, \gamma$. In this case $\epsilon_1 = -(3/2)\alpha k$, $\epsilon_2 = -(1/2)\alpha k$, $\epsilon_3 = (1/2)\alpha k$, and $\epsilon_4 = (3/2)\alpha k$. Here k = 0 possesses the required properties. Hence it follows that ranges of values of the coefficients exist for which the point k = 0 retains these properties.

The group T_d is in a special position. The solution of the secular equation leads to the following eigenvalues:

$$\varepsilon_{1,2,3,4}=\pm\gamma\sqrt{3}k[1\pm\sqrt{3}/_2\sqrt{1-(n_x^4+n_y^4+n_z^2)}], \quad (14)$$

where n denotes a unit vector in the direction **k**. In the direction [111] two of the roots vanish in this approximation. In analogy to the preceding discussion, by adding the invariant δk^2 one can also verify here that the point k = 0 can possess the necessary properties. In the case when the little group contains a center of inversion (T_h or O_h), the invariant must be quadratic in **k** and **J**. The general form of the invariant is again given by formula (9). Here, however, the matrix J_i corresponds to J = 3/2. Solving the secular equation we obtain the well-known formula

$$\varepsilon_{1,2}=\alpha k^2\pm[9\beta^2k^4\pm3(\gamma^2-9\beta^2)(k_x^2k_y^2+k_y^2k_z^2+k_z^2k_x^2)]^{1/2}. \quad (15)$$

Each level is doubly degenerate. For the symmetric directions we obtain

$$\begin{aligned}
&[100]: \quad \varepsilon_{1,2}=(\alpha\pm3\beta)k^2,\\
&[110]: \quad \varepsilon_{1,2}=(\alpha\pm{}^1\!/_2\sqrt{9\beta^2+3\gamma^2})k^2,\\
&[111]: \quad \varepsilon_{1,2}=(\alpha\pm\gamma)k^2.
\end{aligned} \quad (16)$$

From here it follows that contact of the valence band and the conduction band can actually occur at the point **k** = 0. We note that precisely this representation is realized in α-Sn (usually denoted by Γ_8.

5. MORE COMPLICATED SYMMETRY

The analysis carried out above exhausts all cases involving the appearance of a "fermi point" with a little group equivalent to a point group. We shall not give such a complete analysis of the cases of more complicated symmetry, and we only show that here "Fermi points" can also appear. The appearance of nontrivial screw axes and glide planes frequently increases the multiplicity of the degeneracy of the energy levels. In the literature this phenomena is sometimes called coalescence of the energy bands. It is essential that in the majority of the cases the bands coincide on intact pieces of the surface of the Brillouin zone, and consequently there cannot be any "Fermi points" here. However, in a number of cubic groups (namely O^8 and O^{10}, and also O_h^{10}, using the notation from[9]) the little groups of the vectors $\mathbf{k}_0$ with a point at a corner of the cubic Brillouin zone have four-dimensional and six-dimensional single-valued representations.

Let us present the invariant for the four-dimensional representations:

$$Q(\mathbf{k};\mathbf{r},\mathbf{r}')=iv[\{u_1(\mathbf{r})u_2^*(\mathbf{r}')-u_2(\mathbf{r})u_1^*(\mathbf{r}')\}k_z+\{u_2(\mathbf{r})u_3^*(\mathbf{r}')-u_3(\mathbf{r})u_2^*(\mathbf{r}')\}k_x+\{u_3(\mathbf{r})u_1^*(\mathbf{r}')-u_1(\mathbf{r})u_3^*(\mathbf{r}')\}k_y$$

$$+ \{u_3(\mathbf{r})u_4^*(\mathbf{r}') - u_4(\mathbf{r})u_3^*(\mathbf{r}')\}k_x + \{u_4(\mathbf{r})u_2^*(\mathbf{r}') - u_2(\mathbf{r})u_4^*(\mathbf{r}')\}k_y + \{u_1(\mathbf{r})u_4^*(\mathbf{r}') - u_4(\mathbf{r})u_1^*(\mathbf{r}')\}k_z]. \quad (17)$$

The matrix $\hat{D}$ may be written in the form $\hat{D}(\mathbf{k}) = v\alpha\mathbf{k}$, where the matrices α_i possess the commutation relations of the Dirac matrices. In our approximation the levels remain doubly degenerate and $\epsilon_{1,2}(\mathbf{k}) = \pm v k$.

The six-dimensional representations also admit realization of a point Fermi surface but, as in the case of three-dimensional representations of the simple cubic groups, part of the energy bands begins with terms which are quadratic in **k**.

Thus, we have shown that cases when contact of the valence band with the conduction band occurs in the electron spectrum are not unusual, and consequently such substances may be encountered in nature. Of course, we did not exhaust all possible cases involving complicated symmetry, but here it is apparently reasonable to consider specific examples for which the experimental data will indicate the existence of a "Fermi point."

Our conclusions have been reached in the one-electron approximation, i.e., we have not taken the interaction of the electrons into consideration. This problem merits very serious attention because it is quite possible that it may turn out that the interaction leads to instability of the obtained spectrum.

6. COULOMB INTERACTION ASSOCIATED WITH A LINEAR SPECTRUM

We consider, first of all, the simplest case of two energy branches which are described by formula (12). In this case the matrix D can be written in the form

$$\hat{D} = v(\sigma\mathbf{k}), \quad (18)$$

where the σ_i are the Pauli matrices. In accordance with this the Green's function for noninteracting electrons may be written in the form

$$G_0 = [\omega - v(\sigma\mathbf{k}) + i\delta\,\mathrm{sign}\,\omega]^{-1}. \quad (19)$$

Now let us introduce the Coulomb interaction between the electrons, which we shall write in the usual form. Let us consider how screening occurs. The screening interaction has the form

$$\Gamma(\mathbf{k},\omega) = \frac{4\pi e^2}{\varepsilon_0 k^2 - 4\pi e^2 \Pi(\mathbf{k},\omega)}, \quad (20)$$

where $\Pi(\mathbf{k},\omega)$ denotes the sum of all possible electron loops. In the first approximation

$$\Pi(\mathbf{k},\omega) = -i\,\mathrm{Sp}\int \hat{G}_0(\omega_1,\mathbf{k}_1)\hat{G}_0(\omega+\omega_1,\mathbf{k}+\mathbf{k}_1)\frac{d\omega_1\,d^3k_1}{(2\pi)^4}.$$

Substituting here the function $\hat{G}_0$, which it is more convenient to write in the form

$$\hat{G}_0 = \frac{\omega + v\sigma\mathbf{k}}{\omega^2 - (vk - i\delta)^2}, \quad (19')$$

we obtain

$$\Pi_0 = -\frac{k^2}{12\pi^2 v}\ln\frac{k_{max}}{\sqrt{k^2 - \omega^2/v^2}}, \quad (21)$$

where k_{max} is the limit of integration, which is of the order of the period of the reciprocal lattice.

Thus, in the case under consideration the Coulomb interaction is practically not screened. As we see below, the evaluation of the self-energy corrections to the electronic spectrum with the aid of formula (20) also leads to a logarithmic integral. From here it follows that at the point $\mathbf{k} = 0$ the energy spectrum has a singularity, and the matrix D cannot be expanded in a series in powers of **k**. However, in view of the fact that the singularity is weak, we may, without violating the symmetry conditions, assume that the matrix $\hat{D}$ is multiplied by a slowly varying invariant function, i.e., by a function which depends on the absolute value of k and on ω. In order to find this function we apply the method of article[10].

The equation for the Green's function is written in the form

$$G^{-1} = G_0^{-1} - \Sigma = \omega - v(\sigma\mathbf{k}) - \Sigma(\mathbf{k},\omega). \quad (22)$$

We shall seek the complete Green's function in the form

$$G = \frac{\omega + v(\sigma\mathbf{k})f(k,\omega)}{\omega^2 - [vkf(k,\omega) - i\delta]^2}, \quad (23)$$

where f is a slowly varying function of **k** and ω. The first approximation for the polarization operator, evaluated with the aid of this function instead of (19'), is

$$\Pi(\mathbf{k},\omega) = -\frac{k^2}{12\pi^2 v}\int_0^{\xi}\frac{d\xi_1}{f(\xi_1)}, \quad (24)$$

where ξ is a logarithmic variable;

$$\xi = \ln\frac{k_{max}}{\max(k,\omega/v)}, \quad \xi_1 = \ln\frac{k_{max}}{k_1}$$

(the integral is taken over the region where $\omega_1 \sim v k_1$). Consequently, according to (20),

$$\Gamma(\mathbf{k},\omega) = \frac{4\pi e^2}{\varepsilon_0(k^2)}d(\mathbf{k},\omega), \quad d(\mathbf{k},\omega) = \left[1 + \frac{e^2}{3\pi v\varepsilon_0}\int_0^{\xi}\frac{d\xi_1}{f(\xi_1)}\right]^{-1}. \quad (25)$$

In general, besides the simplest loops of two Green's functions, more complicated diagrams (Fig. 1) enter into Π. One can join all of them together into a diagram of the first type, but in this connection one of the triple-vertices is replaced by the appropriate total vertex. For simplicity we make the assumption: $k \gtrsim \omega/v$. The logarithmic integral in Π goes over the region $\omega/v \sim k_1 \gg k$. Consequently the triple-vertex $\gamma(\mathbf{k},\mathbf{k}_1)$ is necessary for us for $k \ll k_1$. Since the interaction is assumed to be weak, we are interested in terms containing the maximum power of the logarithm for a given power of the coupling constant. Considering the first correction to γ, we actually obtain terms with logarithmic integrals. They have the form

$$\frac{k^2}{k_1^2}d(k_1)\int_{\eta}^{\xi}\frac{d\xi_1}{f(\xi_1)} \quad \text{or} \quad \frac{\omega\sigma\mathbf{k}}{vk_1^2}d(k_1)\int_{\eta}^{\xi}\frac{d\xi_1}{f(\xi_1)};$$

$$\xi = \ln\frac{k_{max}}{k}, \quad \eta = \ln\frac{k_{max}}{k_1}.$$

But upon substitution of these expressions into the integral:

$$\Pi^{(1)} = -i\,\mathrm{Sp}\int G(\mathbf{k}_1,\omega_1)G(\mathbf{k}_1+\mathbf{k},\omega_1+\omega)\gamma(\mathbf{k}\omega,\mathbf{k}_1\omega_1)\frac{d^3k_1\,d\omega_1}{(2\pi)^4}$$

we obtain a convergent integral instead of a logarithmically divergent one. Therefore, the corrections to the obtained form of Π will be small, and one can neglect them.

FIG. 1

FIG. 2

Now let us go on to the self-energy of an electron. In the first approximation we have

$$\hat{\Sigma}(\omega, \mathbf{k}) = i\int \Gamma(-\omega_1, -\mathbf{k}_1)\hat{G}(\omega_1+\omega, \mathbf{k}_1+\mathbf{k})\frac{d^3k_1\, d\omega_1}{(2\pi)^4}.$$

By means of a renormalization of the chemical potential we can always eliminate an arbitrary constant from Σ. Therefore we subtract $\Sigma(0, 0)$ and as a result we obtain

$$\Sigma(\omega, \mathbf{k}) = i\int [\hat{G}(\omega_1+\omega, \mathbf{k}_1+\mathbf{k}) - \hat{G}(\omega_1, \mathbf{k}_1)]\Gamma(-\omega_1, -\mathbf{k}_1)\frac{d^3k_1\, d\omega_1}{(2\pi)^4}. \tag{26}$$

Aside from this term, the total self-energy contains terms corresponding to more complicated diagrams (see Fig. 2), but one can show that they give a small contribution. In fact, let us consider the next correction containing two Γ-lines. In order for this term to be important, it must contain two logarithmic integrals with respect to the momenta of the Γ-lines: k_1 and k_2. Consequently there should be two logarithmic regions, $k_1 \gg k_2 \gg k$ and $k_2 \gg k_1 \gg k$, but as already mentioned earlier the vertex γ may contain logarithmic integrals only in that case if its arguments differ appreciably in magnitude, where these integrals correspond to the regions between the values of the arguments. In view of this, in the region $k_1 \gg k_2 \gg k$ the logarithmic integral with respect to k_1 is not present, but in the second region there is no logarithmic integral with respect to k_2. Thus, we may confine our attention to expression (26). Substituting expressions (23) and (25) here, we find

$$\Sigma = -\frac{2e^2}{3\pi\varepsilon_0}\sigma k\int_0^{\xi} d\xi_1\, d(\xi_1). \tag{27}$$

Here $\xi_1 = \ln(k_{max}/k_1)$, $\xi = \ln(k_{max}/k)$, and it is assumed that $vk \gtrsim \omega$. In the case when $vk \ll \omega$,

$$\Sigma = -\frac{2e^2}{3\pi\varepsilon_0}\sigma k\left[\int_0^{\eta} d\xi_1\, d(\xi_1) + d(\eta)(\xi-\eta)\right]. \tag{27'}$$

where $\eta = \ln(vk_{max}/\omega)$ and $\xi = \ln(k_{max}/k)$. The most interesting case is $\omega \sim vk$, i.e. formula (27). It is necessary to explicitly specify the case $\omega \approx \pm vk$. Strictly speaking, at such values of ω a singularity must appear in Σ because the spectrum linear in k is unstable with respect to the decay of quasi-particles into many other particles traveling in the same direction. Such singularities actually exist, but they appear only in the next approximation, in terms proportional to

$$\frac{e^4}{\varepsilon_0^2}\ln\frac{(vk_{max})^2}{(vk)^2-\omega^2}.$$

In those terms which we are considering, the lower limit of the integrals is simply k, and the proximity of ω to $\pm vk$ does not manifest itself in any way. Thus, in the approximation under consideration the case $\omega \approx \pm vk$ is described by formula (27).

If formulas (22), (23), (25), and (27) are compared, then we obtain an equation for the function $f(\xi)$ (we assume $vk \gtrsim \omega$)

$$f(\xi) = 1 + \frac{2e^2}{3\pi v\varepsilon_0}\int_0^{\xi}\left[1 + \frac{e^2}{3\pi v\varepsilon_0}\int_0^{\xi_1}\frac{d\xi_2}{f(\xi_2)}\right]^{-1} d\xi_1. \tag{28}$$

By means of two differentiations this equation reduces to a differential equation. Solving the latter equation we obtain

$$f(\xi) = \left(1 + \frac{e^2}{\pi v\varepsilon_0}\xi\right)^{2/3} \tag{29}$$

and substituting into the expression for $d(\xi)$ we find

$$d(\xi) = \left(1 + \frac{e^2}{\pi v\varepsilon_0}\xi\right)^{-1/3}. \tag{30}$$

Thus, for the case under consideration one obtains the result that G has a pole for $\omega = \pm vkf(\xi)$. The imaginary part which determines the attenuation is small in comparison with the real part ($\mathrm{Im}\, G^{-1}/\mathrm{Re}\, G^{-1} \sim e^2/v\epsilon_0$). For momenta of the order of or smaller than $k_{max}e^{-v\epsilon_0/e^2}$, the energy spectrum differs from a purely linear spectrum by a logarithmic factor.

We note that the sign in front of the logarithm for $d(\xi)$ is positive, which corresponds to a weakening of the interaction in the region of small momenta and frequencies. This situation is analogous to quantum electrodynamics and may be called the "zero-charge" situation (in electrodynamics the situation involves, of course, not a decrease of the momentum but a decrease of the radius over which the interaction is smeared out, i.e., an increase of the upper limit on the logarithmically divergent integrals).

It is of interest to clarify whether this situation is common for all types of linear spectra or if it is unique only for the model under consideration. In general a sum of the invariants of k and $\hat{J}$ with arbitrary coefficients appears in the matrix $\hat{D}$. Under the influence of the interaction each of these coefficients changes into a slowly varying function. Ultimately one must obtain a system of integral equations for these functions. Let us confine our attention to such a generalization of the case under consideration, for which the matrix $\hat{D}$ contains on only one invariant of k and J, i.e., there is only one unknown function. An examination of the problem with several arbitrary invariants would be of significant interest.

In a noninteracting model the matrix $\hat{D}$ has the form $\hat{D} = \hat{v}k$. From symmetry conditions it follows that the Green's function of the electrons has the form

$$G = [\omega - \hat{v}kf(k, \omega) + i\delta\,\mathrm{sign}\,\omega]^{-1}, \tag{31}$$

where $f(k, \omega)$ is an invariant function which, in analogy with the preceding discussion, we assume to be logarithmic (this is confirmed by the following calculation). Since factors of the order of unity are not important inside the logarithm, then one can assume that the invariant function f always depends only on k^2.

Let us evaluate the polarization operator Π. It is given by

$$\Pi(k, \omega) = -i\,\mathrm{Sp}\int \hat{G}(\mathbf{k}_1, \omega_1)\hat{G}(\mathbf{k}_1+\mathbf{k}, \omega_1+\omega)\frac{d^3k_1\, d\omega_1}{(2\pi)^4}.$$

Let us assume that the integral runs over the region $k_1 \geq k$. In the first approximation one could neglect k. However, if this is done, then the integral turns out to be equal to zero. In fact, if the trace is taken with respect to the eigenfunctions of the operator $\hat{D}(k_1)$, then we obtain the integral

$$\int \sum_i [\omega - \varepsilon_i(k) + i\delta \operatorname{sign} \varepsilon_i]^{-2} d\omega,$$

which is equal to zero, since the poles always lie on one side.

In view of this fact, it is necessary to use the next terms of the expansion in powers of **k**. A nonvanishing contribution can appear only from terms of the second order in **k**. It has the form

$$\Pi(\mathbf{k}, \omega) = -i \operatorname{Sp} \int \hat{G}(\mathbf{k}_1, \omega_1) \hat{G}(\mathbf{k}_1, \omega_1 + \omega) \hat{\mathbf{v}}_1 \mathbf{k} \times \hat{G}(\mathbf{k}_1, \omega_1 + \omega) \hat{\mathbf{v}}_1 \mathbf{k} \hat{G}(\mathbf{k}_1, \omega_1 + \omega) f^2(\mathbf{k}_1, \omega_1) \frac{d^3k_1\, d\omega_1}{(2\pi)^4}, \tag{32}$$

where $\hat{\mathbf{v}}_1 = \partial D(\mathbf{k}_1)/\partial \mathbf{k}_1$. Having taken the trace with respect to the eigenfunctions of the operator $\hat{D}(\mathbf{k}_1)$, we obtain

$$\Pi(\mathbf{k}, \omega) = -i \sum \int [\omega_1 - \varepsilon_i(\mathbf{k}_1) + i\delta \operatorname{sign} \varepsilon_i]^{-1} [\omega_1 + \omega - \varepsilon_l(\mathbf{k}_1) + i\delta \operatorname{sign} \varepsilon_l]^{-1} \times (\hat{\mathbf{v}}_1 k)_{il} [\omega_1 + \omega - \varepsilon_l(\mathbf{k}_1) + i\delta \operatorname{sign} \varepsilon_l]^{-1} (\hat{\mathbf{v}}_1 \mathbf{k})_{li} \times (\omega_1 + \omega - \varepsilon_i(\mathbf{k}_1) + i\delta \operatorname{sign} \varepsilon_i)^{-1} f^2(k_1, \omega_1) \frac{d^3k_1\, d\omega_1}{(2\pi)^4}.$$

Since $\hat{\mathbf{v}}_1$ is a self-adjoint operator, then $(\hat{\mathbf{v}}_1 \cdot \mathbf{k})_{il} (\hat{\mathbf{v}}_1 \cdot \mathbf{k})_{li} = |(\hat{\mathbf{v}}_1 \cdot \mathbf{k})_{il}|^2$. In order for the integral to be different from zero, it is necessary that the poles ϵ_i and ϵ_l should have different signs. Carrying out the integration over ω_1 and assuming $\epsilon_i(\mathbf{k}_1) - \epsilon_l(\mathbf{k}_1) \gg \omega$, we obtain

$$\Pi(\mathbf{k}, \omega) = -2 \int \sum_{\substack{v_i > 0 \\ v_l < 0}} (v_{1i} - v_{1l})^{-1} |(\hat{\mathbf{v}}_1 \mathbf{k})_{il}|^2 f^{-1}(k_1) \frac{d^3k_1}{k_1^2 (2\pi)^3}.$$

(Here we have introduced the notation $\epsilon_i(\mathbf{k}_1) = v_{1i} k_1 f(k_1)$, i.e., $v_{1i}(\mathbf{n})$ is the coefficient in the appropriate branch of the energy spectrum, and **n** is a unit vector in the direction of $\mathbf{k}_1$.) Introducing logarithmic variables we obtain

$$\Pi(\mathbf{k}, \omega) = -\frac{1}{\pi^2} \overline{\sum_{\substack{v_{1i} > 0 \\ v_{1l} < 0}} (v_{1i} - v_{1l})^{-2} |(\hat{\mathbf{v}}_1 \mathbf{k})_{il}|^2} \int_0 \frac{d\xi_1}{f(\xi_1)},$$

where the bar denotes averaging over the directions of $\mathbf{k}_1$. The function Γ will have the form

$$\Gamma = \frac{4\pi e^2}{\varepsilon_0 k^2} d(\xi), \quad d(\xi) = \left[1 + a \int_0^{\xi} \frac{d\xi_1}{f(\xi_1)}\right]^{-1}, \tag{33}$$

$$a = \frac{4e^2}{\pi \varepsilon_0} \overline{\sum_{\substack{v_{1i} > 0 \\ v_{1l} < 0}} (v_{1i} - v_{1l})^{-2} \left| \left(\frac{\hat{v}_1 k}{k} \right)_{il} \right|^2}. \tag{34}$$

In the same way as earlier, one can show that the corrections of higher order in e^2 are small.

Let us go on to the self-energy. In analogy to the previous discussion, it is sufficient to consider only the first diagram. Since the integral over k_1 is logarithmic, then we may expand the difference $\hat{G}$ in the integral (26) with respect to k. Having made use of the fact that Γ is an even function of k and ω, we obtain

$$\hat{\Sigma}(\omega, \mathbf{k}) = i \int \left[\hat{G}(\omega_1, \mathbf{k}_1) \hat{\mathbf{v}}_1 \mathbf{k} \hat{G}(\omega_1, \mathbf{k}_1) \cdot f(k_1, \omega_1) \Gamma(\omega_1, \mathbf{k}_1) \frac{d^3k_1\, d\omega_1}{(2\pi)^4} \right].$$

From considerations of symmetry it follows that $\Sigma \sim \hat{\mathbf{v}} \cdot \mathbf{k}$. According to formula (31), $\hat{\Sigma} = \hat{\mathbf{v}} \cdot \mathbf{k}(f - 1)$. Having substituted this expression, we multiply by $\hat{\mathbf{v}}\mathbf{k}$ and take the trace on both sides. In this connection we obtain

$$\sum_{i,l} |(\hat{\mathbf{v}}\mathbf{k})_{il}|^2 [f(\xi) - 1] = i \sum_{i,l} \int (\hat{\mathbf{v}}_1 \mathbf{k})_{il} (\omega - \varepsilon_i + i\delta \operatorname{sign} \varepsilon_i)^{-1} \times (\hat{\mathbf{v}}_1 \mathbf{k})_{li} (\omega - \varepsilon_l + i\delta \operatorname{sign} \varepsilon_l)^{-1} f(k_1) \Gamma(\omega_1, k_1) \frac{d^3k_1\, d\omega_1}{(2\pi)^4}$$

$$= 2 \int \sum_{\substack{v_{1l} > 0 \\ v_{1i} < 0}} |(\hat{v}_1 k)_{il}|^2 (v_{1i} - v_{1l})^{-1} \frac{1}{k_1} \frac{4\pi e^2}{\varepsilon_0 k_1^2} d(k_1) \frac{d^3k_1}{(2\pi)^3}.$$

Carrying out the last integration, we obtain the relation

$$f(\xi) = 1 + b \int_0^{\xi} d(\xi_1) d\xi_1, \tag{35}$$

where

$$b = \frac{4e^2}{\pi \varepsilon_0} \overline{\sum_{\substack{v_{1i} > 0 \\ v_{1l} < 0}} \left| \left(\frac{\hat{\mathbf{v}}_1 \mathbf{k}}{k} \right)_{il} \right|^2 (v_{1i} - v_{1l})^{-1} \left(\sum_{i,l} \left| \left(\frac{\hat{\mathbf{v}}\mathbf{k}}{k} \right)_{il} \right|^2 \right)^{-1}}. \tag{36}$$

Equations (33) and (35) differ from (28) only by the coefficients. The solution has the form

$$f(\xi) = [(a + b)\xi + 1]^{a/(a+b)}, \tag{37}$$

$$d(\xi) = [(a + b)\xi + 1]^{-b/(a+b)}. \tag{38}$$

According to the definitions (34) and (36), both coefficients a and b are positive and of the order of $e^2/v\epsilon_0$. Thus, the conclusions reached for the simplest example considered at the beginning are completely valid for the more general case of a "Fermi point" with a linear spectrum.

7. QUADRATIC SPECTRUM

The situation is completely different in the case of a quadratic spectrum. First of all we note that if we set k = 0 in the polarization operator, then we obtain zero independently of the form of the spectrum. A nonvanishing contribution appears in second order in k. But in the present case the integral analogous to (32) is convergent and for $\omega \ll k$ has the form mk, where m is the order of the mass of a free electron. Therefore the function Γ will have the form

$$\Gamma = 4\pi e^2 / (\varepsilon_0 k^2 + m e^2 k a),$$

where $a \sim 1$. Thus, for $k < me^2/\epsilon_0$ the behavior of Γ changes radically. One can use the obtained expression in order to evaluate the following corrections to the triple-vertex γ.

Let us consider the first such correction (Fig. 3). The corresponding expression has the form

$$\gamma(\mathbf{k}_1, \omega_1; \mathbf{k}_1 + \mathbf{k}, \omega_1 + \omega) = \int \hat{G}(\mathbf{k}_2, \omega_2) \hat{G}(\mathbf{k}_2 + \mathbf{k}, \omega_2 + \omega) \Gamma(\mathbf{k}_1 - \mathbf{k}_2, \omega_1 - \omega_2) \frac{d^3k_2\, d\omega_2}{(2\pi)^4}.$$

We shall assume $k_1 \gg k$, and we shall use a representation which diagonalizes $D(k_1)$. However since $\hat{G}(k_2)$ stands inside the integral, then we arrive at a representation which diagonalizes $D(k_2)$ by introducing conversion coefficients $C_{il}(k_1 k_2)$. If we assume $k_2 \gg k$, then in the first approximation one can neglect k inside the integral. But in this connection we obtain

FIG. 3

$$\int \sum_n C_{in}(k_1k_2)\,(\omega_2-\varepsilon_n(k_2)+i\delta\,\mathrm{sign}\,\varepsilon_n)^{-1}$$
$$\times(\omega_2+\omega-\varepsilon_n(k_2)+i\delta\,\mathrm{sign}\,\varepsilon_n)^{-1}C_{nl}(k_2k_1)\frac{d\omega_2}{2\pi}.$$

ı each of the terms of the sum, the poles with respect o ω_2 lie on the same side of the real axis. Therefore he corresponding integral is equal to zero. A finite ontribution appears only in first order with respect to :. This introduces an extra factor k/k_2 into the integ-rand. After this one can easily see that the integral will onverge, and in the region $k < me^2/\epsilon_0$ it turns out that $^{(1)} \sim 1$. One can verify that this is valid for the correc-ions to γ of arbitrary order.

Here we encounter a new situation: at small momenta ıll of the terms of the perturbation-theoretic series ecome of the same order, i.e., weak coupling changes nto strong coupling. Such a situation also arises in ertain versions of field theory, and also in the problem of a phase transition of the second kind. This situation vas considered in the well-known work by V. N. Gribov ınd A. A. Migdal,[11] where it was shown that the be-ıavior of the fundamental quantities is determined by power laws which are established with the aid of the heory of scaling. We shall use the method of article[11].

We make the assumption that the dependence of all quantities on k and ω is determined by the following relations (we do not present the detailed matrix struc-ture, which depends on the specific representation):

$$G=c_1k^{-\alpha}g(\omega/sk^{\nu}),\qquad \gamma=c_2k^{\beta}q(\omega/sk^{\nu}),\qquad \Gamma=c_3k^{-\delta}d(\omega/sk^{\nu}),\tag{39}$$

and we choose the powers α, β, δ, ν and the constants c_1, c_2, c_3, and s in order to satisfy all the relations of the theory (here a symmetric γ is to be understood, i.e., having arguments of the same order).

First of all, on the basis of the previous discussion one can conclude that for small momenta $4\pi e^2\Pi \gg \epsilon_0 k^2$ and $\Sigma \gg G_0^{-1}$, i.e., the behavior of the G- and Γ-func-tions is determined by the self-energy parts. But the corresponding integrals only depend on k if k is suffi-ciently large, and they only depend on ω if ω dominates. This is guaranteed only in that case when in (39)

$$\begin{aligned} g(0)&=1, & g(x\gg 1)&\sim x^{-\alpha/\nu},\\ q(0)&=1, & q(x\gg 1)&\sim x^{\beta/\nu},\\ d(0)&=1, & d(x\gg 1)&\sim x^{-\delta/\nu} \end{aligned}\tag{40}$$

(the values at zero are normalized by the appropriate choice of the constant).

Let us consider the diagrams of higher orders for γ. The transition to the next order means the appearance in the diagram of two additional G-lines, two γ vertices, one Γ-line, and integration over $d^3k\,d\omega$. We require that this operation preserve the form of γ. For this it is necessary that

$$\int G^2\gamma^2\Gamma\, d^3k_1\, d\omega_1 \sim 1.$$

Substituting (22) and proceeding to the integration with respect to $x_1 = \omega/sk_1^{\nu}$, from here we obtain

$$c_1{}^2c_2{}^2c_3sk^{-2\alpha+2\beta-\delta+\nu+3}\int g^2q^2\,ddx \sim 1.$$

Hence it follows that

$$-2\alpha+2\beta-\delta+\nu+3=0,\tag{41}$$

$$c_1{}^2c_2{}^2c_3s \sim 1.\tag{42}$$

Here it is necessary to add a condition in order that the integral in the correction for γ should converge. Con-sidering the first correction, but with complete lines and vertices, we obtain

$$-2\alpha+3\beta-\delta+\nu+3<0.\tag{43}$$

This guarantees the convergence of the integral over k. It is not difficult to see that here the integral over x will automatically converge.

Then let us consider the G-function. First of all we note that it satisfies the so-called Ward identity:

$$\partial G^{-1}/\partial\omega=\gamma.\tag{44}$$

Substituting formula (39) here, we obtain

$$sc_1c_2k^{\beta-\alpha+\nu} \sim 1,$$

from where it follows that

$$\beta-\alpha+\nu=0,\tag{45}$$

$$sc_1c_2 \sim 1.\tag{46}$$

Now we introduce a new function[1)] γ_2 which we define as

$$\gamma_2=\partial^2G^{-1}/\partial k^2.\tag{47}$$

We assume that this function has a form analogous to (39), that is,

$$\gamma_2=c_4k^{\mu}\varphi(\omega/sk^{\nu}).\tag{48}$$

Substituting (39) and (48) into relation (47), we obtain

$$\mu-\alpha+2=0,\tag{49}$$

$$c_4c_1 \sim 1.\tag{50}$$

Now let us consider $\Sigma(\mathbf{k},\omega)-\Sigma(0,0)$. In the expansion in terms of k in the integral for the self-energy, the contribution of second order in k is essential. Taking into consideration that according to our condition all of the diagrams give a contribution of the same order of magnitude, it is sufficient to consider only the first diagram (Fig. 4). From it we obtain ($\omega \ll sk^{\nu}$)

$$\Sigma(\mathbf{k},\omega)-\Sigma(0,0) \sim k^2\int G^2\gamma_2\gamma^2\Gamma\, d^3k_1\, d\omega_1.$$

If this expression is assumed to be of order G^{-1}, then conditions (41) and (42) are again obtained here together with (49) and (50). From this point of view the introduc-tion of the quantity γ_2 does not give anything new. It is necessary, however, so that the integral $\Sigma(\mathbf{k},\omega)-\Sigma(0,0)$ converges. Hence follows the requirement

$$3+\mu+2\beta-\delta-2\alpha+\nu>0.\tag{51}$$

Let us go on to the polarization operator. In the previous Section it was shown that in the first approxi-mation $\Pi = 0$ for $k = 0$. In the general case it is not difficult to see that $\Pi(0, 0) = 0$. This follows from the

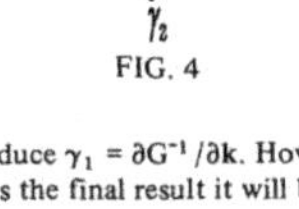

FIG. 4

---

1)One can also introduce $\gamma_1 = \partial G^{-1}/\partial\mathbf{k}$. However, in view of its odd-ness with respect to $\mathbf{k}$, as the final result it will be proportional to $\mathbf{k}\gamma_2$.

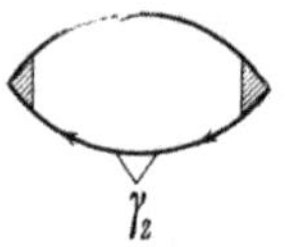

FIG. 5

fact that $\Pi(0, 0)$ determines the change of the number of particles due to the action of the electrostatic potential, which does not depend on the coordinates. Mathematically this is written as:

$$\Pi(0,0) = -i\int \mathrm{Sp}\, G^2(\omega_1, \mathbf{k}_1)\gamma(\omega_1, \mathbf{k}_1)\frac{d\omega_1\, d\mathbf{k}_1}{(2\pi)^4} = i\int \mathrm{Sp}\frac{\partial G}{\partial\omega_1}\frac{d\omega_1\, d\mathbf{k}_1}{(2\pi)^4} = 0.$$

Expanding the integrand up to second order in k, we obtain for $\omega \ll sk^\nu$ (see Fig. 5)

$$\Pi \sim k^2\int d^3k_1\, d\omega_1 G^3\gamma_2\gamma^2.$$

Assuming $\Gamma \sim 1/\Pi$ and using (39) and (48), we again obtain the old relations for the powers and for the coefficients. However, the condition for convergence of the right hand side gives a new restriction:

$$3 + \mu + 2\beta + \nu - 3\alpha < 0. \qquad (52)$$

Relations (41), (45), and (49) give the possibility to express all of the coefficients α, β, δ, μ, and ν in terms of any two of them, for example:

$$\delta = 3 - \nu, \quad \beta = \alpha - \nu, \quad \mu = \alpha - 2. \qquad (53)$$

From conditions (43), (51), and (52) with the aid of (53) we obtain the following restrictions on the possible values of the coefficients:

$$\nu > 1, \quad \alpha < 2, \quad \alpha < \nu, \quad \delta < 2, \quad \beta < 0, \quad \mu < 0. \qquad (54)$$

The question may arise whether new restrictions cannot appear if we enclose several vertices of type γ_2 and γ by Coulomb lines. If we have n_1 vertices of type γ_2 and n_2 vertices of type γ, then the condition for convergence is written in the form

$$n_1\mu + n_2\beta - \alpha(n_1 + n_2 - 1) < 0.$$

Substituting (45) and (49), we obtain

$$\alpha < 2n_1 + \nu n_2.$$

This condition is ensured by the inequalities (54).

It is of interest that upon fulfillment of the restrictions (54), the behavior of G and Γ for small momenta and frequencies actually is determined by the self-energy parts. For $\omega \ll sk^\nu$ this is ensured by the conditions $\alpha < 2$, $\delta < 2$, but for $\omega \gg sk^\nu$ this follows from the condition $\alpha < \nu$.

For the constants c_1, c_2, c_3, and s we have, according to (42) and (46),

$$sc_1c_2 \sim 1, \quad c_1c_2c_3 \sim 1. \qquad (55)$$

The orders of magnitude of these constants are obtained from the conditions for joining with the region of large momenta. According to the previous discussion, joining should occur for $k \sim me^2/\epsilon_0$. In the region of large momenta the electron spectrum is given by $\omega \sim k^2/m$. Therefore, with regard to ω, joining should occur for $\omega \sim me^4/\epsilon_0^2$. It is natural to expect that in the joining region $\omega \sim sk^\nu$. From here it follows that

$$s \sim (e^2/\varepsilon_0)^{2-\nu}m^{1-\nu}. \qquad (56)$$

In the joining region the function $\mathbf{\Gamma}$ is of the order of $(e^2/\epsilon_0k^2) \sim (\epsilon_0/m^2e^2)$. Comparing with expression (39) we have

$$c_3 \sim (e^2/\varepsilon_0)^{\delta-1}m^{\delta-2} \sim (e^2/\varepsilon_0)^{2-\nu}m^{1-\nu}. \qquad (57)$$

In the joining region the function G is of the order of $1/\omega$ or $(me^4/\epsilon_0^2)^{-1}$. Hence it follows that

$$c_1 \sim (e^2/\varepsilon_0)^{\alpha-2}m^{\alpha-1}. \qquad (58)$$

Finally, for $k \sim me^2/\epsilon_0$, γ should be of the order of unity. Hence

$$c_2 \sim (me^2/\varepsilon_0)^{-\beta}. \qquad (59)$$

The values (56)–(59) satisfy the relations (55).

From the obtained results it follows that in the region $k < me^2/\epsilon_0$ the electron's Green's function does not have such a simple form as for large momenta. Its poles, if such exist in general, may be complex, where in general their imaginary part is of the order of the real part. Therefore, in this region it is impossible to talk about quasi-particles in the usual sense of the word.

However, the value of the Green's function makes it possible to determine various physical effects. Of course, it is necessary to bear in mind that the theory developed above does not give the exact value of the constant, and also of the functions g, q, and d in the region $x \sim 1$. However, even what is known makes it possible to make a number of predictions.

Let us consider the momentum distribution of the particles. It is expressed by the integral

$$n(k) = -i\int_{-\infty}^{+\infty} G(\omega, k)e^{i\omega\tau}\frac{d\omega}{2\pi},$$

where $\tau \to +0$. Using formula (22) we obtain

$$n(k) \sim c_1sk^{-\alpha+\nu}\int g(x)\,dx.$$

But as $x \to \infty$ the quantity $g \sim x^{-\alpha/\nu}$ and since $\alpha < \nu$ according to what was said previously, the integral diverges. Hence for small momenta n(k) is actually determined by the value of G in the region of large momenta, and we do not have the right to use formula (22) directly. One can, however, evaluate the difference

$$n(k) - n(0) \sim -i\int_{-\infty}^{+\infty}[G(\omega, k) - G(\omega, 0)]e^{i\omega\tau}\frac{d\omega}{2\pi}.$$

In this integral the region of large values of ω is contracted, and the integral converges for $x \sim 1$. In this connection we obtain

$$n(k) - n(0) \sim c_2sk^{\nu-\alpha} \sim (k\varepsilon_0/me^2)^{\nu-\alpha}. \qquad (60)$$

According to Eq. (37) the power $\nu - \alpha$ is positive. In the region of large momenta we again obtain a constant.

Now let us consider the heat capacity at low temperatures. According to[2] we have

$$\Omega(T) - \Omega(0) = 2\left(T\sum_\omega - \frac{1}{2\pi}\int d\omega\right)\int\frac{d^3k}{(2\pi)^3}e^{i\omega\tau}\ln G(\omega, k). \qquad (61)$$

It is necessary to take $G(\omega)$ for $T = 0$. Replacing the summation over ω by an integration over real frequencies and differentiating with respect to T, we obtain the entropy

$$S(T) = -\frac{\partial\Omega}{\partial T} = -\frac{2}{\pi}\int_{-\infty}^{+\infty}\frac{\omega}{T}\frac{\partial n_F}{\partial\omega}\int\frac{d^3k}{(2\pi)^3}\operatorname{Im}\ln G_R(\omega, k),$$

where n_F denotes the Fermi function. It is impossible to evaluate this integral by the same method as in the presence of the Fermi surface. However, we obtain a result correct to within a constant coefficient if we assume that only small values of ω are essential. Taking into account the evenness of $\partial n_F/\partial\omega$, we obtain

$$S(T) \sim T\int\frac{d^3k}{(2\pi)^3}\left[\frac{\partial}{\partial\omega}\operatorname{Im}\ln G_R(\omega, k)\right]_{\omega\sim T}.$$

Regarding the imaginary part of G to be of the order of the real part, we may remove the Im sign and write the derivative inside the square brackets as

$$G(\omega, k)\frac{\partial G^{-1}(\omega, k)}{\partial\omega} = G\gamma.$$

Substituting G according to formula (22) we obtain

$$S \sim C \sim c_1c_2T\int d^3k k^{-\alpha+\beta} g\left(\frac{T}{sk^\nu}\right) q\left(\frac{T}{sk^\nu}\right).$$

Changing to the variable $x = T/sk^\nu$, we verify that the integral over x is convergent. Substituting the values of the coefficients c_1, c_2, and s, we finally obtain

$$C \sim \left(\frac{me^2}{\varepsilon_0}\right)^3\left(\frac{T\varepsilon_0^2}{me^4}\right)^{3/\nu}. \quad (62)$$

Taking the restriction (37) into account, we find the power of the temperature:

$$3 > 3/\nu > 0. \quad (63)$$

we note that one can also obtain the temperature dependence of the heat capacity from the following simple argument. The heat capacity is proportional to $T_\rho(T)$, where ρ is the density of the electronic states. The polarization operator Π is proportional to ρ. Therefore $\rho(T) \sim k^\delta(T)$, where k(T) denotes the momenta characteristic for a given temperature. It is obvious that $k(T) \sim T^{1/\nu}$. Also keeping in mind that $\delta = 3 - \nu$, we obtain $C \sim T^{3/\nu}$.

In conclusion the authors express their gratitude to L. P. Gor'kov, A. A. Migdal, and I. E. Dzyaloshinskiĭ for valuable discussions.

[1] L. D. Landau, Zh. Eksp. Teor. Fiz. 30, 1058 (1956) [Sov. Phys.-JETP 3, 920 (1957)].
[2] J. M. Luttinger, Phys. Rev. 119, 1153 (1960).
[3] Steven Groves and William Paul, Phys. Rev. Letters 11, 194 (1963).
[4] V. I. Ivanov-Omskiĭ, B. T. Kolomiets, L. I. Kleshchinskiĭ and K. P. Smekalova, Fiz. Tverd. Tela 10, 3106 (1968) [Sov. Phys.-Solid State 10, 2447 (1969)].
[5] A. A. Abrikosov, L. P. Gor'kov, and I. E. Dzyaloshinskiĭ, Metody kvantovoĭ teorii polya v statisticheskoĭ fizike (Methods of Quantum Field Theory in Statistical Physics), Fizmatgiz, 1962 (English Transl., Prentice-Hall, Inc., 1963).
[6] A. A. Abrikosov, Zh. Eksp. Teor. Fiz. 43, 1083 (1962) [Sov. Phys.-JETP 16, 765 (1963)].
[7] P. Nozières and J. M. Luttinger, Phys. Rev. 127, 1423 (1962); J. M. Luttinger and P. Nozières, Phys. Rev. 127, 1431 (1962).
[8] G. E. Pikus, Zh. Eksp. Teor. Fiz. 41, 1258, 1507 (1961) [Sov. Phys.-JETP 14, 898, 1075 (1962)].
[9] O. V. Kovalev, Neprivodimye predstavleniya prostranstvennykh grupp (Irreducible Representations of the Space Groups), Kiev, 1961.
[10] L. D. Landau, A. A. Abrikosov, and I. M. Khalatnikov, Dokl. Akad. Nauk SSSR 95, 1177 (1954).
[11] V. N. Gribov and A. A. Migdal, Zh. Eksp. Teor. Fiz. 55, 1498 (1968) [Sov. Phys.-JETP 28, 784 (1969)].

Translated by H. H. Nickle
148

Journal of Low Temperature Physics, Vol. 10, Nos. 1/2, 1973

Transition of a Bismuth-Type Semimetal to an Excitonic Insulator in a Strong Magnetic Field

A. A. Abrikosov

L. D. Landau Institute for Theoretical Physics
Academy of Sciences of the USSR, Moscow, USSR

(Received January 5, 1972)

The metal–insulator transition in a strong magnetic field is analyzed for a model composed of three electron groups and one hole group, corresponding to the energy spectrum of bismuth in a magnetic field oriented along the principal axis. Two possible types of carrier pairing are found: (a) electrons with holes, and (b) electrons with electrons within a single group. Both are pairings of the insulator type. Nevertheless, free carriers remain in the metal, even in the "insulator" state. The thermodynamic properties, dielectric constant at T = *0, and conductivity as a function of the temperature are investigated. The absence of superthermal conductivity is demonstrated.*

1. INTRODUCTION

Brandt and Chudinov[1] have recently discovered in experiments with bismuth–antimony alloys in a strong magnetic field that the properties of the material undergo significant changes at temperatures of the order of a few degrees Kelvin. The nature of these changes (diminution and strong temperature dependence of the conductivity) affords a basis for hypothesizing a transition of the material into an excitonic insulator.

We have analyzed[2,3] the possibility of such a transition for a model composed of equal numbers of electrons and holes with spherically symmetrical energy spectra and a large momentum difference in momentum space. This spectrum differs appreciably from the true spectrum of bismuth-type semimetals, which contains three groups of electrons and one group of holes with anisotropic spectra.

The most fundamental difference from the model considered previously is the fact that the number of holes does not equal the number of electrons in each group. In zero magnetic field the numbers of electrons in all the groups are equal, while the number of holes is thrice the number of electrons in a single group. In a strong magnetic field, in which case the electrons and holes are transferred to a single Landau level, the numbers of electrons in different groups and, accordingly, the limiting Fermi momenta differ, in general, on account of the anisotropy of the spin and orbital splitting of the levels. In particular, one or two electron groups can become completely "frozen out." An exception arises in the case of a symmetrical field direction, such that either all three electron groups ($\mathbf{H} \parallel C_3$) or two of them ($\mathbf{H} \parallel C_2$) have identical limiting momenta.

The number of holes is equal to the number of electrons. Consequently, the limiting momentum for holes is equal to the sum of the limiting momenta for all the electron groups. Equality of the limiting momenta of electrons and holes will take place only in the event that, due to "freeze-out," only one electron group remains, a situation that is clearly very difficult to realize experimentally.

A second defect of the previous approach[2,3] is the replacement of the actual problem by a one-dimensional model in which no allowance is made for the transverse motion of electrons as described by the oscillator wave function $\psi_0(x - cp_y/eH)$. This aspect of the problem was recently considered by Brazovskii[4,5] in application to our earlier model.[2] The result of this analysis may be summed up as follows. A pole singularity appears only in those scattering channels which correspond to the coupling of carriers having different charge signs. For this kind of complex there is a conservative variable corresponding to the transverse momentum of the paired complex. The scattering amplitude pole depends on this transverse momentum. Near the pole a portion of diagrams entering into the "parquet" turns out to be nonsingular insofar as in these diagrams the integration is taken over the transverse momentum. Actually, the only essential channels are those in which the transverse momentum runs through the diagram. These are diagrams of the ladder type.

The replacement of the parquet by a ladder circumvents the issue of whether the phase transition is realistic or not, because the problem becomes analogous in this case to the superconductivity problem. In the presence of a parquet the appearance of a pole corresponds to a transition into a domain of strong coupling at low energies, implying not so much the inception of a new phase as a singular behavior of the variables in the vicinity of zero energies and temperatures. A representative example is the familiar Kondo effect.[6]

The discarded nonsingular diagrams have the same order as the "bare" interaction constant and, therefore, determine the exact position of the singularity. This differentiates the given problem from the superconductivity problem, the specifics of the interaction in the latter being such that the corrections for the bare interaction have a relative order $\omega_D/\varepsilon_F \ll 1$ (where ω_D is the Debye energy and ε_F is the Fermi energy). However, inasmuch as the integrals in these terms cover a large domain, they depend little on the energy in the scales that we require. The inclusion of such terms, therefore, merely alters the numerical value of the bare constant without changing its order. In view of the fact that the exact value of the bare constant is still unknown, we can reject these additional terms and assume that they are included in the bare constant.

Another conclusion that can be drawn from Brazovskii's work[4,5] is the theoretical impossibility of the pairing of carriers having equal charge, i.e., the absence of superconductivity. It is essential to note that this conclusion was already obtained in our simplified one-dimensional model,[2,3] but at that time it was not apparent beforehand.

In the present paper we investigate another model, which is more nearly compatible with the real semimetal. The new model comprises three isotropic groups of electrons and one group of holes, where the limiting momentum of the latter is thrice that of the former. This case corresponds to the magnetic field along the axis C_3. All the masses and coupling constants of the different groups will clearly be the same. Since only the immediate neighborhood of the Fermi momentum is essential in the integrals, the assumption of isotropy of the electron and hole spectra is not significant. As before,[2] we shall assume point interaction between particles.

At the end of the paper we discuss the qualitative characteristics of the pairing process for other orientations of the magnetic field.

2. SCATTERING AMPLITUDES

We begin with the simpler amplitude of the scattering of an electron by a hole. Inasmuch as the limiting momentum of the electron is equal to p_0 and that of the hole is $3p_0$, any scattering event in which an electron goes from the vicinity of p_0 to within the vicinity of $-p_0$ moves a hole away from the vicinity of its limiting momentum. Consequently, of the singular elements represented in Fig. 3 of our earlier paper,[2] only 3*a* and 3*e* remain. We have reconstructed them here in Fig. 1. Also possible, of course, are the analogous diagrams in which an electron is situated in the vicinity of p_0 and the hole in the vicinity of $-3p_0$. However, these diagrams do not intermingle with the diagrams of Fig. 1, and the result for the corresponding amplitude does not differ from the given case. In the simplified one-dimensional model, Figs. 1a and 1b give logarithmic integrals of the same absolute value but different signs. Consequently, the second-order correction vanishes. It is easily verified by inspection of the total parquet that the entire correction to the scattering amplitude vanishes in the logarithmic approximation, so that the total amplitude turns out in this approximation to be equal to the bare interaction constant.

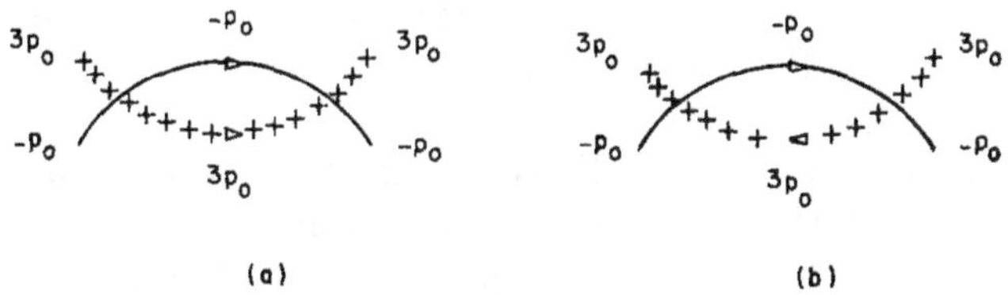

Fig. 1

The situation is altered when the transverse-motion wave functions are taken into account. According to Brazovskii,[4,5] the transverse momentum does not pass through the channel to which Fig. 1b corresponds. In the corresponding terms we find integration over the transverse momentum. In exact analogy with earlier work[2–5] we obtain

$$\Gamma(x, \mathbf{P}) = \pm e^{-P^2/2} - \int_0^x dy \Gamma^2(y, \mathbf{P}) + \int_0^x dy \int (d\mathbf{P}_1/2\pi) \int (d\mathbf{P}_2/2\pi) \Gamma(y, \mathbf{P}_1) \Gamma(y, \mathbf{P}_2) K(\mathbf{P}_1, \mathbf{P}_2, \mathbf{P}) \tag{1}$$

in which Γ is the total vertex divided by $|\lambda_1|$ (λ_1 is the bare interaction constant; we assume a point interaction); the sign $\pm$ in the first term corresponds to the sign of the interaction constant, $x = |\lambda_1|\{\pi p_0[(1/m_e) + (3/m_h)]\}^{-1} \ln(\varepsilon_F/\omega)$; y is the analogous variable of integration; $\mathbf{P}$ is the two-component transverse momentum, expressed in units of $\sqrt{eH/c}$; and the kernel $K(\mathbf{P}_1, \mathbf{P}_2, \mathbf{P})$ is equal to

$$K(\mathbf{P}_1, \mathbf{P}_2, \mathbf{P}) = \cos\{PP_2 \sin(\phi_2 - \phi) + P_1P_2 \sin(\phi_1 - \phi_2) + PP_1 \sin(\phi - \phi_1)\} \tag{2}$$

where P_1, P_2, and P are the absolute values, and ϕ_1, ϕ_2, and ϕ are the polar angles.

The poles of Γ must clearly depend on $\mathbf{P}$; consequently, the pole vanishes in the terms involving integration. In the first approximation, therefore, we can neglect close to the pole the nonsingular free term and terms with integrals over $\mathbf{P}$. In this case $\mathbf{P}$ becomes involved as a parameter. We then obtain

$$\Gamma(x, \mathbf{P}) \approx -\int_0^x dy \Gamma^2(y, \mathbf{P}) \tag{3}$$

This is a ladder-type equation. Its solution is

$$\Gamma = 1/[x - x_0(\mathbf{P})] \tag{4}$$

The quantity $x_0(\mathbf{P})$ is determined from the condition at $x = 0$, i.e., by matching with the domain far from the pole. All terms of Eq. (1) become essential in this domain.

If the third term in Eq. (1) were absent, we would have obtained $x_0 = \mp e^{-P^2/2}$. In this case, therefore, for a positive constant λ there is no pole in Γ, whereas if the constant is negative there is a pole, as expected, because the case in question is exactly analogous to the super-conductivity case. The inclusion of the third term in (1), however, prevents us from associating the occurrence of a pole in one-to-one fashion with the sign of λ_1.

There is one other hazard, in principle. The third term discarded from Eq. (1) in the determination of Γ near a pole might be singular, even though weaker than the saved terms. In this event it would no longer be admissible to state that Γ has a simple pole. However, the substitution of expression (4) into the last term of Eq. (1) shows that the singular part of this term is proportional to $\sqrt{x - x_0(\mathbf{P})}$, i.e., does not affect the character of the singularity of Γ. It is this fact which enables us to assert that the last term in Eq. (1), like the analogous terms in the other equations given below, essentially serves merely to renormalize the bare interaction, i.e., influence the positions of the poles, but does not alter the character of the singularity per se.

We now consider the scattering amplitude of electrons of different groups.

As already stated, in our model there are three electron groups with a limiting momentum p_0. They are described by the operators ψ, ϕ, and η. We first consider a one-dimensional model analogous to our previous one.[2]* The electron interaction Hamiltonian in this case takes the form

$$H_{\text{int}} = \lambda_2 \int \psi^+(z)\varphi^+(z)\varphi(z)\psi(z)\,dz + \lambda_2 \int \varphi^+(z)\eta^+(z)\eta(z)\varphi(z)\,dz + \lambda_2 \int \eta^+(z)\psi^+(z)\psi(z)\eta(z)\,dz \quad (5)$$

The first diagrams containing logarithmic integrals are shown in Fig. 2, in which the solid curve corresponds to the Green function of the operators ψ, the broken curves to the same for ϕ, and the wavy curve to the same for η; the plus and minus signs denote the neighborhood of $p_z = p_0$ or $p_z = -p_0$, respectively.

Considering the symmetry of the problem, it is logical to assume from the outset that there is a total of three different scattering amplitudes, which we denote by Γ_1 (Fig. 3a), Γ_2 (Fig. 3b), and Γ_3 (Fig. 3c); the lines in Fig. 3

*The calculations for the one-dimensional model were carried out in collaboration with E. V. Klishin, to whom the author conveys his gratitude.

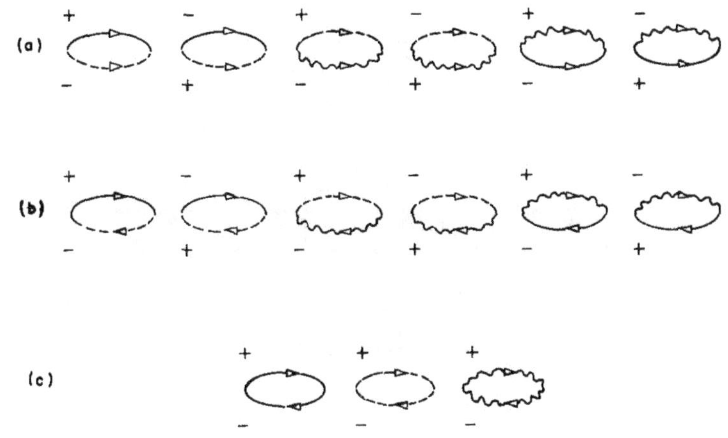

Fig. 2

can belong to different particles; it matters only that all the lines in Γ_1 are of one kind, while those in Γ_2 and Γ_3 are of two different kinds.

The equations for Γ_1 and Γ_3 do not differ in principle from those derived earlier.[2] However, to facilitate what follows we shall reiterate the derivation. The amplitude Γ_1 has two channels, as illustrated in Fig. 4. In the first channel the diagrams can be cut by two parallel lines for the same particle. In the second channel the diagrams can be cut by antiparallel lines, which can correspond to all three kinds of particles. In analogy with Ref. 2 we have the equation

$$\Gamma_1(x) = 2\int_0^x \Gamma_2^2(y)\,dy \tag{6}$$

in which Γ_i are the vertices divided by $|\lambda_2|$ and $x = (|\lambda_2|/2\pi v)\ln \varepsilon_F/\omega$.

The vertex Γ_3 corresponds to two channels (see Fig. 5): Λ_3 with parallel and Λ_4 with antiparallel lines. In Λ_3 cuts are possible which preserve the initial signs of the lines (in the sense of the neighborhood of p_0 or $-p_0$)

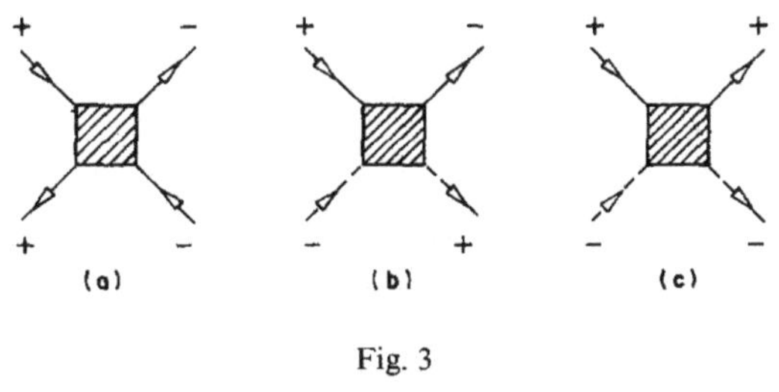

Fig. 3

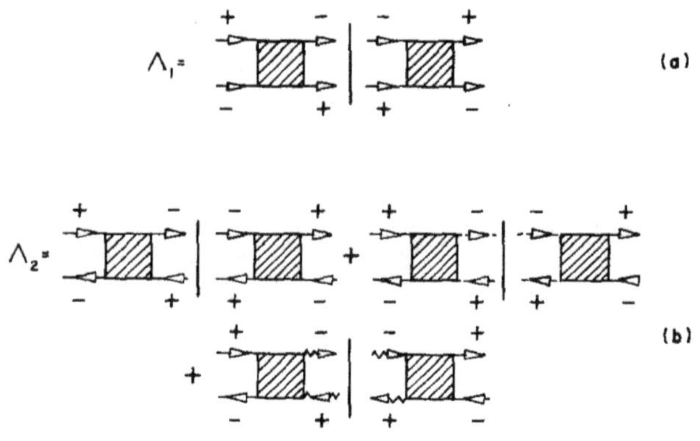

Fig. 4

as well as those which change sign. The corresponding equation has the form

$$\Gamma_3 = \pm 1 - \int_0^x \Gamma_2^2(y)\, dy \tag{7}$$

where $\pm$ corresponds to the sign of λ_2. Finally, for Γ_2 we have the channels illustrated in Fig. 6. As a result we have the equation

$$\Gamma_2(x) = \pm 1 + 2\int_0^x [\Gamma_1(y) - \Gamma_3(y)]\Gamma_2(y)\, dy - \int_0^x \Gamma_2^2(y)\, dy \tag{8}$$

The occurrence of the last term, corresponding to the last diagram in Fig. 6b, is associated with the presence of three kinds of particles and does not have any analog in the earlier model.[2]

The set of equations (6)–(8) has the simple solution

$$\begin{aligned} \Gamma_1 &= -\tfrac{2}{9}[1/(x \pm \tfrac{1}{3})] \pm \tfrac{2}{3} \\ \Gamma_2 &= \tfrac{1}{3}[1/(x \pm \tfrac{1}{3})] \\ \Gamma_3 &= \pm\tfrac{2}{3} + \tfrac{1}{9}[1/(x \pm \tfrac{1}{3})] \end{aligned} \tag{9}$$

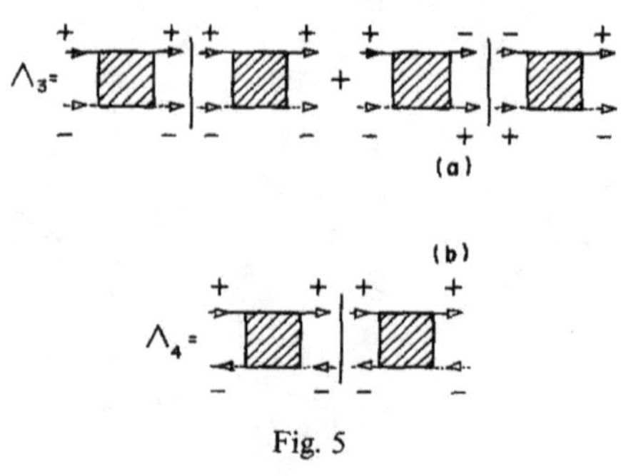

Fig. 5

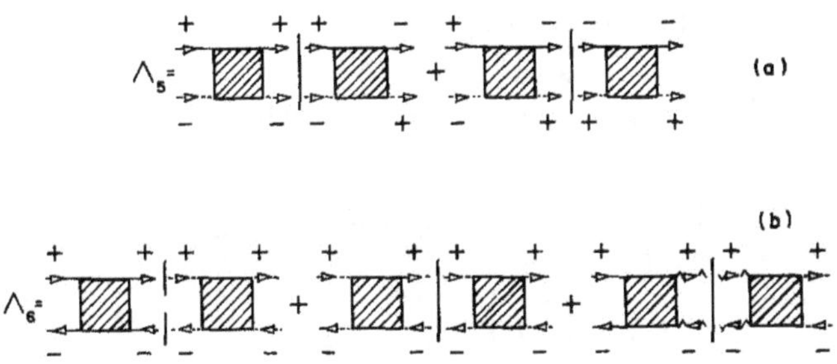

Fig. 6

Inasmuch as we are concerned with the pole for $x = (|\lambda_2|/2\pi v) \ln(\varepsilon_F/|\omega|) > 0$, it follows from (9) that it exists only for $\lambda_2 < 0$ and corresponds to $x = \frac{1}{3}$.

We now inquire what happens when the transverse motion of electrons is taken into account. According to Brazovskii,[4,5] the transverse momentum does not run through the channels corresponding to Λ_1 (Fig. 4a), Λ_3 (Fig. 5a), and Λ_5 (Fig. 6a).

The corresponding terms contain integration over the transverse momentum, and in place of Eqs. (6)–(8) we obtain

$$\begin{aligned}
\Gamma_1(x, \mathbf{P}) &= \int_0^x dy[\Gamma_1^2(y, \mathbf{P}_1) + 2\Gamma_2^2(y, \mathbf{P})] \\
&\quad - \int_0^x dy \int (d\mathbf{P}_1/2\pi) \int (d\mathbf{P}_2/2\pi)\Gamma_1(y, \mathbf{P}_1)\Gamma_1(y, \mathbf{P}_2)K(\mathbf{P}_1, \mathbf{P}_2, \mathbf{P}) \\
\Gamma_2(x, \mathbf{P}) &= \pm e^{-P^2/2} + \int_0^x dy[2\Gamma_1(y, \mathbf{P})\Gamma_2(y, \mathbf{P}) - \Gamma_2^2(y, \mathbf{P})] \\
&\quad - 2\int_0^x dy \int (d\mathbf{P}_1/2\pi) \\
&\quad \cdot \int (d\mathbf{P}_2/2\pi)\Gamma_3(y, \mathbf{P}_1)\Gamma_2(y, \mathbf{P}_2)K(\mathbf{P}_1, \mathbf{P}_2, \mathbf{P}) \\
\Gamma_3(x, \mathbf{P}) &= \pm e^{-P^2/2} + \int_0^x \Gamma_3^2(y, \mathbf{P})\, dy \\
&\quad - \int_0^x dy \int (d\mathbf{P}_1/2\pi) \int (d\mathbf{P}_2/2\pi)[\Gamma_3(y, \mathbf{P}_1)\Gamma_3(y, \mathbf{P}_2) \\
&\quad + \Gamma_2(y, \mathbf{P}_1)\Gamma_2(y, \mathbf{P}_2)]K(\mathbf{P}_1, \mathbf{P}_2, \mathbf{P})
\end{aligned} \tag{10}$$

[All of the notations here correspond to (1) and (2).]

As before, in the first approximation we can neglect nonsingular free terms and terms with integrals over **P** near a pole. Then, **P** is automatically involved as a parameter, and we obtain

$$\Gamma_1(x) \approx \int_0^x dy[\Gamma_1^2(y) + 2\Gamma_2^2(y)]$$

$$\Gamma_2(x) \approx \int_0^x dy[2\Gamma_1(y)\Gamma_2(y) - \Gamma_2^2(y)]$$

$$\Gamma_3(x) \approx \int_0^x dy\Gamma_3^2(y) \tag{11}$$

The last equation is of the ladder type. Combining the first two equations, we obtain two more ladder equations:

$$\Gamma_1(x) + \Gamma_2(x) = \int_0^x dy[\Gamma_1(y) + \Gamma_2(y)]^2$$

$$\Gamma_1(x) - 2\Gamma_2(x) = \int_0^x dy[\Gamma_1(y) - 2\Gamma_2(y)]^2 \tag{12}$$

Equations of this type have as their solution

$$\Gamma_i = 1/[x_{0i}(\mathbf{P}) - x] \tag{13}$$

The quantity $x_{0i}(\mathbf{P})$ is determined from the condition at $x = 0$, i.e., by matching with the domain far from the pole, wherein all terms of Eqs. (10) are significant. The following, however, warrants our attention. At $x = 0$ we have $\Gamma_1 = 0$, whereas $\Gamma_2, \Gamma_3 = \pm e^{-P^2/2}$. It is readily perceived that the relation $\Gamma_3(x) = \Gamma_1(x) + \Gamma_2(x)$ holds. Indeed, if we substitute this condition into the integral terms of the last equation of (10) and the sum of the first two equations, they coincide.

3. PAIRING

We now investigate the possibility of pairing. According to our earlier paper,[2] we need to find asymmetric vertices $\Gamma(x, y)$ in each of the possible channels. Every such vertex is a product $P(x)Q(y)$. Pairing is possible if there is a value of $x = s$ such that $P(s) = \infty$. In the general case the equations for $\Gamma(x, y)$ have the form

$$\Gamma(x, y) = \Gamma(y) \pm \int_y^x \Gamma(u)\Gamma(u, y)\, du \tag{14}$$

As $x \to s$ the function $P(x)$ and, together with it, $\Gamma(x, y) \to \infty$, so that we can neglect the free term and regard the lower limit of the integral as arbitrary, say, equal to zero. Then, $\Gamma(x, y)$ depends on y as on a parameter, and we can write for $P(x)$

$$P(x) = \pm \int_0^x \Gamma(u)P(u)\,du \tag{15}$$

This type of equation has as its solution

$$P(x) \sim C \exp\left[\pm \int_0^x \Gamma(u)\,du\right] \tag{16}$$

It is a simple matter to verify that the pairing of electrons with holes does not occur in the one-dimensional model, because the corresponding function P(x) is equal to $Ce^{\mp x}$. Taking transverse motion into account, we have

$$P(x) \sim 1/[x - x_0(\mathbf{P})] \sim \Gamma(x) \tag{17}$$

Consequently, pairing takes place when the function $x_0(\mathbf{P})$ is not everywhere negative.

The method of computing the asymmetric vertices $\Gamma_i(x, y)$ is inconvenient for studying the pairing of electrons with electrons, because it does not enable one to account properly for the symmetry between separate electron groups. It is better from this point of view to proceed as follows.[7] Assuming that we have certain infinitesimal quasiaverages of the type $\langle\psi_+^+\psi_-\rangle$, we find linear equations for those quantities directly from the diagrams. It is easily verified that all such equations will have the form (15). In the given instance, however, we know the combinations of the quasiaverages to which the function $P(x)$ corresponds. Some of them must be identically equal to zero by virtue of the symmetry between electrons of different groups. This reduces the number of possibilities.

We first consider the one-dimensional model with $\lambda_2 < 0$. The channel Λ_1 in Fig. 4 corresponds to the Cooper quasiaverage $\langle\psi_+\psi_-\rangle$. The appropriate equation has the form

$$P_1(x) = -\int_0^x \Gamma_1(u)P_1(u)\,du$$

Hence,

$$P_1(x) = C \exp\left[-\int_0^x \Gamma_1(u)\,du\right] = C\,e^{\frac{2}{3}x}(x - \tfrac{1}{3})^{2/9}$$

There is no pole, i.e., Cooper pairing does not take place.

In the channel Λ_4 (Fig. 5), which corresponds to the quasiaverage $\langle\psi_+\eta_-^{\pm}\rangle$, we obtain

$$P_2(x) \sim \exp\left[\int_0^x \Gamma_3(u)\,du\right] \sim e^{-\frac{2}{3}x}(x-\tfrac{1}{3})^{1/9}$$

Again there is no pole, and pairing is impossible.

In a channel corresponding to Λ_3, Λ_5, i.e., to a quasiaverage of the Cooper type $\langle\psi_+^+\eta_-^+\rangle$, using the symmetry between electrons of different groups we obtain

$$P_3(x) \sim \exp\left[-\int_0^x (\Gamma_3-\Gamma_2)\,du\right] \sim e^{\frac{2}{3}x}(x-\tfrac{1}{3})^{2/9}$$

Again pairing does not occur.

Finally, in the channel Λ_2, Λ_6 (quasiaverages of the type $\langle\psi_+^+\psi_-\rangle$) we again make use of the symmetry of electrons to find

$$P_4(x) \sim \exp\left[\int_0^x (\Gamma_1-2\Gamma_2)\,du\right] \sim e^{-\frac{2}{3}x}(x-\tfrac{1}{3})^{-8/9}$$

This function goes to infinity at $x = s = \frac{1}{3}$. Consequently, pairing of the type $\langle\psi_+^+\psi_-\rangle$ is possible.

As mentioned in our earlier paper[2] and in Section 1 of the present article, in the parquet situation we can only eliminate certain possibilities, but we cannot assert conclusively that wherever pairing is obtained with logarithmic accuracy it does in fact take place. The situation is otherwise when the transverse motion is taken into account, in which case the equations for Γ reduce to simple chains, and it can definitely be stated that for a particular sign of the renormalized interaction constant Γ_i will necessarily have a pole and, as we shall see presently, certain types of pairing are possible.

It was noted above that in the given case Cooper-type pairings, corresponding to the channels Λ_1 and Λ_3, Λ_5, are absent. In Λ_4 we obtain

$$P_2(x) \sim \exp\left[\int_0^x \Gamma_3(u)\,du\right] \sim 1/[x_{03}(\mathbf{P}) - x]$$

Therefore, it appears at first glance that pairing of the type $\langle\psi_+^+\eta_-\rangle$ is possible.

Next we consider the channel Λ_2, Λ_6, which corresponds to quasiaverages of the type $\langle\psi_+^+\psi_-\rangle$. Equations of the type (15) are obtained for the combinations $\langle\psi_+^+\psi_-\rangle + \langle\phi_+^+\phi_-\rangle + \langle\eta_+^+\eta_-\rangle$ and $\langle\psi_+^+\psi_-\rangle - \langle\eta_+^+\eta_-\rangle$.

It clearly follows from the electron symmetry that the second combination is equal to zero.* For the first combination we have

$$P_4(x) \sim \exp\left[\int_0^x (\Gamma_1 - 2\Gamma_2)\,du\right] \sim 1/[x_{04}(\mathbf{P}) - x]$$

i.e., pairing of the type $\langle\psi_+^+\psi_-\rangle$ is possible. It is readily verified that the combination $\langle\psi_+^+\psi_-\rangle - \langle\eta_+^+\eta_-\rangle$ corresponds to $\Gamma_1(x, y) + \Gamma_2(x, y)$. It is easily shown that, as in the case of symmetrical vertices, the complete equations imply $\Gamma_1(x, y) + \Gamma_2(x, y) = \Gamma_3(x, y)$. But the quasiaverage $P_2(x)$ found above exactly corresponds to $\Gamma_3(x, y)$. This justifies the assumption that pairing of the type $\langle\psi_+^+\eta_-\rangle$ is in fact absent.

Thus, both approaches indicate that only electron pairings of the type $\langle\psi_+^+\psi_-\rangle = \langle\phi_+^+\phi_-\rangle = \langle\eta_+^+\eta_-\rangle$, corresponding to the onset of a variation of the charge density with period $1/2p_0$ and the occurrence of an energy gap on the Fermi distribution, take place in the given model.

The difference between the two approaches lies in the problem of electron–hole pairing. Inasmuch as the one-dimensional model is less realistic, it must be assumed that the pairing of electrons with holes is possible. Since the diagrams corresponding to the pairing of electrons with holes and the pairing of electrons with each other are not intermingled, only one type of pairing occurs. Although we cannot determine the sign of the renormalized coupling constant, we can nevertheless consider it physically probable that it corresponds to the sign of the bare constant. If this is the case, then with prevalence of the Coulomb forces the pairing of electrons with holes takes place, and with prevalence of the phonon forces electron–electron pairing takes place.†

One more problem confronts us, namely, the transverse momentum $\mathbf{P}$. It is fairly obvious that pairing will take place in the state corresponding to positive $x_{0i}(\mathbf{P})$ having minimum absolute value [the quantity $|\lambda|/x_{0i}(\mathbf{P})$ acts as an effective interaction constant]. If the integral terms were absent from Eqs. (10), $x_{0i}(\mathbf{P})$ would be proportional to $e^{P^2/2}$, i.e., pairing would occur for $\mathbf{P} = 0$. We shall assume hereafter that this is in fact true for renormalized

*Suppose that this is not true. The symmetry with respect to rotations about C_3 then implies

$$\langle\eta_+^+\eta_-\rangle = e^{4\pi i/3}\langle\psi_+^+\psi_-\rangle$$

or

$$\langle\eta_+^+\eta_-\rangle = e^{2\pi i/3}\langle\psi_+^+\psi_-\rangle$$

But the symmetry with respect to the plane σ_d implies simultaneously that $\langle\psi_+^+\psi\rangle = \pm\langle\eta_+^+\eta_-\rangle$. These conditions are incompatible.

†Even if the renormalized constants had the same sign, they would have different magnitudes, and this would still imply only one type of pairing.

interaction. This assumption is the most natural and straightforward, although of course it should be subjected to detailed analysis.

4. TYPE I: SPECTRUM AND THERMODYNAMIC PROPERTIES

We propose to investigate various properties in both types of pairing: electrons with holes (type I) and electrons with electrons (type II). In both cases our procedure will be analogous to the theory of superconductivity (see Abrikosov *et al.*,[8] Chap. 7). In the case of type I, leaving only terms corresponding to electron–hole pairing in the Hamiltonian, we obtain (assuming $P = 0$) in the absence of external fields and impurities

$$\begin{aligned} H_I = \sum_k \{ & (\psi^+_{p_0+k}\psi_{p_0+k} + \psi^+_{-p_0-k}\psi_{-p_0-k} + \varphi^+_{p_0+k}\varphi_{p_0+k} \\ & + \varphi^+_{-p_0-k}\varphi_{-p_0-k} + \eta^+_{p_0+k}\eta_{p_0+k} + \eta^+_{-p_0-k}\eta_{-p_0-k})v_e k \\ & + (\rho^+_{3p_0+k}\rho_{3p_0+k} + \rho^+_{-3p_0-k}\rho_{-3p_0-k})v_h k \\ & - \Delta_1^*\rho_{-3p_0-k}(\psi_{p_0+k} + \phi_{p_0+k} + \eta_{p_0+k}) \\ & - \Delta_1(\psi^+_{p_0+k} + \varphi^+_{p_0+k} + \eta^+_{p_0+k})\rho^+_{-3p_0-k} \\ & - \Delta_1'^*\rho_{3p_0+k}(\psi_{-p_0-k} + \varphi_{-p_0-k} + \eta_{-p_0-k}) \\ & - \Delta_1'(\psi^+_{-p_0-k} + \varphi_{-p_0-k} + \eta_{-p_0-k})\rho^+_{3p_0+k} \end{aligned} \tag{18}$$

where

$$v_e = p_0/m_e \qquad v_h = 3p_0/m_h$$

$$\begin{aligned} \Delta_1 &= -\lambda_1 \sum_{k_1} \langle \rho_{-3p_0-k_1}\psi_{p_0+k_1}\rangle = -\lambda_1 \sum_{k_1} \langle \rho_{-3p_0-k_1}\varphi_{p_0+k_1}\rangle \\ &= -\lambda_1 \sum_{k_1} \langle \rho_{-3p_0-k_1}\eta_{p_0+k_1}\rangle \\ \Delta_1' &= -\lambda_1 \sum_{k_1} \langle \rho_{3p_0+k_1}\psi_{-p_0-k_1}\rangle = \ldots \end{aligned} \tag{19}$$

λ_1 is the renormalized interaction constant, assumed to be negative. Here we consider $k, k_1 \ll p_0$. These are the only values that are significant below.

It is readily seen that the Hamiltonian is separable into two independent parts, one of which includes ρ_{-3p_0-k}, ψ_{p_0+k}, ϕ_{p_0+k}, η_{p_0+k}, while the other includes ρ_{3p_0+k}, ψ_{-p_0-k}, ϕ_{-p_0-k}, η_{-p_0-k}. One part is obtained from the other by a simple change of notation. Hence it follows that we have two independent subsystems, so that the thermodynamic variables can be calculated from one of them and then doubled.

We introduce the thermodynamic Green functions

$$
\begin{aligned}
F^{+}(z-z',\tau-\tau') &= \langle T(\rho_{-}^{+}\psi_{+}^{+})\rangle \equiv \sum_{k}\langle T_{\tau}(\rho_{-3p_0-k}^{+}(\tau)\psi_{p_0+k}^{+}(\tau'))\rangle\, e^{ik(z-z')} \\
&= \langle T(\rho_{-}^{+}\varphi_{+}^{+})\rangle = \langle T(\rho_{-}^{+}\eta_{+}^{+})\rangle \\
G &= -\langle T(\psi_{+}\psi_{+}^{+})\rangle = -\langle T(\varphi_{+}\varphi_{+}^{+})\rangle = \ldots \\
D &= -\langle T(\psi_{+}\varphi_{+}^{+})\rangle = \ldots \\
\mathscr{G} &= -\langle T(\rho_{-}\rho_{-}^{+})\rangle
\end{aligned}
\tag{20}
$$

We obtain the following equations for the Fourier components:

$$
\begin{aligned}
(i\omega_n-\xi_e)G-\Delta_1F^{+} &= 1 \\
(i\omega_n-\xi_e)D-\Delta_1F^{+} &= 0 \\
(-i\omega_n-\xi_h)F^{+}+\Delta_1^{*}(G+2D) &= 0 \\
(-i\omega_n-\xi_h)\mathscr{G}(-\omega_n)-3\Delta_1F^{+} &= 1
\end{aligned}
\tag{21}
$$

in which $\xi_e = v_e k$ and $\xi_h = v_h k$.

The solution of these equations has the form

$$
\begin{aligned}
G &= \frac{1}{i\omega_n-\xi_e}\cdot\frac{(i\omega_n-\xi_e)(-i\omega_n-\xi_h)+2|\Delta_1|^2}{(i\omega_n-\xi_e)(-i\omega_n-\xi_h)+3|\Delta_1|^2} \\
D &= \frac{1}{i\omega_n-\xi_e}\cdot\frac{-|\Delta_1|^2}{(i\omega_n-\xi_e)(-i\omega_n-\xi_h)+3|\Delta_1|^2} \\
F^{+} &= -\frac{\Delta_1^{*}}{(i\omega_n-\xi_e)(-i\omega_n-\xi_h)+3|\Delta_1|^2} \\
\mathscr{G}(-\omega) &= \frac{i\omega_n-\xi_e}{(i\omega_n-\xi_e)(-i\omega_n-\xi_h)+3|\Delta_1|^2}
\end{aligned}
\tag{22}
$$

The poles of the Green functions with respect to the variable $i\omega$ yield the energy spectrum. From (22) we find three types of poles:

$$
\omega_1 = \xi_e \qquad \omega_{2,3} = \frac{\xi_e-\xi_h}{2} \pm \sqrt{\left(\frac{\xi_e+\xi_h}{2}\right)^2 + 3|\Delta_1|^2} \tag{23}
$$

Consequently, in addition to the two branches having an energy gap, there is a gapless branch having the same spectrum as for free electrons. The corresponding residue of the G function is equal to 2/3. It may therefore be assumed that only a third of the electrons is involved in pairing, the other $\frac{2}{3}$ remaining free. This interpretation is confirmed by our later results. It may clearly be attributed to the fact that pairing involves only electrons and holes in a small neighborhood of the limiting momentum, and the number of such neighborhoods for electrons is three times (one-dimensional problem) the number for holes.

From the definition of Δ_1 we obtain

$$1 = |\lambda_1| T \sum_{\omega} \int (dk/2\pi)\{1/[(i\omega_n - \xi_e)(-i\omega_n - \xi_h) + 3|\Delta_1|^2]\}$$

$$= (|\lambda_1|/2\pi) \int_0^{L/(v_e+v_h)} dk\{[\tanh(\omega_2/2T) - \tanh(\omega_3/2T)]/(\omega_2 - \omega_3)\} \quad (24)$$

where $L \sim \varepsilon_F$ for electrons is the upper limit of integration. Setting $T = 0$, we obtain

$$\Delta_1(0) = (L/\sqrt{3}) \exp\{-\pi p_0[(3/m_h) + (1/m_e)]/|\lambda_1|\} \quad (25)$$

The argument of the exponential function is of the order $\pi p_0 \varepsilon_0 / e^2 m_e$, where ε_0 is the dielectric constant. Conversely, assuming $\Delta_1 = 0$, we find

$$T_c = \sqrt{3}\Delta_1(0)(2\gamma/\pi)(3m_e m_h)^{1/2}/(3m_e + m_h) \quad (26)$$

where $\gamma = e^c$.†

Next we calculate the thermodynamic functions. Let us find $\partial H_I/\partial\lambda_1$. It is important to realize that the initial Hamiltonian contained terms of the type $\lambda_1 \psi^+ \rho^+ \rho \psi$ and that two terms of Eq. (18) were obtained from each term of the type indicated, namely, $-\Delta_1^* \rho\psi - \Delta_1 \psi^+ \rho^+$. Bearing this fact in mind, along with the presence of the two subsystems, we obtain

$$\frac{\partial\langle H_I\rangle}{\partial\langle\lambda_1\rangle} = 6\frac{|\Delta_1|^2}{|\lambda_1|^2} \quad (27)$$

From this we obtain for the thermodynamic potential

$$\Omega - \Omega_n = 6(eH/2\pi c)\int_0^{|\lambda_1|} |\Delta_1|^2\, d(1/|\lambda_1|)$$

$$= 6(eH/2\pi c)\int_0^{\Delta_1} [d(1/|\lambda_1|)/d|\Delta_1|]|\Delta_1|^2\, d|\Delta_1| \quad (28)$$

†It follows from Eq. (23) for $\omega_{2,3}$ that the minimum value of ω_2 corresponds to $k_0 = -[(v_e - v_h)/(v_e + v_h)\cdot\sqrt{3}\Delta_1(0)/\sqrt{v_e v_h}]$ and is equal to $\delta = \sqrt{3}\Delta_1(0)[2\sqrt{v_e v_h}/(v_e + v_h)] = \sqrt{3}\Delta_1(0)[2\sqrt{3m_e m_h}/(3m_e + m_h)]$ (the corresponding quantities for ω_3 are $-k_0$ and $-\delta$). The relationship between T_c and δ has the same form as for superconductors: $T_c = (\gamma/\pi)\delta$. We observe that whereas the number of carriers and related variables, such as the conductivity, are determined by the "gap" δ, the threshold for direct optical transitions between ω_2 and ω_3 corresponds to $2\sqrt{3}\Delta_1(0)$ [see Eq. (39) below].

The factor $eH/2\pi c$ is the density of states. With the help of Eqs. (23) and (24) we establish the relationship between $|\Delta_1|$ and $1/|\lambda_1|$ and then use (28) to find the difference between the free energies. Inasmuch as $|\Delta_1|$ decreases with increasing $1/|\lambda_1|$, we find that $\Omega(T) - \Omega_n(T) < 0$ up to the critical temperature, i.e., pairing is favorable energywise. In particular, at $T = 0$ we find

$$1/|\lambda_1| = (1/\pi p_0)[(3/m_h) + (1/m_e)]^{-1} \ln (L/\sqrt{3}\Delta_1)$$

We have as a result

$$\Omega(0) - \Omega_n(0) = -\tfrac{3}{2}(eH/\pi^2 p_0 c)[(3/m_h) + (1/m_e)]^{-1}|\Delta_1(0)|^2 \qquad (29)$$

Of particular interest is the low-temperature heat capacity. Unlike the superconductivity problem, in the given case the energy spectrum has a gapless branch. As we shall verify shortly, it induces a linear heat capacity. We write Eq. (24) in the form

$$1/|\lambda_1| = \frac{1}{\pi}\int_0^{L/(v_e+v_h)} [dk/(\omega_2 - \omega_3)] - \frac{1}{\pi}\int_0^\infty [n_F(\omega_2)/(\omega_2 - \omega_3)]\, dk$$

$$-\frac{1}{\pi}\int_0^\infty [n_F(-\omega_3)/(\omega_2 - \omega_3)]\, dk \qquad (30)$$

in which $n_F(\omega) = (e^{\omega/T} + 1)^{-1}$. It is quickly perceived from this equation that the difference $\Delta_1(T) - \Delta_1(0)$ depends exponentially on the temperature. Substituting $1/|\lambda_1|$ into Eq. (28), we find that the first term of Eq. (30) yields expression (29) with $\Delta_1(T)$ in place of $\Delta_1(0)$. The corresponding temperature increment depends exponentially on the temperature and is small in comparison with the linear terms. By substituting the remaining terms of Eq. (30) into (28) we can extend the integration over Δ_1 to ∞, because the additional part $\int_{\Delta_1}^\infty d\Delta_1$ introduced by this is also exponentially dependent on the temperature.

Substituting Eq. (23) for ω_2 and ω_3 and integrating over Δ_1, we find

$$\Omega(T) - \Omega_n(T) = T(eH/\pi^2 c)\left[\int_0^\infty dk \ln (1 + e^{-\xi_e/T}) + \int_0^\infty dk \ln (1 + e^{-\xi_h/T})\right]$$

$$- \tfrac{3}{2}(eH/\pi^2 p_0 c)[(3/m_h) + (1/m_e)]^{-1}|\Delta_1(0)|^2 \qquad (31)$$

At the same time,

$$\Omega_n(T) - \Omega_n(0) = -T(eH/\pi^2 c)\left[3\int_0^\infty dk \ln(1 + e^{-\xi_e/T}) + \int_0^\infty dk \ln(1 + e^{-\xi_h/T})\right] \tag{32}$$

Hence it follows that in the event of pairing two-thirds of the linear contribution of electrons to the heat capacity is retained. On the whole, for $T \to 0$ we can write

$$C(T)/C_n(T) \approx 2m_e/[3m_e + (m_h/3)] \tag{33}$$

Normally in semimetals of the bismuth type the electron effective mass is considerably smaller than the hole mass. Consequently, the reduction of the linear term in the heat capacity can be a considerable effect.

It is also important to note that, as shown in Ref. 5, near the critical temperature itself, clearly, there is a singularity in the thermodynamic variables, and the equations obtained here are inapplicable.*

5. TYPE I: DIELECTRIC CONSTANT

We now calculate the longitudinal dielectric constant at $T = 0$ for frequencies $\omega_0 \gg 1/\tau$ (τ is the mean free time). In this case it is more convenient to work in the time technique for $T = 0$. Determining

$$G = -i\langle T(\psi\psi^+)\rangle \qquad F^+ = \langle T(\rho^+\psi^+)\rangle \qquad F = \langle T(\rho\psi)\rangle$$
$$D = -i\langle T(\psi\varphi^+)\rangle \qquad \mathscr{G} = -i\langle T(\rho\rho^+)\rangle \tag{34}$$

we obtain

$$\begin{aligned}
G &= \{[(\omega + \xi_h)(\omega - \xi_e) - 2|\Delta_1|^2]/[(\omega + \xi_h)(\omega - \xi_e) - 3|\Delta_1|^2]\} \cdot [1/(\omega - \xi_e)] \\
D &= \{|\Delta_1|^2/[(\omega + \xi_h)(\omega - \xi_e) - 3|\Delta_1|^2]\} \cdot [1/(\omega - \xi_e)] \\
F^+ &= \{-i\Delta_1^*/[(\omega + \xi_n)(\omega - \xi_e) - 3|\Delta_1|^2]\} \\
F(-\omega) &= \{i\Delta_1/(\omega + \xi_h)(\omega - \xi_e) - 3|\Delta_1|^2]\} \\
\mathscr{G}(-\omega) &= -\{(\omega - \xi_e)/[(\omega + \xi_h)(\omega - \xi_e) - 3|\Delta_1|^2]\}
\end{aligned} \tag{35}$$

*Due to the condition $e^2/\pi\varepsilon_0 v \ll 1$ and hence $T_c \ll \varepsilon_F$, the singular region must be very small ($|(T - T_c)/T_c| \sim (T_c/\varepsilon_F)^4$).

Let the electric field depend only on the frequency and be directed along the constant magnetic field. Then the current operator has the form

$$\hat{j} = [ev_e - (e^2A/m_ec)](\psi_+^+\psi_+ + \varphi_+^+\varphi_+ + \eta_+^+\eta_+) + [ev_h - (e^2A/m_hc)]\rho_-^+\rho_-$$
$$+[-ev_e - (e^2A/m_ec)](\psi_-^+\psi_- + \varphi_-^+\varphi_- + \eta_-^+\eta_-)$$
$$+[-ev_h - (e^2A/m_hc)]\rho_+^+\rho_+ \tag{36}$$

where ψ_+ and ψ_- denote the neighborhoods of p_0 and $-p_0$, respectively, and A is the vector potential, which is proportional to $e^{-i\omega_0 t}[E = (i\omega_0/c)A]$.

Assuming that the Hamiltonian is augmented with a term accounting for interaction with the field, $-(1/c)\hat{j}\hat{A}$, and expanding in A, we obtain the following for $\omega_0 > 0$ by the standard technique:

$$j = \langle \hat{j} \rangle = QA$$

$$Q = 2(eH/2\pi c)\int_{-\infty}^{\infty}(dk/2\pi)\int_{-\infty}^{\infty}(d\omega/2\pi)(ie^2/c)[3v_e^2G(\omega+\omega_0)G(\omega)$$
$$+ 6v_e^2D(\omega+\omega_0)D(\omega) + v_h^2\mathscr{G}(-\omega-\omega_0)\mathscr{G}(-\omega)$$
$$- 6v_ev_hF^+(\omega+\omega_0)F(-\omega)] - (Ne^2/c)[1/m_e) + (1/m_h)] \tag{37}$$

where $N = (3p_0eH/2\pi^2c)$ is the number of holes or, equal to it, the number of electrons in all three groups.

Substituting expressions (35) and gathering terms, we obtain

$$Q = 2(eH/2\pi c)\cdot 12(e^2/c)\Delta^2\int_0^\infty \frac{dk}{2\pi}\{(v_e+v_h)^2/(\omega_2-\omega_3)[(\omega_2-\omega_3)^2-\omega_0^2]\}$$
$$-(Ne^2/c)[(1/m_e) + (1/m_h)] \tag{38}$$

Substituting into the latter the expressions for ω_2 and ω_3 and taking account of the relationship between Q and the dielectric constant, $\varepsilon = \varepsilon_0 + 4\pi Q/\omega_0^2$, we finally obtain

$$\varepsilon = \varepsilon_0 - (4\pi Ne^2/\omega_0^2)[(1/m_e) + (1/m_h)]$$
$$+(4\pi Ne^2/\omega_0^2)[(1/3m_e) + (1/m_h)]f(\omega_0/2\sqrt{3}|\Delta_1|)$$

$$f(x) = \begin{cases} \dfrac{1}{x\sqrt{1-x^2}}\sin^{-1}x & x < 1 \\ -\dfrac{1}{x\sqrt{x^2-1}}\ln(x+\sqrt{x^2-1}) + \dfrac{i\pi/2}{x\sqrt{x^2-1}} & x > 1 \end{cases} \tag{39}$$

For $x \ll 1$ we find that $f \to 1$. Hence it follows that at low frequencies

$$\varepsilon \approx \varepsilon_0 - (4\pi e^2/m_e\omega_0^2)(\tfrac{2}{3}N) \tag{40}$$

i.e., as before in the heat capacity two thirds of the electrons remain "free."

Near $\omega = 2\sqrt{3}|\Delta_1|$ there is an absorption threshold, near which

$$f(x) \approx \begin{cases} (\pi/2\sqrt{2})(1/\sqrt{1-x}) & x = 1 - 0 \\ (i\pi/2\sqrt{2})(1/\sqrt{x-1}) & x = 1 + 0 \end{cases}$$

At high frequencies $f \sim 1/x^2$, i.e., the last term in Eq. (39) is nonessential.

6. TYPE I: INTERACTION WITH IMPURITIES

Consider now the interaction with impurities. We assume that the impurity potential is screened at distances much larger than the interatomic spacing. Therefore we neglect intervalley transitions.* The terms appearing in the Hamiltonian are

$$\sum_{k,k'} \Big(U_1 \psi^+_{p_0+k} \psi_{p_0+k'} + U_2 \psi^+_{p_0+k} \psi_{-p_0+k'} e^{-2ip_0z} + U_3 \rho_{3p_0+k} \rho^+_{3p_0+k'} + U_4 \rho_{3p_0+k} \rho^+_{-3p_0+k'} e^{-6ip_0z} + \cdots \Big) e^{i(k'-k)z} \tag{41a}$$

We did not write explicitly the corresponding terms for the other electron groups and the complex conjugate.

Limiting ourselves with the Born approximation we need the diagrams of the type of Fig. 7, where the crosses denote the scattering by an impurity and the joining line with a point denotes that both of them refer to the same impurity atom (see Ref. 8, Chap. 7). After averaging over the positions of the impurity it appears that the momentum transfers at the two crosses must exactly compensate each other. This imposes a hard restriction on the possible combination of the crosses of different types since the momentum transfers such that $\pm 2p_0$ and $\pm 6p_0$ must be compensated.

It is easy to see that the following kinds of combinations are possible:

$$\begin{array}{ll} \psi^+_+\psi_\pm | \psi^+_\pm \psi_+ & \rho_+\rho^+_\pm | \rho_\pm \rho^+_+ \\ \psi^+_+\psi_+ | \varphi^+_\pm \varphi_\pm & \psi^+_+\psi_+ | \rho_\pm \rho^+_\pm \\ \psi^+_+\psi_+ | \psi^+_- \psi_- & \psi^+_+\psi_- | \varphi^+_- \varphi_+ \end{array} \tag{41b}$$

*Actually, the amplitude of such transitions is $4\pi e^2/[\varepsilon_0(\mathbf{K})\mathbf{K}^2]$, where $\mathbf{K}$ is a large momentum. The quantity $\varepsilon_0(\mathbf{K})$ is of the order of unity and not 100 as $\varepsilon_0(0)$, so the possibility of such a neglection is not evident and must be considered as a simplifying assumption. It is justified by the experimental results (see Conclusion).

Fig. 7

Let us suppose (it is justified by the result) that the scattering by the impurities does not change the kinds of coupling, i.e. the functions G, D, etc., become more complicated but no new functions appear, as, for instance, $\langle \psi_+ \varphi_-^+ \rangle$. Therefore, we again have two independent equivalent subsystems (in fact, not completely independent since the line in the middle of Fig. 7 can belong to either subsystem).

It will be convenient instead of several functions $G, D, F, \mathscr{G}$ to introduce a matrix $\hat{G}$. For that we numerate the operators:

$$\psi_{p_0+k} \equiv \psi_{1k}, \varphi_{p_0+k} \equiv \psi_{2k}, \eta_{p_0+k} \equiv \psi_{3k}, \overset{+}{\rho}_{-3p_0-k} = \psi_{4k}$$

The connection of the new matrix with the previous functions has the form [we use the definition $F = \langle T(\rho_- \psi_+) \rangle$]

$$\hat{G}_{\alpha\beta}(\omega, k) = - \langle T(\psi_\alpha(z, \tau)\psi_\beta^+(z', \tau)) \rangle_{\omega,k}$$

$$= \begin{pmatrix} G & D & D & F(-\omega) \\ D & G & D & F(-\omega) \\ D & D & G & F(-\omega) \\ -F^+ & -F^+ & -F^+ & -\mathscr{G}(-\omega) \end{pmatrix} \tag{42}$$

Taking into account the possible kinds of lines between the two crosses at Fig. 7 and the Hamiltonian (18), we obtain the following matrix $(i\omega - \hat{H})$:

$$(i\omega - \hat{H})_{\alpha\beta}$$

$$= \begin{pmatrix} i\omega - 2\bar{G} - \xi_e & -2\bar{D} & -2\bar{D} & \tilde{\Delta}_1 \\ -2\bar{D} & i\omega - 2\bar{G} - \xi_e & -2\bar{D} & \tilde{\Delta}_1 \\ -2\bar{D} & -2\bar{D} & i\omega - 2\bar{G} - \xi_e & \tilde{\Delta}_1 \\ \tilde{\Delta}_1^* & \tilde{\Delta}_1^* & \tilde{\Delta}_1^* & i\omega + 2\bar{\mathscr{G}}(-\omega) + \xi_h \end{pmatrix} \tag{43}$$

where

$$\tilde{\Delta}_1 = \Delta_1 - \bar{F}(-\omega)$$

$$\tilde{\Delta}_1^* = \Delta_1^* + \bar{F}^+(\omega)$$

$$\bar{G}(\omega) = (1/\tau_1)(v_e/2\pi) \int_{-\infty}^{\infty} dk\, G(\omega, k)$$

$$\bar{D}(\omega) = (1/\tau_1)(v_e/2\pi) \int_{-\infty}^{\infty} dk\, D(\omega, k)$$

$$\bar{\mathscr{G}}(-\omega) = (1/\tau_1)(v_h/2\pi) \int_{-\infty}^{\infty} dk\, \mathscr{G}(-\omega, k)$$

$$\bar{F}(-\omega) = (1/\tau_3)(\sqrt{v_e/v_h}/2\pi)\int_{-\infty}^{\infty} dk\, F(-\omega, k)$$

$$\bar{F}^+(\omega) = (1/\tau_3)(\sqrt{v_e v_h}/2\pi)\int_{-\infty}^{\infty} dk\, F^+(\omega, k)$$

$$1/\tau_1 = N_i(eH/2\pi c)[(U_1^2 + U_2^2)/2v_e]$$

$$1/\tau_2 = N_i(eH/2\pi c)[(U_3^2 + U_4^2)/2v_h]$$

$$1/\tau_3 = N_i(eH/2\pi c)(U_1 U_3/\sqrt{v_e v_h}) \tag{44}$$

Solving the equation $(i\omega - \hat{H})\hat{G} = 1$ we find

$$G = \tfrac{1}{3}\{(i\tilde{\omega}_2 + \xi_h)/[(i\tilde{\omega}_1 - \xi_e)(i\tilde{\omega}_2 + \xi_h) - 3|\tilde{\Delta}_1|^2]\} + \tfrac{2}{3}[1/(i\tilde{\omega}_1' - \xi_e)]$$

$$D = \tfrac{1}{3}\{(i\tilde{\omega}_2 + \xi_h)/[(i\tilde{\omega}_1 - \xi_e)(i\tilde{\omega}_2 + \xi_h) - 3|\tilde{\Delta}_1|^2]\} - \tfrac{1}{3}[1/(i\tilde{\omega}_1' - \xi_e)]$$

$$F^+ = \tilde{\Delta}_1^*/[(i\tilde{\omega}_1 - \xi_e)(i\tilde{\omega}_2 + \xi_h) - 3|\tilde{\Delta}_1|^2] = -F(-\omega)\tilde{\Delta}_1^*/\tilde{\Delta}_1$$

$$\mathscr{G}(-\omega) = -(i\tilde{\omega}_1 - \xi_e)/[(i\tilde{\omega}_1 - \xi_e)(i\tilde{\omega}_2 + \xi_h) - 3|\tilde{\Delta}_1|^2] \tag{45}$$

where

$$i\tilde{\omega}_1 = i\omega - 2(\bar{G} + 2\bar{D})$$

$$i\tilde{\omega}_1' = i\omega - 2(\bar{G} - \bar{D})$$

$$i\tilde{\omega}_2 = i\omega + 2\bar{\mathscr{G}}(-\omega)$$

All the following mathematics resembles very much the theory of superconductors with magnetic impurities.[9]† According to the definition (44) we get

$$\tilde{\omega}_1 = \omega + (1/\tau_1)(u/\sqrt{u^2+1}) \qquad \omega_1' = \omega + (1/\tau_1)\,\mathrm{sign}\,\omega$$

$$\tilde{\omega}_2 = \omega + (1/\tau_2)(u/\sqrt{u^2+1}) \qquad \tilde{\Delta}_1 = \Delta_1 - (1/2\sqrt{3})(1/\tau_3)(1/\sqrt{u^2+1}) \tag{46}$$

where

$$u = (v_e\tilde{\omega}_2 + v_h\tilde{\omega}_1)/2\sqrt{3}\tilde{\Delta}_1\sqrt{v_e v_h}$$

From (46) we find the equation for u:

$$\omega/\delta = u[1 - (1/\delta\tau)(1/\sqrt{u^2+1})] \tag{47}$$

where

$$\delta = \Delta_1 \cdot 2\sqrt{3v_e v_h}/(v_e + v_h)$$

$$1/\tau = [(\sqrt{v_e v_h}/\tau_3) + (v_e/\tau_2) + (v_h/\tau_1)]/(v_e + v_h) \tag{48}$$

†Zittartz has shown[10] that a similar situation arises in the three-dimensional isotropic Keldish–Kopajev model.[11]

Note that in a pure substance the quantity $\delta_0 = \Delta_1 2\sqrt{3v_e v_h}/(v_e + v_h)$ is the minimal energy gap (see the footnote on p. 17).

Equation (47) exactly corresponds to Eq. (42) of Ref. 9. Therefore, one can immediately derive that when impurities are present, the energy gap in the spectrum becomes $[1 - (\tau\delta)^{-2/3}]^{3/2}$ and it vanishes at $\tau^{-1} = \delta$. The quantity δ at $T = 0$ is defined by the relation

$$\ln [\delta_0(0)/\delta(0)] = \ln (u_0 + \sqrt{u_0^2 + 1}) + [1/2\delta(0)\tau]\{(\pi/2) - \arctan u_0 - [u_0/(1 + u_0^2)]\} \quad (49)$$

where δ_0 is δ for a pure substance, $u_0 = 0$ if $\delta\tau > 1$ and $[(\delta\tau)^{-2} - 1]^{1/2}$ if $\delta\tau < 1$.

When $1/\tau = 1/\tau_0 = \frac{1}{2}\delta_0, \delta(0) = 0$. The condition $\delta\tau = 1$ gives $\delta(0) = \delta_0(0)\, e^{-\pi/4}$, i.e., $1/\tau = \delta_0(0)\, e^{-\pi/4} = 0.91/\tau_0$. Hence the concentration region corresponding to $0.91/\tau_0 < 1/\tau < 1/\tau_0$ is a gapless state. With small concentrations of the impurity we have

$$\delta(0) = \delta_0(0) - (\pi/4\tau)$$

The critical temperature depends on the impurity concentration according to

$$\ln (T_c/T_{c0}) = \psi(\tfrac{1}{2}) - \psi[\tfrac{1}{2} + (1/2\tau\pi T_c)] \qquad \psi(x) = \Gamma^1(x)/\Gamma(x) \quad (50)$$

Particularly with small impurity concentration we have

$$T_c = T_{c0} - (\pi/4\tau)$$

So in our case scattering by ordinary impurities destroys the coupling in the same way as the scattering by magnetic impurities destroys superconductivity. This is due to the fact the electron and hole having different charges and p_0, anisotropic spectra interact differently with the impurities.

The most interesting property for experiment is the conductivity. The current operator (36) written in the matrix form corresponding to (42) is

$$\hat{j} = \begin{pmatrix} ev_e - (e^2A/m_e c) & & & \\ & ev_e - (e^2A/m_e c) & & \\ & & ev_e - (e^2A/m_e c) & \\ & & & -ev_h + (e^2A/m_h c) \end{pmatrix}$$

For the other subsystem, described by ψ_-, φ_-, η_-, ρ_+^+, the v_e, v_h have the signs reversed. To the first-order interaction we have in the thermodynamic technique

$$j = \langle \hat{j} \rangle = QA$$

$$Q = -(e^2/c)(eH/2\pi c)\cdot 2\int_{-\infty}^{\infty}(dk/2\pi)T\sum_{\omega}\mathrm{Tr}\,[\hat{G}(\omega)\hat{j}(A=0)\hat{G}(\omega+\omega_0)$$
$$\times\,\hat{j}(A=0)] - (Ne^2/c)[(1/m_e)+(1/m_h)]$$
$$= -(e^2/c)(eH/2\pi c)\cdot 2\int_{-\infty}^{\infty}(dk/2\pi)T\sum_{\omega}[3v_e^2G(\omega)G(\omega+\omega_0)$$
$$+\,6v_e^2D(\omega)D(\omega+\omega_0) + v_h^2\mathcal{G}(-\omega)\mathcal{G}(-\omega-\omega_0)$$
$$+\,6v_ev_hF(-\omega)F^+(\omega+\omega_0) - (Ne^2/c)[(1/m_e)+(1/m_h)]$$

where $N = 3p_0eH/2\pi^2c$ is the total number of electrons. Substituting in the first two terms in formulas (45) for G and D and writing them as $G = \frac{1}{3}G_1 + \frac{2}{3}G_2, D = \frac{1}{3}G_1 - \frac{1}{3}G_2$, we get $G_1(\omega)G_1(\omega+\omega_0) + 2G_2(\omega)G_2(\omega+\omega_0)$. This we shall have in mind in subsequent transformations.

In order to obtain the current as a function of the real frequency ω_0 we must continue this expression analytically to the real axis.[8] Here all the functions: G, D, $\mathcal{G}$, and F have the analytic properties of a Green function. Using the general rules we have

$$T\sum_n G(i\omega_n)G[i(\omega_n+\omega_0)] \to (1/2\pi)\int_{-\infty}^{\infty}\tanh(\omega/2T)\,\mathrm{Im}\,G_R(\omega)$$
$$\times\,[G_R(\omega+\omega_0) + G_A(\omega-\omega_0)]\,d\omega$$

We are interested in the static conductivity given by the imaginary part of Q at small frequencies ω_0. Taking the imaginary part of the last expression we get

$$(1/2\pi)\int_{-\infty}^{\infty}\tanh(\omega/2T)\,\mathrm{Im}\,G_R(\omega)[\mathrm{Im}\,G_R(\omega+\omega_0) - \mathrm{Im}\,G_R(\omega-\omega_0)]\,d\omega$$
$$= (1/2\pi)\int_{-\infty}^{\infty}[\tanh(\omega/2T) - \tanh[(\omega+\omega_0)/2T]]\,\mathrm{Im}\,G_R(\omega)\,\mathrm{Im}\,G_R$$
$$\times\,(\omega+\omega_0)\,d\omega \approx -(\omega_0/2T)\int_{-\infty}^{\infty}[\cosh(\omega/2T)]^{-2}$$
$$\times\,[\mathrm{Im}\,G_R(\omega)]^2\,d\omega/2\pi$$

Applying this to the equation above and using the connection between σ

and Q, $\sigma = (c/\omega_0)\,\mathrm{Im}\,Q_k(\omega_0)$, we get*

$$\sigma = e^2\frac{eH}{2\pi c}\frac{1}{T}\int_{-\infty}^{\infty}\frac{dk}{2\pi}\int_{-\infty}^{\infty}\frac{d\omega}{2\pi}\left(\cosh\frac{\omega}{2T}\right)^{-2}$$
$$\times\left\{v_e^2\left[\mathrm{Im}\frac{\tilde\omega_2+\xi_h}{(\tilde\omega_1-\xi_e)(\tilde\omega_2+\xi_h)-3|\tilde\Delta_1|^2}\right]^2+2v_e^2\right.$$
$$\times\left[\mathrm{Im}\frac{1}{\tilde\omega_1'-\xi_e}\right]^2-6v_ev_h$$
$$\times\left[\mathrm{Im}\frac{\tilde\Delta_1(\omega)}{(\tilde\omega_1-\xi_e)(\tilde\omega_2+\xi_h)-3|\tilde\Delta_1|^2}\right]^2$$
$$\left.+\,v_h^2\left[\mathrm{Im}\frac{\tilde\omega_1-\xi_e}{(\tilde\omega_1-\xi_e)(\tilde\omega_2+\xi_h)-3|\tilde\Delta_1|^2}\right]^2\right\} \tag{51}$$

Here $\tilde\omega$ and $\tilde\Delta_1$ are expressed in terms of ω and Δ_1 by relations similar to (46) and (47) in which, however, $1/\sqrt{u^2+1}\rightarrow i/\sqrt{u^2-1}$.

According to (46) the second term in (51) has exactly the same form as for electrons in the absence of any pairing with collision time $\tau_1/2$. This term is easily calculated and is equal to

$$\sigma(0) = \tfrac{2}{3}[Ne^2(\tau_1/2)/m_e]$$

(This result is rather evident.)

Suppose that the impurity concentration is not too large, i.e., $1/\tau \ll \delta_0$. In this case, as we shall see, the contribution of the remaining terms to σ is proportional to $\exp[-\delta_0/T]$. Hence $\sigma(0)$ is the conductivity at $T = 0$. Again we see that two-thirds of the electrons remain free.

Now we calculate the other terms of the conductivity. Let us consider the most interesting case for experiment $T \ll \delta_0$. We shall develop the expressions in powers of $1/\tau$ and leave the most important terms. It is easy to see that this demands

$$1/\tau \ll \overline{(\omega^2-\delta_0^2)^{3/2}/\delta_0^2} \sim T^{3/2}/\delta_0^{1/2} \qquad \delta_0 \gg T \gg \delta_0^{1/3}\tau^{-2/3}$$

After some calculations we get for the corresponding part of the conductivity

$$\tfrac{8}{3}(Ne^2\tau/m_e)(T/\delta_0)[v_h/(v_e+v_h)]\,e^{-\delta_0/T}$$

The conductivity in the normal state evidently has the form

$$\sigma_n = [Ne^2(\tau_1/2)/m_e] + [Ne^2(\tau_2/2)/m_h] \tag{52}$$

*Strictly speaking this expression is not exact since in averaging over the impurity position $\overline{GG} \neq \overline{G}\cdot\overline{G}$. To simplify the calculations we suppose that the Born amplitudes within every group do not depend on the momentum transfer, i.e., $U_1 = U_2$, $U_3 = U_4$. In this case the contribution of the additional terms arising in the correct averaging turns out to be zero.

[this follows also from (51)]. Hence the ratio $\sigma(T)/\sigma_n$ becomes

$$\sigma(T)/\sigma_n = \tfrac{2}{3}\{1 + (8T\tau/\delta_0\tau_1)[v_h/(v_e + v_h)]\, e^{-\delta_0/T}\}/[1 + (\tau_2 m_e/\tau_1 m_h)] \quad (53)$$

If $m_h \gg m_e$, then $\tau_2 \sim (m_e/m_h)\tau_1$ and $m_e\tau_2/m_h\tau_1 \sim (m_e/m_h)^2 \ll 1$. In this case for $T \to 0$, $\sigma(0)/\sigma_n = \frac{2}{3}$.

7. TYPE II: ENERGY SPECTRUM AND THERMODYNAMICS

We now investigate the pairing of electrons with electrons within a single group, as occurs whenever the effective interaction between carriers of the same sign is one of attraction. If we assumed that the sign of the effective interaction is determined by the bare interaction, type II pairing is realized if the phonon attraction exceeds the Coulomb repulsion.

In the given case electrons pair within a single group, i.e., three identical independent subsystems occur with respect to pairing. Let us pick one of the subsystems, say ψ. Its Hamiltonian has the form

$$H_{\text{II}} = \sum_k \psi^+_{p_0+k} v_e k \psi_{p_0+k} - \sum_k \psi^+_{-p_0+k} v_e k \psi_{-p_0+k} - \Delta_2 \sum_k \psi^+_{p_0+k}\psi_{-p_0+k} - \Delta_2^* \sum_k \psi^+_{-p_0+k}\psi_{p_0+k} \quad (54)$$

in which

$$\Delta_2 = -\lambda_2 \sum_{k_1} (\langle \varphi^+_{-p_0+k_1}\varphi_{p_0+k_1}\rangle + \langle \eta^+_{-p_0+k_1}\eta_{p_0+k_1}\rangle) = -2\lambda_2 \sum_{k_1} \langle \psi^+_{-p_0+k_1}\psi_{p_0+k_1}\rangle \quad (55)$$

(we assume that $\lambda_2 < 0$).

We can simplify things by invoking matrix notation at the outset. We introduce $\psi_{1k} = \psi_{p_0+k}$ and $\psi_{2k} = \psi_{-p_0+k}$. Here,

$$\hat{H} = \begin{pmatrix} \xi & -\Delta_2 \\ -\Delta_2^* & -\xi \end{pmatrix} = \xi\sigma_z - \Delta_2\sigma_x \quad (56)$$

(we assume that Δ_2 is real). In the temperature base $G_{\alpha\beta} = -\langle T_\tau \psi_\alpha(x)\psi^+_\beta(x')\rangle$. From the equation $(i\omega - \hat{H})\hat{G} = 1$ we obtain

$$\hat{G} = (i\omega_n - \xi\sigma_z + \Delta_2\sigma_x)^{-1} = -(i\omega + \xi\sigma_z - \Delta_2\sigma_x)/(\omega_n^2 + \xi^2 + \Delta_2^2) \quad (57)$$

Thus, the energy spectrum is expressed by the following equation:

$$\omega = \pm\sqrt{\xi^2 + \Delta_2^2} \quad (58)$$

as in superconductors.

From the definition of Δ_2 (55) we find

$$1 = (|\lambda_2|/\pi v_e)\int_0^{\varepsilon_F} d\xi[\tanh(\sqrt{\xi^2 + \Delta_2^2}/2T)/\sqrt{\xi^2 + \Delta_2^2}] \qquad (59)$$

Thus, the dependence $\Delta_2(T)$ is the same as for superconductors with[8]

$$\Delta_2(0) = 2\varepsilon_F \exp(-\pi v_e/|\lambda_2|) \qquad (60)$$

In order to obtain the part of the thermodynamic potential associated with electrons we differentiate $\langle H_{II} \rangle$ [in the unreduced form (5)] with respect to λ_2. This gives us

$$\partial\Omega/\partial|\lambda_2| = -(3/\lambda_2^2)\Delta_2^2(eH/2\pi c)$$

or

$$\Omega - \Omega_n = (3eH/2\pi c)\int_0^{\Delta_2} [d(1/|\lambda_2|)/d\Delta_2]\Delta_2^2\, d\Delta_2 \qquad (61)$$

Once this has been done, the derivation simply follows the theory of superconductivity. In particular, for $T \ll T_c = (\gamma/\pi)\Delta_2(0)$ the electron heat capacity depends exponentially on the temperature.

It is important to note at this point that the hole group is not directly involved in the given type of pairing. On the other hand, the loop shown in Fig. 8a has a logarithmic singularity. Within the framework of the given point interaction such a loop should not occur, because the point interaction of holes with holes cannot take place, and the transition to an electron loop is rendered impossible by the transfer of momentum $3p_0$. The interaction of holes, however, can be communicated by several electron loops, as in Fig. 8b. In this case, of course, the logarithms are lost [if we assume that $(e^2/\varepsilon_0 v)\ln(\varepsilon_F/\omega) \sim 1$, the vertex in Fig. 8b will have a relative magnitude of order $(e^2/\varepsilon_0 v)^2 \ll 1$]. Hence, pairing of this nature will not occur at realistic temperatures.

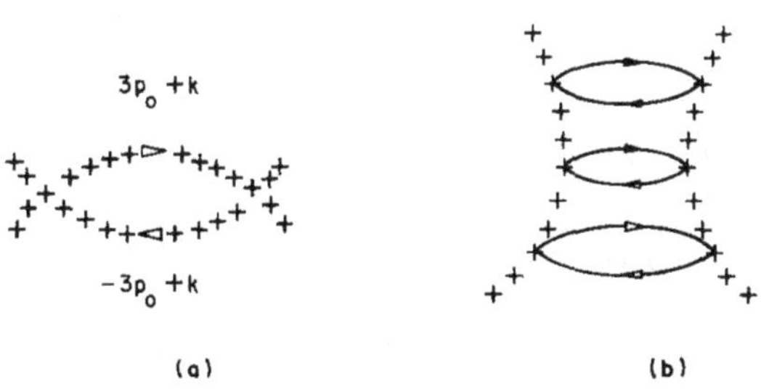

Fig. 8

On the other hand, however, electron pairing of the type $\langle\psi^+_{p_0}\psi_{-p_0}\rangle$ connotes the onset of a periodic variation of the electron density with period $1/2p_0$ and a corresponding component of the electrostatic potential acting on the holes. This potential undoubtedly contains higher harmonics, specifically one with period $1/6p_0$ (it is readily seen that its amplitude is of the order $\Delta_2^3/\varepsilon_F^2$). This potential will necessarily create an energy gap in the hole spectrum, although certainly of much smaller size than in the electron case (i.e., $\sim\Delta_2^3/\varepsilon_F^2$). In principle it will persist up to T_c, but, because of its smallness, it will not be felt at any realistic temperatures.

Consequently, the linear hole term is retained in the heat capacity. The ratio of the linear contributions to the heat capacity, as can readily be seen, is equal to

$$C/C_n = (1/v_h)/[(3/v_e) + (1/v_h)] = m_h/(m_h + 9m_e) \tag{62}$$

The value of this ratio depends on the ratio of the masses, but if $m_h \gg m_e$, unlike type I pairing, the given ratio can be close to unity.

As in the case of type I pairings, Eqs. (59) and (60) are clearly inapplicable near T_c.

8. TYPE II: DIELECTRIC CONSTANT

As we did for type I, we first calculate the longitudinal dielectric constant at $T = 0$ for frequencies $\omega_0 \gg 1/\tau$. The time technique matrix Green function has the form

$$G = -i\langle T(\psi_\alpha(x)\psi^+_\beta(x'))\rangle_{\omega,k} = (\omega - \xi_e\sigma_z + \Delta_2\sigma_x)^{-1} \tag{63}$$

The electron current operator is obviously equal to

$$\hat{j}_e = ev_e\psi^+_\alpha\sigma^z_{\alpha\beta}\psi_\beta + ev_e\varphi^+_\alpha\sigma^z_{\alpha\beta}\varphi_\beta + ev_e\eta^+_\alpha\sigma^z_{\alpha\beta}\eta_\beta - (e^2A/m_ec)N$$

Recognizing that the Hamiltonian includes interaction with the field, $-(1/c)jA$. and expanding in A, we obtain

$$j_e = \langle\hat{j}_e\rangle = Q_eA$$

$$Q_e = 3(eH/2\pi c)\int_{-\infty}^{\infty}(dk/2\pi)\int_{-\infty}^{\infty}(d\omega/2\pi)(ie^2/c)v_e^2$$
$$\cdot \operatorname{Tr}[\sigma_z\hat{G}(\omega)\sigma_z\hat{G}(\omega+\omega_0)] - (e^2N/m_ec)$$
$$= 12(eH/2\pi c)(ie^2/c)(v_e/2\pi)\int_{-\infty}^{\infty}d\xi\int_{-\infty}^{\infty}(d\omega/2\pi)$$
$$\cdot\{[\omega(\omega+\omega_0)+\xi^2-\Delta_2^2]/(\omega^2-\xi^2-\Delta_2^2)[(\omega+\omega_0)^2-\xi^2-\Delta_2^2]\}$$

Evaluating the integrals with regard for the relation $\varepsilon = 4\pi Q/\omega_0^2$ and adding the contribution from free holes, $-4\pi Ne^2/m_h\omega_0^2$, we find

$$\varepsilon = \varepsilon_0 - (4\pi Ne^2/m_h\omega_0^2) + (4\pi Ne^2/m_e\omega_0^2)[f(\omega_0/2\Delta_2) - 1] \tag{64}$$

where f is the same function as in (39).

9. TYPE II: INTERACTION OF ELECTRONS WITH IMPURITIES

We shall consider now the interaction of electrons with impurities. In the case under investigation the holes do not participate in the coupling, and we can consider only electrons. Taking into account, as before, only the intravalley transitions we get the same combinations of crosses in Fig. 7 [Eq. (41b)] and finally obtain

$$\hat{G} = (i\omega - \hat{H})^{-1} = [i\tilde{\omega} - \xi_e\sigma_z + \tilde{\Delta}_2\sigma_x]^{-1} \tag{65a}$$

where

$$i\tilde{\omega} = i\omega - (1/\tau_1)\int_{-\infty}^{\infty} (d\xi_e/2\pi)\,\delta p\hat{G}(\omega, k)$$

$$\tilde{\Delta}_2 = \Delta_2 - (1/2\tau_1')\int_{-\infty}^{\infty} (d\xi_e/2\pi)\delta p[\sigma_x\hat{G}(\omega, k)] \tag{65b}$$

Here $1/\tau_1$ is defined by Eq. (44) and $1/\tau_1'$ is obtained from $1/\tau$ by replacing $(U_1^2 + U_2^2)/2$ by U_1^2.

Substituting Eq. (65a) into the definition (65b) we get

$$\tilde{\omega} = \omega + (1/\tau_1)(u/\sqrt{1 + u^2})$$

$$\tilde{\Delta}_2 = \Delta_2 - (1/2\tau_1')(1/\sqrt{1 + u^2}) \tag{66}$$

where $U = \tilde{\omega}/\tilde{\Delta}_2$ and satisfies the equation

$$\omega/\Delta_2 = u[1 - (1/\tau\Delta_2)(1/\sqrt{1 + u^2})] \tag{67}$$

This time $1/\tau$ is defined by

$$1/\tau = (1/2\tau_1') + (1/\tau_1)$$

The equations for Δ_2, T_c, and the energy gap are the same as in Section 6.

Let us now calculate the conductivity under the same conditions as in Section 6, i.e., with $1/\tau$, $T \ll \Delta_2$. Since the holes do not participate in the pairing their contribution is always the same as in the normal state, i.e.,

$$\sigma(0) = Ne^2(\tau_2/2)/m_h \tag{68}$$

We denote it by $\sigma(0)$ since if $\Delta_2 \gg 1/\tau$, T, the electronic conductivity drops exponentially with decreasing temperature.

In the same way as (51) we get for the electronic part*

$$\sigma = \frac{Ne^2}{m_e T}\int_{-\infty}^{\infty}\frac{d\xi}{2\pi}\int_0^{\infty} d\omega\left(\cosh\frac{\omega}{2T}\right)^{-2}\left[\left(\mathrm{Im}\frac{\tilde{\omega}}{\tilde{\omega}^2-\xi^2-\tilde{\Delta}_2^2}\right)^2 + \xi^2\left(\mathrm{Im}\frac{1}{\tilde{\omega}^2-\xi^2-\tilde{\Delta}_2^2}\right)^2 - \left(\mathrm{Im}\frac{\tilde{\Delta}_2}{\tilde{\omega}^2-\xi^2-\tilde{\Delta}_2^2}\right)^2\right] \quad (69)$$

where $N = 3(eHp_0/2\pi^2 c)$ is the total number of electrons.

Applying Eqs. (66) and (67), developing them in powers of $1/\tau$, and calculating the integral (69) we obtain the electronic contribution to the conductivity:

$$2(Ne^2\tau/m_e)(T/\Delta)\,e^{-\Delta_2/T}$$

This calculation is correct if

$$\frac{1}{\tau} \ll \overline{(\omega^2-\Delta_2^2)^{3/2}}/\Delta_1^2 \sim \frac{T^{3/2}}{\Delta_2^{1/2}} \quad \text{or} \quad \Delta_2 \gg T \gg \Delta_2^{1/3}\tau^{-2/3}$$

The relation of the total conductivity to the conductivity in the normal state becomes

$$\sigma(T)/\sigma_n = [(\tau_2 m_e/\tau_1 m_h) + 4(\tau/\tau_1)(T/\Delta_2)\,e^{-\Delta_2/T}]/[1 + (\tau_2 m_e/\tau_1 m_h)]$$

If $m_e \ll m_h$, then $\tau_2/\tau_1 \sim m_e/m_h$ and hence in this case $\sigma(0)/\sigma_n \sim (m_e/m_h)^2 \ll 1$.

10. THE SUPERTHERMAL CONDUCTIVITY PROBLEM

In the event that pairing takes place in the system and the pairs form a "condensate" it is logical to inquire whether the motion of the condensate will produce a nondissipative transport process. Inasmuch as the complexes in any type of pairing do not have any charge, they obviously cannot transfer current. Another type of transport is possible, namely entropy transfer, but since the complexes form a condensate, they do not transfer entropy, so that entropy transfer must be associated with excitations in the system, i.e., with detached complexes. However, the latter are subjected to scattering by impurities, so that "nondissipative" super-thermal conductivity is impossible.

We wish to point out the difference from liquid helium. In the latter heat transfer in the superfluid state is also realized by excitations. In the normal state the transfer mean free path is of the order of the interatomic

*As in Section 6, this is true, strictly speaking, only for $U_1 = U_2$.

spacings, while in the superfluid state the entire gas of excitations is in motion as a whole, and dissipation is associated only with wall friction. In the case of a metal dissipation occurs due to collisions of the excitations with impurities, which do not move together with the excitation gas and thus play the same role in both the normal and paired states.

In both of our cases the situation is even simpler. In order to grasp this fact, we derive an equation relating the electric current and the heat flow. We choose μ and T as our independent thermodynamic variables. We then have

$$(\partial S/\partial \mu)_T = (\partial N/\partial T)_\mu$$

whence

$$S = \int (\partial N/\partial T)_\mu \, d\mu$$

Differentiating with respect to the time, we obtain

$$\partial S/\partial t = \operatorname{div} \mathbf{q}/T = \int d\mu(\partial/\partial T)(\partial N/\partial t) = \int d\mu(\partial/\partial T) \operatorname{div} \mathbf{j}_N$$

where $\mathbf{q}$ is the heat flux and $\mathbf{j}_N$ is the particle flux. However, inasmuch as each electron carries a charge e, $\mathbf{j}_N$ coincides, correct to a factor e, with the electric current even in the case when there are hole carriers. Integrating, we obtain the following relation between the heat and particle fluxes:

$$\mathbf{q} = T(\partial/\partial T)\int d\mu \cdot \mathbf{j}_N(\mu)$$

The introduction of a moving condensate means replacing the corresponding Δ by $\Delta\, e^{2ilz}$. But since Δ enters into the Hamiltonian in combination with certain operators ψ, ρ, etc., this substitution is equivalent to the operator transformations $\psi \to \psi\, e^{i\alpha lz}$, where α represents certain constants. It is easily shown that in both types of pairing treated above, as well as in those types which occur in the model of Ref. 2, these transformations are simply equivalent to a variation of the chemical potential.

Thus, in the case of type II pairing we set

$$\psi_+ \to \psi_+\, e^{ilz} \qquad \psi_- \to \psi_-\, e^{-ilz}$$

but this is equivalent to the substitution $p_0 \to p_0 + l$ or

$$\delta\mu = v_e l$$

In a semimetal of the first type we require that $\psi_+^+ \rho_-^\pm \to \psi_+^+ \rho_-^\pm\, e^{-2ilz}$. However, with a variation of μ the limiting momentum for electrons (i.e., p_0) is shifted by $\delta\mu/v_e$, the limiting momentum for holes (i.e., $3p_0$) by $-\delta\mu/v_h$. It is required, therefore, that

$$\delta\mu[(1/v_e) + (1/v_h)] = 2l$$

Thus, we have in both cases $\delta\mu = vl$, where v is a constant. But this implies that in the linear approximation on l

$$j_N(l, \mu) = [\partial j_N(0, \mu)/\partial\mu]vl$$

Substituting into (82), we obtain

$$q = vlT(\partial j_N/\partial T)$$

where j_N corresponds to the true μ and $l = 0$. But for $l = 0$ in every case $j_N = 0$. Therefore, $q = 0$.

One remark is in order. In each of the pairing types analyzed we have several independent subsystems and, therefore, several condensates. A variation of the chemical potential, strictly speaking, implies the occurrence of the factor e^{2ilz} for all Δ. The question naturally arises as to whether it is impossible to obtain j_N by means of different l for different Δ. Actually, in the event of several subsystems Eq. (82) contains $\int j_N\, d\mu = \sum_i \int j_{N_i}\, d\mu$. But it was established above that $\int j_{N_i}\, d\mu = 0$ for each subsystem separately.

11. CONCLUSION

Now we shall briefly note some special features of the cases with other field directions. In Ref. 1 the case $\mathbf{H} \parallel C_2$ was investigated where one electron group was completely frozen out and the two others were symmetric. This case is very close to that considered above. Also, two kinds of pairing are possible, and in either case the substance does not become a true insulator. In the experiments of Ref. 1 the mass ratio was small: $m_e/m_h \sim 0.02$. In such geometry for coupling of the type I one-half of the electrons must freeze out, i.e., for $m_e \ll m_h$, $\sigma_n/\sigma(0) = 2$. For coupling of type II it should be $\sigma_n/\sigma(0) \sim (m_h/m_e)^2 \gg 1$. Experimentally this relation for two samples was close to two, and for the third it varied with magnetic field between three and four. Evidently one must conclude that in the alloys BiSb coupling of the type I takes place. Some increase of $\sigma_n/\sigma(0)$ compared to the theoretic prediction is most probably due to the fact that the condition $e^2/\varepsilon_0\hbar v \ll 1$, assumed in these calculations, was not fulfilled in experiment.

In the case $e^2/\varepsilon_0\hbar v \sim 1$ the coupling takes place not in a narrow vicinity of p_0 but over the whole electron distribution. In the low-density limit every electron participates in the pairing and $\sigma(0) \to 0$, i.e., $\sigma_n/\sigma(0) \to \infty$. Hence the increase of this ratio compared to two, is to be expected.

There are also other arguments for the preference of the electron-hole pairing type. This is the observation in Ref. 1 that the transition takes place only for $H > 30\,\text{kOe}$. This is the field magnitude when $eH/m_h^* c \sim \varepsilon_{Fh}$ for holes, since for the electrons much smaller fields suffice. Furthermore, a slight inclination of the field direction which makes the two electron groups inequivalent should have destroyed the pairing of type II. In experiment[12] no essential changes are observed.

For a general field direction all the carrier groups will have different momenta (if more than one electron group remains). In this case only the pairing of type I between some two groups can take place. Note that independently of the fact of whether electrons are paired with holes or with electrons of another group (e.g., $\langle\psi^+_{p_{01}}\eta_{-p_{02}}\rangle$) the interaction sign must be that of Coulomb interaction.

In the case when all three electron groups exist and all of them are inequivalent, two subsequent phase transitions are to be expected (e.g., one electron group is paired with holes and then the two remaining with each other). Finally, the substance becomes a true insulator.

A separate case is a single group of electrons. It corresponds to the model considered in Refs. 2–5. In this case different types of pairing are possible and a true insulator always arises. But this case is difficult to produce experimentally.

ACKNOWLEDGMENTS

I would like to take this opportunity to express my gratitude to N. B. Brandt, S. M. Chudinov, and S. A. Brazovskii for valuable discussions.

REFERENCES

1. N. B. Brandt and S. M. Chudinov, *Zh. Eksperim. i Teor. Fiz. Pis. Red.* **13**, 146 (1971); *J. Low Temp. Phys.*, in press.
2. A. A. Abrikosov, *J. Low Temp. Phys.* **2**, 37 (1970).
3. A. A. Abrikosov, *J. Low Temp. Phys.* **2**, 177 (1970).
4. S. A. Brazovskii, *Zh. Eksperim. i Teor. Fiz.* **61**, 2401 (1971).
5. S. A. Brazovskii, *Zh. Eksperim. i Teor. Fiz.* **62**, 820 (1972).
6. A. A. Abrikosov and A. A. Migdal, *J. Low Temp. Phys.* **3**, 519 (1970).
7. A. I. Larkin and D. E. Khmelnitskii, *Zh. Eksperim. i Teor. Fiz.* **56**, 2087 (1969).
8. A. A. Abrikosov, L. P. Gor'kov, and I. E. Dzyaloshinski, *Methods of Quantum Field Theory in Statistical Physics* (Prentice-Hall, Englewood Cliffs, N.J., 1963).
9. A. A. Abrikosov and L. P. Gor'kov, *Zh. Eksperim. i Teor. Fiz.* **39**, 1781 (1960).
10. J. Zittartz, *Phys. Rev.* **165**, 605, 612 (1967).
11. L. V. Keldish and Yu. V. Kopajev, *Fiz. Tverd. Tela* **6**, 2791 (1964).
12. S. M. Chudinov, private communication.

Calculation of critical indices for zero-gap semiconductors

A. A. Abrikosov

L. D. Landau Theoretical Physics Institute, USSR Academy of Sciences

(Submitted November 19, 1973)
Zh. Eksp. Teor. Fiz. **66**, 1443–1460 (April 1974)

The indices defining the exponents in the power laws in the Green functions and physical characteristics of zero-gap semiconductors of the second kind in the strong-coupling region (i.e., the region of energies $\omega \ll \omega_0$, where $\omega_0 = m^* e^4/2\kappa_0^2$ is the exciton binding energy, and momenta $k \ll k_0$, where $k_0 = m^* e^2/\kappa_0$ is the inverse Bohr radius of the exction) are calculated. Two methods are applied: expansion in $\epsilon = 4 - d$ (d is the dimensionality of space) and expansion in the "large number of components," i.e., in $1/\Pi$, where Π is the polarization operator. In the first approximation the two procedures give results that are numerically slightly different but qualitatively similar. It follows from the results that there are several energy and momentum regions: (1) the "free region," (2) the region $\omega_1 < \omega < \omega_0$, $k_1 < k < k_0$ ($\omega_1 \sim 2.5 \times 10^{-4}\ \omega_0$, $k_1 \sim 1.6 \times 10^{-2} k_0$), in which the "random-phase approximation" is applicable, i.e., the interaction of the carriers is altered but their spectrum is conserved, and (3) the true strong-coupling region $\omega < \omega_1$, $k < k_1$. In the regions (2) and (3) the results of Abrikosov and Beneslavskiĭ are applicable, with $\alpha = \nu = 2$ in region (2) and $\alpha = 1.76$, $\nu = 1.92$ in region (3).

1. INTRODUCTION

It has been shown earlier (cf.[1,2]) that in zero-gap semiconductors of the second kind (e.g., α-Sn), a strong-coupling regime arises in the vicinity of the point of contact of the branches of the spectrum (a point of the order of $k_0 = m^* e^2/\kappa_0$ along the momentum axis and $\omega_0 = m^* e^4/2\kappa_0^2$ along the energy axis, where m^* is the effective mass and κ_0 is the dielectric permittivity). In this region the Green functions and vertices acquire the self-similar form:

$$G \sim \omega_0^{-1} \left(\frac{k}{k_0}\right)^{-\alpha} g\left[\frac{\omega/\omega_0}{(k/k_0)^\nu}\right],$$
$$\Gamma \sim \frac{\omega_0}{k_0^3} \left(\frac{k}{k_0}\right)^{\nu-3} d\left[\frac{\omega/\omega_0}{(k/k_0)^\nu}\right], \tag{1}$$
$$\gamma \sim \left(\frac{k}{k_0}\right)^{\alpha-\nu} q\left[\frac{\omega/\omega_0}{(k/k_0)^\nu}\right].$$

The indices α and ν determine the behavior of physical quantities in the singular region, e.g., their dependence on the temperature, magnetic field, frequency, etc.

In view of the fact that the problem of zero-gap semiconductors is methodologically similar to the theory of phase transitions, to determine the unknown indices α and ν it is natural to attempt to apply the methods developed in recent years by Wilson and Fisher[3] and Ferrell and Scalapino[4] and successfully used for the analysis of different types of phase transitions. The present article is devoted to this question.

2. CONTINUATION TO NONINTEGER DIMENSIONS

The calculation of the indices should be performed for a specific model. Although the indices do not depend on the interaction constant, they can depend on the geometry (the little group and specific representation; cf.[1]). We shall consider a very realistic situation, namely, the case of α-Sn with a diamond-type lattice. The bands touch at a point with little group O_h. We take a representation corresponding to $J = {}^3/_2$; the corresponding Hamiltonian and spectrum are well known[1]. To simplify matters, we put $3\beta = \gamma$ in the formula in[1]; the spectrum then possesses spherical symmetry. The Hamiltonian in this case can be written in the form

$$H = fk^2 + c[\hat{J}_\mu \hat{J}_\nu - {}^1/_3 \hat{J}^2 \delta_{\mu\nu}] k_\mu k_\nu, \tag{2}$$

whence it can be seen that there is invariance with respect to any rotations. The eigenvalues are doubly degenerate and equal to

$$\varepsilon_{1,2} = (f \pm c) k^2. \tag{3}$$

Hence it can be seen that a zero-gap spectrum is obtained for $c > f$.

It was shown in[1] that the zeroth-approximation term for the polarization operator is proportional to an integral that converges like $1/k$ at the lower limit (k is the external momentum). It is clear that in four-dimensional space we would have a logarithmic integral. This corresponds precisely to the situation in phase transitions. The application of Wilson's method[3] is associated with formulation of the problem in a space with a noninteger number of dimensions $d = 4 - \epsilon$. In the present case, the problem is complicated by the fact that operators J_μ associated with the three-dimensional rotation group appear in the Hamiltonian and form tensor combinations with the momentum components. The continuation to noninteger dimensions in such a situation has been effected for the case $J = 1$[5], but the method used is not applicable for $J = {}^3/_2$. In view of this, we shall proceed in the following way.

We write the second term in the Hamiltonian (2) in the form cH_1,

$$H_1 = k_\mu k_\nu \hat{A}_{\mu\nu}, \tag{4}$$

and analyze the properties of the operators $A_{\mu\nu}$. According to (2), these operators possess the following properties:

$$\sum_\mu \hat{A}_{\mu\mu} = 0, \quad \{A_{\mu\mu}, A_{\nu\nu}\} = -1 \quad (\nu \neq \mu),$$
$$\mathrm{Sp}\,\hat{A}_{\mu\nu} = 0, \quad A_{\mu\mu}^2 = 1, \quad \hat{A}_{\mu\nu}^2 = {}^3/_4 \quad (\nu \neq \mu); \tag{5}$$

$\{\hat{A}_{\mu\nu}, \hat{A}_{\rho\sigma}\} = 0$ in all other cases. Here there is no summation over repeated indices. All these properties can be written in the form of a single formula:

$$\{\hat{A}_{\mu\nu}, \hat{A}_{\rho\sigma}\} = {}^3/_2 (\delta_{\mu\rho}\delta_{\nu\sigma} + \delta_{\mu\sigma}\delta_{\nu\rho}) - \delta_{\mu\nu}\delta_{\rho\sigma}. \tag{6}$$

Hence it follows that the anticommutator of two Hamiltonians H_1 at different points k is equal to

$$\{H_1(\mathbf{k}_1), H_1(\mathbf{k}_2)\} = 3(\mathbf{k}_1\mathbf{k}_2)^2 - k_1^2 k_2^2. \tag{7}$$

In particular, if $k_1 = k_2 = k$ we obtain $2c^2k^4$, and averaging (7) over the directions of k_2 gives zero. These properties are carried over in an obvious way to a space with noninteger dimensions. For this it is sufficient to require that

$$\{H_1(\mathbf{k}_1),\ H_1(\mathbf{k}_2)\} = \frac{2}{d-1}[d(\mathbf{k}_1\mathbf{k}_2)^2 - k_1^2k_2^2]. \tag{8}$$

For this, we can take H_1 in the form (4), where the operator coefficients $\hat{A}_{\mu\nu}$ satisfy the condition

$$\{\hat{A}_{\mu\nu},\ \hat{A}_{\rho\sigma}\} = \frac{d}{d-1}(\delta_{\mu\rho}\delta_{\nu\sigma} + \delta_{\mu\sigma}\delta_{\nu\rho}) - \frac{2}{d-1}\delta_{\mu\nu}\delta_{\rho\sigma}. \tag{9}$$

In the following we shall need to use the property (8) more frequently than any other. However, in the calculation of the polarization operator it is necessary to take the trace, and we need to know the rank of the matrices $\hat{A}_{\mu\nu}$. In a space of three dimensions this is equal to 4. We shall determine the rank from the condition that it be sufficient to enable the conditions (9) to be satisfied. We shall use the following method. For the three-dimensional case, the matrices $A_{\mu\nu}$ can be represented with the aid of direct products of two sets of Pauli matrices:

$$\begin{aligned} &A_{zz} = J_z^2 - \frac{5}{4} = \sigma_z s_z, \quad A_{xy} = \frac{1}{2}\{J_x, J_y\} = \frac{\sqrt{3}}{2}s_y, \\ &A_{xx} = J_x^2 - \frac{5}{4} = \frac{\sqrt{3}}{2}s_x - \frac{1}{2}\sigma_z s_z, \quad A_{zx} = \frac{1}{2}\{J_z, J_x\} = \frac{\sqrt{3}}{2}s_z\sigma_x, \\ &A_{yy} = J_y^2 - \frac{5}{4} = -\frac{\sqrt{3}}{2}s_x - \frac{1}{2}\sigma_z s_z, \quad A_{yz} = \frac{1}{2}\{J_y, J_z\} = \frac{\sqrt{3}}{2}s_z\sigma_y. \end{aligned} \tag{10}$$

A similar representation can also be used in other cases. The operators $A_{\mu\nu}$ are symmetric in μ and ν, and $\Sigma A_{\mu\mu} = 0$. If the dimensionality of space (i.e., the number of values of μ) is d, then the number of such matrices equals $(d-1)(d+2)/2$.

The conditions (9) can be satisfied by means of linear combinations of matrices with zero trace, if these matrices anticommute, their squares are equal to unity, and there are $(d-1)(d+2)/2$ of them. This is proved as follows. We have at our disposal $(d-1)(d+2)/2$ matrices, which we denote by $\gamma_1, \gamma_2, \ldots$. We take any $(d-1)d/2$ of them as the $A_{\mu\nu}$ with $\mu \neq \nu$. There remain $d-1$ matrices, in terms of which we must express the d matrices $A_{\mu\mu}$. We must then have: a) $A_{\mu\mu}^2 = 1$, b) $\{A_{\mu\mu}, A_{\nu\nu}\} = -2/(d-1)\ (\mu \neq \nu)$, c) $\sum_\mu A_{\mu\mu} = 0$. We denote $A_{11} \equiv \gamma_1$. We choose $A_{22} = p\gamma_1 + q\gamma_2$ in such a way that the conditions (a) and (b) are satisfied. This gives $p = -1/(d-1)$, $q = [1 - 1/(d-1)^2]^{1/2}$. We next choose $A_{33} = p_1\gamma_1 + q_1\gamma_2 + r\gamma_3$, again to satisfy the same conditions. This gives

$$p_1 = -\frac{1}{d-1}, \quad q_1 = -\frac{1}{d-2}\left[1 - \frac{1}{(d-1)^2}\right]^{1/2},$$
$$r = \left[1 - \frac{1}{(d-1)^2}\right]^{1/2}\left[1 - \frac{1}{(d-2)^2}\right]^{1/2}.$$

Continuing in the same spirit, we obtain expressions for all the $A_{\mu\mu}$ (for integer d):

$$\begin{aligned} A_{11} &= \gamma_1, \\ A_{22} &= -\frac{1}{d-1}\gamma_1 + \left[1 - \frac{1}{(d-1)^2}\right]^{1/2}\gamma_2, \\ A_{33} &= -\frac{1}{d-1}\gamma_1 - \frac{1}{d-2}\left[1 - \frac{1}{(d-1)^2}\right]^{1/2}\gamma_2 + \\ &+ \left[1 - \frac{1}{(d-1)^2}\right]^{1/2}\left[1 - \frac{1}{(d-2)^2}\right]^{1/2}\gamma_3. \\ &\ldots\ldots\ldots\ldots\ldots\ldots \end{aligned}$$

$$A_{d-1\,d-1} = -\frac{1}{d-1}\gamma_1 - \ldots + \left[1 - \frac{1}{(d-1)^2}\right]^{1/2}\ldots\left[1 - \frac{1}{(d-d+2)^2}\right]^{1/2}\gamma_{d-1},$$
$$A_{dd} = -\frac{1}{d-1}\gamma_1 - \ldots - \left[1 - \frac{1}{(d-1)^2}\right]^{1/2}\ldots\left[\frac{3}{4}\right]^{1/2}\gamma_{d-1}.$$

The condition (c) is fulfilled automatically. This is not surprising, as it already follows from (9) that $\{A_{\rho\sigma}, \sum_\nu A_{\nu\nu}\} = 0$ for any of the matrices $A_{\rho\sigma}$.

If we construct such matrices as direct products of Pauli matrices, we obtain the following cases:

for $n = 1$,

$$\sigma_x,\ \sigma_y,\ \sigma_z;$$

for $n = 2$,

$$s_x,\ s_y,\ s_z\sigma_x,\ s_z\sigma_y,\ s_z\sigma_z;$$

for $n = 3$,

$$\nu_x,\ \nu_y,\ \nu_z s_x,\ \nu_z s_y,\ \nu_z s_z\sigma_x,\ \nu_z s_z\sigma_y,\ \nu_z s_z\sigma_z.$$

Thus, in a similar manner, we find that the number of matrices equals $3 + 2(n-1)$, where n is the number of types of Pauli matrices. This number should be equal to (or greater than) $(d-1)(d+2)/2$. The rank of the matrices equals 2^n. Hence we obtain

$$r(d) = 2^n = 2^{d(d+1)/4-1}. \tag{11}$$

In particular, for $d = 3$ we have $r = 4$; if $d = 4$, $r = 16$. The rule (8) and the rank (11) of the matrices are sufficient in practice for our purposes.

3. FIRST APPROXIMATION IN $\epsilon = 4 - d$

According to the philosophy of Wilson's approach[3], we can argue in the following way. We shall assume that $d = 4 - \epsilon$. If we put $\epsilon = 0$, we obtain logarithmic integrals. This gives us the possibility of choosing principal diagrams in which the power of the interaction constant corresponds to the power of a large logarithm. As a result of summation, certain functions of this logarithm arise. Then, assuming ϵ to be small, we carry out the formal replacement:

$$\ln\frac{k_0}{k} = \frac{1}{2\pi^2}\int_k^{k_0}\frac{d^4k}{k^4} \to \text{const}\int_k^{k_0}\frac{d^{4-\epsilon}k}{k^4} = \text{const}\frac{k^{-\epsilon} - k_0^{-\epsilon}}{\varepsilon}.$$

For $k \ll k_0$ we are left with $k^{-\epsilon}/\epsilon$. Putting $\epsilon = 1$, we obtain the critical exponents for all the quantities.

We shall carry out the above program for our model. We put the external frequency equal to zero. Then according to[1], $G \sim \omega_0^{-1}(k/k_0)^{-\alpha}$. First of all we shall consider the polarization operator (Fig. 1). In the first approximation, this is equal to

$$\begin{aligned} \Pi^{(1)}(\omega, \mathbf{k}) = -i\,\mathrm{Sp}\int G^{(0)}\left(\omega_1 - \frac{\omega}{2},\ \mathbf{k}_1 - \frac{\mathbf{k}}{2}\right) \\ \times G^{(0)}\left(\omega_1 + \frac{\omega}{2},\ \mathbf{k}_1 + \frac{\mathbf{k}}{2}\right)\frac{d^dk_1}{(2\pi)^d}\frac{d\omega}{2\pi}. \end{aligned} \tag{12}$$

We put $d = 4$. According to the arguments expressed in[1], the integral over ω_1 goes to zero for $k = 0$. Consequently, we must expand in k under the integral. We then obtain an integral of the type $k^2\int dk_1/k_1$. The situation turns out to be very similar to that which was found in an zero-gap semiconductors of the first kind (with a linear spectrum). This is confirmed by the fact that an estimate of the correction to γ represented in Fig. 2 gives only unimportant terms. In view of this, we can confine ourselves to summing the logarithmic terms in the electron Green function G and Coulomb-quantum Green function Γ, taking $\gamma = 1$.

FIG. 1

FIG. 2

We call attention to the following circumstance. The Hamiltonian $fk^2 + cH_1$ is a linear combination of two invariants of the rotation group. According to[1], we must assume that in the logarithmic approximation the constants f and c are replaced by functions of $\ln(k_0/k)$. However, it is not difficult to see that logarithmic corrections arise only in the term with H_1. In fact, assuming that $\omega \ll ck^2$, we have for the self-energy in the zeroth approximation

$$\Sigma^{(1)}(0,\mathbf{k}) = i\frac{4\pi e^2}{\varkappa_0}\int \frac{1}{k_1^2} G^{(0)}(\mathbf{k}+\mathbf{k}_1,\omega_1)\frac{d\omega_1}{2\pi}\frac{d^dk_1}{(2\pi)^d}. \quad (13)$$

The function $G^{(0)}(k_1)$ can be written in the form

$$G^{(0)}(\mathbf{k}_1) = (\omega_1 - fk_1^2 - cH_1)^{-1} = \frac{\omega_1 - fk_1^2 + cH_1}{(\omega_1 - fk_1^2)^2 - (ck_1^2 - i\delta)^2}. \quad (14)$$

Substituting this expression into (13) and taking the integral over ω_1, we find

$$\Sigma^{(1)}(0,\mathbf{k}) = \frac{2\pi e^2}{\varkappa_0}\int \frac{1}{k_1^2}\frac{H_1(\mathbf{k}+\mathbf{k}_1)}{(\mathbf{k}+\mathbf{k}_1)^2}\frac{d^dk_1}{(2\pi)^d}. \quad (15)$$

But it immediately follows from this that $\mathrm{Tr}\,\Sigma = 0$ (since $\mathrm{Tr}\,H_1 = 0$). Consequently, a logarithmic correction to the first invariant fk^2 does not arise, and only the correction to H_1 remains. This means that the diagonal term in G^{-1} for $\omega \ll ck^2$ has the power α_1, equal to $2 - O(\epsilon^2)$. But if the term of the type H_1 has a lower power (of the type $\alpha = 2 - O(\epsilon)$), and below we shall verify that this is precisely the case, then in the approximation under consideration it is sufficient to retain only this term. In view of this, we shall assume below that $f = 0$.

Earlier[1], the general case of a logarithmic situation with one invariant in the Hamiltonian was analyzed. The result reduces to the following. It is necessary to determine the quantities a and b from the relations

$$a\ln\frac{k_0}{k} = -\Gamma^{(0)}\Pi^{(1)}, \quad b\ln\frac{k_0}{k} = \frac{\{\Sigma^{(1)}(k), H_1(k)\}}{2ck^4}. \quad (16)$$

In the region where the logarithms are large,

$$G(0,k) \sim G^{(0)}\left(\ln\frac{k_0}{k}\right)^{-b/(a+b)}, \quad \Gamma(0,k) \sim \Gamma^{(0)}\left(\ln\frac{k_0}{k}\right)^{-a/(a+b)}. \quad (17)$$

Replacing $\ln(k_0/k) \to k^{-1}$, we obtain

$$\alpha = 2 - \frac{b}{a+b}, \quad \nu - 3 = -2 + \frac{a}{a+b}, \quad (18)$$

i.e., $\alpha = \nu$, which, incidentally, was obvious, since $\gamma \sim (k/k_0)^{\alpha-\nu} = 1$ in the approximation under consideration. ν can differ from α in second order in ϵ. But in the first approximation, according to (18), it is necessary to determine the ratio b/a, and from it

$$\alpha = \nu = 2 - \frac{b/a}{1+b/a}. \quad (18a)$$

According to (12), (14) (with $f = 0$) and (11), we have

$$\Pi^{(1)}(\omega,\mathbf{k}) = -ir(d)\int\{\omega_+\omega_- + c^2[d(\mathbf{k}_+\mathbf{k}_-)^2 - k_+^2k_-^2]/(d-1)\}$$
$$\times[(\omega_- - ck_-^2 + i\delta)(\omega_- + ck_-^2 - i\delta)(\omega_+ - ck_+^2 + i\delta)(\omega_+ + ck_+^2 - i\delta)]^{-1}\frac{d\omega}{2\pi}\frac{d^dk}{(2\pi)^d},$$

where $k_\pm = k_1 \pm k/2$ and $\omega_\pm = \omega_1 \pm \omega/2$. Integrating over the frequencies, we obtain

$$\Pi^{(1)}(\omega,\mathbf{k}) = -\frac{r(d)}{2c}\frac{d}{d-1}$$
$$\times\int\frac{(k_+^2 + k_-^2)[k_+^2k_-^2 - (\mathbf{k}_+\mathbf{k}_-)^2]d^dk_1}{[(k_+^2 + k_-^2 - i\delta) - (\omega/c)^2]k_+^2k_-^2(2\pi)^d}. \quad (19)$$

We shall put $\omega = 0$ and confine ourselves to logarithmic accuracy. Assuming that $k \ll k_1$, we find

$$\Pi^{(1)}(0,\mathbf{k}) = -\frac{r(d)}{4c}\frac{d}{d-1}\int\frac{k_1^2k^2 - (\mathbf{k}_1\mathbf{k})^2}{k_1^6}\frac{d^dk_1}{(2\pi)^d}. \quad (20)$$

We put $d = 4$; in the integral,

$$\frac{d^dk_1}{(2\pi)^d} \to 4\pi\int_0^\pi \sin^2\theta\, d\theta\int\frac{k_1^3dk_1}{(2\pi)^4}.$$

We then obtain

$$a = \frac{1}{c}\frac{4\pi\cdot 2\pi^2}{(2\pi)^4}\frac{e^2}{\varkappa_0}4. \quad (21)$$

We turn now to the self-energy. It follows from symmetry considerations that $\Sigma^{(1)} = qcH_1$, where q can depend only on k^2. Taking the anticommutator with $H_1(k)$, according to (9) we obtain

$$2qk^4 = \frac{4\pi e^2d}{c\varkappa_0(d-1)}\int\frac{1}{k_1^2}\left\{\frac{(\mathbf{k}_1\mathbf{k}+k^2)^2}{(\mathbf{k}_1+\mathbf{k})^2} - \frac{(\mathbf{k}_1\mathbf{k})^2}{k_1^2}\right\}\frac{d^dk_1}{(2\pi)^d}$$

(here we have made use of the relation $\overline{(\mathbf{k}_1\cdot\mathbf{k})^2} = k_1^2k^2/d$). Assuming that $k \ll k_1$ and expanding in k, we obtain

$$q = \frac{2\pi e^2 d}{c\varkappa_0(d-1)}\int(1 + 4\cos^4\theta - 5\cos^2\theta)\frac{d^dk_1}{(2\pi)^dk_1^4}.$$

Since $q = b\ln(k_0/k)$ (cf. 17), we obtain from this ($d = 4$)

$$b = \frac{1}{c}\frac{4\pi\cdot 2\pi^2}{(2\pi)^4}\frac{e^2}{\varkappa_0}\frac{1}{6}. \quad (22)$$

Substituting this expression into (18), we have

$$\alpha = \nu = 2 - 1/25. \quad (23)$$

It is interesting to note that the indices α and ν for the electron G-function differ little from the "free indices." As regards the function $\Gamma(k)$, it becomes close to the quantity $\sim k^{-1}$ obtained from the "random-phase approximation," i.e., obtained on the basis of a calculation of the first-approximation diagram for the polarization operator[6].

Furthermore, in the approximation under consideration it turns out that the term with fk^2 in the Hamiltonian is not renormalized, and $\alpha = \nu$. The next approximations would undoubtedly introduce corrections to these statements. However, it is extremely difficult to calculate the corresponding integrals and we shall apply another approximate method, which, moreover, gives an idea of how the transition to the strong-coupling regime occurs.

4. THE APPROXIMATION OF A LARGE NUMBER OF COMPONENTS

In the preceding section, it was noted that indices are obtained that are close to those given by taking into account only the first approximation for the polarization operator. This circumstance is characteristic of the situation in phase transitions when the "order parameter" has a large number of components. Although this case is rare in true physical models, we can nevertheless count on the fact that expansion in the number of components gives a series that is not too slowly convergent even for $n = 3$. Such an approximation was proposed by Ferrell and Scalapino[4]. The special role of the first-order diagram for the polarization operator is explained by the fact that, as a consequence of taking the trace, it is proportional to n. Since $\Gamma \sim 1/\pi$ appears in the self-energy Σ and, at the same time, there is no summation over the components, Σ will be of order $1/n$. In our case, the number of components is equal to four and we can count on the good convergence of this procedure.

Although, as we shall see below, the $1/n$ expansion leads in first order to more complicated calculations than does the ϵ expansion, it has important merits. First of all, the calculations are performed in the real three-dimensional space and questions associated with the nonuniqueness of the continuation to noninteger d do not arise. It is then found that, even in the first approximation, $\alpha \neq \nu$. Having a real model, we shall perform the calculation only for this model, and not for arbitrary n.

The expression for the first approximation for the polarization operator is obtained from (19), if we put $d = 3$. Introducing the variables $x = \cos\theta$, $z = 2k_1/k$ and $s = 2\omega/ck^2$, we obtain

$$\Pi^{(1)} = -\frac{3k}{2c\pi^2}\int_0^1 dx \int_0^\infty dz \frac{z^4(z^2+1)(1-x^2)\,dx}{[(z^2+1-i\delta)^2-s^2][(z^2+1)^2-4z^2x^2]}. \tag{24}$$

Integrating over x, we obtain

$$\Pi^{(1)} = -\frac{3k}{8c\pi^2}\int_0^\infty \frac{z^2 dz}{(z^2+1-i\delta)^2-s^2}\left[z^2+1-\frac{1}{2}\frac{(z^2-1)^2}{z}\ln\frac{z+1}{|z-1|}\right]. \tag{25}$$

The integral over z can be taken if we note that

$$\int_0^\infty dz \to \frac{1}{2}\int_{-\infty+i\delta}^{\infty+i\delta} dz,$$

with

$$\ln\frac{z+1}{|z-1|} \to \ln\frac{z+1}{z-1}.$$

We then obtain

$$\Pi^{(1)} = -\frac{3k}{32c\pi}\varphi(s), \tag{26}$$

$$\varphi(s) = \frac{(2+s)^2}{s}\operatorname{arctg}(1+s)^{-1/2} - \frac{(2-s)^2}{s}\operatorname{arctg}(1-s)^{-1/2} - (1+s)^{1/2} - (1-s)^{1/2} \tag{27a}$$

for $0 < s < 1$, and

$$\varphi(s) = \frac{(2+s)^2}{s}\operatorname{arctg}(1+s)^{-1/2} - (1+s)^{1/2} - \frac{\pi(2-s)^2}{2s}\theta(2-s) + i\left[(s-1)^{1/2} - \frac{(2-s)^2}{2s}\ln\left|\frac{1+(s-1)^{1/2}}{(s-1)^{1/2}-1}\right|\right] \tag{27b}$$

for $s > 1$;

$$\varphi(-s) = \varphi(s); \tag{27c}$$

the asymptotic values are

$$\varphi(s) \to 4\left(\frac{\pi}{2}-1\right), \quad s \to 0;$$
$$\varphi(s) \to \frac{8}{3\sqrt{s}}(1+i), \quad s \to \infty; \tag{28}$$

the singular values are

$$\varphi(1) = 9\operatorname{arctg}\frac{1}{\sqrt{2}} - \frac{\pi}{2} - \sqrt{2},$$
$$\varphi(2) = \frac{4\pi}{3} - \sqrt{3} + i. \tag{29}$$

Graphs of the real and imaginary parts of the function φ ($\varphi = \varphi_1 + i\varphi_2$) are given in Fig. 3.

It will be important for the following that $\varphi(s)$ is the value on the real axis of a function of the complex variable s, defined by the formula (27a), analytic in the plane of the complex variable s with cuts along the real axis for $1 < |s| < \infty$, and symmetric with respect to $s \to -s$. The formulas (27b) and (27c) correspond to taking the value on the upper side along the right cut $1 < s < \infty$ and on the lower side along the left cut $-\infty < s < -1$.

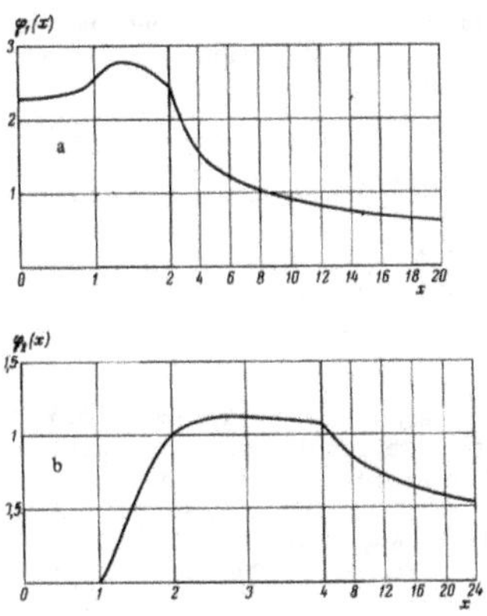

FIG. 3

On decrease of k and ω in the approximation under consideration, we enter first of all a region in which $\Pi^{(1)} \gg \Gamma^{(0)}$. However, the self-energy correction to the electron Green function in this region is still small, and only on further decrease of ω and k does it begin to play a fundamental role. By assumption (cf.[1]), it is precisely then that there arise the power dependences $G \sim \omega^{-\alpha/\nu}$ for $\omega/\omega_0 \gg (k/k_0)^\nu$ or $G \sim k^{-\alpha}$ for $\omega/\omega_0 \ll (k/k_0)^\nu$ (we recall that $\alpha < \nu$ and $\alpha < 2$). Matching between these laws and the free Green function should occur in the region in which Σ is still small. This is possible only in the case when α/ν is close to 1 and α is close to 2.

In fact, for $\omega/\omega_0 \gg (k/k_0)^\nu$,

$$G^{-1} \infty \omega^{\alpha/\nu} \infty \omega\left(\frac{\omega}{\omega_0}\right)^{-(1-\alpha/\nu)} \approx \omega\left[1+\left(1-\frac{\alpha}{\nu}\right)\ln\frac{\omega_0}{\omega}\right].$$

On the other hand,

$$G^{-1} = \omega - cH_1 - \Sigma.$$

Hence it follows that in the matching region, for $\omega/\omega_0 \gg (k/k_0)^\nu$,

$$\Sigma(\omega) \approx -\left(1-\frac{\alpha}{\nu}\right)\omega\ln\frac{\omega_0}{\omega}. \tag{30}$$

In exactly the same way, if we assume that $\Sigma = qcH_1$ for $\omega/\omega_0 \ll (k/k_0)^\nu$, we have

$$G^{-1} \infty k^\alpha \infty k^2\left(\frac{k}{k_0}\right)^{-(2-\alpha)} \approx k^2\left[1+(2-\alpha)\ln\frac{k_0}{k}\right]$$

and

$$q = (2-\alpha)\ln(k_0/k). \tag{31}$$

Thus in the approximation under consideration, determination of Σ with logarithmic accuracy is sufficient to determine α and ν. Of course, all these arguments are valid only in the case when $1 - \alpha/\nu \ll 1$ and $2 - \alpha \ll 1$. The self-energy is obtained from (13) by replacing $\Gamma^{(0)} = 4\pi e^2/\kappa_0 k^2$ by $\Gamma = -1/\pi$. In addition, it is necessary to take into account that $\Sigma(0, 0)$ must be subtracted from $\Sigma(\omega, k)$ (renormalization of the chemical potential). For finite ω we have

$$\Sigma(\omega, k) = -i\int \Pi^{-1}(\omega_1, \mathbf{k}_1) G(\omega-\omega_1, \mathbf{k}-\mathbf{k}_1)\, d\omega_1 \frac{d^3k_1}{(2\pi)^4}. \tag{32}$$

The function $G^{(0)}$ can be written in the form

$$\begin{aligned} G^{(0)}(\omega-\omega_1, \mathbf{k}-\mathbf{k}_1) = {} & \frac{1}{2}\left[1+\frac{H_1(\mathbf{k}-\mathbf{k}_1)}{\varepsilon_1(\mathbf{k}-\mathbf{k}_1)}\right][\omega-\omega_1-\varepsilon_1(\mathbf{k}-\mathbf{k}_1)+i\delta]^{-1} \\ & +\frac{1}{2}\left[1-\frac{H_1(\mathbf{k}-\mathbf{k}_1)}{\varepsilon_1(\mathbf{k}-\mathbf{k}_1)}\right][\omega-\omega_1+\varepsilon_1(\mathbf{k}-\mathbf{k}_1)-i\delta]^{-1}, \end{aligned} \tag{33}$$

where $\epsilon_1(\mathbf{k}) = ck^2$.

For Π we substitute the expression (26), (27). Owing to the fact that this is the value on the real axis of an analytic function with two cuts (with an appropriate definition on the cut), the integral over ω_1 can be represented as a contour integral, where the contour C is depicted in Fig. 4a, in which the poles from $G^{(0)}(\omega - \omega_1)$ are also shown. Since the integral over ω_1 for $\Sigma(\omega, k) - \Sigma(0, 0)$ converges in any case, the contour can be closed and deformed in such a way that it takes the form C′ (Fig. 4b). Since the function has complex-conjugate values on the two sides of the cut, we obtain (we do not write the integral to be subtracted for $\Sigma(0, 0)$)

$$\begin{aligned} \Sigma = {} & \frac{8}{3\pi^2}\int_1^\infty ds_1 \int ck_1\, dk_1 \frac{d\Omega}{4\pi} A(s_1) \\ & \times\left\{\left[1+\frac{H_1(\mathbf{k}-\mathbf{k}_1)}{\varepsilon_1(\mathbf{k}-\mathbf{k}_1)}\right]\left[\frac{2\omega}{ck_1^2}-s_1-\frac{2(\mathbf{k}-\mathbf{k}_1)^2}{k_1^2}\right]^{-1}\right. \\ & \left.+\left[1-\frac{H_1(\mathbf{k}-\mathbf{k}_1)}{\varepsilon_1(\mathbf{k}-\mathbf{k}_1)}\right]\left[\frac{2\omega}{ck_1^2}-s_1+\frac{2(\mathbf{k}-\mathbf{k}_1)^2}{k_1^2}\right]^{-1}\right\} \\ & +\frac{8}{3\pi}\int dk_1 ck_1 \frac{d\Omega}{4\pi} B(s_{10})\left[\frac{H_1(\mathbf{k}-\mathbf{k}_1)}{\varepsilon_1(\mathbf{k}-\mathbf{k}_1)}-1\right], \end{aligned} \tag{34}$$

where $s_1 = 2\omega/ck_1^2$, $s_{10} = 2\omega/ck_1^2 + 2(\mathbf{k} - \mathbf{k}_1)^2/k_1^2$ is the value of s_1 at the pole, $A(s_1) = -\mathrm{Im}[1/\varphi(s_1)]$, and $B(s_1) = \mathrm{Re}[1/\varphi(s_1)]$ (we assume that $k \ll k_1$ and $2\omega \ll ck_1^2$; this is valid in the calculation of the logarithmic part of the integrals—cf. below). This expression can be transformed by making use of properties of the function $1/\varphi(s) = B(s) - iA(s)$, namely, the analytic properties and the symmetry in s. Making use of the Cauchy theorem for the contour C″ (Fig. 4c), we obtain

$$B(s_{10}) = -\frac{1}{\pi}\int_1^\infty A(s_1)\left[\frac{1}{s_1+s_{10}}+\frac{1}{s_1-s_{10}}\right] ds_1. \tag{35}$$

Substituting this relation into (37), we obtain

$$\begin{aligned} \Sigma = {} & \frac{8}{3\pi^2}\int_1^\infty ds_1 \int ck_1\, dk_1 \frac{d\Omega}{4\pi} A(s_1)\cdot \\ & \times\left\{\left[1+\frac{H_1(\mathbf{k}-\mathbf{k}_1)}{\varepsilon_1(\mathbf{k}-\mathbf{k}_1)}\right]\left[\frac{2\omega}{ck_1^2}-s_1-\frac{2(\mathbf{k}-\mathbf{k}_1)^2}{k_1^2}\right]^{-1}\right. \\ & \left.+\left[1-\frac{H_1(\mathbf{k}-\mathbf{k}_1)}{\varepsilon_1(\mathbf{k}-\mathbf{k}_1)}\right]\left[\frac{2\omega}{ck_1^2}+s_1+\frac{2(\mathbf{k}-\mathbf{k}_1)^2}{k_1^2}\right]\right\}. \end{aligned} \tag{36}$$

In this form we need no longer imply the subtraction of $\Sigma(0, 0)$. In fact, in view of the fact that $\int d\Omega H_1(\mathbf{k}_1) = 0$, the quantity $\Sigma(0, 0) = 0$.

If $\omega \gg ck^2$ we can put $k = 0$. Expanding in ω and integrating over $\mathbf{k}_1$, we obtain

$$\Sigma(\omega) = -\frac{16}{3\pi^2}\int_1^\infty \frac{A(s)}{(s+2)^2} ds\, \omega \ln\frac{\omega_0}{\omega}, \tag{37}$$

where $\omega_0 \sim ck_0^2$; k_0 is the upper limit of the integration over k. But if $\omega \ll ck^2$, we must, on the contrary, put $\omega = 0$ in the integral (36). We then obtain

$$\Sigma(k) = -\frac{16}{3\pi^2}\int_1^\infty ds_1 \int ck_1\, dk_1 \frac{d\Omega}{4\pi} A(s_1)\frac{H_1(\mathbf{k}-\mathbf{k}_1)}{\varepsilon_1(\mathbf{k}-\mathbf{k}_1)}\frac{1}{s_1+2(\mathbf{k}-\mathbf{k}_1)^2/k_1^2}.$$

Since its trace equals zero, this expression should be proportional to H_1; $\Sigma = qcH_1$. Calculating the anticommutator with H_1 and making use of the rule (7), we have

$$\begin{aligned} q = {} & -\frac{8}{3\pi^2 k^4}\int_1^\infty ds_1 \int k_1\, dk_1 \frac{d\Omega}{4\pi} A(s_1) \\ & \times\frac{3[\mathbf{k}\times(\mathbf{k}-\mathbf{k}_1)]^2-k^2(k-k_1)^2}{(\mathbf{k}-\mathbf{k}_1)^2}\frac{1}{s+2(\mathbf{k}-\mathbf{k}_1)^2/k_1^2} \end{aligned}$$

Expanding in k/k_1, we obtain

$$q = -\frac{8}{3\pi^2}\left[\frac{2}{5}\int_1^\infty\frac{A(s)}{s+2}ds-\frac{16}{5}\int_1^\infty\frac{A(s)}{(s+2)^2}ds+\frac{64}{15}\int_1^\infty\frac{A(s)}{(s+2)^3}ds\right].$$

However, this expression is not completely accurate. This can be seen from the fact that the first term in it diverges as $s \to \infty$. Indeed, according to (28), $A(s) \propto s^{1/2}$ as $s \to \infty$. The inaccuracy has arisen in the use of the transformation (35), which is valid only for the case when $A(s)$ and $B(s)$ fall off at infinity[1)]. We can correct the situation by subtracting from $B - iA$ a symmetric function of s with the same cuts and asymptotic value, in such a way that the result now be a decreasing function. We note that the trouble arises entirely from the terms of second order in k/k_1, which originate from H_1/ϵ_1 (the zeroth approximation is taken from s_{10} and the denominators of the first term in (34)). Therefore, we shall consider only these terms.

Subtracting from $1/\varphi(s)$ a function which we denote by $1/\varphi_\infty(s) = B_\infty - iA_\infty$, we obtain

$$B(2) = B_\infty(2) - \frac{1}{\pi}\int_1^\infty [A(s)-A_\infty(s)]\left(\frac{1}{s+2}+\frac{1}{s-2}\right) ds.$$

As a result, it turns out that in the expression for q we must make the replacement

$$\int_1^\infty \frac{A(s)}{s+2} ds \to -\frac{\pi}{2}B_\infty(2)+\int_1^\infty\left[\frac{A(s)}{s+2}-A_\infty(s)\frac{s}{s^2-4}\right] ds.$$

For $1/\varphi_\infty(s)$ we can take the simplest function:

$$1/\varphi_\infty(s) = {}^3/_{16}[(s+1)^{1/2}+(1-s)^{1/2}];$$

then $B_\infty = (3/16)(s + 1)^{1/2}$ and $A_\infty = (3/16)(s - 1)^{1/2}$. Substituting $1/\varphi_\infty$ into the integrals, we find

$$\int_1^\infty \frac{A(s)}{s+2} ds \to -\frac{3\sqrt{3}\pi}{32}+\int_1^\infty\left(\frac{A(s)}{s+2}-\frac{3}{16}\frac{s(s-1)^{1/2}}{s^2-4}\right) ds.$$

But we can do still better. We add and subtract $3/16\sqrt{s}$ under the integral. We then obtain

$$\int_1^\infty \frac{A(s)}{s+2} ds \to -\frac{3}{8}+\int_1^\infty\left(\frac{A(s)}{s+2}-\frac{3}{16\sqrt{s}}\right) ds.$$

Carrying out these replacements in q and comparing the expressions for q and for $\Sigma(\omega)$ with (31) and (30), we obtain

$$2-\alpha = \frac{2}{5\pi^2}-\frac{16}{15\pi^2}\left(A_1-8A_2+\frac{32}{3}A_3\right), \tag{38}$$

$$1-\frac{\alpha}{\nu} = \frac{16}{3\pi^2}A_2, \tag{39}$$

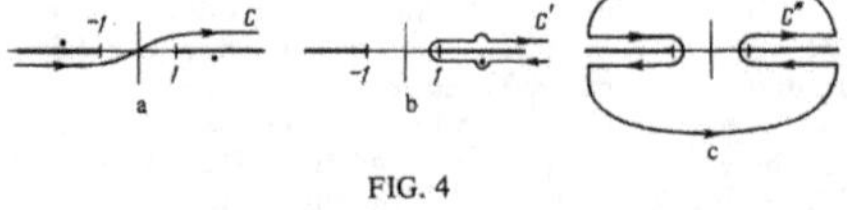

FIG. 4

$$A_1=\int_1^\infty\left(\frac{A(s)}{s+2}-\frac{3}{16\sqrt{s}}\right)ds,\quad A_{n>1}=\int_1^\infty\frac{A(s)}{(s+2)^n}ds. \tag{40}$$

The function $A(s) = -\mathrm{Im}[1/\varphi(s)] = \mathrm{Im}\,\varphi(s)/|\varphi(s)|^2$, and $\varphi(s)$ is given by the formula (27b). Numerical calculation gives:

$$A_1=-0.6976;\quad A_2=0.1570;\quad A_3=0.01207;\quad A_4=0.00183.$$

As a result,

$$2-\alpha=0.238,\quad 1-\alpha/\nu=0.0847 \tag{41a}$$

or

$$\alpha=1.76,\quad \nu=1.92. \tag{41b}$$

The result obtained is evidence that the approach used is highly successful. In fact, both $1-\alpha/\nu$ and $(2-\alpha)/2$ are of the order of 0.1. In view of this, the values obtained for α and ν can be regarded as fairly reliable. But if we take the first approximation of the ϵ expansion, it turns out there that the electron Green function is weakly renormalized, but, in return, the Coulomb-quantum Green function is strongly renormalized: $\delta = 3-\nu \approx 1$, in place of 2 as in Coulomb's law. This shows that, in the given problem, $\epsilon = 4-d$ is not as suitable an expansion parameter as $1/n$, in contrast to the theory of phase transitions. This is also indicated by the fact that, according to Sec. 3, $1-\alpha/\nu$ is a quantity of order ϵ^2, whereas $2-\alpha$ is of order ϵ. In reality, as can be seen from (41a), they are of the same order (nevertheless, it is true that there is a certain tendency for $1-\alpha/\nu$ to be small). In view of this, we must conclude that, although the ϵ-expansion method can be applied in the given problem for certain qualitative estimates (the absence of the term f, and arguments in favor of the $1/n$ expansion; cf. also Sec. 5), for quantitative calculations it is necessary to use the $1/n$ expansion.

In the derivation performed we have assumed the Hamiltonian to be equal to cH_1, on the basis of the result from the preceding Section that the contribution fk^2 in the Hamiltonian is not renormalized. Actually, this is somewhat inconsistent. In fact, this result was obtained in the first approximation of the ϵ expansion, whereas here we are using a completely different approximation. Nevertheless, a certain argument can be adduced in favor of the validity of the procedure used. Suppose that $f \neq 0$. In this case, the replacement

$$s\to s-2\eta zx,$$

where $\eta = f/c$, takes place in the expression (24) for the polarization operator. Let η be small. In this case, expanding in η gives a correction of order η^2 to Π. Substituting Π into Σ leads to the appearance of corrections of order η^2 in all terms. In particular, a term proportional to $\eta^2k^2\ln(k_0/k)$ can arise in Σ. Combining this with the term $-fk^2$ in $G^{(0)-1}$, we arrive at the conclusion that for small η the power $2-\alpha$ in the isotropic term of the Hamiltonian will be proportional to η, i.e., it will be considerably smaller than the constant obtained for the term of the type cH_1. But this means that, in this case, we can neglect the isotropic term in the asymptotic region $\omega, k \to 0$. Consequently, the case we have considered is internally consistent. It is not difficult to see that the opposite assumption, namely, that $f \neq 0$, $c = 0$, leads to $\Pi^{(1)} = 0$, i.e., to the absence of a strong-coupling situation. Consequently, the possibility that the power of the isotropic term is lower than the power of the term of the type cH_1 cannot be realized. Finally, equality of the powers is also impossible, since, if this were the case, it would be true for all f.

One can also discuss all this in a different way. Matching of the power laws with perturbation theory is not always possible. In particular, in matching with ordinary perturbation theory in the ϵ-expansion method[3] it is necessary to select the value of the interaction constant for which this is possible. In our case, for $ck^2 \gg \omega$, there can be terms of the type H_1 and an isotropic term in G^{-1}. We can make three assumptions: a) the power of the term of the type H_1 is lower than the power of the isotropic term. Then matching with the perturbation theory starting from $f = 0$ should exist, and has been effected in this Section; b) the power of the term of the type H_1 is higher than the power of the isotropic term. In this case, there should be matching of the strong-coupling regime with the perturbation theory starting from $c = 0$, and this is not obtained; c) finally, by assuming the powers to be equal we can take any value of the ratio $f/c = \eta$ in the initial Hamiltonian. The two powers should turn out to be equal for any η. But this is not corroborated for small η.

Thus, the possibility (a) considered above is the only one.

5. "ISOTROPIZATION" IN THE SINGULAR REGION

Up to now we have considered only the spherically symmetric model with the Hamiltonian (2). It was found that the power of the second term, which we called cH_1, in the singular region is lower than the power of the term fk^2. This corresponds to the fact that the electron and hole masses are, as it were, equalized in the singular region[2)]. We now show that "isotropization" of the spectrum occurs in the singular region, i.e., even in the case when we do not assume isotropy at large momenta, it arises automatically in the singular region.

For this we note that an anisotropic model corresponds to replacement of cH_1, the part of the Hamiltonian given by formula (4), by $c_1H_1 + c_2H_2$, where H_1 includes the terms with the "non-diagonal" matrices $A_{\mu\nu}$ $(\mu \neq \nu)$ and H_2 includes the terms with the matrices $A_{\nu\nu}$. We call attention to the fact that, according to the rules (5) and (6), we can easily replace the normalization of the matrices $A_{\mu\nu}$ with $\mu \neq \nu$ without damaging the whole scheme. This gives the possibility of generalizing the method proposed in Sec. 1 for continuing to noninteger dimensions for the case under consideration, and for all integer d the number of matrices γ_i and, consequently, their rank are not increased in comparison with the isotropic model.

Without repeating all the arguments, we cite the principal formulas. Writing

$$c_1H_1+c_2H_2=c_2\sum_{\eta\nu}A_{\eta\nu}k_\eta k_\nu, \tag{42}$$

we have in place of (9)

$$\{A_{\eta\nu},A_{\rho\sigma}\}=\frac{d\mu^2}{d-1}(\delta_{\eta\rho}\delta_{\nu\sigma}+\delta_{\eta\sigma}\delta_{\nu\rho})-\frac{2}{d-1}\delta_{\eta\nu}\delta_{\rho\sigma}+\frac{2d}{d-1}(1-\mu^2)\delta_{\eta\nu\rho\sigma}, \tag{43}$$

where $\mu = c_1/c_2$, and $\delta_{\mu\nu\rho\sigma} = 1$ when all the indices coincide and zero otherwise. It follows from (43) that

$$\{c_1H_1(\mathbf{k}_1)+c_2H_2(\mathbf{k}_1),\, c_1H_1(\mathbf{k}_2)+c_2H_2(\mathbf{k}_2)\}=\frac{2c_2^2}{d-1}\Big[d\mu^2(\mathbf{k}_1\mathbf{k}_2)^2 -k_1^2k_2^2+d(1-\mu^2)\sum_\nu k_{1\nu}^2k_{2\nu}^2\Big]. \quad (44)$$

In particular, for $\mathbf{k}_1 = \mathbf{k}_2 = \mathbf{k}$ the right-hand side becomes equal to $2(\epsilon/c_2)^2$;

$$\varepsilon=c_2k^2\left[\mu^2+\frac{2d}{d-1}(\mu^2-1)A_{22}\right]^{1/2} \quad (45)$$

is the pole of the bare electron Green function;

$$A_{22}=\sum_{\mu\neq\nu} n_\mu^2 n_\nu^2.$$

With the same rule (11) for the rank of the matrices, we can solve the problem. To obtain the first approximation in $\epsilon = 4 - d$ it is sufficient to perform the calculations with logarithmic accuracy for $d = 4$. Since in the present case we have not one but two invariants, the scheme we constructed earlier[1] is not directly applicable. However, the basic line of the calculations is the same. We shall find Π and Σ in the logarithmic approximation, assuming c_1 and c_2 (or c_2 and μ) to be functions of $\xi = \ln(k_0/k)$, and then obtain equations from which it will be possible to determine them.

Substituting $G = [\omega - c_1H_1 - c_2H_2]^{-1}$ into (12), taking the integral over ω and transforming the expression obtained, we find

$$\Pi=-4k^2\frac{2\pi^2}{(2\pi)^4}\int_0^\xi\frac{d\xi_1}{c_2(\xi_1)}\left\{\frac{\mu^2+{}^4/_9(\mu^2-1)(4\mu^2-1)A_{22}-{}^8/_3(\mu^2-1)^2A_{222}}{[1+{}^8/_3(\mu^2-1)A_{22}]^{3/2}}\right\}_{av}, \quad (46)$$

where

$$\{\ \}_{av}=(2\pi^2)^{-1}\int_0^\pi\sin^2\theta\,d\theta\int_0^\pi\sin\varphi\,d\varphi\int_0^{2\pi}d\gamma\{\ \},$$

A_{22} was defined earlier,

$$A_{222}=\sum_{\mu\neq\nu\neq\rho} n_\mu^2 n_\nu^2 n_\rho^2,$$

and the projections n_ν of the unit vector are equal to:

$$n_1=\cos\theta,\quad n_2=\sin\theta\cos\varphi,\quad n_3=\sin\theta\sin\varphi\cos\gamma,$$
$$n_4=\sin\theta\sin\varphi\sin\gamma.$$

In (46), μ depends on ξ_1. It follows from (46) that

$$\Gamma=\frac{4\pi e^2}{\varkappa_0(k^2-4\pi e^2\Pi/\varkappa_0)}=\frac{4\pi e^2}{\varkappa_0k^2}d(\xi), \quad (47)$$

where $d(\xi)$ is a logarithmic function.

We turn now to the self-energy. We write, analogously to formula (15),

$$\Sigma^{(1)}(0,\mathbf{k})=\frac{2\pi e^2}{\varkappa_0}\int\frac{d(\xi_1)}{(\mathbf{k}_1+\mathbf{k})^2}\frac{c_1H_1(\mathbf{k}_1)+c_2H_2(\mathbf{k}_1)}{\varepsilon(\mathbf{k}_1)}\frac{d^dk_1}{(2\pi)^d}$$
$$\approx\frac{8\pi e^2}{\varkappa_0}\int\frac{d(\xi_1)(\mathbf{k}_1\mathbf{k})}{k_1^6}\frac{c_1H_1(\mathbf{k}_1)+c_2H_2(\mathbf{k}_1)}{\varepsilon(\mathbf{k}_1)}\frac{d^dk_1}{(2\pi)^d}=q_1H_1(\mathbf{k})+q_2H_2(\mathbf{k}). \quad (48)$$

We take the anticommutator with $A_{\mu\nu}$ $(\mu\neq\nu)$. Comparing the two parts and applying the definition $c_1 = c_{10} + q_1$, we obtain the equation

$$c_1(\xi)=c_{10}+\frac{2\pi^2}{(2\pi)^4}\frac{4\pi e^2}{\varkappa_0}{}^8/_3\cdot\int_0^\xi\left\{\frac{d(\xi_1)\mu(\xi_1)A_{22}}{(1+{}^8/_3A_{22}[\mu^2(\xi_1)-1])^{1/2}}\right\}_{av}. \quad (49)$$

By taking the anticommutator of the formula (48) with $A_{\rho\rho}$, we have, in the same way,

$$c_2(\xi)=c_{20}+\frac{2\pi^2}{(2\pi)^4}\frac{4\pi e^2}{\varkappa_0}2\int_0^\xi\left\{\frac{d(\xi_1)({}^1/_4-{}^2/_3A_{22})}{(1+{}^8/_3A_{22}[\mu^2(\xi_1)-1])^{1/2}}\right\}_{av}. \quad (50)$$

From (46) and (47) we have

$$\frac{1}{d(\xi)}=1+\frac{4\pi e^2}{\varkappa_0}\frac{2\pi^2}{(2\pi)^4}$$
$$\times4\int_0^\xi\left\{\frac{\mu^2+{}^4/_9(\mu^2-1)(4\mu^2-1)A_{22}-{}^8/_3(\mu^2-1)^2A_{222}}{[1+{}^8/_3(\mu^2-1)A_{22}]^{3/2}}\right\}_{av}d\xi_1. \quad (51)$$

From Eqs. (49), (50) and (51) we can find the three functions c_1, c_2, and d. Our purpose is not this, however, but the determination of the laws

$$c_1\sim\xi^{\alpha_1},\quad c_2\sim\xi^{\alpha_2},\quad d\sim\xi^{-\alpha_3} \quad (52)$$

in the asymptotic region. Using the notation $\beta = 2 \times 2\pi^2 \times 4\pi e^2/(2\pi)^4\kappa_0$, introducing $\mu = c_1/c_2$ and differentiating the equations with respect to ξ, we have

$$\frac{d\ln\mu}{d\xi}=\frac{\beta d(\xi)}{c_2(\xi)}\left\{\frac{A_{22}-{}^1/_4}{[1+{}^8/_3A_{22}(\mu^2-1)]^{1/2}}\right\}_{av}, \quad (53a)$$

$$\frac{d\ln c_2}{d\xi}=\frac{\beta d(\xi)}{c_2(\xi)}\left\{\frac{{}^1/_4-{}^2/_3A_{22}}{[1+{}^8/_3A_{22}(\mu^2-1)]^{1/2}}\right\}_{av}; \quad (53b)$$

$$\frac{d\ln d^{-1}}{d\xi}=\frac{\beta d(\xi)}{c_2}\left\{\frac{\mu^2+{}^4/_9(\mu^2-1)(4\mu^2-1)A_{22}-{}^8/_3(\mu^2-1)^2A_{222}}{[1+{}^8/_3A_{22}(\mu^2-1)]^{3/2}}\right\}_{av}. \quad (53c)$$

First of all, we make the assumption that μ = const. For this it is necessary that the right-hand side of (53a) vanish, or, in other words,

$$\frac{1}{2\pi^2}\int_0^\pi\sin^2\theta\,d\theta\int_0^\pi\sin\varphi\,d\varphi\int_0^{2\pi}d\gamma\frac{A_{22}-{}^1/_4}{[1+{}^8/_3A_{22}(\mu^2-1)]^{1/2}}=0. \quad (54)$$

This equation is satisfied for $\mu = 1$, since $\{A_{22}\}_{av} = 1/4$. This solution corresponds to the isotropic case. But in principle (54) could also have other solutions.

We note that the left-hand side of Eq. (54) has the asymptotic forms[3)]

$$\left\{\frac{A_{22}-{}^1/_4}{(1-{}^8/_3A_{22})^{1/2}}\right\}_{av},\quad \mu\to0, \quad (55a)$$

$$\frac{1}{|\mu|}\left\{\frac{A_{22}-{}^1/_4}{({}^8/_3A_{22})^{1/2}}\right\}_{av},\quad \mu\to\infty. \quad (55b)$$

The maximum value of A_{22}, corresponding to $n_1^2 = n_2^2 = n_3^2 = n_4^2 = 1/4$, equals $3/8$, and the minimum value equals zero. Since $\bar{A}_{22} = 1/4$, we should expect that values $A_{22} > 1/4$ will play the principal role in the averaging of (55a), i.e., the result will be positive, and that values $A_{22} < 1/4$ will play the main role in averaging of (55b), i.e., the result will be negative.

These arguments are reinforced by the following estimate. We expand the integrand in (54) in $A_{22} - 1/4$ and confine ourselves to the first nonvanishing term. The integral is then easily taken and we obtain

$$-\frac{4}{3}\left\{\left(A_{22}-\frac{1}{4}\right)^2\right\}_{av}\frac{(\mu^2-1)}{[1+{}^2/_3(\mu^2-1)]^{3/2}}=-\frac{1}{120}\frac{(\mu^2-1)}{[1+{}^2/_3(\mu^2-1)]^{3/2}}. \quad (56)$$

This formula is accurate near $\mu^2 = 1$ and somewhat less accurate at the limits (the expansion parameter is of the order of $1/3$). But it is, however, sufficient for the following conclusions to be drawn: 1) there is only one value $\mu^2 = 1$ at which the expression (54) vanishes; 2) for $\mu \to 0$ this expression is positive and for $\mu \to \infty$ the coefficient of $|\mu|^{-1}$ is negative.

We return now to Eqs. (53) and make an assumption about the powers in (52): $\alpha_1 > \alpha_2$. This means that for $\xi \to \infty$ we have $\mu \to \infty$, i.e., the power of μ is positive. But for this the right-hand side of (53a) must be positive for $\mu \to \infty$ (d cannot become negative and in reality c_2 implies $|c_2|$). But it is proportional to (55b), and, according to the analysis carried out above, the coefficient of $1/|\mu|$ is, in reality, negative. The same contradiction arises if we assume that the powers in (52) satisfy $\alpha_1 < \alpha_2$, i.e., $\mu \to 0$ as $\xi \to \infty$.

Thus, it follows from the arguments given that μ^2

= 1 is the only solution, i.e., "isotropization" arises in the asymptotic region, irrespective of the initial conditions (c_{10} and c_{20}).

6. CONCLUSION

Although the calculations performed above pertained to the singular region, in which all quantities are strongly renormalized because of the interaction, nevertheless, since we have used a matching method, we can analyze qualitatively the transition to this region.

Deviation from the "free" behavior begins for $k < k_0$, $\omega < \omega_0$. So long as the momenta and energies are not too small (we shall denote the lower boundary by k_1, ω_1), the change in the electron Green function can be neglected. In the region $k_1 < k < k_0$, $\omega_1 < \omega < \omega_0$, which we shall call the second region (the first is the "free" region), only the electron-electron interaction describable by the function Γ changes. The inclusion of the zeroth-approximation polarization operator has already been made by a number of authors[6]. Our calculation justifies this approximation in the second region. The Coulomb law in this region is replaced by an interaction proportional to $1/r^2$ (in the coordinates). For $\omega \gg ck^2$, allowance for the polarization leads to the appearance of a dielectric permittivity proportional to $1/\omega^{1/2}$. From formulas (26) and (28) it is not difficult to obtain

$$\varkappa = \varkappa_0 + \frac{e^2(1+i\,\mathrm{sign}\,\omega)}{(2\omega c)^{1/2}}. \quad (57)$$

In this form, this formula is also true for $\omega \gtrsim \omega_0$. It can be shown that formula (57) also remains valid if we take an isotropic model, but with different masses, i.e., $f \neq 0$. In this case, $c = \frac{1}{4}(1/m_e + 1/m_h)$.

Since the "free" Green function and the interaction have the form (1) with $\alpha = \nu = 2$, all the temperature and other dependences found earlier in[2] are operative in the second region. As regards estimating the coefficients, the effect can be appreciably reduced in those cases in which interaction of the electrons is important. For example, the electron-electron scattering probability will be proportional to the number $(1 - \alpha/\nu)$ (41a), which is approximately 0.08. In view of the fact that the second region is very simple for calculations, all the expressions for the physical effects in this region can be found, without great complications, in an explicit form with numerical coefficients (in any case, for the isotropic model)[4)], and this has been done in a number of papers (cf. the literature cited in[7]).

According to Sec. 4, change of the electron Green function begins either when

$$k \lesssim k_1 \sim k_0 e^{1/(\alpha-2)} \sim 1{,}6 \cdot 10^{-2} k_0,$$

or when

$$\omega \lesssim \omega_0 e^{1/(\alpha/\nu-1)} \sim 10^{-5}\omega_0.$$

These limits are somewhat different. In fact, $ck_1^2 \sim 2.5 \times 10^{-4}\omega_0$, whence it follows that, in principle, there is a certain region of energies in which Σ is substantially renormalized and, at the same time, the dependence of Σ on ω can be neglected. But this means in practice that a single-particle spectrum remains for energies $2.5 \times 10^{-4}\omega_0 > \omega > 10^{-5}\omega_0$, although it differs from the "free" spectrum. For $\omega < 10^{-5}\omega_0$, the single-particle spectrum is strongly damped.

However, all these changes occur in the region of extremely low energies. We shall call this the third region. In the known zero-gap semiconductors, because of the large dielectric permittivity κ_0 and the small effective mass, the second region is itself shifted into the region of energies of the order of $10^{-3} - 10^{-4}$ eV, so that even this region, in the purest samples, may be difficult to "seek out." In principle, we can assume that substances with significantly larger values of ω_0 will be found. However, even in a case in which ω_0 reaches 10 eV, the third region will remain extremely difficult to attain.

In conclusion, the author expresses his gratitude to A. A. Migdal and I. E. Dzyaloshinskiĭ for discussion of the work, to Yu. V. Petrov and V. I. Fisher for the numerical calculations, and to S. I. Anisimov for friendly assistance in organizing these calculations.

1) The expression (34) contains no such divergence.

2) We recall that the single-particle spectrum may not exist in the singular region (cf. the Conclusion).

3) A change of sign of μ corresponds only to a change of basis and, therefore, has no physical significance.

4) Our calculation had the purpose of matching with the singular region, and we have therefore considered the isotropic model with $f = 0$. In calculations for the second region it is necessary to use the true "free" Green function.

1 A. A. Abrikosov and S. D. Beneslavskiĭ, Zh. Eksp. Teor. Fiz. 59, 1280 (1970) [Sov. Phys.-JETP 32, 699 (1971)].

2 A. A. Abrikosov and S. D. Beneslavskiĭ, J. Low Temp. Phys. 5, 141 (1971).

3 K. G. Wilson and M. E. Fisher, Phys. Rev. Lett. 28, 240 (1972); K. G. Wilson, Phys. Rev. Lett. 28, 548(1972).

4 R. A. Ferrell and D. J. Scalapino, Phys. Rev. Lett. 29, 413 (1972).

5 M. E. Fisher and A. Aharony, Phys. Rev. Lett. 30, 559 (1973).

6 L. Liu and D. Brust, Phys. Rev. Lett. 20, 651 (1968); Phys. Rev. 173, 777 (1968); D. Sherrington and W. Kohn, Phys. Rev. Lett. 21, 153 (1968); Rev. Mod. Phys. 40, 767 (1968).

7 J. C. Broerman, p. 917 in Proc. Int. Conf. Physics of Semiconductors 2, PWN, Warsaw, 1972.

Translated by P. J. Shepherd
150

Possible mechanism of high-temperature superconductivity

A. A. Abrikosov

L. D. Landau Institute of Theoretical Physics, USSR Academy of Sciences
(Submitted 23 January 1978)
Pis'ma Zh. Eksp. Teor. Fiz. **27**, No. 4, 235–238 (20 February 1978)

A superconductivity mechanism based on the formation of "metallic excitonium", the exciton phase in a semiconductor or in a semimetal with $m_e/m_a \lesssim 0.01$, is proposed.

PACS numbers: 74.10.+v, 74.20.Fg

Recent experiments by Brandt, Kuvshinnikov Rusakov, and Semenov[1] have reawakened interest in the possibility of high-temperature superconductivity. A mechanism explaining the almost complete Meissner effect observed in[1] was proposed in[2]. This explanation is based on the idea of a transition to a new excitonic phase characterized by the presence of a spontaneous current.

In this article we call attention to one principally rather simple mechanism of producing high-temperature superconductivity, which is possibly realized in CuCl at high pressures.

Halperin and Rice, in their well known article,[3] have analyzed various types of excitonic phases and, in particular, have noted C. Herring's idea that "crystalline excitonium" can be created at a large mass difference between the electrons and holes. If the spectra of the electrons and holes are isotropic, then an individual exciton is in this case quite similar to a hydrogen atom, so that one can imagine that condensed phases of these excitons are just like those of hydrogen—molecular liquid, molecular crystal, atomic crystal (metallic hydrogen). Of course, the characteristic parameters of these phases will be different than those of hydrogen: the role of the atomic dimensions will be played by $r_B^*=\epsilon_0/m_e e^2$ (we use $\hbar=1$ throughout), where ϵ_0 is the dielectric constant and m_e is the effective mass of the electron (we assume that $m_e \ll m_h$). In addition to ordinary sound, exciton-lattice vibrations will exist and the corresponding vibration frequencies and sound velocities will be determined by m_h rather than by m_H, i.e., they will be larger by a factor $\sqrt{m_H/m_h} \sim 10$ (at $m_h \sim 20\, m_0$).

In hydrogen, the phase transition from the molecular crystal to the atomic crystal can be realized with the aid of a pressure of several megabars. In the excitonic phases, the role of the "pressure" is played by the width of the gap between the valence and conduction bands, which can give way to band overlap. In some substances the gaps can depend strongly on the external pressure or on the magnetic field. Thus, transitions between different excitonic phases are again ensured in final analysis by the external pressure. But the scale of these pressures in BiSb and CuCl is of the order of not more than several kbar, i.e., smaller by hundreds of times than in hydrogen.

Let us imagine qualitatively how the transitions occur with decreasing gap. When the gap reaches the value $(1/2)\mathrm{Ry}^*=m_e e^4/2\epsilon_0^2$, exciton production sets in. Of course, Van der Waals attraction forces act between the individual "atoms"–excitons–or "molecules"–biexcitons–and lead to a first-order transition into the condensed phase at a gap width somewhat larger than Ry*/2. We, however, will disregard this small effect. If the gap is less than Ry*/2, then the density of the excitons is determined by their repulsion at short distances. Consequently $n_e=n_h\sim(m_e e^2/\epsilon_0)^3$. In view of the exponential dependence of the repulsion forces on the distance, the density will change little with further decrease of the gap, until the band overlap becomes strong enough. Then a metallic phase sets in, which can be called "metallic excitonium." The condition for this is $e^2/\epsilon_0 v_e=e^2 m_e/\epsilon_0 p_0\lesssim 1$, where p_0 is the limiting Fermi momentum determined from the condition $p_0=\sqrt{2m_e(E_v-E_c)}$, where E_v and E_c are the edges of the bands (it is assumed that $E_v>E_c$, i.e., the bands overlap; the formula for p_0, strictly speaking, is valid only at $e^2/\epsilon_0 v_e\ll 1$). Obviously, for this purpose the band overlap must be $E_v-E_c\gtrsim \mathrm{Ry}^*/2$. If $m_h\gg m_e$, then it is still possible to have for holes $e^2/\epsilon_0 v_h\gg 1$, i.e., they can form a lattice.

Knowledge of m_e, m_h, and E_v-E_c makes it possible in principle to obtain all the characteristics of the crystalline excitonic phases, but the numerical calculations are complicated. We confine ourselves therefore to some simple estimates for metallic excitonium in the limiting case $e^2\epsilon_0 v_e\ll 1$. The most interesting is the question of superconductivity. The most effective electron-interaction mechanism is exchange of exciton–lattice vibration quanta. In view of the Coulomb forces, it is easy to estimate the quantities that enter in the well known BCS-theory formula $T_c\sim\omega(2p_0)\exp(-1/\lambda)$. Namely, it turns out that $\omega(2p_0)\sim(p_0^2/m_e)(\sqrt{e^2 m_e/p_0\epsilon_0})\sqrt{m_e/m_h}$ and $\lambda\sim e^2 m_e/\epsilon_0 p_0$. In the metallic region $\lambda\lesssim 1$. With decreasing overlap, λ is increased both by the decrease of p_0 and by the increase of the effective mass of the electron m_e^* when the role of the interaction with the lattice increases (this tendency can be traced in the limiting case $\lambda\ll 1$, but it is preserved in all probability also at $\lambda\sim 1$). It is clear that although the increase of the band overlap contributes to the onset of the metallic excitonium, the critical temperature is decreased in this case. The maximum attainable critical temperature is obtained at $\lambda\sim 1$. In this case there is only one scale for the electron energies, namely Ry*, and thus

$$T_{c\,max}\sim\frac{m_e e^4}{\epsilon_0^2}\sqrt{\frac{m_e}{m_h}}\;.$$

If $m_e\sim 0.2m_0$, $\epsilon_0\sim 1$–3, and $m_e/m_h\sim 0.01$, then this can amount to several hundred degrees.

Thus, we can count on obtaining a rather high critical temperature. As to the critical field, it will not be so strong as might be assumed on the basis of the proportionality of H_c and T_c. According to the BCS theory $H_c\sim T_c\sqrt{m_e p_0}$. This yields

$$H_c\sim\epsilon_0^{-1/2}p_0^2\,e\sqrt{m_e/m_h}\,\exp(-1/\lambda)\;,$$

at $\lambda\ll 1$ and $H_c\sim(e/r_B^*\epsilon_0^{1/2})\sqrt{m_e/m_h}$ at $\lambda\sim 1$ (r_B^* is the effective Bohr radius). Thus, the critical field increases with decreasing overlap. The maximum value of the field at

$m_0=0.2m_0$, $\epsilon_0\approx 2$, and $m_e/m_h\sim 0.01$ is of the order of 10^4 Oe. It is of interest to determine the behavior of κ. It is known that $\kappa=0.96\,\delta/\xi$ where $\delta=\sqrt{m_e c^2/4\pi n_e e^2}$, $\xi=0.18\,v/T_c$, and n_e is the electron density. This yields $\kappa\sim c/v\sqrt{m_e/m_h}\exp(-1/\lambda)$, i.e., κ decreases with increasing band overlap. As $\lambda\to 1$ we get approximately $\kappa\sim\epsilon_0/\alpha\sqrt{m_e/m_h}$, where α is the fine-structure constant ($\alpha=e^2/\hbar c=1/137$). If $m_e/m_h\sim 0.01$, then $\kappa\sim 10$, i.e., if the overlap is not too large even pure metallic excitonium is a type-II superconductor. The corresponding $H_{c2}\sim\kappa H_c\sim 10^5$ Oe.

The increase of m_e^* on account of the Coulomb interaction with the hole lattice contributes to an increase of the maximum values of T_c, H_c, and κ.

Let us see what bearing these arguments can have on CuCl. According to[4], at temperatures below 200–300K this substance has three phases, two dielectric (0–6 and 6–40 kbar) and one metallic (40–100 kbar). These are room-temperature pressures. The energy spectrum has according to[5,6] a valence-band maximum at the point Γ, with a mass $m_h\sim 20m_0$, and conduction-band minima at the points Γ and X. The width of the direct gap at the point Γ is $\Delta_{\Gamma\Gamma}=3.4$ eV, and the corresponding electron mass is $m_e=0.34m_0$. The width of the indirect gap is $\Delta_{\Gamma X}<0.35$ eV and the corresponding electron mass is more likely smaller than the mass of the first electronic minimum. There are data indicating that the indirect gap decreases with increasing pressure. Thus, the ratio of the electron and hole masses is $m_h/m_e\sim 100$, and the width of the gap (or the overlap) can be regulated by external pressure. It is clear that CuCl is a suitable candidate for searches for the described hypothetical situation.

According to[7,8], a smeared transition into the diamagnetic phase ($\chi=-\alpha/4\pi$, $\alpha\sim 1$) is actually observed in the 100 K region, and the transition temperature increases as the pressure is removed.[1)] This agrees qualitatively with the described regularities of T_c, although no superconductivity was observed directly in the experiment. An interesting feature of the metallic phase is also that it is preserved when the pressure is removed, down to ~ 1 kbar at 200 K. Diamagnetic states arise also in other phases,[1,9], but always only under conditions of rapid cooling or heating.

The following hypothesis can be advanced: The two dielectric phases observed at lower pressures are also condensed excitonic phases. What they really are cannot be ascertained without detailed experiments or calculations, but the possibilities are numerous: various crystalline modifications made up of rotating molecules or nonrotating molecules, an atomic antiferromagnetic dielectric, or finally an atomic or molecular exciton liquid. The energy differences between these phases are most likely small, as is evidenced by the small volume effects observed in the transitions. We note also that the transitions are most probably pure electronic, since the principal crystal lattice is isomorphic in all three phases. We can further assume that under nonequilibrium conditions a metallic phases can appear at low pressures and be preserved in a metastable state. This phase will be superconducting in an appropriate temperature region.

If this assumption is correct, then it follows that it is meaningless to study the details of the appearance of diamagnetic states under nonequilibrium conditions. On the contrary metallic superconducting excitonium must be primarily investigated at high pressures, where it is an equilibrium phases, and attempts should then be made to preserve it as a metastable phase as the pressure is removed. It is possible that at a temperature in the 100 K range this phase can be brought down to zero pressure.

A distinguishing feature of CuCl are the very heavy holes. The reason is that the valence band stems mainly from the *d*-state of the copper atoms. It is natural to expect other substances of the same type to be found among the semiconducting or semimetallic compounds with valence bands made up of *d* or *f* atomic cation states.

In conclusion, I am grateful to A. P. Rusakov for numerous discussions.

[1)]A. P. Rusakov, private communication.

[1]N. B. Brandt, S. V. Kuvshinnikov, A. P. Rusakov, and M. V. Semenov, Pis'ma Zh. Eksp. Teor. Fiz. **27**, 37 (1978) [JETP Lett. **27**, 33 (1978)].
[2]B. A. Volkov and Yu. V. Kopaev, Pis'ma Zh. Eksp. Teor. Fiz. **27**, 10 (1978) [JETP Lett. **27**, 7 (1978)].
[3]B. I. Halperin and T. M. Rice, Rev. Mod. Phys. **40**, 755 (1968).
[4]A. P. Rusakov, S. G. Grigoryan, A. V. Omel'chenko, and A. E. Kadyshevich, Zh. Eksp. Teor. Fiz. **72**, 726 (1977) [Sov. Phys. JETP **45**, 380, (1977)].
[5]I. Ringeissen and S. Nikitine, J. Phys. (Paris) **28**, 53, (1967); S. Nikitine, Proc. Tenth Intern. Conf. on Physics of Semiconductors, Cambridge, Mass., 1970, publ. by US Atomic Energy Commission, Washington, DC (1970), p. 196.
[6]A. P. Rusakov, Phys. Status Solidi B **72**, 503 (1975); A. P. Rusakov, N. V. Fistul', M. A. Il'in, A. A. Abdullaev, and S. G. Grigoryan, Fiz. Tverd. Tela (Leningrad) **18**, 3546 (1976) [Sov. Phys. Solid State **18**, 2067 (1976)].
[7]A. P. Rusakov, V. N. Laukhin, and Yu. A. Lisovskii, Phys. Status Solidi B, **71**, K191 (1975).
[8]C. W. Chu, S. Early, T. H.. Geballe, A. P. Rusakov, and R. E. Schwall, J. Phys. C **8**, L241 (1975).
[9]C. W. Chu, A. P. Rusakov, S. Early, and P. Hambourger, Phys. Rev. 1978 (in press).

Journal of Low Temperature Physics, Vol. 39, Nos. 1/2, 1980

Spin-Glass with a Semiconductor As Host

A. A. Abrikosov

L. D. Landau Institute for Theoretical Physics, Academy of Sciences, Moscow, USSR

(Received October 1, 1979)

The indirect exchange interaction of localized spins in a semiconductor due to conduction electrons is calculated. It is alternating in sign, with an exponentially decreasing amplitude. The oscillation period and the decrement are defined by the band structure of the host. Physical peculiarities are studied. Properties of one-dimensional semiconductor-based spin-glasses are considered. Results are compared with experimental data.

1. INTRODUCTION

It is well known that between magnetic impurities placed in a nonmagnetic metal there is an exchange interaction transferred by conduction electrons. In an isotropic metal this interaction is described by the Ruderman–Kittel–Kasuya–Yosida (RKKY) formula.[1] Taking the scattering of electrons at defects into account, this formula is as follows:

$$\mathscr{H}_{12} = V_0(\mathbf{S}_1\mathbf{S}_2)\cos(2p_0 r)r^{-3}\exp(-r/l) \tag{1}$$

where

$$V_0 = (J/n)^2 p_0 m/(4\pi^3) \tag{2}$$

At sufficiently low temperatures this interaction produces a specific magnetic state called a "spin-glass." In Refs. 2 and 3 the properties of such objects were considered for an electron mean free path l much less than the average distance between the spins $n_m^{1/3}$, where n_m is the volume concentration of spins. Under this assumption the interaction (1) becomes short-ranged. In Ref. 2 it was demonstrated that the preparation of such spin-glasses, although possible in principle, is actually very complicated in practice.

Recently Tholence,[4] analyzing his data on the temperature dependence of the magnetic susceptibility of $\mathrm{Eu}_x\mathrm{Sr}_{1-x}\mathrm{S}$,[5] found that at $x \leqslant 3\%$ the dependence $\chi(T)$ in a rather wide temperature interval is described by the relation $\chi(T) \propto \ln(T_0/T)$, which was predicted in Ref. 2. The dependence of

0022-2291/80/0400-0217$03.00/0

T_0 on concentration also follows the theoretical prediction. This causes some surprise, since SrS is not a metal, but a semiconductor.

In this paper it will be shown that in a semiconductor an exchange interaction of spins appears which is close to (1) in may respects. The physical peculiarities of such objects are considered; we call them "semiconductor-based spin-glasses."

2. EXCHANGE INTERACTION

As in a metal, we start from an exchange interaction of electrons with spins of magnetic impurity atoms:

$$\mathscr{H}_{ei} = -(J/n)\sum_i \psi^+(\mathbf{R}_i)\boldsymbol{\sigma}\mathbf{S}_i\psi(\mathbf{R}_i) \tag{3}$$

It produces an interaction between impurity spins

$$\mathscr{H}_{12} = 2\left(\frac{J}{n}\right)^2 \mathbf{S}_1\mathbf{S}_2 \int \exp(i\mathbf{kr})\frac{d^3\mathbf{k}}{(2\pi)^3} \times\left[-i\int G(\omega,\mathbf{p})G(\omega,\mathbf{p}-\mathbf{k})\frac{d\omega\, d^3\mathbf{p}}{(2\pi)^3}\right] \tag{4}$$

where $G(\omega, \mathbf{p})$ is the electron Green's function.

We choose the following energy spectrum: one maximum in the valence band at $\mathbf{K}=0$ and several equivalent minima in the conduction band at $\mathbf{K}=\mathbf{K}_i$. The electrons and holes we assume isotropic, with masses m_e and m_h, respectively. Such a model spectrum is sufficiently general to describe the principally important properties of the interaction and at the same time permits a generalization to other cases. Given such a spectrum, the Green's function is

$$G(\omega,\mathbf{p}) = \left(\omega+\frac{p^2}{2m_h}-i\delta\right)^{-1}+\sum_i\left[\omega-\Delta-\frac{(\mathbf{p}-\mathbf{K}_i)^2}{2m_e}+i\delta\right]^{-1} \tag{5}$$

(we take as origin the top of the valence band).

Substituting $G(\omega, \mathbf{p})$ into the square brackets of (4), we obtain

$$-\sum_i\int\frac{d^3\mathbf{p}}{(2\pi)^3}\left\{\left[\Delta+\frac{(\mathbf{p}-\mathbf{K}_i)^2}{2m_e}+\frac{(\mathbf{p}-\mathbf{k})^2}{2m_h}\right]^{-1}+\left[\Delta+\frac{(\mathbf{p}-\mathbf{K}_i-\mathbf{k})^2}{2m_e}+\frac{p^2}{2m_h}\right]^{-1}\right\}$$

The $\mathbf{p}$ integral is formally divergent. It actually should be limited by the Brillouin zone boundaries, but since we are interested only in its singular part, we subtract from the integrand $4m^*/p^2$ with $m^*=(m_e^{-1}+m_h^{-1})^{-1}$. Then the integral converges. Performing it, we obtain the singular part of the square brackets in (4):

$$(2\pi)^{-1}\sum_i (m^*)^{3/2}(m_e+m_h)^{-1/2}\{[(\mathbf{k}+\mathbf{K}_i)^2+2\Delta(m_e+m_h)]^{1/2}$$
$$+[(\mathbf{k}-\mathbf{K}_i)^2+2\Delta(m_e+m_h)]^{1/2}\}$$

Substituting in (4), we get

$$\mathcal{H}_{12}=4(J/n)^2\mathbf{S}_1\mathbf{S}_2\sum_i \cos(\mathbf{K}_i\mathbf{r})(m^*)^{3/2}(m_e+m_h)^{-1/2}$$
$$\times\int[\exp(i\mathbf{qr})][q^2+2\Delta(m_e+m_h)]^{1/2}\,d^3\mathbf{q}/(2\pi)^3$$

Separating the long-range part of the latter integral, we obtain finally

$$\mathcal{H}_{12}=-\frac{\pi^{-5/2}(\alpha m^*)^{3/2}}{[2(m_e+m_h)]^{1/2}}\left(\frac{J}{n}\right)^2\mathbf{S}_1\mathbf{S}_2\,e^{-\alpha r}r^{-5/2}\sum_i\cos\mathbf{K}_i\mathbf{r} \tag{6}$$

where

$$\alpha=[2\Delta(m_e+m_h)]^{1/2} \tag{7}$$

It should be mentioned that the important $|\mathbf{q}|$ in the integral are of the order of α. This means that the important $\mathbf{p}$ in both the valence and conduction bands are within a distance of the order of α from the corresponding extrema. This justifies the use of a quadratic spectrum if α is less than the reciprocal lattice period.

The interaction (6) reminds one of the RKKY formula (1). The role of l is taken over by α^{-1}, i.e., the interaction range is defined by the energy gap and the effective masses. The oscillating factor is not as simple as in (1). However, if $\mathbf{K}_i\neq 0$, it also produces an interaction of alternating sign at distances of the order of the interatomic spacing. The formula (6) is applicable also in the case $\mathbf{K}_i=0$, i.e., for a direct gap. In this case $\sum_i\cos\mathbf{K}_i\mathbf{r}=1$, i.e., the interaction of spins in such a semiconductor is ferromagnetic. The theory of such a type of random ferromagnet was constructed by Korenblit and Shender.[6]

An interaction of spins very close to (6) was obtained many years ago in Ref. 7. The authors, however, considered a different model: a wide gap and a narrow valence band. In a real insulator most probably an interaction of spins should be expected decaying at distances on the interatomic order, but the derivation in Ref. 7 cannot be taken seriously. The oscillations are due to the integration range of the momentum in the valence band, which is taken as the Brillouin zone, approximated by a sphere (and at the same time a quadratic spectrum is assumed!). For obscure reasons the integration in the conduction band is performed over the whole momentum space. In any case, such a short-ranged interaction is of no interest for the spin-glass problem.

3. SOME PECULIARITIES

In spite of the similarity of the interaction (6) with the RKKY result (1), there are some differences. First, the interaction range, although larger than the interatomic spacing in semiconductors with a narrow gap and small effective masses, can be of the same order in other substances. In the theory of Ref. 2 the assumption $l \gg p_0^{-1}$ was used. Hence it becomes necessary to reconsider whether the theory remains applicable for $l \sim p_0^{-1}$ or $\alpha \sim |\mathbf{K}_i|$. The other point is that the oscillating factor differs from the $\cos(2p_0 r)$ in (1). But as we know from Ref. 2, at low temperatures the details of the behavior of this factor become very important, particularly close to its zeros. Therefore this question also should be studied. We shall begin with it.

According to Ref. 2, in all cases we averaged over coordinates of spins in two steps: first we averaged over distances larger than the interatomic distance but smaller than l, and after that we averaged over larger distances. With that aim we introduced the "density" of values of $\cos(2p_0 r)$. Now we calculate a similar quantity for the sum of cosines in (6),

$$\rho(q) = \lim_{V\to\infty} \frac{1}{V} \int d^3\mathbf{r}\, \delta\Big[\sum_i \cos(\mathbf{K}_i \mathbf{r}) - q\Big] \tag{8}$$

As examples we consider two particular cases: a cubic crystal with the electronic minima placed (a) at the fourth-order axes (this corresponds to Si) and (b) at the third-order axes (as in Ge). In case (a) we obtain

$$\begin{aligned}\rho(q) &= (2\pi)^{-3} \int_0^{2\pi} d\varphi_1 \int_0^{2\pi} d\varphi_2 \int_0^{2\pi} d\varphi_3 \\ &\quad \times \delta[q - \eta(\cos\varphi_1 + \cos\varphi_2 + \cos\varphi_3)] \\ &= (2\pi)^{-4} \int_{-\infty}^{\infty} e^{iqs} \left[\int_0^{2\pi} d\varphi\, e^{-is\cos\varphi}\right]^3 ds \\ &= (1/\pi) \int_0^{\infty} (\cos qs)\, J_0^3(\eta s)\, ds \end{aligned} \tag{9}$$

where $\eta = 1$ if the minima are placed at the Brillouin zone boundaries and $\eta = 2$ for arbitrary points at C_4.

The integral (9) converges with any q and remains finite at $q = 0$. This is seen at once if $\rho(q)$ is represented in the form

$$\rho(q) = \frac{1}{\eta\pi^3}\int_0^1 \frac{ds_1}{(1-s_1^2)^{1/2}}\int_0^1 \frac{ds_2}{(1-s_2^2)^{1/2}}\int_0^1 \frac{ds_3}{(1-s_3^2)^{1/2}}$$
$$\times\left[\delta\left(\frac{q}{\eta}-s_1-s_2-s_3\right)+3\delta\left(\frac{q}{\eta}-s_1-s_2+s_3\right)\right.$$
$$\left.+3\delta\left(\frac{q}{\eta}+s_1+s_2-s_3\right)+\delta\left(\frac{q}{\eta}+s_1+s_2+s_3\right)\right] \quad (10)$$

Putting $q \to 0$, we get

$$\rho(0) = \frac{6}{\eta\pi^3}\int_0^1 \frac{ds_1}{(1-s_1^2)^{1/2}}\int_0^{1-s_1} \frac{ds_2}{(1-s_2^2)^{1/2}[1-(s_1+s_2)^2]^{1/2}} = \frac{6}{\eta\pi^3} \quad (11)$$

Now consider case (b). The components of $\mathbf{K}_i$ are $(\pm K_0, \pm K_0, \pm K_0)$. Scaling the coordinates as $1/K_0$, we obtain

$$\sum_i \cos(\mathbf{K}_i\mathbf{r}) = \eta[\cos(x+y+z)+\cos(x-y+z)$$
$$+\cos(-x+y+z)+\cos(-x-y+z)]$$
$$= 4\eta\cos x\cos y\cos z$$

Integrating, we get

$$\rho(q) = \frac{4}{\pi^3}\int_0^1 \frac{ds_1}{(1-s_1^2)^{1/2}}\int_0^1 \frac{ds_2}{(1-s_2^2)^{1/2}}\int_0^1 \frac{ds_3}{(1-s_3^2)^{1/2}}$$
$$\times[\delta(q-4\eta s_1s_2s_3)+\delta(q+4\eta s_1s_2s_3)]$$
$$= \frac{4}{\pi^3}\int_0^1 \frac{ds_1}{(1-s_1^2)^{1/2}}\int_0^1 \frac{ds_2\,\theta(4\eta s_1s_2-|q|)}{(1-s_2^2)^{1/2}[(4\eta s_1s_2)^2-q^2]^{1/2}}$$
$$= \frac{1}{\pi^3\eta}\int_0^{4\eta/|q|} \frac{du}{(u^2-1)^{1/2}}K\left\{\left[1-\left(\frac{qu}{4\eta}\right)^2\right]^{1/2}\right\} \quad (12)$$

where K is the elliptic integral. At $q \to 0$ the integral becomes logarithmic in the range $1 \ll u \ll 4\eta/|q|$. The calculation gives

$$\rho(|q| \ll 1) = \frac{1}{2\pi^3\eta}\left(\ln^2\frac{32\eta}{|q|}-\frac{\pi^2}{4}\right) \quad (13)$$

For finite $|q| < 4\eta$, $\rho(q)$ is finite and for $|q| > 4\eta$, $\rho(q) = 0$.

According to Ref. 2, at low temperatures $T \ll T_0$ the specific heat is proportional to $\rho(q)$ taken at $|q| \sim q(T)$, corresponding to an interaction of the order of T at a distance $r_0 = (4\pi n_m l)^{-1/2}$ or $r_0 = (4\pi n_m \alpha^{-1})^{1/2}$. The magnetic susceptibility is proportional to $\int_{q(T)}^1 dq\, q^{-1}\rho(q)$. If ρ tends to a finite limit at $q \to 0$, then the temperature dependences are the same as in Ref. 2. However, in the case that $\rho(q)$ at $q \to 0$ tends to zero or infinity, these dependences will be entirely different. Formula (13) can serve as an example; it gives the following laws:

$$C \propto T \ln^2 (T_0/T), \qquad \chi \propto \ln^3 (T_0/T) \tag{14}$$

One sees how important it is to know at least qualitatively the band structure of the semiconductor in order to explain the physical properties of the spin-glass formed in it.

Now we shall study the validity of the theory in the case $\alpha \sim |\mathbf{K}_i|$. For definiteness we assume an interaction of the simplified form

$$\mathcal{H}_{12} = V_0 \mathbf{S}_1 \mathbf{S}_2 (\cos Kr)\, e^{-\alpha r} r^{-5/2} \tag{15}$$

and calculate the specific heat. According to Ref. 2, we get

$$C = n_m \int 4\pi n_m r^2 \exp\left(-\frac{4\pi}{3} n_m r^3\right) \times \left\{ \varphi\left[\frac{V_0 S |\cos Kr|\, e^{-\alpha r}}{2Tr^{5/2}}\right] - \varphi\left[\frac{V_0 S(2S+1)|\cos Kr|\, e^{-\alpha r}}{2Tr^{5/2}}\right]\right\} \tag{16}$$

where $\varphi(x) = x^2/\text{sh}^2\, \chi$. At low temperatures the vicinities of the zeros of $\cos Kr$ are important. Define

$$r = (\pi/2K)(2j+1) + \delta r \tag{17}$$

j being an integer. Then $|\cos Kr| = \sin (K\,\delta r) \approx K\,\delta r$ if $K\,\delta r \ll 1$. With this assumption we get

$$C = n_m \sum_j 4\pi n_m r_j^2 \exp\left(-\frac{4\pi}{3} n_m r_j^3\right) \int d(\delta r) \times \left\{ \varphi\left[\frac{V_0 SK\,\delta r\, e^{-\alpha r_j}}{2Tr_j^{5/2}}\right] - \varphi\left[\frac{V_0 S(2S+1)K\,\delta r\, e^{-\alpha r_j}}{2Tr_j^{5/2}}\right]\right\}$$

where $r_j = (\pi/2K)(2j+1)$. The argument of the functions in the brackets should be of the order of 1. Our assumptions are justified if this is compatible with $K\,\delta r \ll 1$, i.e., if

$$V_0 S^2 e^{-\alpha r_j}/Tr_j^{5/2} \gg 1 \tag{18}$$

With this we perform the δr integration and obtain

$$C = \frac{4\pi^2}{3K} \frac{T}{V_0} \frac{1}{2S+1} n_m \sum_j 4\pi n_m r_j^{9/2} \exp\left(-\frac{4\pi}{3} n_m r_k^3 + \alpha r_j\right)$$

We make the further assumption that only r_j with large j are important. This permits us to pass from summation to integration

$$\sum_j = \frac{K}{\pi}\int dr_j$$

The dr_j integral is similar to the dr integral in Ref. 2. The steepest descent method is used and the result is

$$C = \frac{\pi^{1/2}}{3}\frac{T}{V_0}\frac{(4\pi n_m)^{-1/2}}{2S+1}\alpha^2 \exp\left[\frac{2}{3}\left(\frac{4\pi n_m}{\alpha^3}\right)^{-1/2}\right] \tag{19}$$

The saddle point is $r_0 = (4\pi n_m \alpha^{-1})^{-1/2}$. The integration range in r_j is $|r_j - r_0| \sim (4\pi n_m \alpha)^{-1/4}$. The interval between adjacent r_j is $K(4\pi n_m\alpha)^{-1/4}$. If $\alpha \lessdot K$, $K \gg n_m^{1/3}$, this is large, and this justifies the transformation from the sum to an integral by dr_j.

Now we can prove condition (18). Substituting $r_j \sim r_0$, we obtain

$$T \ll V_0 S^2 r_0^{-5/2} e^{-\alpha r_0}$$

A more precise inequality is obtained from a comparison of the results for different temperature regions, as in Ref. 2. The result is $T \ll T_0$, where

$$T_0 = V_0 S \alpha^{5/2}(4\pi n_m/\alpha^3)^{3/2} \exp[-(4\pi n_m/\alpha^3)^{-1/2}] \tag{20}$$

We also present the expression for the magnetic susceptibility for the interaction (15) at $T \ll T_0$:

$$\chi = \frac{1}{3\pi^{3/2}}\frac{\mu^2\alpha^{1/2}}{V_0}\left(\frac{4\pi n_m}{\alpha^3}\right)^{-1/2} \exp\left[\frac{2}{3}\left(\frac{4\pi n_m}{\alpha^3}\right)^{-1/2}\right]\ln\frac{T_0}{T} \tag{21}$$

4. "COMMENSURATE" CASE

A special case can arise if every magnetic impurity occupies a definite position in the crystal cell although the choice of cells remains random. For the metal-based spin-glass this case does not differ from the general one, since π/p_0 is incommensurate with the lattice periods and hence the δr in (17) can have arbitrary values with equal probability. For the semiconductor-based spin-glass the situation is similar if the minima of the conduction band are at some arbitrary $\mathbf{K}_i$. If, however, $\mathbf{K}_i$ are equal to rational parts of the reciprocal lattice periods, then the properties can be drastically changed.

We consider what happens in the models for which we calculated $\rho(q)$ in the previous section. We assume a lattice with the zincblende structure and a substitutional alloy, i.e., the magnetic impurity atoms can be located at sites of a fcc lattice. There are the following possible $\mathbf{r}$:

$$(n_1 a, n_2 a, n_3 a); \qquad ((n_1+\tfrac{1}{2})a, (n_2+\tfrac{1}{2})a, n_3 a)$$

$$(n_1 a, (n_2+\tfrac{1}{2})a, (n_3+\tfrac{1}{2})a); \qquad ((n_1+\tfrac{1}{2})a, n_2 a, (n_3+\tfrac{1}{2})a)$$

(a is the edge of the basic cube). For $\mathbf{K}_l$ we consider the most typical case when the conduction band minima are located at the Brillouin zone boundaries. In this case the $\mathbf{K}_l$ are one-half of the corresponding reciprocal lattice periods.

For the first model we consider the case when

$$\mathbf{K}_1 = (2\pi/a, 0, 0); \qquad \mathbf{K}_2 = (0, 2\pi/a, 0); \qquad \mathbf{K}_3 = (0, 0, 2\pi/a)$$

Substituting $\mathbf{r}$ of the first type, we get $\sum_i \cos(\mathbf{K}_i\mathbf{r}) = 3$. For the other three types of $\mathbf{r}$, $\sum_i \cos(\mathbf{K}_i\mathbf{r}) = -1$. The distribution function $\rho(q)$ is

$$\rho(q) = \tfrac{1}{4}\delta(q-3) + \tfrac{3}{4}\delta(q+1) \tag{22}$$

For the second model the $\mathbf{K}_i$ are

$$\mathbf{K}_1 = (\pi/a, \pi/a, \pi/a); \qquad \mathbf{K}_2 = (-\pi/a, \pi/a, \pi/a)$$

$$\mathbf{K}_3 = (\pi/a, -\pi/a, \pi/a); \qquad \mathbf{K}_4 = (-\pi/a, -\pi/a, \pi/a)$$

Substituting $\mathbf{r} = (n_1 a, n_2 a, n_3 a)$, we obtain $\sum_i \cos(\mathbf{K}_i\mathbf{r}) = \pm 4$ according to the parity of the n's. For all the other r, $\sum_i \cos(\mathbf{K}_i\mathbf{r}) = 0$. This is a very peculiar situation: there are four quasiindependent spin subsystems; in every system we have an alternating exchange interaction; the interaction between subsystems is achieved by weaker forces of magnetic dipole origin. The ρ function for every subsystem has the form

$$\rho(q) = \tfrac{1}{2}\delta(q-4) + \tfrac{1}{2}\delta(q+4) \tag{23}$$

We now consider the principal physical characteristics of these cases.

The main difference from the incommensurate glasses is the absence of small q. As we know from Ref. 2, of greatest importance are pairs of spins with interaction energy of the order of T. As $T > T_0$ this corresponds to the choice of interspin distance $r(T)$ and at $T < T_0$ in incommensurate glasses to the choice of small q. In commensurate glasses even in the latter case $r(T)$ has to be chosen. But $T = T_0$ corresponds to distances $r(T) \sim r_0 = (4\pi n_m \alpha^{-1})^{-1/2}$. This means that distances larger than r_0 become important at $T < T_0$. As already mentioned in Refs. 2 and 3, at such distances a given spin interacts not with one but with several closest neighbors which manage to get into a layer of thickness α^{-1}. It was mentioned also that these spins have different orientations, since they belong to different domains of the infinite cluster. Hence a commensurate short-range glass at $T \ll T_0$ is similar to the ordinary long-range spin-glass which appears in a metal at $n_m^{-1/2} \ll l$.

In the range $\theta \gg T \gg T_0$ formulas of the type (29) and (30) of Ref. 2 are valid, namely

$$C = n_m 4\pi n_m \alpha^{-3} \ln^2 (V_0 S n_m^{5/6}/T) \ln (2S+1) \times \exp [-\tfrac{4}{3}\pi n_m \alpha^{-3} \ln^3 (V_0 S n_m^{5/6}/T)] \tag{24}$$

$$\chi = [n_m \mu^2 S(S+1)/3T] \exp [-\tfrac{4}{3}\pi n_m \alpha^{-3} \ln^3 (V_0 S n_m^{5/6}/T)] \tag{25}$$

The main exponential factor remains the same at $T < T_0$ since it describes the probability of the absence of a neighboring spin at distances less than $r(T) = \alpha^{-1} \ln (V_0 S n_m^{5/5}/T)$. Assuming that the formulas for C and χ keep their main features even for $T < T_0$ [with corrections of the order of $n_m \alpha^{-3} \ln^2 (V_0 S n_m^{5/6}/T)$ in the exponent and preexponential factor, see Ref. 6], it is possible to judge qualitatively the temperature dependence of these quantities. Differentiating (21) with respect to T, we see that at $T < T_0$ [where $r(T) > r_0$] the change with temperature of the exponential factor dominates over the changes of the factor T^{-1}, i.e., in this temperature range $\chi(T)$ decreases with decreasing temperature. Hence the curve $\chi(T)$ must have two maxima: one at $T = \theta$ and the other at $T \sim T_0$. This behavior distinguishes very clearly the commensurate case from the incommensurate one, where, according to Ref. 2, there exists only one maximum at θ and a minimum at a slightly lower temperature and then χ increases with decreasing temperature.

We assumed, as in Ref. 2, that the dipole interaction at distances $r(T)$ is still small compared with the exchange interaction. The condition for that is

$$T \gg \mu^2 \alpha^3 \ln^{-3} (V_0 S/\mu^2 n_m^{1/6}) \tag{26}$$

5. ONE-DIMENSIONAL CASE

In Ref. 8 we considered a one-dimensional spin-glass with short-range interaction. It is known, however, that most quasi-one-dimensional compounds transform at low temperatures to nonmetallic phases. Therefore a one-dimensional semiconductor-based spin-glass is a realistic object for investigation.

We shall start from (3), where in the present case n is the number of atoms per unit length of the chain and $\mathbf{R}_i \to z_i$. The derivation follows Section 2 with the substitution of three-dimensional vectors by one-dimensional quantities. Inserting the G function similar to (5) into the analog of (4), we obtain a momentum integral

$$-\eta \int \frac{dp}{2\pi} \left\{ \left[\Delta + \frac{(p-K)^2}{2m_e} + \frac{(p-k)^2}{2m_h} \right]^{-1} + \left[\Delta + \frac{(p-K-k)^2}{2m_e} + \frac{p^2}{2m_h} \right]^{-1} \right\}$$

Here we took into account that in the conduction band there is one minimum at $p = K$ if $K = 0$ or π/a, and two equivalent minima otherwise. In the last case the result is doubled. This is reflected by the factor η.

The p integral is convergent. Taking it, we get

$$\eta(m_e m_h)^{1/2}\{[(k-K)^2+2\Delta(m_e+m_h)]^{-1/2} + [(k+K)^2+2\Delta(m_e+m_h)]^{-1/2}\}$$

Substituting into (4), we get

$$\mathscr{H}_{12} = -4\eta(J/n)^2\mathbf{S}_1\mathbf{S}_2(\cos Kz)(m_e m_h)^{1/2} \times \int e^{iqz}[q^2+2\Delta(m_e+m_h)]^{-1/2}\, dq/2\pi$$

The q integral is a Hankel function $K_0(\alpha z)$, where $\alpha = [2\Delta(m_e+m_h)]^{1/2}$. Since we are interested only in large z, we shall use the asymptotic form

$$\mathscr{H}_{12} = -[4\eta/(2\pi)^{1/2}](J/n)^2\mathbf{S}_1\mathbf{S}_2(m_e m_h)^{1/2}(\cos Kz)\, e^{-\alpha z}/(\alpha z)^{1/2} \quad (27)$$

The result is qualitatively the same as in Ref. 8 with the difference that the preexponential factor contains $z^{-1/2}$ instead of z^{-1}.

As was demonstrated in Ref. 8, in the one-dimensional case small values of $\cos Kz$ play no special role at any temperature and therefore the commensurate one-dimensional spin-glass does not differ qualitatively from the incommensurate glass.

At $K = 0$ the interaction (27) is ferromagnetic. Hence there is another type of random one-dimensional magnet which differs both from a one-dimensional spin-glass and from a one-dimensional random antiferromagnet. Its most interesting property is of course the susceptibility. We use the cluster approach described in Ref. 8.

Since the spins in every cluster are parallel, the total spin of the cluster is equal to nS, where n is the number of spins entering the cluster. The magnetic susceptibility is [we use formulas (22) and (23) from Ref. 8]

$$\chi = \frac{n_m\mu^2}{3T}\sum_{n=1}^{\infty} a(n)Sn(Sn+1) = \frac{n_m\mu^2}{3T}\left[S^2\left(\frac{2}{\lambda}-1\right)+S\right] \quad (28)$$

where $a(n)$ is the cluster distribution function, $\lambda = \exp[-n_m z(T)]$, and $z(T)$ is the thermal interaction radius. In the present case

$$z(T) = \alpha^{-1}\ln(V_0 n_m^{1/2}/T) \quad (29)$$

and

$$V_0 = 4\eta(2\pi)^{-1/2}(J/n)^2(m_e m_h)^{1/2}S^2$$

Substituting λ with this value of $z(T)$ into (28), we obtain χ. At low temperatures $[z(T) \gg n_m^{-1}]$

$$\chi = (n_m \mu^2/3T)[2S^2(V_0 n_m^{1/2}/T)^{n_m/\alpha} - S^2 + S] \tag{30}$$

At high temperatures $[z(T) \ll n_m^{-1}]$, $\lambda \to 1$. Therefore

$$\begin{aligned}\chi &\approx (n_m\mu^2/3T)[S(S+1) + 2S^2(1-\lambda)] \\ &= (n_m\mu^2/3T)[S(S+1) + 2S^2 n_m \alpha^{-1} \ln (V_0 n_m^{1/2}/T)]\end{aligned} \tag{31}$$

Now we find the magnetic moment at low temperatures ($\lambda \ll 1$) in a finite field h limited by the condition $\mu hS \ll T$. Such a field polarizes the clusters but does not change their composition. Using formula (41) from Ref. 8 with $B(n, m) = \delta_{n,m}$, we get

$$M = n_m\mu \sum_{n=1}^{\infty} a(n)\left\{\left(Sn + \frac{1}{2}\right) \text{cth}\left[\frac{(Sn + \frac{1}{2})\mu h}{T}\right] - \frac{1}{2}\text{cth}\left(\frac{\frac{1}{2}\mu h}{T}\right)\right\} \tag{32}$$

Taking into account that the main role is played by large n, using the asymptotic form $a(n) = \lambda^2 e^{-\lambda h}$, and remembering that $\mu hS/T \ll 1$, we find

$$M = n_m \mu S \gamma^{-2} \int_0^{\infty} dy \, (y \text{ cth } y - 1) \, e^{-\gamma y} \tag{33}$$

where

$$\gamma = \mu hS/T\lambda \tag{34}$$

The asymptotic forms are

$$\begin{aligned}M &\approx n_m \mu S\left(\frac{2}{3}\gamma - \frac{8}{15}\gamma^3\right) \\ &= \frac{2n_m\mu^2 S^2 h}{3T\lambda}\left(1 - \frac{4}{5}\gamma^2\right), \qquad \gamma \ll 1\end{aligned} \tag{35}$$

$$M \approx n_m \mu S(1 - \gamma^{-1}), \qquad \gamma \gg 1 \tag{36}$$

Formula (35) coincides with (28) in the limit $\lambda \ll 1$. The latter condition is taken as granted, since otherwise we would not be able to assume $n \gg 1$.

From (36) and (28) it follows that the ferromagnetic transition takes place only at $T = 0$. Therefore such one-dimensional magnets can be called ferromagnets only conventionally denoting the type of spin interaction.

6. COMPARISON WITH EXPERIMENT

The dependence $\chi \propto \ln (T_0/T)$ observed in $Eu_xSr_{1-x}S$ alloys makes it plausible that this compound is an incommensurate spin-glass with $\rho(0) =$ const. From experimental data it is possible to define the concentration dependence of T_0. According to formula (20), we get

$$\ln T_0 = \ln (V_0 S\alpha^{5/2}) + \tfrac{3}{2}\ln (4\pi n_m \alpha^{-3}) - (4\pi n_m \alpha^{-3})^{-1/2} \qquad (37)$$

The experimental data can be indeed presented in the form*

$$\ln T_0 = a + b \ln y - y^{-1/2} \qquad (38)$$

where $y = dx$, with x the atomic concentration, and a, b, and d are constant coefficients; we obtain the following values of the constants:

$$a = -1.057, \qquad b = 0.345, \qquad d = 100$$

Of course the accuracy of the preexponential factor in T_0 [i.e., the terms with a and b in (38)] is poor and therefore we shall discuss only the last term. Comparison of (37) and (38) shows that $y = 4\pi n_m \alpha^{-3}$ and hence the constant d is

$$d = 4\pi n \alpha^{-3} \qquad (39)$$

where n is the number of Sr or S atoms per unit volume. The measurements were performed at concentrations $x = 0.0085$, 0.017, and 0.03. The theory applies at $y \ll 1$. Since $d = 100$ this condition was in fact not fulfilled. Nevertheless the dependence $\ln T_0 = A - Bx^{-1/2}$ can be clearly traced on the experimental graph and this permits us to define the constant α from the result $d = 100$. Inserting $n \sim 10^{22}$ in (39), we obtain $\alpha \sim 10^7\ \text{cm}^{-1}$. Comparing this result with formula (7), we see that it agrees very well with our ideas, since it corresponds to $\Delta \sim 1$ eV and m_e, $m_h \sim 0.1 m_0$. For a more detailed comparison, data for smaller concentrations of Eu are necessary, together with a precise knowledge of the band structure of SrS.

It should be mentioned that formula (21) predicts a decrease of the steepness $\chi/\ln (T_0/T)$ with increasing concentration, contrary to the experimental trend. This can be a consequence of $y \sim 1$, since in the theory terms of higher order in y were neglected. The logarithmic law itself is probably less sensitive to the violation of the condition $y \ll 1$.

Experimentally there is a deviation from the law $\chi \propto \ln (T_0/T)$ with a trend to saturation at the lowest temperatures. If $y \sim 1$, this is hardly due to dipole forces and should be probably attributed to the "effective field" of more distant neighbors. The temperature at which the saturation begins should rise with concentration on the average, but a wide scatter for different samples with the same concentration should be expected. This agrees qualitatively with experimental data.

*This has kindly been done by J. L. Tholence at our request

ACKNOWLEDGMENT

I would like to express my gratitude to Dr. J. L. Tholence for directing my attention to the problem and for a very valuable discussion of various spin-glass properties.

REFERENCES

1. M. A. Ruderman and C. Kittel, *Phys. Rev.* **96**, 99 (1954); T. Kasuya, *Prog. Theor. Phys.* **16**, 45 (1956); K. Yosida, *Phys. Rev.* **106**, 863 (1957).
2. A. A. Abrikosov and S. I. Moukhin, *J. Low Temp. Phys.* **33**, 207 (1978).
3. A. A. Abrikosov, *J. Low Temp. Phys.* **33**, 505 (1978).
4. J. L. Tholence, private communication.
5. J. L. Tholence, F. Holtzberg, H. Godfrin, H. V. Löhneysen, and R. Tournier, *J. Phys.* (*Paris*) **39**, C6-928 (1978).
6. I. Ya. Korenblit and E. F. Shender, *Usp. Fiz. Nauk* **126**, 233 (1978).
7. N. Bloembergen and T. J. Rowland, *Phys. Rev.* **97**, 1679 (1955).
8. A. A. Abrikosov, *J. Low. Temp. Phys.* **37**, 111 (1979).

VOLUME 73, NUMBER 24 PHYSICAL REVIEW LETTERS 12 DECEMBER 1994

Observation of an "Extended" Van Hove Singularity in $YBa_2Cu_4O_8$ by Ultrahigh Energy Resolution Angle-Resolved Photoemission

K. Gofron,[1,2] J. C. Campuzano,[1,2] A. A. Abrikosov,[1] M. Lindroos,[3,4] A. Bansil,[3] H. Ding,[1,2] D. Koelling,[1] and B. Dabrowski[1,5]

[1]*Argonne National Laboratory, 9700 S. Cass Ave, Argonne, Illinois 60439*
[2]*Department of Physics, University of Illinois at Chicago, Chicago, Illinois 60607*
[3]*Physics Department, Northeastern University, Boston, Massachusetts 02115*
[4]*Physics Department, Tampere University of Technology, Tampere, Finland*
[5]*Physics Department, Northern Illinois University, DeKalb, Illinois 60115*

(Received 4 March 1994)

We characterize the nature of the Van Hove singularity near the Fermi energy (E_F) in $YBa_2Cu_4O_8$ using ultrahigh energy resolution (10 meV) angle-resolved photoemission spectroscopy, together with corresponding first-principles spectra intensity computations. The singularity, related to CuO_2 planes, is located about 19 meV below E_F at the Y point in the Brillouin zone and arises from a band which is nearly dispersionless along the Γ to Y direction. Such an "extended" saddle point would render the band structure quasi-1D and yield a power law divergence in the density of states, allowing the occurrence of T_c's of order 100 K even in a weak coupling BCS scheme.

PACS numbers: 74.72.Bk, 74.25.Jb, 79.60.Bm

Ever since the discovery of high-T_c superconductivity, one of the most intensely debated questions has been: What is the reason for the critical temperature in the new materials to be much higher than in the earlier superconductors such as Hg, Pb, Nb, and the A15's? Many mechanisms have been proposed to explain these high transition temperatures, some exotic and others of a more conventional nature [1]. Of interest to this Letter is the so-called Van Hove scenario [2–9], in which the occurrence of high-T_c's is tied to the presence of Van Hove singularities in the density of states (DOS) close to the Fermi energy. So far, there has only been indirect evidence for the existence of such peaks in the DOS, for example, from the discontinuity of the specific heat at T_c [6], studies of the isotope effect [7], of the thermopower [8], and of the quasiparticle lifetime broadening [9].

In this Letter, for the first time, we give direct evidence for the presence of a Van Hove singularity in $YBa_2Cu_4O_8$ near the Fermi energy. The nature of the singularity is characterized by determining its location in the Brillouin zone and the dispersion of the associated spectral density peak by ultrahigh resolution angle-resolved photoemission spectroscopy (ARPES). Here we show data at an energy resolution better than 10 meV. Most of the previously published synchrotron-based work has involved resolutions of 25–30 meV. Since the Van Hove singularity lies very close to the Fermi energy, a high energy resolution is essential for these measurements. The experimental spectra are compared with corresponding first-principles ARPES intensity calculations, and it is clearly established that the measured spectral feature arises from bulk energy bands of the crystal related to the CuO_2 planes. The conclusions are therefore generally relevant to the cuprate superconductors, particularly since we have found similar behavior in $YBa_2Cu_3O_{7-\delta}$, $Bi_2CaSr_2Cu_2O_8$, and $Tl_2Ba_2CuO_6$ [10].

The common Van Hove scenario assumes that the diverging DOS arises from a saddle point in the underlying 2D CuO_2 plane bands (i.e., a minimum along one direction and a maximum along the perpendicular direction). Our results provide a new twist to this picture in that the bands are measured to be nearly flat (within 1 meV) in one direction, so that the situation may be described as a family of simple saddle points (critical line), or an *extended saddle point,* with profound consequences for the normal and superconducting properties of the material.

Van Hove [11] showed that in a two-dimensional system a simple saddle point yields a logarithmic divergence in the DOS. However, for an extended saddle point, the DOS singularity would diverge as the much stronger $(E - E_c)^{-1/2}$, where E_c is the critical point energy [12,13]. It is then straightforward to show that when the distance between the Fermi energy and the saddle point energy is less than T_c, the phonon energy scale is replaced by an electronic energy scale in a weak coupling BCS scheme. In this case, the transition temperature scales as $T_c \sim w\lambda^2$ [12], where w denotes the band width and λ the usual coupling constant. For a reasonable value of w of order 1 eV, a coupling constant of 0.1 would already yield $T_c \sim 100$ K. Thus any coupling mechanism, even phonons, could easily explain the occurrence of high T_c's.

The specific system investigated in this study is $YBa_2Cu_4O_8$ (Y124), which is closely related to the more familiar $YBa_2Cu_3O_{6.9}$ (Y123). The crystal structure of Y124 is the same as that of Y123, except that the former contains two CuO_2 chains per unit cell. The Brillouin zones (BZ) in the two cases are also similar, with the Γ-Y line being parallel to the direction of the CuO chains.

 0031-9007/94/73(24)/3302(4)$06.00

The symmetry points at the zone boundary are $X(\pi, 0)$, $Y(0, \pi)$, and $S(\pi, \pi)$. The crystals of Y124 are, however, naturally untwinned and more stable compared to Y123, making Y124 attractive from an experimental viewpoint. We have compared the detailed spectra in Y124 with the corresponding results for Y123 in a number of cases and find the results to be quite similar, indicating that our conclusions are relevant for Y123.

The measurements were carried out at the Aladdin synchrotron. The high energy resolution was obtained using a commercial spectrometer, but with improved shielding and electronics and a 4 m normal-incidence monochrometer. The measurements employed untwinned Y124 single crystals with T_c of 82 K and transition widths of 1 K as determined by a SQUID magnetometer. The crystals were cleaved at 12 K in a vacuum better than 4×10^{-11} Torr and the ARPES spectra taken below T_c in order to avoid difficulties of oxygen loss and sample degradation. A number of experiments were carried out above T_c but did not show the presence of a superconducting energy gap, consistent with previous studies, even though peak shifts bigger than 1 meV are detectable by our instrument [14]. The specimens were oriented to an accuracy better than 1° by Laue diffraction and the orientation further confirmed in the experimental chamber via the observed symmetry of sharp photoemission features around high symmetry points.

In order to carry out first-principles computations of ARPES intensities in structures of the complexity of the high-T_c superconductors, we have generalized the one-step Green's function based photoemission approach which considers a semi-infinite solid of nonoverlapping muffin-tin potentials separated from the bulk by a potential step [15]. Our specific muffin-tin potentials are obtained via the semi-relativistic self-consistent Korringa-Kohn-Rostoker Green's-function (KKR) band structure methodology [16] and yield the bulk band structures and Fermi surfaces of Y124 and Y123 in reasonable accord with the well-known results [17]. Given the close similarity of the electronic structures and ARPES spectra of Y124 and Y123, the present computations on Y124 follow along the lines of our earlier study of Y123 [15], where an excellent accord was obtained between the computed and measured shape, polarization, and k dependence of the calculated and measured spectral peaks associated with the CuO_2 plane bands. Accordingly, we have assumed the surface to possess an ideal BaO/CuO_2 termination (i.e., the topmost layer is BaO, followed by the CuO_2 layer) and have chosen the imaginary part of the self-energy parameter to reflect an energy dependent broadening of initial and final states representative of the experimental spectra.

Figure 1 shows ARPES measurements in two perpendicular directions centered on the Y symmetry point. The spectra in Fig. 1(a) clearly show an intense peak, nearly dispersionless along the Γ-Y-Γ_1 direction (Γ_1 is in the second BZ). The peak position remains fairly constant for a region of $k_\parallel$ values centered on Y extending to

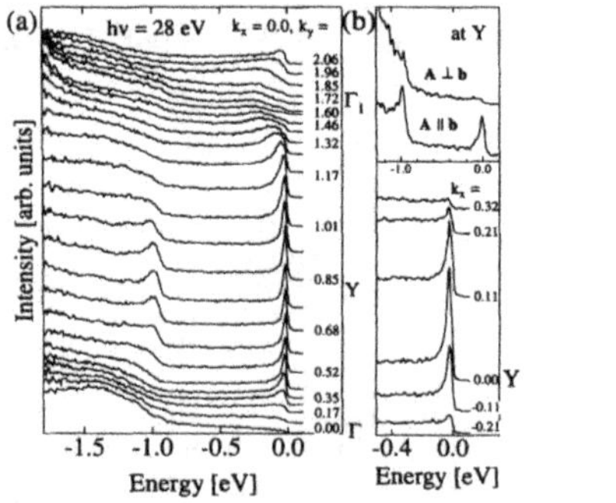

FIG. 1. Experimental ARPES spectra from untwinned $YBa_2Cu_4O_8$ along (a) Γ-Y and (b) Y-S directions. Note the large peak in the spectra very close to E_F in the vicinity of the Y point. The spectra are labeled with momenta at E_F in units of $Å^{-1}$. The inset in (b) shows that the band is excited only when the light is polarized along the **b** axis, having *even* symmetry with respect to the mirror plane containing the **b** and **c** axes (Γ-Y plane).

about $\frac{1}{3}$ of the Γ-Y-Γ_1 distance [18]. When moving from Y toward S [Fig. 1(b)], the band loses intensity as it crosses the Fermi level. This can be seen more clearly if we expand the energy scale by 1 order of magnitude, as shown in Figs. 2 and 3. Figure 2 shows the measured and computed spectra for $k_\parallel$ values varying along the same direction as in Fig. 1(a). The experimental spectra [Fig. 2(a)] show the peak lying 19 meV below the Fermi energy at the Y point. The characteristic features of the measured and computed spectral peak are very similar, in that both become more intense around the Y point, with the intensity remaining quite high as we move away from Y towards either Γ or Γ_1, but decreasing rapidly and broadening as the feature disperses to higher binding energies. The theoretical peak, however, displays more dispersion than the experiment, consistent with the notion that the associated band in the real material becomes flatter, presumably due to electron-electron correlations beyond the local density approximation used in the present calculations. Figure 3(a) shows that as we vary $k_\parallel$ in a perpendicular direction to that of Fig. 2(a) from the Y point (to the BZ corner S, as opposed to Γ), the spectral peak rapidly moves above the Fermi energy. As noted above, a convincing demonstration of this behavior is crucially dependent on the ultrahigh energy resolution achieved in this experiment. As in Fig. 2 above, the dispersion of the theoretical peak in Fig. 3(b) is greater than that of the experimental feature, presumably also due to electron-electron correlations.

The peak in the theoretical spectra of Figs. 2(b) and 3(b) originates in a bulk band related to the CuO_2 planes. We have established this via additional computations which show that, unlike the case of surface related features, this feature is insensitive to changes in the position of the

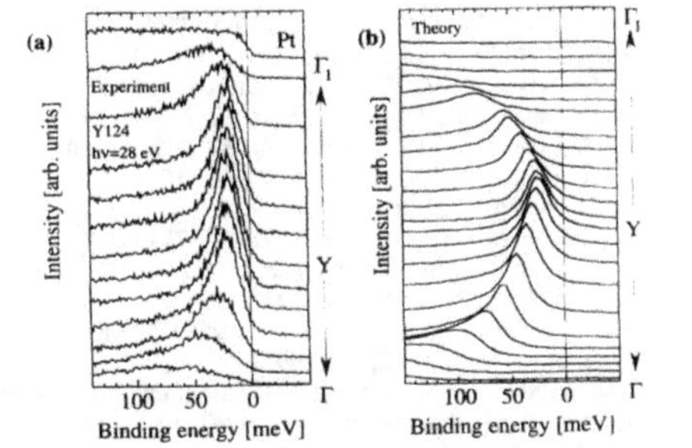

FIG. 2. (a) Same as Fig. 1, except with the energy scale expanded by 1 order of magnitude. The data show the peak at the minimum binding energy of 19 meV at Y. The emission angles vary by 2°. The Pt edge at the top shows the ultrahigh energy resolution (10 meV) of the spectrometer. This figure emphasizes spectra shapes. The quantitative E-k values obtained from these spectra are given in Fig. 4. (b) Corresponding first-principles computed intensities at $\frac{1}{2}$ the momentum interval (= 1°) used in (a) in order to clearly delineate the $k_{\|}$ dependence of the spectral shape.

surface barrier and also to the z component of the incident light. We have measured the detailed dependence of the intensity on polarization and find the observed behavior to be similar to the calculation. Bearing these considerations in mind, we conclude that the experimental feature in Figs. 2(a) and 3(a) is of *bulk* origin.

Figure 4(a) summarizes the dispersion of the state around the Y point obtained from the peak positions in the spectra of Figs. 2(a) and 3(a). A surface plot resulting from additional measurements [19], in which spectra were taken along several lines parallel to Y-S but offset along the Γ-Y direction, is shown in Fig. 5. The data show a maximum around Y as one moves towards Γ by varying k_y, but a minimum when moving towards S by varying k_x, and establish unambiguously an *extended saddle point* behavior of the highest occupied CuO_2 band around the Y point in Y124.

Additional information concerning the k_z dispersion is provided by changing the photon energy at fixed $k_{\|}$. Our studies show the k_z dispersion to be small in the vicinity of the Y point, indicating that the band structure is quasi-two-dimensional. On changing the photon energy, we also find a feature at a binding energy of 115 meV at the Y point [19]. This feature also displays a saddle point behavior and may arise from the second CuO_2-related plane band. We omit a detailed discussion here since

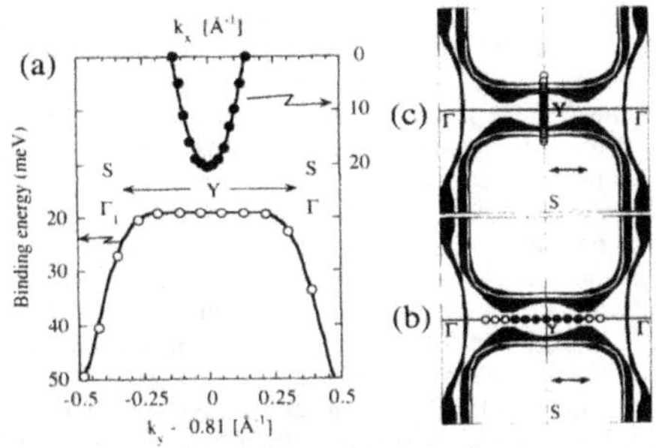

FIG. 4. (a) Band dispersion around Y obtained from the experimental spectra of Figs. 2(a) and 3(a). (b) Positions in momentum space where the spectra in Fig. 2(a) were taken superposed on a plot of the Fermi surface projected onto the basal plane in the zone. Filled dots represent positions where the band is flat. (c) Positions in momentum space where the spectra in Fig. 3(a) were taken. Dots are filled when the band is below the Fermi level. The arrows in (b) and (c) indicate the direction of polarization of the photons.

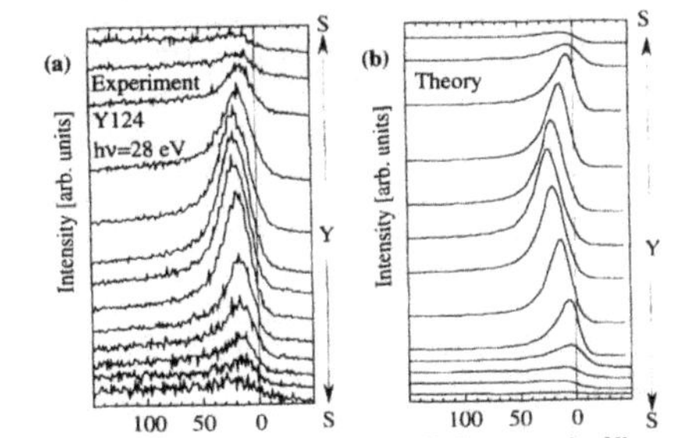

FIG. 3. Same as Fig. 2, except that here the $k_{\|}$ values are varied from Y towards the S point in a direction perpendicular to that of Fig. 2, showing the spectral feature dispersing towards E_F. Experimental and theoretical spectra are at the same momentum intervals (the emission angles vary by 1°).

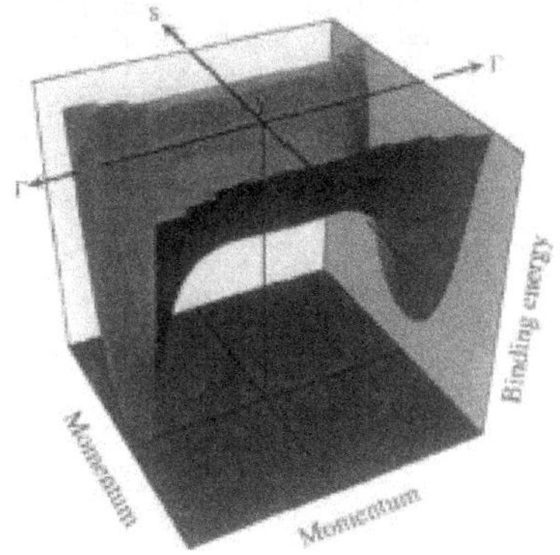

FIG. 5. A surface plot of the measured band dispersion around the Y point. The k_x and k_y axes cover an interval of 1 Å^{-1} centered on the Y point. The binding energy axis spans 0 to 60 meV.

this second saddle point lies at a much higher binding energy compared to the feature in Fig. 1, making it less interesting from the viewpoint of superconductivity. The peak nearest E_F is excited strongly at 28 eV in the first and second BZ, while both peaks are excited at 21.1 and 17 eV in the second BZ.

As already noted, the saddle point singularities appear to be present in all the cuprates studied so far. They lie at the zone boundaries, extended along the direction of the Cu-O bonds. In materials such as Y124, Y123, Tl2212 [10], and Bi2212 [20], which possess T_c's of ~100 K, the saddle points lie within 20 meV of E_F. These cuprates also display linear resistivity in the normal state, temperature independent thermopower, and other "unusual" properties. On the other hand, in NdCeCuO ($T_c = 23$ K), the saddle point binding energy is a few hundred meV [21], too high to significantly enhance T_c, and this material has normal state resistivity exhibiting the usual T^2 dependence. These observations suggest that the binding energy of the saddle point may be a useful parameter in classifying the cuprates.

In summary, we have directly observed an extended saddle point around the Y symmetry point in $YBa_2Cu_4O_8$ and shown that it arises from bulk electronic states related to the CuO_2 planes. Such an extended Van Hove singularity can easily explain the occurrence of high transition temperatures even in a conventional weak coupling BCS scheme. If this scenario is correct, one could conceivably obtain higher T_c's by "engineering" the electronic structure so as to increase the number and extension of saddle point singularities.

We thank R. S. Markiewicz for a discussion. This work was supported by the DOE under contract No. W-31-109-ENG-38, including a subcontract to Northeastern University; NSF Grants No. DMR 8914120, DMR 8809854, and DMR 91-20000, through the Science and Technology Center for Superconductivity; and benefited from a travel grant under the US-Finland program of the NSF, an ASLA-Fullbright travel grant, and the allocation of supercomputer time on the NERSC and Pittsburgh supercomputer centers.

[1] For a recent discussion see, *Spectroscopies in Novel Superconductors*, edited by F. M. Mueller, A. Bansil, and A. J. Arko, J. Phys. Chem. Solids, Vol. 54 (1993).
[2] J. E. Hirsh and D. J. Scalapino, Phys. Rev. Lett. **56**, 2735 (1986).
[3] J. Friedel, J. Phys. Condens. Matter **1**, 7757 (1989).
[4] R. S. Markiewicz, Int. J. Mod. Phys. B **5**, 2037 (1991).
[5] D. M. Newns *et al.*, Phys. Rev. Lett. **69**, 1264 (1992).
[6] C. C. Tsuei *et al.*, Phys. Rev. Lett. **69**, 2134 (1992).
[7] C. C. Tsuei *et al.*, Phys. Rev. Lett. **65**, 2724 (1990); C. C. Tsuei *et al.*, Phys. Rev. Lett. **68**, 1091 (1992).
[8] D. M. Newns *et al.* (to be published).
[9] P. C. Pattnaik *et al.*, Phys. Rev. B **45**, 5714 (1992).
[10] K. Gofron *et al.* (to be published).
[11] L. Van Hove, Phys. Rev. **89**, 1189 (1953).
[12] A. A. Abrikosov, J. C. Campuzano, and K. Gofron, Physica (Amsterdam) **214C**, 73 (1993).
[13] The experiment with finite resolution can, of course, only establish that the effective mass in the vicinity of the Y point is substantially greater than that near the Γ point. Under this assumption, Ref. [12] shows that the DOS peak will possess a power law divergence $(E - E_c)^{-1/2}$, even though the band dispersion is not absolutely flat. More generally, extended saddle points will possess higher power law divergences, which are nonetheless qualitatively different from logarithmic ones.
[14] A number of representative spectra taken above T_c indicate, however, that the present results are not affected significantly by the superconducting transition.
[15] A. Bansil *et al.*, J. Phys. Chem. Solids **53**, 1541 (1992); **54**, 1185 (1993); M. Lindroos et al., Physica (Amsterdam) **212C**, 347 (1993).
[16] Specifically, the muffin-tin potentials are the same as used previously in our work on the Y123 (see Ref. [15]), except that we have placed these on the base centered Y124 lattice. For the BaO/CuO_2 terminated surface used here we expect this to be a reasonable approximation, the Cu-O chain region where the Y123 and Y124 structures differ being six layers deep below the surface.
[17] D. Koelling *et al.* (unpublished); J. Yu *et al.*, Phys. Lett. **A 122**, 203 (1987); W. E. Pickett, R. E. Cohen, and H. Krakauer, Phys. Rev. B **42**, 8764 (1990); O. K. Andersen *et al.*, Physica (Amsterdam) **185C**, 147 (1991); J. Yu, K. T. Park, and A. J. Freeman, Physica (Amsterdam) **172C**, 467 (1991).
[18] The existence of a narrow feature in the ARPES spectra of Y123 has previously been reported by Tobin *et al.* [Phys. Rev. B **45**, 5563 (1992)], and Manzke *et al.* [Surf. Sci. **269**, 1066 (1992)]. However, the dispersion of the feature and in particular its interpretation in terms of an extended Van Hove scenario has not been considered before.
[19] K. J. Gofron, Ph.D. thesis, University of Illinois at Chicago 1993 (unpublished); K. Gofron *et al.*, J. Phys. Chem. Solids **54**, 1193 (1993).
[20] D. S. Dessau *et al.*, Phys. Rev. Lett. **71**, 2781 (1993).
[21] D. M. King *et al.*, Phys. Rev. Lett. **70**, 3159 (1993).

PHYSICAL REVIEW B VOLUME 58, NUMBER 5 1 AUGUST 1998-I

Quantum magnetoresistance

A. A. Abrikosov
Materials Science Division, Argonne National Laboratory, 9700 South Cass Avenue, Argonne, Illinois 60439
(Received 26 September 1997; revised manuscript received 9 March 1998)

An explanation is proposed of the unusual magnetoresistance, linear in magnetic field and positive, observed recently in nonstoichiometric silver chalcogenides. The idea is based on the assumption that these substances are basically gapless semiconductors with a linear energy spectrum. Most of the excess silver atoms form metallic clusters which are doping the remaining material to a very small carrier concentration, so that even in a magnetic field as low as 10 Oe, only one Landau band participates in the conductivity.
[S0163-1829(98)07829-1]

I. INTRODUCTION

Recently an interesting phenomenon has been observed by Xu *et al.*[1] in two doped silver chalcogenides $Ag_{2+\delta}Se$ and $Ag_{2+\delta}Te$, with $\delta\sim0.01$, the magnetoresistance depended linearly on magnetic field starting from approximately 10 Oe and going on to 55 kOe without any sign of saturation. This took place at temperatures ranging from 4.5 to 300 K. The slope of the linear field dependence decreased with increasing temperature in this range approximately three times. The Hall constant did not depend on temperature at $T<100$ K and decreased at higher temperatures.

With ideal stoichiometry, $\delta=0$, or small nonstoichiometry, at low temperatures these substances are intrinsic semiconductors with a narrow direct gap.[2,3] The gap is sample dependent and lies in the range of several tens of meV. The effective mass of carriers is of the order of $10^{-2}m_0$, where m_0 is the free electron mass.[4] At higher temperatures ($T=133$ °C for Ag_2Se) these substances undergo a phase transition into a phase which behaves more similar to a metal. Contrary to that, for nonstoichiometric (or doped) compounds, from the finite value of the Hall constant we can conclude that at any temperature they are more similar to the high-temperature phase of the stoichiometric compounds.

According to conventional theory (see, e.g., Ref. 5) the magnetoresistance in a metal with nonequal densities of electrons and holes behaves as

$$\Delta\rho\sim\begin{cases}\rho_0(\Omega\tau)^2, & \Omega\tau\ll1\\ \rho_0, & \Omega\tau\gg1\end{cases} \tag{1}$$

where $\Omega=eH/(m_1c)$ is the Larmor frequency, m_1 is the cyclotron mass, τ is the collision time, and ρ_0 is the resistance at zero field. This means that first it grows quadratically with field and then reaches saturation. With equal densities of electrons and holes (this can hardly correspond to the present case), the limit at $\Omega\tau\gg1$ is quadratic in Ω. None of these laws resembles the observations. There exists a certain possibility of a linear magnetoresistance, namely, a polycrystalline metal with an open Fermi surface (see in Ref. 5) but this is definitely not the present case. Therefore it is clear that the conventional approach, based on the quasiclassical treatment of electrons, cannot explain the phenomenon. Exotic mechanisms, such as constant and variable range hopping in semiconductors (see Ref. 6) also do not lead to a linear dependence of the resistivity on magnetic field. Besides, the substances in question behave as metals, and not as semiconductors.

In an old and thoroughly forgotten (including the author himself) paper,[7] published almost 30 years ago, the present author developed a more general approach to the theory of galvanomagnetic phenomena in metals, based on the exact quantum description of electrons in a magnetic field. The real goal of that paper was to establish the truth concerning the amplitude of the Shubnikov-deHaas oscillations (this problem caused great confusion in the scientific community of those days), however, for generality, the limiting quantum case, with electrons only in one Landau band, was also considered. The formulas obtained in that work were

$$\rho_{xx}=\rho_{yy}=\frac{N_iH}{\pi n^2ec},\quad \rho_{xy}=RH=\frac{H}{nec}, \tag{2}$$

where N_i is the density of scattering centers, n is the electron density, R is the Hall constant, and it was assumed that $T=0$. To my knowledge, this was the only possible case of a linear positive magnetoresistance.

In formula (2) $\rho_{xx}=\rho_{yy}$ is the total resistance, whereas the experimental result[1] can be described as

$$\rho=\rho_0+aH, \tag{3}$$

and this formula is valid not only at magnetic fields of several T but down to 10 Oe. According to what was said before [formula (1)], there is no way to explain such a behavior assuming that the metal is homogenous. Indeed the condition for only one Landau level to participate is (we include $\hbar$ in the numerical estimates; otherwise, units are used with $\hbar=1$)

$$n<(eH/c\hbar)^{3/2}. \tag{4}$$

If we substitute the electron concentration obtained from the Hall measurements $n\sim10^{17}$ cm^{-3}, we get $H>2$ T. The only possibility is to assume that the real samples used in the experiments are highly inhomogenous, so that they contain small regions with a large concentration of excess silver atoms and, correspondingly, higher electron concentration, imbedded into regions with a much smaller electron concentration where the extremal quantum situation takes place. For such a structure one could get the formula (3), however, one would have to assume that the concentration in the poorly doped regions has the order of 10^{12} cm^{-3}, i.e., $\sim 10^{-10}$ per atom, or less. This can be obtained from the condition (4) and the experimental result that the dependence of the magnetoresistance deviates from linearity only at $H<10$ Oe. Since the atomic concentration of excess silver atoms is of the order of 1%, it seems that these atoms form metallic clusters, which are doping the rest of the material with electrons.

For an inhomogenous system the calculation of the Hall constant requires a more precise characterization of the material, and therefore the experimentally obtained "effective concentrations" of the order of 10^{17} cm^{-3} cannot be calculated unambiguously. However, even these concentrations corresponding to $\sim 10^{-5}$ electrons per atom are much less than the concentration of the excess silver atoms; this shows the trend.

There are, however, other concerns. At small fields the Landau level spacing becomes smaller than T. For a quadratic energy spectrum at 10 Oe it is (in K)

$$\frac{\hbar eH}{m_1 c k_B} \sim 10^{-3}\left(\frac{m_0}{m_1}\right) \text{ K},$$

where k_B is the Boltzmann constant and m_0 is the mass of the free electron. If one uses the value of $m_1/m_0 \sim 10^{-2}$ given in the literature,[4] one gets 10^{-1} K, which is too small, since the linear magnetoresistance at 10 Oe was observed at $T=4.5$ K. The other assumption in Ref. 7, namely, that the electrons are degenerate, is even more difficult to achieve at the required densities.

II. MODEL

Since the problem seems completely hopeless for an ordinary approach, some very unusual path has to be taken. In Ref. 3 an idea was proposed about the high-temperature α phase of Ag_2Se. Due to the increased mobility of silver ions a substantial disorder appears which creates "tails" in both the conduction band and the valence band. Eventually the bands overlap, and the substance becomes metallic. It should be mentioned that according to experimental data the gap is direct. In Ref. 8 it was concluded that in Ag_2Te with increasing temperature a phase transition from a narrow-gap semiconductor to a gapless semiconductor takes place. One could guess that since the main reason for this change of the spectrum is disorder, the same could happen as result of doping and not of increased mobility (the ions move slowly, and they are always static from the viewpoint of electrons); this conclusion is supported by the finite Hall constant. Under these conditions the nonstoichiometric compound cannot be treated as a semiconductor with carriers in the bands resulting from doping such as, e.g., Ge and Si, but the start must be done from a different phase, which is closer to a gapless semiconductor.[9,10] The latter is a substance where at $T=0$ a completely filled valence band matches an empty conduction band. From the small value of the effective mass[4] it looks more likely that the energy spectrum in both bands is linear. Indeed, since a small effective mass can appear only in some restricted regions of the momentum space, it has to grow with energy (momentum). If the valence band contacts the conduction band, or slightly hybridizes with it, the spectrum becomes even closer to linearity.

The possibility of a linear spectrum was analyzed in Ref. 11 (type I). It can be a consequence of cubic symmetry without an inversion center and can also happen at some random point in the reciprocal space at proper "tuning", e.g., by pressure or doping. An example of the latter was analyzed in Ref. 12.

We will consider here the simplest spectrum of this kind, just in order to have an example of what can happen in such a substance. The spectrum corresponds to a double representation of cubic group T or O, and its Hamiltonian can be written as

$$H=\int \psi^+ v\left[\sigma\left(\mathbf{p}-\frac{e}{c}\mathbf{A}\right)\right]\psi dV, \tag{5}$$

where σ are the Pauli matrices and $\mathbf{p}$ is the momentum operator. The velocity v can be assumed of the usual order of magnitude $v \sim 10^8$ cm/sec. In the absence of magnetic field we get two branches of the spectrum with energies $\varepsilon = vp$, $-vp$. Suppose that the magnetic field is along z, and we chose the vector potential $A_y = Hx$. The electronic wave functions will have two components satisfying the equations

$$-i\frac{\partial}{\partial z}\psi_1+\left(-i\frac{\partial}{\partial x}-\frac{\partial}{\partial y}+i\frac{eH}{c}x\right)\psi_2=\frac{\varepsilon}{v}\psi_1,$$

$$\left(-i\frac{\partial}{\partial x}+\frac{\partial}{\partial y}-i\frac{eH}{c}x\right)\psi_1+i\frac{\partial}{\partial z}\psi_2=\frac{\varepsilon}{v}\psi_2. \tag{6}$$

Since the equations contain explicitly only x, we will search for solutions of the usual form

$$\psi_{1,2}=\psi_{1,2}(x)e^{(ip_y y+ip_z z)}. \tag{7}$$

The eigenvalues of Eqs. (6) are

$$\varepsilon_n^{(+)}=v\left(p_z^2+\frac{2eHn}{c}\right)^{1/2},$$

$$\varepsilon_n^{(-)}=-v\left(p_z^2+\frac{2eHn}{c}\right)^{1/2}. \tag{8}$$

Corresponding normalized eigenfunctions have the form

$$\psi_{n1}^{(+)}=\frac{1}{\sqrt{2}}\left(1+\frac{p_z}{(p_z^2+2neH/c)^{1/2}}\right)^{1/2}\psi_n,$$

$$\psi_{n2}^{(+)}=-\frac{i}{\sqrt{2}}\left(1-\frac{p_z}{(p_z^2+2neH/c)^{1/2}}\right)^{1/2}\psi_{n-1},$$

$$\psi_{n1}^{(-)}=\frac{1}{\sqrt{2}}\left(1-\frac{p_z}{(p_z^2+2neH/c)^{1/2}}\right)^{1/2}\psi_n,$$

$$\psi_{n2}^{(-)}=\frac{i}{\sqrt{2}}\left(1+\frac{p_z}{(p_z^2+2neH/c)^{1/2}}\right)^{1/2}\psi_{n-1}, \tag{9}$$

where ψ_n are the usual normalized eigenfunctions of a free electron in a magnetic field:

$$\psi_n=(2^n n!)^{-1/2}(\beta/\pi)^{1/4}e^{-(\beta/2)[x-(p_y/\beta)]^2}\times H_n[\sqrt{\beta}(x-p_y/\beta)]. \tag{10}$$

Here $\beta=eH/c$ and H_n are Hermite polynomials.

Formulas (9) describe the eigenfunctions for $n\neq 0$. In the case $n=0$ the eigenfunctions are

$$\psi_{01}^{(+)}=\theta(p_z)\psi_0,$$
$$\psi_{02}^{(+)}=0,$$
$$\psi_{01}^{(-)}=\theta(-p_z)\psi_0,$$
$$\psi_{02}^{(-)}=0. \tag{11}$$

We assume that in the undoped substance at zero temperature all the negative bands are filled and the positive bands are empty (see Ref. 9). We will assume also that at the temperatures and levels of doping under consideration only the bands $\varepsilon_0^{(+)}=v|p_z|$ and $\varepsilon_0^{(-)}=-v|p_z|$ will contain charge carriers. The necessary condition will be established later. Imagine that the doping corresponds to the electron density n_0. This means that the density of electrons in the band $\varepsilon_0^{(+)}$ minus the density of holes in the band $\varepsilon_0^{(-)}$ must be equal to n_0:

$$\frac{eH}{\pi c}\int_0^\infty([e^{(vp_z-\mu)/T}+1]^{-1}-[e^{(vp_z+\mu)/T}+1]^{-1})\frac{dp_z}{2\pi}=n_0. \tag{12}$$

Here we took into account two projections of the electron spin and the fact that the spin splitting $\mu_B H$ can be neglected. It is also important to remember that according to Eq. (11) only electron states with $p_z>0$ and hole states with $p_z<0$ are available. From condition (12) we obtain

$$\mu=\frac{2\pi^2 n_0 cv}{eH}. \tag{13}$$

We see that the chemical potential does not depend on temperature, which means that at all temperatures the electron system is described by the Fermi distribution.

Now we can obtain the condition for one Landau level to participate in kinetics. The first and rather evident condition is (in ordinary units)

$$T<v\sqrt{eH\hbar/c}. \tag{14}$$

If we substitute $H\sim 10$ Oe, we get $T<10$ K, which corresponds to the measurement conditions of Ref. 1. With larger H the lower temperature boundary also rises, so that at 1 T we get $T<300$ K, which is also quite satisfactory. The contribution of the Landau band with $n=1$ to the electron density is

$$n^{(1)}=\frac{eH}{2\pi^2 c}\int_0^\infty[e^{[v(p_z^2+2eH/c)^{1/2}-\mu]/T}+1]^{-1}dp_z.$$

The contribution will be small if condition (14) is fulfilled, provided that $\mu<(eH/c)^{1/2}$. Using formula (13), we get the condition (4) for n_0.

We must be aware that for the model under consideration this quantity is only the difference between densities of electrons and holes, entering, as we will see below, the Hall constant. Both types of carriers contribute to the conductivity, and their densities can be considerably larger than n_0. For example, the density of electrons is

$$n^{(0)}=\frac{eH}{2\pi^2 c}\int_0^\infty[e^{(vp_z-\mu)/T}+1]^{-1}dp_z=\frac{eHT}{2\pi^2 cv}\ln(1+e^{\mu/T}).$$

At low temperatures, when $\mu\gg T$, the density is equal to n_0 but at higher temperatures and magnetic fields it becomes much larger.

III. HALL CONSTANT

For the Hall constant we need only "bare" Green functions of the type defined in Ref. 7:

$$G_{\alpha\beta}^{(0)}(p_z,p_y,x,x',\omega_m)=\sum_n\frac{\psi_{n\alpha}(x-cp_y/eH)\psi_{n\beta}^*(x'-cp_y/eH)}{i\omega_m+\mu-\varepsilon_n(p_z)}, \tag{15}$$

where the summation over n means different energy levels, positive and negative, and $\alpha,\beta=1,2$. We use the temperature technique since we want to consider finite temperatures.

The current operator can be obtained from the Hamiltonian (1) and the relation

$$\delta H=-\frac{1}{c}\int \mathbf{j}\delta\mathbf{A}dV. \tag{16}$$

From here we obtain

$$\mathbf{j}(\mathbf{r},t)=e\psi^+(\mathbf{r})v\sigma\psi(\mathbf{r},t). \tag{17}$$

In the usual way we get for the linear response (see Ref. 13)

$$j_i(\omega_0)=-Q_{ik}^R(\omega_0)A_k(\omega_0), \tag{18}$$

where $Q_{ik}^R(\omega_0)$ is the analytical continuation from the upper half plane of complex frequencies of the corresponding thermodynamic average

$$Q_{ik}(i\omega_0)=\frac{2e^2v^2}{c}T\sum_m\int\frac{dp_y}{2\pi}\int\frac{dp_z}{2\pi}\int dx'\times\mathrm{Tr}[\sigma_i G(p_y,p_z,x,x',\omega_m+\omega_0)\times\sigma_k G(p_y,p_z,x',x,\omega_m)], \tag{19}$$

where the trace is taken over the components 1,2 and two spin projections were taken into account. Performing the analytical continuation we get

$$Q^R_{xy}(\omega_0)=\frac{2ie^2v^2}{c}\int_{-\infty}^{\infty}\frac{d\omega}{2\pi}\int\frac{dp_y}{2\pi}\int\frac{dp_z}{2\pi}\int dx'\bigg\{\tanh\frac{\omega}{2T}[G^R_2(p_y,p_z,x,x',\omega+\omega_0)\mathrm{Im}\,G^R_1(p_y,p_z,x',x,\omega)$$
$$-G^R_1(p_y,p_z,x,x',\omega+\omega_0)\mathrm{Im}\,G^R_2(p_y,p_z,x',x,\omega)]+\tanh\frac{\omega+\omega_0}{2T}[\mathrm{Im}\,G^R_2(p_y,p_z,x,x',\omega+\omega_0)$$
$$\times G^A_1(p_y,p_z,x',x,\omega)-\mathrm{Im}\,G^R_1(p_y,p_z,x,x',\omega+\omega_0)G^A_2(p_y,p_z,x',x,\omega)]\bigg\}. \tag{20}$$

Now we substitute the Green functions (15) and use the eigenvalues and eigenfunctions (8)–(11). Here Im G^R means that the imaginary part of the factor $[\omega+\mu-\varepsilon_n(p_z)+i\delta]^{-1}$ in Eq. (15) has to be taken equal to $-\pi\delta[\omega+\mu-\varepsilon_n(p_z)]$. We can limit ourselves with the states $\varepsilon_0^{(\pm)}$, $\varepsilon_1^{(\pm)}$ since the Green function of the state $n=0$ has no large magnetic field in the denominator, and the orthogonality of the eigenfunctions permits the connection between n's differing not more than by 1. Then every one of the Green functions contains the product $\psi_0(x-cp_y/eH)\psi_0(x'-cp_y/eH)$. Integrating in Eq. (20) over x' and p_y, we obtain the factor $eH/(2\pi c)$. After that, instead of the eigenfunctions we can substitute only the p_z-dependent prefactors in Eq. (9). Taking into account that every Green function can correspond to a (+) state, as well as to a (−) state, we obtain for small ω_0

$$Q^R_{xy}(\omega_0)=\frac{2ie^2v^2}{c}\frac{eH}{2\pi c}\frac{c\omega_0}{2eHv^2}\int_0^{\infty}\frac{dp_z}{2\pi}\bigg(\tanh\frac{vp_z-\mu}{2T}-\tanh\frac{vp_z+\mu}{2T}\bigg). \tag{21}$$

The connection between $Q^R_{xy}(\omega_0)$ and the static conductivity σ_{xy} is

$$\sigma_{xy}=\left[\frac{ic}{\omega_0}Q^R_{xy}(\omega_0)\right]_{\omega_0\to 0} \tag{22}$$

From here we obtain

$$\sigma_{xy}=\frac{e^2}{\pi}\int_0^{\infty}([e^{(vp_z-\mu)/T}+1]^{-1}-[e^{(vp_z+\mu)/T}+1]^{-1})\frac{dp_z}{2\pi}.$$

Comparing with Eq. (12), we obtain

$$\sigma_{xy}=\frac{n_0ec}{H},\quad \rho_{xy}=\frac{H}{n_0ec},\quad R_H=\frac{\rho_{xy}}{H}=\frac{1}{n_0ec}, \tag{23}$$

i.e., the usual result for the Hall constant R_H.

IV. CONDUCTIVITY

To obtain the conductivity we use the same approach except that we calculate this time the diagonal component of the Q matrix. We obtain

$$Q^R_{xx}(\omega_0)=Q^R_{yy}(\omega_0)=\frac{2e^2v^2}{c}\int_{-\infty}^{\infty}\frac{d\omega}{2\pi}\int\frac{dp_y}{2\pi}\int\frac{dp_z}{2\pi}\int dx'\times\bigg\{\tanh\frac{\omega}{2T}[G^R_2(p_y,p_z,x,x',\omega+\omega_0)\mathrm{Im}\,G^R_1(p_y,p_z,x',x,\omega)$$
$$+G^R_1(p_y,p_z,x,x',\omega+\omega_0)\mathrm{Im}\,G^R_2(p_y,p_z,x',x,\omega)]+\tanh\frac{\omega+\omega_0}{2T}[\mathrm{Im}\,G^R_2(p_y,p_z,x,x',\omega+\omega_0)G^A_1(p_y,p_z,x',x,\omega)$$
$$+\mathrm{Im}\,G^R_1(p_y,p_z,x,x',\omega+\omega_0)G^A_2(p_y,p_z,x',x,\omega)]\bigg\}. \tag{24}$$

The connection between $Q^R_{xx}(\omega_0)$ and σ_{xx} is similar to Eq. (22), and hence we need the imaginary part of $Q^R_{xx}(\omega_0)$. From Eq. (24) we obtain

$$\mathrm{Im}\,Q^R_{xx}(\omega_0)=\frac{2e^2v^2}{c}\int_{-\infty}^{\infty}\frac{d\omega}{2\pi}\int\frac{dp_y}{2\pi}\int\frac{dp_z}{2\pi}\int dx'\bigg(\tanh\frac{\omega}{2T}-\tanh\frac{\omega+\omega_0}{2T}\bigg)[\mathrm{Im}\,G^R_2(p_y,p_z,x,x',\omega+\omega_0)$$
$$\times\mathrm{Im}\,G^R_1(p_y,p_z,x',x,\omega)+\mathrm{Im}\,G^R_1(p_y,p_z,x,x',\omega+\omega_0)\mathrm{Im}\,G^R_2(p_y,p_z,x',x,\omega)] \tag{25}$$

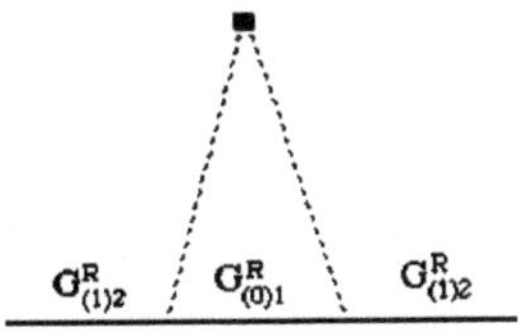

FIG. 1. Correction to the Green function due to scattering, entering the expression (28).

(we used here that Im $G^R=-\text{Im}\, G^A$). Since, according to Eq. (22), we are interested in small ω_0, we expand to the lowest order and obtain

$$\text{Im}\, Q_{xx}^R(\omega_0)$$
$$=-\frac{e^2v^2\omega_0}{cT}\int_{-\infty}^{\infty}\frac{d\omega}{2\pi}\int\frac{dp_y}{2\pi}\int\frac{dp_z}{2\pi}\int dx'$$
$$\times\cosh^{-2}\frac{\omega}{2T}\,[\text{Im}\, G_2^R(p_y,p_z,x,x',\omega)$$
$$\times\text{Im}\, G_1^R(p_y,p_z,x',x,\omega)$$
$$+\text{Im}\, G_1^R(p_y,p_z,x,x',\omega)\text{Im}\, G_2^R(p_y,p_z,x',x,\omega)].$$

Again, since all the Green functions contain products $\psi_0(x-cp_y/eH)\psi_0(x'-cp_y/eH)$, the integrations over x' and p_y can be performed resulting in a factor $eH/(2\pi c)$, and we are left with the prefactors in formulas (9) and energy denominators. We can write this as

$$\text{Im}\, Q_{xx}^R(\omega_0)$$
$$=-\frac{2e^2v^2\omega_0}{cT}\frac{eH}{2\pi c}\int_{-\infty}^{\infty}\frac{d\omega}{2\pi}\int_{-\infty}^{\infty}\frac{dp_z}{2\pi}\cosh^{-2}\frac{\omega}{2T}$$
$$\times\text{Im}\, G_{(1)2}^R(p_z,\omega)\text{Im}\, G_{(0)1}^R(p_z,\omega), \tag{26}$$

where $G_{(1)2}^R$, $G_{(0)1}^R$ mean that in the sums over n [see Eq. (15)] only terms with $n=1$ or $n=0$ are left.

Taking the scattering into account we can conclude that the product of Green functions contains

$$\frac{1/(2\tau_0)\theta(\pm p_z)}{(\omega+\mu\mp v|p_z|)^2+(1/2\tau_0)^2}\,\frac{1/(2\tau_1)}{[\omega+\mu\mp\varepsilon_1(p_z)]^2+(1/2\tau_1)^2}. \tag{27}$$

Both products are close to δ functions. Since, however, the second factor would be $\pi\delta[\omega+\mu\mp\varepsilon_1(p_z)]$ and $\varepsilon_1(p_z)$ contains the large magnetic field, this would substantially reduce the contribution to Eq. (26) due to the factor $\cosh^{-2}(\omega/2T)$ in the integrand. Therefore, we can replace only the first factor by a δ function and get from Eq. (27)

$$\frac{\pi\delta(\omega+\mu\mp v|p_z|)\theta(\pm p_z)(1/2\tau_1)}{[\omega+\mu\mp\varepsilon_1(p_z)]^2}. \tag{28}$$

Here $1/(2\tau_1)$ is $-\text{Im}\,\Sigma$, where Σ is the self-energy due to scattering. The expression (28) contains, actually, the first order scattering correction to the Green function, which is presented in Fig. 1. The imaginary part of Σ comes from integration along the half circle around the pole in the central Green function, and hence, it will impose a limitation $\omega>\min[\varepsilon(p_z')-\mu]$. Again, in order not to reduce the contribution to the integral over ω, the central Green function has to belong to $n=0$.

The impurity lines contribute after averaging over positions a factor

$$N_i\int d^3\mathbf{k}(2\pi)^{-3}|U(\mathbf{k})|^2\int\psi_{12}^*(x_1-cp_y/eH)e^{ik_xx_1}$$
$$\times\psi_{0j}[x_1-c(p_y-k_y)/eH]dx_1$$
$$\times\int\psi_{0j}^*[x_2-c(p_y-k_y)/eH]e^{-ik_xx_2}$$
$$\times\psi_{12}(x_2-cp_y/eH)dx_2. \tag{29}$$

We have included here integrals coming from the "envelope" ψ functions. $U(\mathbf{k})$ is the Fourier component of the electron-impurity interaction potential. It is natural to assume that it is a screened Coulomb interaction

$$U(\mathbf{k})=\frac{4\pi e^2}{\varepsilon_\infty(k^2+\kappa^2)}, \tag{30}$$

where ε_∞ is the background dielectric constant, which may be large; κ is the reciprocal Debye screening radius

$$\kappa^2=\frac{8\pi e^2}{\varepsilon_\infty}\frac{eH}{2\pi c}T\sum_m\int_{-\infty}^{\infty}\frac{dp_z}{2\pi}\frac{1}{(i\omega_m+\mu-vp_z)^2}$$
$$=\frac{2e^3H}{\pi\varepsilon_\infty cv}. \tag{31}$$

Here we took into account both branches + and − at $n=0$ and integrated in the same way as before.

Since $e^2/\hbar v\sim 1$ for $v\sim 10^8$, $\kappa^2\ll eH/c$, if $\varepsilon_\infty\gg 1$. We will assume this, since in semimetals, where the situation is close to the one under consideration, $\varepsilon_\infty\sim 100$. According to this, in the integral over $d^3\mathbf{k}$ in Eq. (29) small $\mathbf{k}$'s are the most important ($k\sim\sqrt{eH/c}$ is the upper limit). However, if we put $k_x=k_y=0$ in Eq. (29), the integral will vanish, since $\psi_{02}=0$, and there are no nondiagonal matrix elements. Therefore we must expand the matrix elements in powers of k_x and k_y. In this way we get

$$\int\psi_{12}^*(x_1-cp_y/eH)e^{ik_xx_1}\psi_{0j}[x_1-c(p_y-k_y)/eH]dx_1$$
$$\approx ik_x(\hat{x})_{0j}^{12}+(c/eH)k_y(\partial/\partial\hat{x})_{0j}^{12}. \tag{32}$$

The matrix elements, entering here we can obtain from two relations. Since the group velocity in the current is $\hat{v}=v\sigma$, we can write

$$\hat{v}_x=\partial\hat{x}/\partial t=i[H,\hat{x}];\quad(\hat{x})_{0j}^{12}=\frac{(\hat{v}_x)_{0j}^{12}}{i(\varepsilon_1-\varepsilon_0)}. \tag{33}$$

Similarly, the matrix elements of $\partial/\partial\hat{x}$ can be obtained from the commutation relation $[\partial/\partial\hat{x},\hat{x}]=1$. One must also take into account that every one of the three Green functions entering the diagram in Fig. 1 can be + or −, and the matrix elements depend on that. So we obtain

$$(\hat{x})_{01+}^{12+}=(\hat{x})_{01+}^{12-}=(\hat{x})_{12+}^{01+}=(\hat{x})_{12-}^{01+}=\frac{\theta(p_z)}{\sqrt{2}(p_z^2+2eH/c)^{1/2}},$$

$$(\hat{x})_{01-}^{12+}=(\hat{x})_{01-}^{12-}=(\hat{x})_{12+}^{01-}=(\hat{x})_{12-}^{01-}=\frac{\theta(p_z)}{\sqrt{2}(p_z^2+2eH/c)^{1/2}}, \tag{34}$$

$$(\partial/\partial\hat{x})_{12+}^{01+}=(\partial/\partial\hat{x})_{12-}^{01+}=-(\partial/\partial\hat{x})_{01+}^{12+}=-(\partial/\partial\hat{x})_{01+}^{12-}=\frac{\theta(p_z)(p_z^2+2eH/c)^{1/2}}{2\sqrt{2}},$$

$$(\partial/\partial\hat{x})_{12+}^{01-}=(\partial/\partial\hat{x})_{12-}^{01-}=-(\partial/\partial\hat{x})_{01-}^{12+}=-(\partial/\partial\hat{x})_{01-}^{12-}=\frac{\theta(-p_z)(p_z^2+2eH/c)^{1/2}}{2\sqrt{2}}. \tag{35}$$

The matrix elements do not depend on the + or − index of external Green functions. Therefore from the internal Green function we obtain

$$\int_0^\infty \frac{dp_z'}{2\pi}\frac{1}{\omega+\mu-vp_z'+i\delta}+\int_{-\infty}^0 \frac{dp_z'}{2\pi}\frac{1}{\omega+\mu+v|p_z'|+i\delta}$$
$$=\int_{-\infty}^\infty \frac{dp_z'}{2\pi}\frac{1}{\omega+\mu-vp_z'+i\delta}=-\frac{i}{2v}.$$

In the last factor of Eq. (29) the operator $\partial/\partial\hat{x}$ is acting on the internal function, and so integration by parts has to be performed.

We will see below that under condition (14) the p_z, entering the matrix elements will be much less than $\sqrt{eH/c}$ and can be neglected. Then the terms with k_x and k_y will be equal, and since the $k_z=p_z-p_z'$ is small, the integral over **k** becomes

$$N_i\int d^2\mathbf{k}(2\pi)^{-2}(k_x^2+k_y^2)|U(\mathbf{k})|^2=N_i\frac{4\pi e^4}{\varepsilon_\infty^2}\ln\varepsilon_\infty$$

[we have substituted formulas (30) and (31)]. Actually, this result is true with logarithmic accuracy. Higher order powers of k_x and k_y would add a term of the order of unity to $\ln\varepsilon_\infty$.

Collecting all the factors together, using Eqs. (22) and (28), and taking into account that

$$\delta(\omega+\mu-v|p_z|)\theta(p_z)+\delta(\omega+\mu+v|p_z|)\theta(-p_z)$$
$$=\delta(\omega+\mu-vp_z)$$

we obtain finally

$$\sigma_{xx}=\sigma_{yy}=\frac{1}{2\pi}\left(\frac{e^2}{\varepsilon_\infty v}\right)^2\ln\varepsilon_\infty\frac{ecN_i}{H}. \tag{36}$$

The corresponding component of the resistivity tensor is

$$\rho_{xx}=\frac{\sigma_{xx}}{\sigma_{xx}^2+\sigma_{xy}^2}$$

and σ_{xy} is given by Eq. (23). It is most likely that $(e^2/\varepsilon_\infty v)^2N_i\ll n_0$, and hence

$$\rho_{xx}=\frac{1}{2\pi}\left(\frac{e^2}{\varepsilon_\infty v}\right)^2\ln\varepsilon_\infty\frac{N_i}{ecn_0^2}H. \tag{37}$$

V. DISCUSSION

As we argued in the Introduction, the resistivity (37) is only a part of the total resistivity and is due to regions with a very small concentration of electrons [term aH in Eq. (3)]. The resistivity indeed depends linearly on H. The interesting feature is that it formally does not depend on temperature. From experimental data[1] we see that it, actually, has to depend on temperature, decreasing approximately three times, as the temperature varies from 4.5 to 300 K. This rather small decrease of ρ, instead of a marked increase, demonstrates that it is not due to the appearance of phonon scattering but is most likely associated with the change of doping (n_0). This is qualitatively confirmed by the fact that the decrease with temperature of the observed Hall constant and of the resistivity starts at the same temperature: 60–70 K in $Ag_{2+\delta}Se$, and 100–130 K in $Ag_{2+\delta}Te$. As we argued already, in an inhomogeneous sample with an unknown internal structure the "effective" Hall constant cannot be calculated. The electron concentration per atom obtained from it has to be somewhere between $\delta\sim10^{-2}$ and the limiting value from our estimate (4), i.e., $\sim10^{-10}$. As we mentioned, the experimental value was $\sim10^{-5}$.

The fact that both the resistivity and the Hall constant decrease with temperature correspond qualitatively to formulas (23) and (37). Since the doping mechanism is rather unusual (doping of pure regions by electrons from metallic clusters) it is difficult to predict the temperature dependence of n_0. Most likely, it will grow with temperature but much slower than the familiar exponent; this corresponds to the observations. Since the resistance depends not only on n_0 but also on N_i, it is hard to compare at this stage samples with different δ's.

In our model we made several unusual assumptions. The gapless spectrum with a linear momentum dependence in both bands, making the electrons similar to charged neutrinos, can be checked by diamagnetic (cyclotron) resonance. The main feature of this spectrum is that contrary to more conventional cases, the quantum condition persist to rather small magnetic fields and rather high temperatures. In order to distinguish this phenomenon from other kinds of unusual magnetiresistance phenomena (they all lead to negative magnetoresistance) I suggest that it be called *quantum magnetoresistance*.

ACKNOWLEDGMENT

I would like to thank the authors of Ref. 1, Dr. M.-L. Saboungi, Professor J. E. Enderby, and Professor P. B. Littlewood for communicating to me their results prior to publication and for useful discussions. This work was supported by the Department of Energy under Contract No. W-31-109-ENG-38.

[1] R. Xu, A. Husmann, T. F. Rosenbaum, M.-L. Saboungi, J. E. Enderbya, and P. B. Littlewood, Nature (London) **390**, 57 (1997).
[2] R. Dalven and R. Gill, Phys. Rev. **159**, 645 (1967).
[3] P. Junod, H. Hediger, B. Kilchör, and J. Wullschleger, Philos. Mag. **36**, 941 (1977).
[4] S. A. Aliev and F. F. Aliev, Izv. Akad. Nauk SSSR, Neorg. Mater. **21**, 1869 (1985).
[5] A. A. Abrikosov, *Fundamentals of the Theory of Metals* (North-Holland, Amsterdam, 1988).
[6] B. I. Shklovskii and A. L. Efros, *Electronic Properties of Doped Semiconductors* (Springer Verlag, Berlin, 1984).
[7] A. A. Abrikosov, Sov. Phys. JETP **29**, 476 (1969).
[8] S. A. Aliev and Z. A. Agaev (unpublished); this paper is briefly described in Ref. 4.
[9] This reminds me of the situation in semimetals. The theory can be constructed, if we start not from a simple cubic metal but from a specific type of gapless semiconductor, which looks quite artificial and definitely has a larger energy. However, it becomes energetically favorable at finite deformations (see Ref. 10).
[10] A. A. Abrikosov and L. A. Fal'kovsky, Sov. Phys. JETP **16**, 769 (1962).
[11] A. A. Abrikosov and S. D. Beneslavsky, Sov. Phys. JETP **32**, 699 (1971).
[12] A. A. Abrikosov, J. Low Temp. Phys. **8**, 313 (1972).
[13] A. A. Abrikosov, L. P. Gor'kov, and I. Ye. Dzyaloshinsky, *Quantum Field Theoretical Methods in Statistical Physics* (Pergamon, Oxford, 1965).

www.ingramcontent.com/pod-product-compliance
Lightning Source LLC
LaVergne TN
LVHW081930200926
848484LV00009B/2402
* 9 7 8 9 8 1 1 2 9 1 7 8 4 *